Seitz

WEED ECOLOGY

WEED ECOLOGY

Implications for Management

Second Edition

Steven Radosevich
Oregon State University

Jodie Holt
University of California, Riverside

Claudio Ghersa
Universidad de Buenos Aires

John Wiley & Sons, Inc.
New York • Chichester • Weinheim • Brisbane • Singapore • Toronto

This text is printed on acid-free paper.

Library of Congress Cataloging in Publication Data

Radosevich, Steven R.
Weed ecology : implications for management / Steven Radosevich,
Jodie Holt, Claudio Ghersa.—2nd ed.
p. cm.
ISBN 0-471-11606-8 (cloth : alk. paper)
1. Weeds—Ecology. 2. Weeds—Control. I. Holt, Jodie S.
II. Ghersa, Claudio. III. Title.
SB611.R33 1996
581.6′52—dc20 96-19470
CIP

Printed in the United States of America
10 9 8 7 6 5 4 3 2

Contents

Preface

Few people will deny that weed scientists, often operating in the separate disciplines of agronomy, horticulture, forestry, or range management, have collected vast amounts of information about weeds, the problems they cause, and the tools used to control them. In this respect, Weed Science is largely an empirical, descriptive discipline and therefore we had little trouble finding a wealth of information with which to update this book. For this we thank our colleagues around the world. We also recognize, however, that the theoretical framework of any scientific discipline is important. In this case, the conceptual underpinnings of our discipline are necessary to understand how weed invasions occur, how weed communities continue to exist, and even how agroecosystems work. It is with such an understanding that we believe new knowledge about the biological and human systems in which weeds exist will develop, along with expectations for improvement both in technology and in the condition of humans and nature.

In this new edition of *Weed Ecology: Implications for Vegetation Management*, we have added three chapters about the technology of weed control. Much of this information has been developed over the past four decades and has become a part of our general knowledge about dealing with weeds. Much new information about weed biology has emerged since the first edition of this book was published; thus we have also greatly expanded topics dealing with weed biology that were covered in the first edition and incorporated many references to original research leading to this new information. We attempt to guide and inform readers about the biology of weeds by relying on theories, concepts, and principles of basic plant ecology, physiology, and genetics. We consider the role of weeds in human systems and introduce new ideas about systems thinking and the role that hierarchical structure plays in both natural and human interrelationships. We believe that such principles ultimately will influence how scientific problems—even those about weeds—will be addressed in the future.

As with the first edition of this book, it became apparent that we would learn much more than our readers would from this project. Therefore, we are grateful for the research of its many contributors and to such people as H. G. Baker, J. L. Harper, E. J. Salisbury, J. P. Grime, and D. Tilman, who have written so thoughtfully about the subject of weed biology. Several people also graciously reviewed portions of this book as it was being prepared. In particular, we sincerely appreciate the efforts of Arnold Appleby, Robert Zimdahl, Richard Meilam, and Peter List.

As you begin this new edition, we ask you once again to consider these words by Emerson:

> *But these young scholars, who invade our hills,*
> *And traveling often in the cut he makes,*
> *Love not the flower they pluck, and know it not,*
> *And all their Botany is Latin names.*

Steven R. Radosevich
Jodie S. Holt
Claudio M. Ghersa

Corvallis, Oregon
Riverside, California
Buenos Aires, Argentina
January 1996

PART

I

Introduction

In one of his early texts on weed control, A.S. Crafts begins by saying, "In the beginning there were no weeds." What Dr. Crafts meant was that even though plants have existed for a long time, weeds did not exist before humankind. Weeds exist because of our human ability to judge and select among the various species of the plant kingdom. However, this anthropomorphic perspective on weeds provides little insight into weed evolution, ecological characteristics, or interactions that occur so markedly with crops. In this text, our focus is on these interactive features of weeds, especially as they occur in agricultural, forest, and rangeland situations. By considering weeds foremost as plants and by relying heavily on the concepts of plant ecology, we hope to provide a better understanding of the vegetation for which our discipline is named. Perhaps a better understanding of weeds will lead to better crop and weed management.

Chapter 1

Weeds and Weed Science

Weeds exist as a category of vegetation because of the human ability to select for desirable traits among the various members of the plant kingdom. Just as some plants are valued for their utility or beauty, others are reviled for their apparent lack of those traits. Weeds are recognized worldwide as an important type of undesirable, economic pest. However, the value of any plant is unquestionably determined by the perceptions of its viewers. Those perceptions also influence the human activities directed at this category of vegetation.

WEEDS

A plant growing out of place, that is, a plant growing where it is not wanted, at least by some people, is a common, accepted explanation of what a weed is. This notion of what is undesirable imparts so much human value to the idea of weediness that it is usually necessary to recognize who is making the determination as well as the characteristics of the plants themselves. For example, certain plants growing in a cereal field, pasture, or along a fence row may be unwanted by a farmer or rancher, but they also may be considered to be wildflowers or a valuable wildlife cover by other people. Vine maple is a valued source of deer browse in the spring and a spectacular source of coloration in the Cascade Mountains of Oregon and Washington during autumn, but it also is known to compete with young conifers, thus hampering forest regeneration activities. Thus, it has been argued that many weeds in agricultural fields, forest plantations, and rangelands are not "out of place" at all, but are simply not wanted there by some people.

In Table 1.1 we list many of the "human" characteristics that have been used to describe weeds. Most of these characteristics are based on some judgment of worth, success, or other human attribute, like aggressiveness, harmfulness, or being unsightly or ugly. Since this anthropomorphic view of weeds is so prevalent (Table 1.1.), it may be that weeds are little more than plants that have aroused a level of human dislike at some particular place or time. Unfortunately, the anthropomorphic view of weeds provides little insight into why and where they exist, their interactions and associations with crops and other organisms, or even how to manage them effectively. Weeds have proven to be efficient and successful organisms in the environments that they inhabit. Therefore, it is important to explore whether weeds possess common traits that distinguish them from other plants or whether they are only set apart by local notions of usefulness.

A list of more biological characteristics that describe weeds has been developed (Table 1.2), but it seems unlikely that any plant species could possess all of those "ideal" weedy traits. However, Herbert Baker, the botanist and originator of the list, suggests that a plant species might possess various combinations of the characteristics in Table 1.2, resulting in a range of weediness from minor to major weeds. In the latter case, Baker believes that evolutionary processes

TABLE 1.1 Definitions and Descriptions of Weeds

Definition	*Description*
Growing in an undesirable location	A plant growing where it is not desired (Terminology Committee of the Weed Science Society of America, 1956).
Competitive and aggressive behavior	A plant that grows so luxuriantly or plentifully that it chokes all other plants that possess more valuable properties (Brenchley, 1920).
Persistence and resistance to control	The predominance and pertinacity of weeds (Gray, 1879).
Useless, unwanted, undesirable	A plant not wanted and therefore to be destroyed (Bailey and Bailey, 1941). A plant whose virtues have not yet been discovered (Emerson, 1878).
Appearing without being sown or cultivated	Any plant other than the crop sown (Brenchley, 1920). A plant that grows spontaneously in a habitat greatly modified by human action (Harper, 1944).
Unsightly	A very unsightly plant of wild growth, often found in land that has been cultivated (Thomas, 1956).

Source: Adapted from King (1966).

TABLE 1.2 Ideal Characteristics of Weeds

Germination requirements fulfilled in many environments
Discontinuous germination (internally controlled) and great longevity of seed
Rapid growth through vegetative phase to flowering
Continuous seed production for as long as growing conditions permit
Self-compatibility but not complete autogamy or apomixy
Cross-pollination, when it occurs, by unspecialized visitors or wind
Very high seed output in favorable environmental circumstances
Production of some seed in a wide range of environmental conditions; tolerance and plasticity
Adaptations for short-distance dispersal and long-distance dispersal
If perennial, vigorous vegetative reproduction or regeneration from fragments
If perennial, brittleness, so as not to be drawn from the ground easily
Ability to compete interspecifically by special means (rosettes, choking growth, allelochemicals)

Source: Baker (1974). Reproduced with permission from the *Annual Review of Ecology and Systematics*, Volume 5. Copyright 1974 by Annual Reviews, Inc. Palo Alto, CA.

would compound specific adaptations into highly successful (weedy) individuals that possess an "all purpose genotype." It must be stressed, however, that ecological success in the form of weediness cannot be measured solely from the agricultural perspective of noxiousness. The number of individuals, the range of habitats occupied, and the ability to continue the species through time must be considered foremost when evaluating success of a species. The obvious limitation of the list in Table 1.2 is that almost every plant species has some "weedy" characteristics; however, not all plants are weeds.

Definitions of Weeds

As we have just observed (in Tables 1.1 and Table 1.2), weeds can be described in either anthropomorphic or biological terms. Weeds emerge from such descriptions as organisms that may possess a particular suite of biological characteristics but also have the distinction of negative human selection. Thus, a definition of a weed as *any plant that is objectionable or interferes with the activities or welfare of man* (Weed Science Society of America, 1994) seems to describe sufficiently this category of vegetation. However, other authors—for example, Zimmerman (1976) and Aldrich (1984)—define weeds in more specific terms than the simple definition given above. Zimmerman believes that the term "weed" should be used to describe plants that display all the following characteristics: they (1) colonize disturbed habitats, (2) are not members of the original plant community, (3) are locally abundant, and (4) are economically of little value [or are costly to control]. A definition by Aldrich describes weeds as plants that originated under a natural environment and, in response to [human] imposed or natural conditions, are now interfering associates of crops and human activities. All of these definitions imply that weeds have some common biological traits but also a level of relative undesirability as determined by particular humans. Whether or not a plant is a weed depends on the context in which someone finds it and on the perspectives and objectives of those involved in dealing with it.

The most important criterion for weediness is interference at some place or time with the values and activities of people—farmers, foresters, and many other segments of society. However, the abundance of weeds is often of more concern than their mere presence. For instance, farmers are usually less concerned about the occurrence of a few isolated plants in a field, even noxious ones, than they are about the occupation of land by vast numbers of weeds. Therefore, the relative abundance of plants, their location, and the potential use of the land they occupy should be considered in weed definitions. When abundance is applied as a criterion for weediness, it suggests a condition of the land (Figure 1.1), as well as a class of vegetation (Table 1.2) and a form of human discrimination (Table 1.1). Weed abundance also may be an indicator or symptom of land mismanagement, neglect, or abuse.

FIGURE 1.1 Leafy spurge: a weedy rangeland condition. (Courtesy of B.D. Maxwell, Montana State University.)

Weed Classification

Botanical classification is based on established criteria that distinguish among types of vegetation, and then the systematic arrangement of plants according to their similarities. Some common methods used to classify weeds are by taxonomic characteristics, life history, habitat, physiology, degree of undesirability, and evolutionary strategy.

BOTANICAL (TAXONOMIC) CLASSIFICATION

All organisms are classified systematically according to simple morphological characteristics. The basis of a biological classification system is the presumed genetic relationships among the organisms. The accepted system used today classifies organisms into kingdom, division (phylum), class, order, family, genus, and species. While biologists disagree about the number of kingdoms that should be recognized, particularly for algae, fungi, and bacteria, all land plants are placed in the plant kingdom. Most weeds occur in the division Anthophyta (angiosperms), although notable exceptions occur (some ferns and conifers are considered weeds). Angiosperms are further divided into the classes Dicotyledones (dicots) and Monocotyledones (monocots).

The next level of classification is the order. There are fifty-six orders of dicots and thirteen orders of monocots. The orders are divided further into families, which like classes and orders, are composed of plants whose morphological similarities are greater than their differences. The level of genus includes plants that have common characteristics, especially of the flower, and that are genetically related. The narrowest category of classification is the species, which consists of plants that are morphologically similar and that interbreed freely. Occasional matings may also occur between members of different species in the same genus. At this point, the plant group is given a name, called a *Latin binomial,* which consists of both the genus and species names of the plant and which also briefly describes it. Table 1.3 is a list of some weed species by Latin binomial. This method of classification is the basis for organization of all taxonomic texts and books used to identify weeds. Some examples for the western United States are the *Grower's Weed Identification Handbook* (Fischer et al., 1978), *Weeds of the West* (Whitson, 1991), and *Weeds of California* (Robbins et al., 1941, 1951).

There are approximately 250,000 species of plants in the world. However, less than 250 plant species, about 0.1 percent, are troublesome enough to be called weeds universally throughout the world. A list of these weeds considered to be the world's worst was developed by Holm and his associates (1977); some of these are shown in Table 1.3. It is surprising that relatively few plant species make up the majority of agricultural weed problems worldwide. Furthermore, nearly 70 percent of these weed species occur in only twelve plant families and over 40 percent are found in only two families: *Poaceae* (grass)

TABLE 1.3 Scientific and Common Names of Certain Annual Weed Species Considered to be Among the World's 76 Worst

Species	*Common Name*
Ageratum conyzoides	tropical ageratum
Amaranthus hybridis	smooth pigweed
Amaranthus spinsous	spiny amaranthus
Anagallis arvensis	scarlet pimpernel
Argemone mexicana	Mexican prickle poppy
Avena fatua	wild oat
Bidens pilosa	hairy beggar-tick
Capsella bursa-pastoris	shepherdspurse
Cenchrus echinatus	southern sandbur
Chenopodium album	common lambsquarters
Cyperus difformis	smallflowered umbrella sedge
Cyperus iria	rice flatsedge
Dactyloctenium aegyptium	crowfootgrass
Digitaria adscendens	Southern crabgrass
Digitaria sanguinalis	large crabgrass
Echinochloa colonum	junglerice
Echinochloa crus-galli	barnyardgrass
Eleusine indica	goosegrass
Euphorbia hirta	garden spurge
Fimbristylis miliacea	globe fringerush
Galinsoga parviflora	small-flowered galinsoga
Galium aparine	catchweed bedstraw
Heliotropium indicum	Indian heliotrope
Ischaemum rugosum	saramollagrass
Leptochloa filiformis	red sprangletop
Lolium temulentum	darnel
Polygonum convolvulus	wild buckwheat
Portulaca oleracea	purslane
Rottboellia exaltata	itchgrass
Setaria verticillata	bristly foxtail
Setaria viridis	green foxtail
Solanum nigrum	black nightshade
Sonchus oleraceus	annual sowthistle
Spergula arvensis	corn spurry

TABLE 1.3 (Continued)

Species	*Common Name*
Sphenoclea zeylanica	gooseweed
Stellaria media	chickweed
Striga lutea	witchweed
Tribulus terrestris	puncturevine
Xanthium spinosum	spiny cocklebur
Xanthium strumarium	common cocklebur

Source: Adapted from Holm et al., 1977.

and *Asteraceae* (sunflower or composite). Although these observations are fruitful areas of speculation for plant evolutionary biologists, it must be realized that about 75 percent of world food production is provided by only a dozen crops: barley, maize, millet, oats, rice, sorghum, sugarcane, wheat, white potato, sweet potato, cassava, and soybean. Eight of these crops also are members of the grass family. The distribution of both the world's worst agricultural weeds and its crops is quite taxonomically restricted, again pointing to the extreme discrimination and selection that humans apply to vegetation.

It is sometimes necessary to distinguish only broadly among weed species. In such situations, distinction among grasses, sedges, and broadleaf (dicot) plants may be sufficient, and a much abbreviated system of classification is satisfactory. Such a system was once in common use by weed control specialists and has been described by Ross and Lembi (1985) as dicots and monocots.

Dicots. Plants whose seedlings produce two cotyledons or seed leaves. Dicots are usually typified by netted leaf venation (Figure 1.2) and flowering parts in fours, fives, or multiples thereof. Examples include mustards, nightshades, and morningglory. They are commonly called broadleaved plants.

Monocots. Plants whose seedlings bear only one cotyledon. Monocots are typified by parallel leaf venation (Figure 1.2) and flower parts in threes or multiples of three. Most monocot weeds are found in only two groups, grasses and sedges, although other groups exist.

Grasses. Leaves usually have a ligule or at times an auricle. The leaf sheaths are split around the stem, with the stem being round or flattened in cross section with hollow internodes.

Sedges. Leaves lack ligules and auricles, and the leaf sheaths are continuous around the stem. In many species the stem is triangular in cross section with solid internodes.

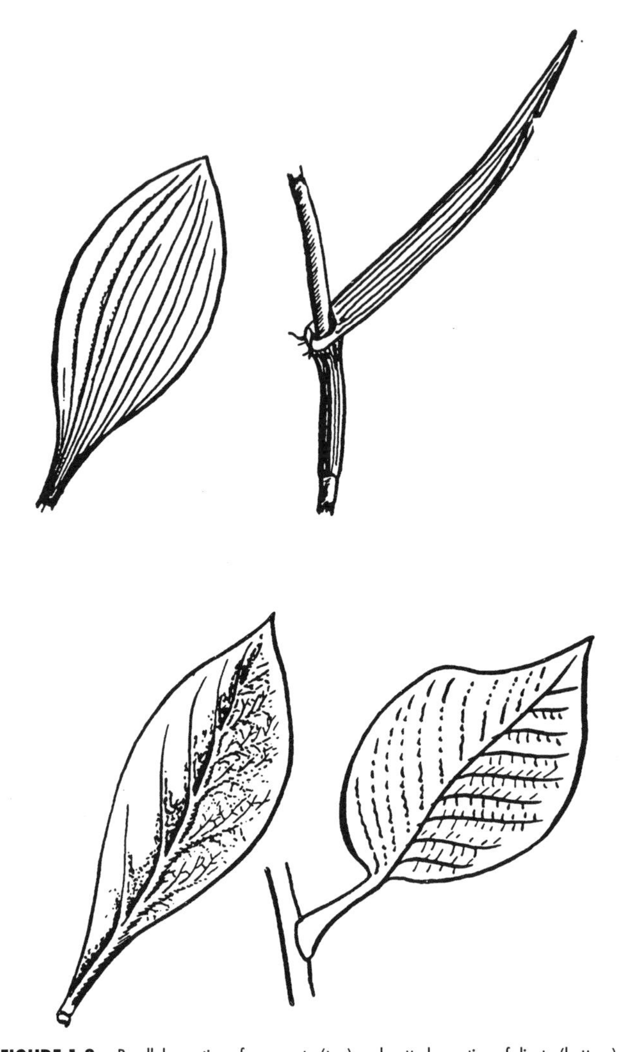

FIGURE 1.2 Parallel venation of monocots (top) and netted venation of dicots (bottom).

CLASSIFICATION BY LIFE HISTORY

Another method used to classify weeds is by the life cycle of the plant. The length of life, season of growth, and time and method of reproduction are used to classify weeds in this way.

Annuals. An annual plant completes its life cycle from seed to seed in one year or less (Figure 1.3). Annuals often are divided into two groups, winter and summer, according to the plant's time of germination, maturation, and death.

Winter annuals. Usually germinate in the fall or winter, grow throughout the spring, and set seed and die by early summer.

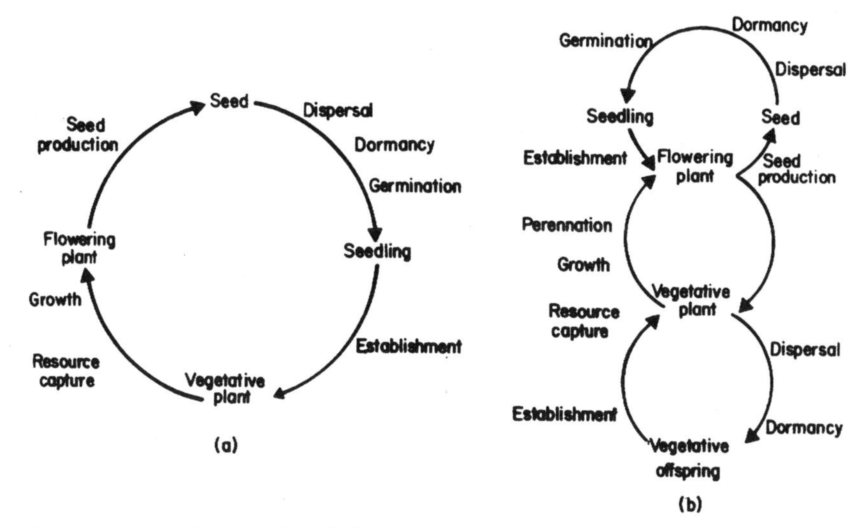

FIGURE 1.3 Schemes illustrating the life cycle of an annual flowering plant (a) and a perennial plant producing both seed and vegetative progeny (b). Adapted from Grime, 1979, *Plant Strategies and Vegetation Processes*. Copyright 1979 by John Wiley and Sons, Ltd., Chichester, U.K.

Summer annuals. Germinate in the spring, grow throughout the summer, set seed by autumn, and die before winter.

In mild climates, however, it is usual for some winter annuals to germinate in late summer or autumn, and for some summer annuals to live throughout the winter. Annual plants are the largest single category of weeds.

Biennials. These plants live longer than one year but less than two years. During the first growth phase, biennials develop vegetatively from a seedling into a rosette. Because of this growth habit, biennials sometimes can be confused with winter annuals. After a cold period, vegetative growth resumes, and floral initiation, seed production, and death occur. Biennials often are large plants when mature, and have thick fleshy roots. Relatively few weed species are biennials, but some annual plants may behave as biennials under certain conditions and some biennials may behave as short-lived perennials in mild climates.

Perennials. Perennial plants live for longer than two years and may reproduce several times before dying (Figure 1.3). These plants are characterized by renewed vegetative growth year after year from the same root system.

Simple herbaceous perennial weeds. Reproduce almost exclusively from seed and normally do not reproduce vegetatively. However, if the root system of these plants is injured or cut, each piece usually regenerates into another plant. Dandelion (*Taraxacum officinale*) and plantain (*Plantago lanceolata*) are examples of simple herbaceous perennials.

Creeping herbaceous perennials. Over-winter and produce new vegetative structures (ramets) from asexual reproductive organs such as rhizomes, tubers, stolons, bulbs, corms, and roots (Figure 1.4). These plants also reproduce sexually from seed (genets). Most aquatic weeds, except algae, are creeping perennial plants.

Woody plants. A special category of perennial weed. Plants in this group are characterized by stems that undergo secondary thickening and an annual growth increment. Some trees and shrubs and many vine species are considered to be woody weeds.

CLASSIFICATION BY HABITAT

Weeds can be classified according to where they grow. For example, most weeds are terrestrial (that is, they are found on land), but some are restricted to the aquatic environment. Some weeds only infest a particular crop or appear during a specific growing condition. Therefore, it is common to find lists and descriptions of weeds that are usually found in a particular habitat, like arable land, pastures, rights-of-way, forests, rangelands, and so on.

Aquatic weeds. These plants have structural modifications that enable them to live in water. They have been categorized further based on their location in the aqueous environment. These categories are depicted in Figure 1.5 as: *floating*, *emergent*, and *submerged*. Algae also are considered to be aquatic weeds.

Floating weeds. Rest upon the water surface. Their roots hang freely into the water or sometimes attach to the bottom of shallow ponds or streams.

Emergent weeds. Typical plants of natural marshlands and are often found along the shorelines of ponds and canals. These plants stand erect and are always rooted into very moist soil.

Submerged weeds. Grow completely under water, although a few floating stems or leaves may exist on the water surface.

PHYSIOLOGICAL CLASSIFICATION

Plant biologists have discovered that plants differ in their response to temperature, light, daylength, and other factors of the environment. These differences in plant physiology also have been used as a basis for weed classification.

Photosynthetic pathway. Most plants, called C_3 plants, use the Calvin cycle exclusively as a method of fixing carbon dioxide, water, and light energy into sugars. This terminology is used because the first stable product of photosynthesis in such plants (phosphoglyceric acid) has three carbon atoms. In some plants, however, the first stable photosynthetic products are 4-carbon atom sugars, such as oxaloacetate, malate, and aspartate. These are called C_4 plants. This physiological distinction may not seem significant as a means of categorizing weeds. However, differences in photosynthetic pathway result in sub-

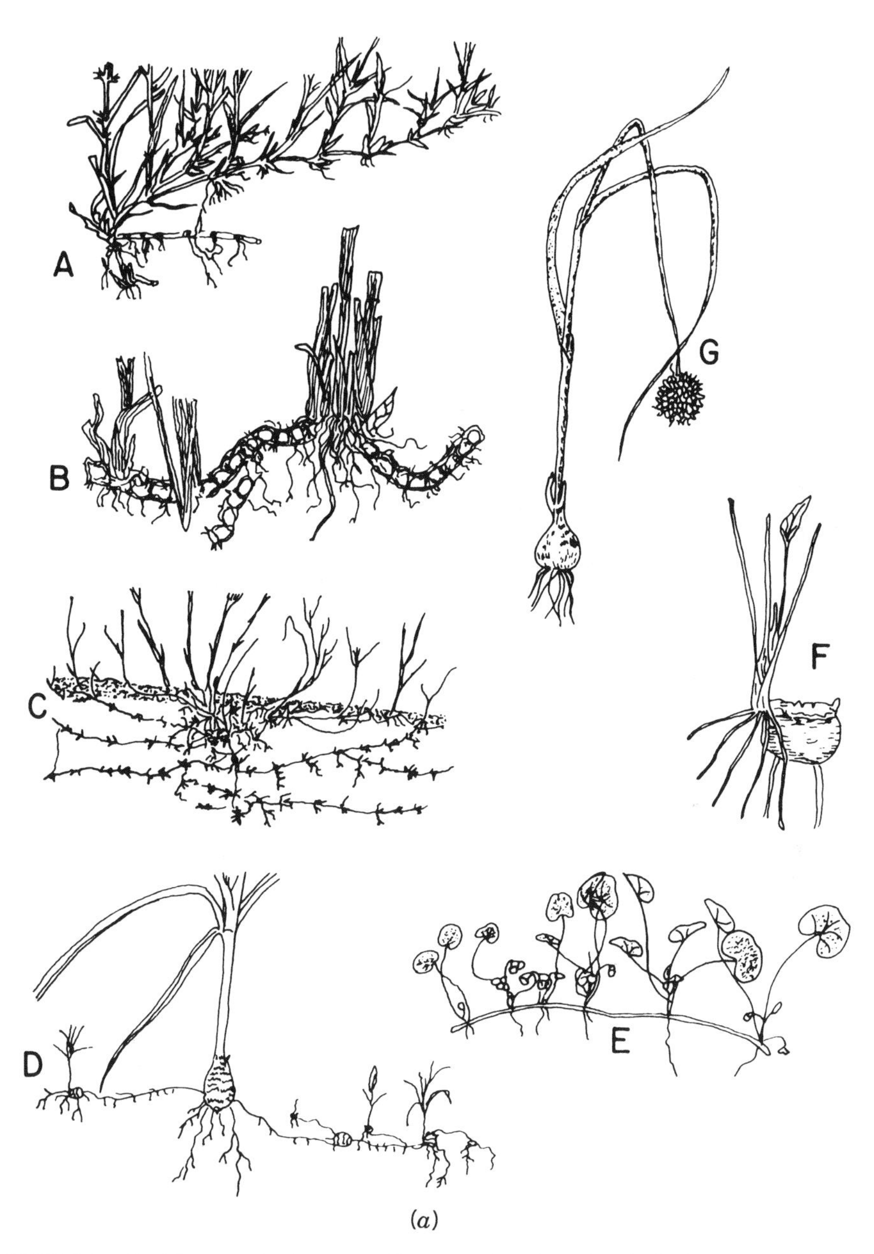

FIGURE 1.4a Vegetative propagation of weeds by modified stems. A, Bermudagrass (*Cynodon dactylon*), stolon; B, Johnsongrass (*Sorghum halepense*), rhizome; C, quackgrass (*Agropyron repens*), rhizome; D, purple nutsedge (*Cyperus rotundus*), tuber; E, dichondra (*Dichondra repens*), creeping stem; F, bulbous buttercup (*Ranunculus bulbosus*), corm; G, wild onion (*Allium bolanderi*), bulb. (Compiled from Robbins et al., 1941, *Weeds of California*.)

(*continued*)

(*b*)

FIGURE 1.4b Vegetative reproduction of weeds by roots. A, whitetop (*Cardaria draba*); B, sheep sorrel (*Rumex acetocella*); C and D, Canada thistle (*Cirsium arvensis*); E, western yarrow (*Achillea millefolium*); F, leafy spurge (*Euphorbia esula*). (Compiled from Robbins et al., 1941, *Weeds of California* and Fischer et al., 1978, *Grower's Weed Identification Handbook*.)

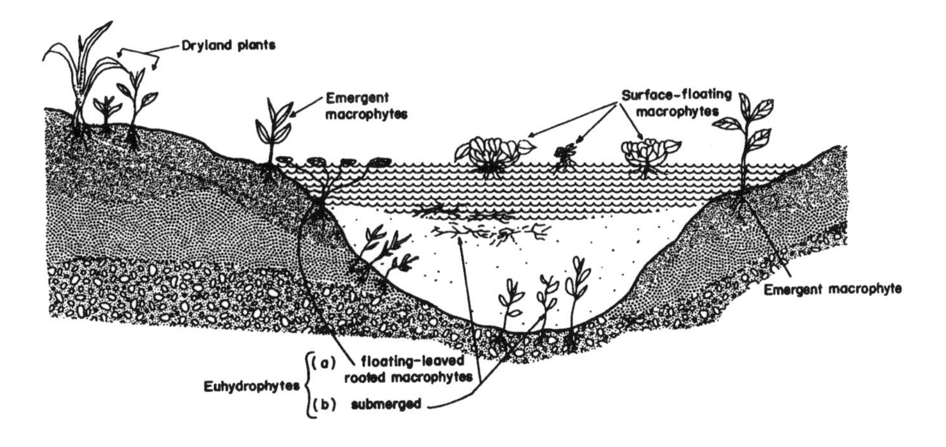

FIGURE 1.5 Habitats of aquatic weeds. (From Akobundu, *Weed Science in the Tropics: Principles and Practices*,1987. Copyright 1987, John Wiley and Sons, New York).

stantial biochemical and anatomical variation among plant species. Because of these differences, C_4 weeds often are more efficient at photosynthesis and can be more competitive than C_3 weeds and crops, especially in hot climates. Of the eighteen worst weeds of the world noted by Holm et al. (1977), fourteen have the C_4 pathway of carbon fixation.

Daylength. Classification by daylength is based on a photoperiodic response of flower inititation in plants. Three distinct classes of daylength response are known: short-day, long-day, and day-neutral. Weeds that have a short-day response to daylength, such as lambsquarters (*Chenopodium album*) and cocklebur (*Xanthium* spp.), are stimulated to flower when days are short and maintain vegetative growth when days are long. Long-day weeds, like black henbane (*Hyoscyamus niger*) and dogfennel (*Eupatorium capillifolium*), maintain vegetative growth when days are short but are induced to flower under long-day conditions. Other weeds, for example, nightshades (*Solanum* spp.), remain vegetative or flower irrespective of the photoperiodic condition.

CLASSIFICATION ACCORDING TO UNDESIRABILITY

The term *noxious weed* refers to any plant species that is capable of becoming detrimental, destructive, or difficult to control. Many states, provinces, and countries maintain at least one official list of such weeds so that their introduction can be prevented or restricted (see Chapter 8). Such weeds usually create a particularly undesirable situation in crops. As a result, the presence of noxious weed seeds in seed crops can prevent the sale and distribution of that crop across national and international boundaries. Poisonous weeds represent a special kind of undesirability since they can be a direct threat to human or animal health. The list of poisonous weeds is extensive; therefore, only an abbreviated list of some prevalent poisonous weeds and a brief description of their effects is presented in Table 1.4.

CLASSIFICATION BY EVOLUTIONARY STRATEGY

Weed species have been classified according to evolutionary strategies that are based on genetically determined patterns of carbon resource allocation. One useful theory holds that two basic external factors limit the amount of plant material that can accumulate within an area. These factors are *stress* and *disturbance* (see Chapter 2). When the extremes of the two factors are considered

TABLE 1.4 Characteristics of Some Poisonous Weeds

Name	*Toxic Principle*	*Source*	*Clinical Signs*
Arrowgrass	Hydrocyanic	Leaves	Nervousness, trembling, spasms or convulsions
Bouncing bet	Saponin	Whole plant; seeds are most toxic	Nausea, vomiting, rapid pulse, dizziness
Bracken fern	Unknown	Fronds	Fever, difficulty in breathing, salivation, congestion
Buffalo bur	Solanine	Foliage and green berries	Most serious in nonruminants
Buttercups	Protoanemonin	Green shoots	Loss of condition, production drops, and milk is often red, diarrhea, nervousness, twitching, labored breathing
Chokecherry and other cherries	Glucoside-amygdalin, a cyanogenic compound	Leaves	Rapid breathing, muscle spasms, staggering, convulsions, and coma
Cocklebur	Hydroquinone	Seeds and seedlings	Nausea, depression, weakness especially in swine
Corn cockle	A glucoside githagin and a saponin	Seeds	Poultry and pigs are most affected; inability to stand, rapid breathing, coma
Horsetail	Thiaminase activity, an alkaloid	Shoots	Loss of condition, excitability, staggering, rapid pulse, difficult breathing, emaciation

TABLE 1.4 Continued

Name	*Toxic Principle*	*Source*	*Clinical Signs*
Indian tobacco	Alkaloids similar to nicotine	Leaves and stems	Ulcers in mouth, salivation, nausea, vomiting, nasal discharge, coma
Jimsonweed	Alkaloids	All parts	Rapid pulse and breathing, coma
Larkspur	Alkaloid	All parts	Staggering, nausea, salivation, quivering, respiratory paralysis
Nightshades	Solanine, a glycoalkaloid	Foliage and green berries	Most cases in sheep, goats, calves, pigs, and poultry; anorexia, nausea, vomiting, abdominal pain, diarrhea
Ohio buckeye	Alkaloid	Sprouts, leaves and nuts	Uneasy or staggering gait, weakness, trembling
Water hemlock	Cicutoxin	Young leaves and roots	Convulsions
Whorled milkweed	A resinoid, galitoxin	Shoots especially near top	Poor equilibrium, muscle tremors, depression, and then nervousness, slobbering, mini-convulsions

Source: Zimdahl, 1993.

(Table 1.5), several possible strategies of evolutionary development emerge: stress-tolerators, competitors, and ruderals (Figure 1.6).

Stress-tolerators. Plants that survive in unproductive environments by reducing their biomass allocation for vegetative growth and reproduction. They exhibit characteristics that ensure the endurance of relatively mature individuals in harsh, limited environments. The environmental limitation may be caused by physical factors, such as recurring drought or flood, or biotic factors, such as use of resources by neighboring plants or herbivory. Species with these characteristics are prevalent in continually unproductive environments or during the late stages of succession in fertile environments.

Competitors. Have evolved characteristics that maximize the capture of environmental resources in productive but relatively undisturbed conditions. These

TABLE 1.5 Plant Evolutionary Strategies Resulting from Disturbance and Stress

	Intensity of Stress	
Intensity of Disturbance	*High*	*Low*
High	Plant mortality	Ruderals
Low	Stress tolerators	Competitors

Source: Grime (1979).

plants have extensive vegetative growth and are abundant during the early and intermediate stages of succession.

Ruderals. Found in highly disturbed but potentially productive environments. They are usually herbs and characteristically have a short life span and high seed production. They occupy the earliest stages of succession.

Grime (1979) suggests that most herbaceous weed species fall into two combined strategies of *competitive ruderals* and *stress-tolerant competitors*. Plants possessing the competitive ruderal strategy have rapid early growth rates and competition between individual plants occurs before flowering. Such plants occupy fertile sites, and periodic disturbance (for example, annual tillage) favors their abundance and distribution. Many annual, biennial, and herbaceous perennial weed species found on arable land fit the criteria for the competitive ruderal tactic (Figure 1.7; see also Chapter 2).

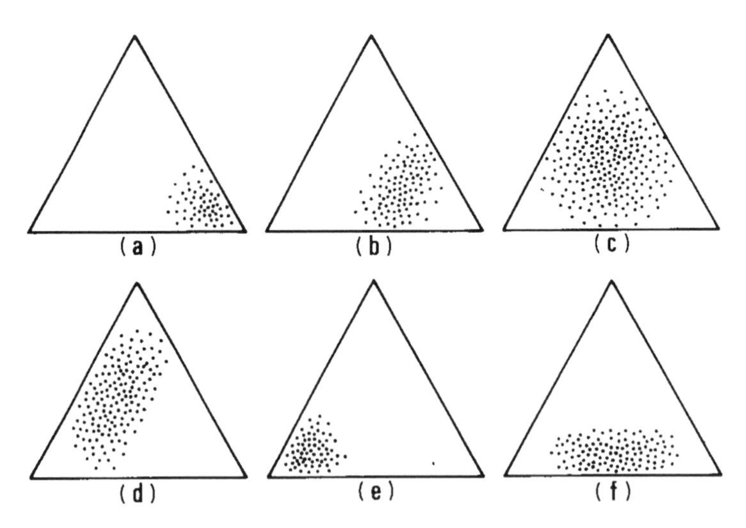

FIGURE 1.6 Diagram describing the range of strategies encompassed by (a) annual herbs, (b) biennial herbs, (c) perennial herbs and ferns, (d) trees and shrubs, (e) lichens, and (f) bryophytes. For the distribution of strategies within a triangle see Figure 2.7. (From Grime, 1977, *American Naturalist* 111:1169–1194. Copyright 1977 by the University of Chicago.)

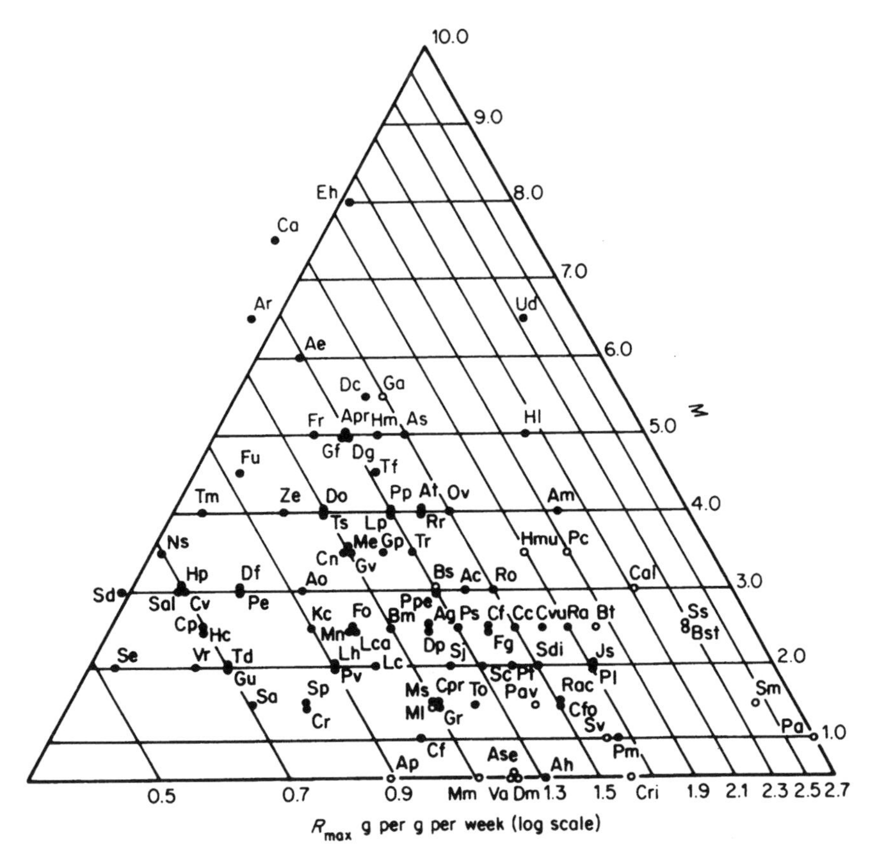

FIGURE 1.7 A triangular ordination of herbaceous species. ○, annuals; •, perennials (including biennials). The morphology index (M) was calculated from the formula $M = (a + b + c)/2$ where a is the estimated maximum height of leaf canopy (1, < 12 cm; 2, 12–25 cm; 3, 25–37 cm; 4, 37–50 cm; 5, 50–62 cm; 6, 62–75 cm; 7, 75–87 cm; 8, 87–100 cm; 9, 100–112 cm; 10, >112 cm); b is the lateral spread (0, small therophytes; 1, robust therophytes; 2, perennials with compact unbranched rhizome or forming small (<10 cm diameter) tussock; 3, perennials with rhizomatous system or tussock attaining diameter 10–25 cm; 4, perennials attaining diameter 26–100 cm; 5, perennials attaining diameter >100 cm); c is the estimated maximum accumulation of persistent litter (0, none; 1, thin discontinuous cover; 2, thin, continuous cover; 3, up to 1 cm depth; 4, up to 5 cm depth; 5, >5 cm depth) (Grime, 1974). Key to species: Ac, *Agrostis canina*, ssp. *canina*; Ae, *Arrhenatherum elatius*; Ag, *Alopecurus geniculatus*; Ah, *Arabis hirsuta*; Am, *Achillea millefolium*; Ao, *Anthoxanthum odoratum*; Ap, *Aira praecox*; Apr, *Alopecurus pratensis*; Ar, *Agropyron repens*; As, *Agrostis stolonifera*; Ase, *Arenaria serpyllifolia*; At, *Agrostis tenuis*; Bm, *Briza media*; Bs, *Brachypodium sylvaticum*; Bst, *Bromus sterilis*; Bt, *Bidens tripartita*; Ca, *Chamaenerion angustifolium*; Cal, *Chenopodium album*; Cc, *Cynosurus cristatus*; Cf, *Carex flacca*; Cfl, *Cardamine flexuosa*; Cfo, *Cerastium fontanum*; Cn, *Centaurea nigra*; Cp, *Carex panicea*; Cpr, *Cardamine pratensis*; Cr, *Campanula rotundifolia*; Cri, *Catapodium rigidum*; Cv, *Clinopodium vulgare*; Cvu, *Cirsium vulgare*; Dc, *Deschampsia cespitosa*; Df, *Deschampsia flexuosa*; Dg, *Dactylis glomerata*; Dm, *Draba muralis*; Do, *Dryas octopetala*; Dp, *Digitalis purpurea*; Eh, *Epilobium hirsutum*; Fg, *Festuca gigantea*; Fo, *Festuca ovina*; Fr, *Festuca rubra*; Fu, *Filipendula ulmaria*; Ga, *Galium aparine*; Gf, *Glyceria fluitans*; Gp, *Galium palustre*; Gr, *Geranium robertianum*; Gu, *Geum urbanum*; Gv, *Galium verum*; Hc, *Helianthemum chamaecistus*; Hl, *Holcus lanatus*; Hm, *Holcus mollis*; Hmu, *Hordeum murinum*; Hp. 1, *Helictotrichon pratense*; Js, *Juncus squarrosus*; Kc, *Koeleria cristata*; Lc, *Lotus corniculatus*; Lca, *Luzula campestris*; Lh, *Leontodon hispidus*; Lp, *Lolium perenne*; Me, *Milium effusum*; Ml, *Medicago lupulina*; Mm, *Matricaria matricarioides*; Mn, *Melica nutans*; Ms, *Myosotis sylvatica*; Ns, *Nardus stricta*; Ov, *Origanum vulgare*; Pa, *Poa annua*; Pav, *Polygonum aviculare*; Pc, *Polygonum convolvulus*; Pe, *Potentilla erecta*; Pl, *Plantago lanceolata*; Pm, *Plantago major*; Pp, *Poa pratensis*; Ppe, *Polygonum persicaria*; Ps, *Poterium sanguisorba*; Pt, *Poa trivialis*; Pv, *Prunella vulgaris*; Ra, *Rumex acetosa*; Rac, *Rumex acetosella*; Ro, *Rumex obtusifolius*; Rr, *Ranunculus repens*; Sa, *Sedum acre*; Sal, *Sesleria albicans*; Sc, *Scabiosa columbaria*; Sd, *Sieglingia decumbens*; Sdi, *Silene dioica*; Sj, *Senecio jacobaea*; Sm, *Stellaria media*; Sp, *Succisa pratensis*; Ss, *Senecio squalidus*; Sv, *Senecio vulgaris*; Td, *Thymus drucei*; Tf, *Tussilago farfara*; Tm, *Trifolium medium*; To, *Taraxacum officinalis*; Tr, *Trifolium repens*; Ts, *Teucrium scorodonia*; Ud, *Urtica dioica*; Va, *Veronica arvensis*; Vr, *Viola riviniana*; Ze, *Bromus erectus*. Estimates of R_{max} are based on measurements during the period 2–5 weeks after germination in a standardized productive controlled environment conducted on seedlings from seeds collected from a single population in Northern England. (From Grime, 1974. Reproduced by permission of Macmillan (Journals) Ltd.)

Stress-tolerant competitors are primarily trees or shrubs, although some perennial herbs also fall into this category (Figure 1.7). Common characteristics of these weeds are rapid dry matter production, large stem extension, and high leaf area production (see Chapter 6).

Weeds as a Product of Human Activity

There are many books that describe and identify weeds. Some weed species have even achieved worldwide prominence (Holm, 1978; Holm et al., 1977; see Table 1.3). Most weeds are important, however, from a more local to regional perspective. The local distribution of weeds is influenced by environmental and biological factors that determine habitat types and human activities. Environmental factors that affect weed occurrence are soil type, pH, soil moisture, light quantity and quality, precipitation pattern, and variation in air, soil, and water temperatures. Biological factors, such as the incidence of insects and diseases on either weeds or associated crops, grazing activities of animals, and plant competition also can influence the distribution of weeds. Human activities, such as farming or forestry, also are a major reason for the local and regional patterns of weed distribution. DeWet and Harlan (1975) indicate that plant species are affected and undergo changes when their habitats are disturbed. Disturbance favors some species and disfavors others, which eventually may be reduced in number or replaced. These are all topics of discussion in later chapters of this text.

WEEDS ON AGRICULTURAL LAND

In his book, *Weeds and Aliens*, E.J. Salisbury (1961) provides several enlightening examples of the shifting patterns of the past and present agricultural weed flora in Great Britain resulting from human activities. He believes that some plants presently regarded as weeds were initially cultivated as crops—a contention supported by Harlan and his co-workers. These species include chess (*Bromus secalinus*) and wild oat (*Avena fatua*) as cereals, lambsquarters (*Chenopodium album*) as poultry feed, and large-seeded false flax (*Camelina sativa*) as an oil crop. Salisbury also describes the introduction of what is considered the native weed flora into Great Britain, primarily by the Romans. Using historical evidence of weeds that have diminished over time, he suggests that weed evolution is a highly dynamic process. Based on that evidence, he predicts that some "casual" weeds of today are likely to become the noxious, troublesome weeds of tomorrow owing to the continual modification of cultural practices and "modernization" of agriculture. The present increase of herbicide resistance in formerly susceptible plant species suggests that he was right.

In the *Boke of Husbandry* (Fitzherbert, 1523) a number of weeds that "doe moche harme" are enumerated. Salisbury infers that they were some of the most common weeds in England 400 years ago. Some are still common but others, such as darnel (*Lolium temulentum*) and corncockle (*Agrostemma githago*), would rarely be listed as modern-day weeds. At that time, the grains of both darnel (Figure 1.8) and wheat were probably harvested together and subsequently sown together, especially since the wheat varieties then more closely resembled darnel in size than do modern varieties. The invention of the seed drill by Jethro Tull in 1730 caused a major revolution in agricultural practices at the time because, in addition to other benefits, it allowed hand hoeing of weeds between crop rows (Tull, 1733). The author of *Hortus Graminus Woburnenis* (1826; see Salisbury, 1961) regarded the diminution of darnel as mainly due to the "new

FIGURE 1.8 Illustration of one of the most ancient weeds, darnel (*Lolium temulentum*), the "tares" of the Bible. From the "Herbal of Dioscorides" (Vienna manuscript, illustrated by a Byzantine in A.D. 512), where it is termed "Aira" and also "Thyaron," and is noted as growing among wheat. The Romans called it "Lolium." (From King, 1966.)

husbandry" for, according to Salisbury, "It is never found on arable land that is cultivated under drill husbandry except when introduced by using seed from other farms where broadcast sowing is practiced." Apparently, darnel diminished as an important weed in cereal culture because of orderly sowing of cereal in rows and the concomitant use of hand hoes made possible by the widespread use of the grain drill.

Corncockle, also a prominent weed in sixteenth- through eighteenth-century England, declined as seed cleaning technologies were improved during the nineteenth century—an observation supported by Harper (1977). Both Salisbury and Harper indicate that corncockle seed is able to remain viable in soil for only a short time; therefore, improved techniques to "clean" crop seed had an almost immediate effect on the prevalence of this species. In a comprehensive study that spanned more than sixty years, Haas and Streibig (1982) also documented changes in weed species composition that resulted from various cultural practices in Danish cereal fields. Table 1.6 lists the weed species that were displaced and enhanced by the agricultural practices in Denmark over this period. Cultural practices believed to influence weed species shifts were fertility management, soil amendments, tillage operations, herbicide application, and methods of harvesting.

Perhaps the most extreme example of how human activities influence weed species distribution and composition are *crop mimics*. These are weed species that have evolved life cycles or morphological features so similar to a crop that the two species cannot be distinguished or separated easily. Chapter 3 considers the influence of humans on the evolution of weed species, including crop mimicry, in much more depth.

Holzner and Immonen (1982) indicate that human action is the most important factor in determining the occurrence and distribution of agricultural weed species. They indicate that many agrestals that accompanied crops for centuries in Europe have now become locally extinct, retreating to their climatic optimum where most survive outside cultivated fields. Other plants, they note, have increased in both prominence and abundance. Holzner and Immonen suggest several causes for these recent changes in weed species composition:

- Improved seed cleaning, which results in the local eradication of "specialists" that are unable to grow outside arable land and depend on being sown with the crop.
- Abandonment of crops, which leads to loss of specialized weeds.
- The "leveling" of environmental conditions, which results in a uniform weed flora.
- Increased reliance on monocultures, which tends to simplify the weed flora.
- Combine harvesting, which allows some weed species to shed seed in the field and distributes the seeds of others.
- Reduced tillage and "no tillage" operations, which promote perennial species.

TABLE 1.6 Displacement of Weed Species in Danish Spring Cereals

	Occurrence in Fields (%)			Responds to[a]		Susceptible to[b]	
Weed Species	*1918*	*1945*	*1970*	*Nitrogen/Soil Amendment*	*Combine Harvest*	*Harrow*	*Phenoxy Herbicide*
Increasing Species							
Matricaria inodora (scentless chamomile)	18	34	27	+	+	S	M
Matricaria matricarioides (pineappleweed)	0	1	18	-	+		M
Plantago major (broadleaf plantain)	36	49	75	-	-		M
Poa annua (annual bluegrass)	18	34	46	+	+		T
Stellaria media (chickweed)	73	94	92	+	+	S	M
Species Unchanged or Retreating							
Agropyron repens (quackgrass)	55	47	60	+	-		T
Atriplex patula (spreading orach)	27	60	21	+	-		S
Capsella bursa-pastoris (shepherdspurse)	46	53	22	+	+		S
Chenopodium album (common lambsquarters)	64	85	68	+	+	S	S
Myosotis arvensis (field forget-me-not)	36	45	50	-	+		M
Polygonum convolvulus (wild buckwheat)	91	93	82	+	-		S
Polygonum aviculare (prostrate knotweed)	82	85	64	-	+	S	M
Rumex acetosella[c] (sheep sorrel)	55	20	15				M
Scleranthus annuus	55	23	21				
Sinapis arvensis (wild mustard)	75	83	12				S
Sonchus arvensis (perennial sow thistle)	73	60	19				M
Taraxacum spp. (dandelion)	73	35	27	-	-		S

Source: Haas and Streibig, 1982, in LeBaron and Gressel (Eds.), *Herbicide Resistance in Plants*. Copyright © 1982 by John Wiley & Sons, Inc., New York.

[a]+ indicates that the species responds favorably to nitrogen fertilization, other soil amendment, or combine harvesting (compaction), and - indicates an unfavorable response.

[b]S indicates that the species is susceptible to harrowing or MCPA or dichlorprop herbicides. M and T refer to moderately tolerant or tolerant to MCPA or dichlorprop. No symbol under harrow indicates some tolerance to harrowing.

[c]Calcifuge species, responds adversely to lime.

- Reduced competitive ability of short-statured crops and crops treated with chemical growth regulators.
- Extensive use of herbicides, which causes sensitive species to become locally extinct or to evolve resistance to the chemical. (They note that herbicide-resistant species usually fill niches rapidly, increase their density and invade areas they were formerly unable to occupy—a phenomenon called compensation.)

WEEDS IN FORESTS

There are many natural conditions, such as climate, soil type and fertility, topography, and events, like hurricanes and wildfire, that shape the forested landscapes of the world. Following "catastrophic" disturbances, it is common for forests to undergo a sequence of vegetation changes that result in a forest nearly identical to the one previously destroyed. This process of natural forest re-establishment through successive changes in vegetation composition is called secondary succession (Figure 1.9). Following a radical disturbance, a new patch in the physical environment is once again available for colonization by plants. In such situations, "pioneer" tree (e.g., poplar, birch, alder and some conifers) or shrub species (e.g., ceanothus or manzanita) (Figure 1.10) are quick to colonize the disturbed areas and can dominate them for years and even for decades. This rapid recolonization by pioneer species, although a normal stage in succession (see Chapter 2), can seriously delay the replacement of disturbed sites with more economically desirable trees.

The major disturbance to forests of any region is the harvesting of wood by humans. It is estimated that each year the world loses 37 million acres of forest in this manner (Perlin, 1989). In temperate conifer forests, especially those with-

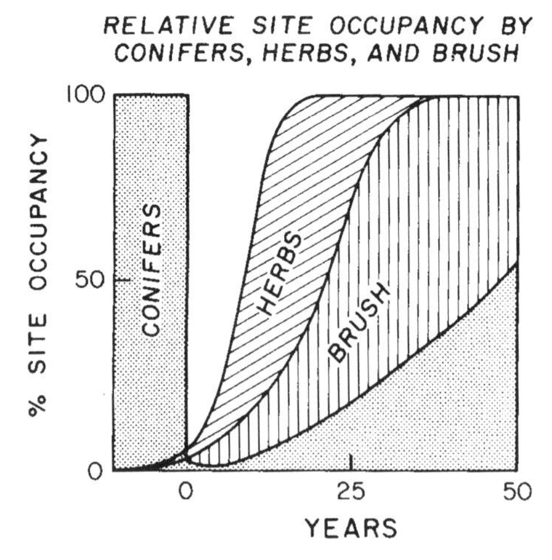

FIGURE 1.9 A hypothetical example of early forest succession following disturbance. (From Radosevich and Conard, 1981. By permission of Academic Press, New York.)

FIGURE 1.10 Forest succession in the Sierra Nevada mountains, California. Reversion to an earlier stage as a result of frequent fires is indicated. (Photograph by S.R. Radosevich, in Radosevich and Conard, 1981. Reproduced by permission of Academic Press, New York.)

out any follow up reforestation activities, logging has led to the gradual replacement of conifers by less desirable herbaceous, shrub, or hardwood species. Sutton (1985) summarizes this situation for much of Canada by saying:

> The extent of the weed problem is influenced greatly by not only the nature of stands that were cut but also by the season and method of harvesting, the intensity of utilization, and the time that has been allowed to elapse between harvesting and the attempt to regenerate. Such factors are ignored in exploitation forestry, which strives to minimize the unit cost of extracted wood and maximize immediate profits. The result has been the virtual complete separation of harvesting and regeneration operations in Canadian forestry. This separation, together with the inadequacy of the regeneration effort by industry and government alike, has produced weed problems on a vast scale.

Although the ability to regenerate forests varies by region or locality, the conditions described by Sutton, unfortunately, are not unique to Canada. For example, Waldstad and his associates (1987) indicate that hardwoods, principally red alder, occupy 32 percent of the prime timberland in western Oregon that was once dominated by conifers.

Forest regeneration. Most forests regenerate naturally following disturbance, given enough time. However, logging activities and land clearing are the principal disturbance factors that both set up and modify the natural patterns and time frames of succession so that "weed" species are favored, and sometimes even dominate, in many forest types. The ability of a site to regenerate following such disturbances, as well as the composition of species that result afterward, is most dependent on the type, frequency, and severity of the tree removal operation.

Brender (1952, in Gjerstad and Barber, 1987) provides a historical account of the impact of European settlers on Georgia forests, a pattern of land use that was similar throughout the southern United States. In so doing, he also describes a pattern of forest regeneration called *old-field succession* (Figure 1.11). The frontier advance across Georgia began in 1773 and reached the Alabama border in 1826. Land was cleared so rapidly for farming that one county open to settlement in 1812 contained only 200 acres of virgin forest by 1847. Following the American Civil War (mid-1800s), most of the suitable land was devoted to continuous cotton production, which in a relatively short time resulted in severe soil erosion and depletion. Except for small scattered areas, virtually none of the original hardwood and pine forests remained. A decline in cotton prices during the 1880s resulted in the abandonment of many fields that were then invaded by pines. When World War I (early 1900s) brought an increased demand for cotton, second-growth pine was cut as the land was again cleared for farming.

The introduction of the cotton boll weevil and the Great Depression (1930s) again resulted in the abandonment of many farms. Natural seeding from fence rows and other adjacent pines, in addition to millions of seedlings planted by the U.S. Civilian Conservation Corps in the 1930s, were the beginning of the

FIGURE 1.11 Old-field succession in western Oregon. (Photograph by S.R. Radosevich, Oregon State University.)

current southern forests that now cover approximately 65 percent of that area's land base. Pines are currently the preferred timber resource in the region. However, Gjerstad and Barber (1987) indicate that hardwood encroachment is occurring each year on about 600,000 acres previously considered to be "pine-type" in the Southeast.

Techniques, collectively known as *artificial regeneration* have been used successfully to replant many logged-over areas in the United States and Canada, including those mentioned above (Figure 1.12) Artificial regeneration usually involves collecting seed of preferred tree species, germinating and growing the seedling trees in nurseries, out-planting them to field sites, followed by intensive weed control. Despite these apparent successes, however, important questions still remain about the ecological, social, and economic desirability of converting vast acreages of virgin and naturally regenerated forests into tree farms.

WEEDS IN RANGELANDS

The destruction and replacement of vegetation by humans is now a common occurrence over most of the world, with a loss in primary productivity and floristic diversity often being the result. The invasion of exotic species also can be both a cause and a consequence of such environmental manipulation. However, it is rare that invaders cause the replacement of most or all of the plant and animal species in a disturbed ecosystem (Billings, 1990). A possible

FIGURE 1.12 A young Douglas-fir plantation following logging and artificial regeneration. (Photograph by S.R. Radosevich, Oregon State University.)

exception to this generalization is rangeland weeds. In this system of production, the species replacement following disturbance often seems so complete that only a sketchy picture of pre-disturbance conditions remain. We offer downy brome, also called cheatgrass (*Bromus tectorum*), as an example.

Cheatgrass (Figure 1.13) is a desert annual grass that is exotic to North America. Woodwell (1986) suggests that it is "one of those small bodied, rapidly reproducing, hardy plants that finds a variety of open niches around the world and [in so doing] changes the world." In this case, the chance introduction of cheatgrass before the turn of the century to the Great Basin of North America has altered the entire native shrub ecosystem of that region. Billings (1990) indicates that this introduction also provides us with a classical case of biological impoverishment and the concomitant environmental change that allows successful replacement of an indigenous vegetation, that is, native perennial bunchgrasses and shrubs. The particular example in this case is sagebrush (*Artemisia tridentata*) rangeland, which was first grazed by large herbivores, then invaded by cheatgrass, and subsequently subjected to range fires.

Original vegetation and early land use history of the Great Basin. Billings (1990) and others (Klemmedson and Smith, 1964; Mack, 1984) indicate that the western Great Basin was not part of the bison range of the North American

FIGURE 1.13 Cheatgrass (*Bromus tectorum*) in former sagebrush-bunchgrass range. (Photograph by S.R. Radosevich, Oregon State University.)

FIGURE 1.13 Continued.

Great Plains because the rhizomatous C_4 grasses on which the bison thrived cannot grow on the summer-dry steppes of this region. Rather, perennial C_3 bunchgrasses of the genera *Poa, Festuca, Agropyron*, and *Stipa* dominated the grass stratum of this sagebrush ecological formation (Billings, 1990). Apparently, the native bunchgrasses of the region also did not carry fire well because range fires in the sagebrush-bunchgrass steppe, in contrast to the Great Plains, were rare.

The native ungulate herbivores were antelope, deer, desert bighorn sheep, and elk, which, because of their smaller size and numbers than bison, created a relatively light impact on the sagebrush-grass community. During the 1840s and 1850s, the first overland wagon trains to Oregon and California introduced domestic livestock to the region. Thus, the first grazing impacts in the area appeared along the Oregon and California trails. For example, Beckwith, an early explorer, noted the following in June, 1854:

> Fine droves of cattle, which had been overwintered near Great Salt Lake passed today on their way to California, and one or two large flocks of sheep are but a few miles behind them. . . . The more experienced stock drovers send their cattle back from the river to feed on the nutritious grass of the hills (Beckwith, 1854, in Billings, 1990).

Watson (1871) made one of the first good botanical descriptions of the area, listing 59 species of Poaceae. Cheatgrass was not among the species listed, suggesting that it had not yet arrived in the intermountain region of North America. In the summer of 1902, Kennedy (1903) made the first survey of range conditions in northern Nevada and fifty years later Robertson (1954) retraced Kennedy's route. Billings (1990) compared their writings and noted the following changes in range conditions over that 50-year period:

- Desirable livestock browse shrubs decreased.
- *Agropyron spicatum*, a prime forage bunchgrass, decreased from "abundant" to "generally absent" or less than 5% density.
- Annuals, notably cheatgrass, not present in 1902, had increased to an "extreme degree."
- Burn scars were "absent or unimportant" in 1902. In 1952 much of the route was bordered or crossed by "burned off range" and covered by cheatgrass or little rabbit-brush (*Chrysothamnus viscidiflorus*).
- Big sagebrush replaced "bluegrass meadows" at lower elevations.
- "Stream channels had eroded deeper and wider."

All of the conditions in the above list indicate heavy grazing, cheatgrass invasion, and occurrence of repeated fires.

Introduction of cheatgrass and fire. According to Mack (1981, 1986), the first collections of cheatgrass in the Great Basin were from Spense's Ridge, British Columbia, 1889; Ritzville, Washington, 1893; and Provo, Utah, 1894. Each location is in a wheat growing area, which suggests that cheatgrass seeds may have arrived as a contaminant of crop seed. From these beginnings, the species spread throughout eastern Washington and Oregon, southern Idaho, northern Nevada and Utah. By the 1930s, it was abundant throughout the entire sagebrush steppe. Billings (1990) believes that the rapid spread of the grass across the region was aided by railroad stock cars and grazing animals that were subsequently driven onto the rangelands. In addition, the climate of the Great Basin was ideal for the new weed. As a winter annual, it requires moist soils during the cold season and cold winter weather while it is vegetative in order to flower the following spring.

Billings (1990) indicates that once cheatgrass became established, the region was set for wildfires. Cheatgrass usually sets seed, dies and dries up by June in most areas of the region. Thus, a supply of fuel that was nonexistent in the original open sagebrush-bunchgrass ecosystem becomes available. Without fire, cheatgrass simply invades the overgrazed sagebrush range, where it forms an ephemeral annual stratum in that community. However, once this plant community experiences either lightning- or human-caused fire, the sagebrush is killed. Since this shrub cannot sprout following fire, the native shrublands of the Great Basin have been replaced by vast expanses of annual grassland. Billings (1990) points out that as cheatgrass has become more and more dominant, upland areas of the Great Basin, notably the pinyon pine-juniper biome, are now increasingly threatened by a similar process of vegetative change.

Management of cheatgrass. Mack (1984) notes that the assemblage of features that allowed cheatgrass to invade the Great Basin has also largely prevented its eradication. For example, a plant capable of germinating over an eight-month period, as cheatgrass can, is nearly impossible to control completely in the seedling stage. Mowing or grazing in early spring makes little difference since developing seeds of the species are viable and readily capable of germinating the following autumn. Even when fire removes all vegetative plants, new ones emerge from seed reserves in the soil and, of course, further reentry is always possible. Furthermore, as we will see in Chapter 3, the plant has little problem adapting to the wide variety of environmental conditions of both rangeland and cultivated fields.

On the other hand and from the standpoint of volume of herbage produced and extent of area covered, cheatgrass is unquestionably the most important forage plant in the Great Basin now (Klemmedson and Smith, 1964). It provides the bulk of early spring grazing for all classes of livestock on millions of acres in the arid West. Thus, it is not easy to comprehend the economic importance of this ecological change, although the extent and permanence of it are readily comprehensible.

Weeds in Production Systems

Weeds are first and foremost part of a complex biological system. They also are a component in several complex socio-economic systems that we call agriculture, forestry, or range management. Figure 1.14 depicts a general conceptual diagram of how agriculture interacts with all other components of the total human ecosystem (Nevah, 1980; Roush et al 1989b). Obviously, weeds are only one component of the agricultural and semi-agricultural landscapes, and they interact with, influence, and are influenced by many other components of the entire human ecosystem. The simplest approach to dealing with weeds has been to control them, to reduce their abundance. However, this tactic has sometimes been inadequate, especially when all biological, social, and economic factors are considered (Chapter 11). In order to understand how weeds fit into production systems, it is necessary to determine how the processes and factors within the

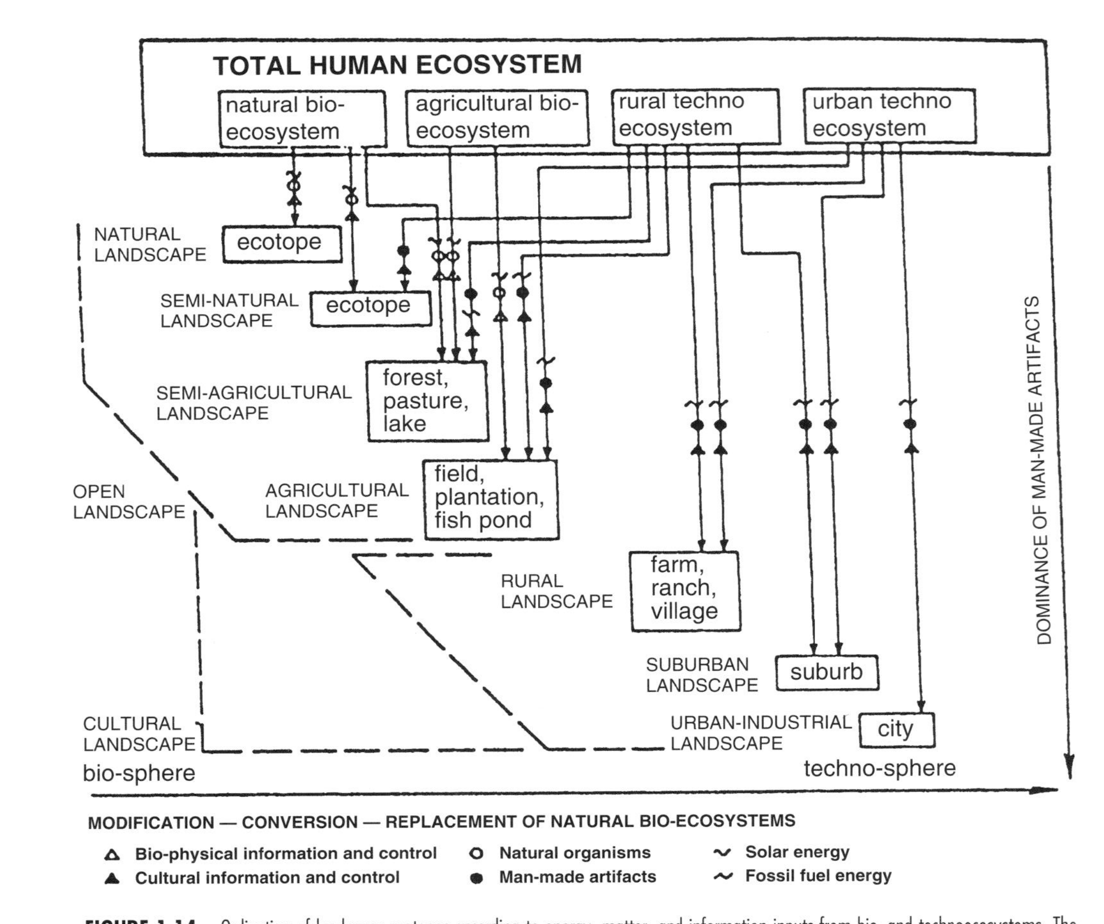

FIGURE 1.14 Ordination of landscape ecotopes according to energy, matter, and information inputs from bio- and technoecosystems. The achievement of a new balance between the left and the right poles of this ordination is the major goal of landscape ecology. (From Naveh, 1980.)

system interact and then to organize these processes into a logical framework. These and other approaches to dealing with weeds form the basis of the field of Weed Science. Thus, we examine the development and evolution of the discipline of Weed Science in the remainder of this chapter.

WEED SCIENCE

Weed Science is an integrative, applied scientific discipline typical of most other pest-management and production-oriented disciplines of modern agriculture. Weed scientists have been extremely successful in developing tools and tactics to reduce weeds in crops, rangelands, and forest plantations. Here we examine the history, interdisciplinary composition, and activities of Weed Science.

History of Weed Science

Many textbooks and most historical papers about Weed Science are quick to point out that weeds have been with us since settled agriculture began, perhaps ten thousand years ago. Weeds must have been known to early farmers because hoes and other "grubbing" implements, artifacts of those ancient times, have been found at archeological sites. In addition, many references and accounts of weeds and their detrimental effects on crop yields are found in early writings, from Theophrastus and the Bible to more recent books. These writings have shaped our ideas and definitions of weeds as we saw earlier in this chapter. However, the existence of weeds in agriculture does not necessarily coincide with the evolution of Weed Science as a scientific discipline. In fact, even today, weeds are just an incidental part of food production in most parts of the world and farmers are simply people with hoes. Rather, the history of Weed Science parallels the history of modern-day agriculture and is probably less than a century old, even though weeds have been with us since agriculture began.

In many respects, the history of Weed Science is the history of weed control. Control has obviously progressed from hand pulling and primitive hand tools to animal- and fossil-fuel-powered implements (Chapter 8). The earliest weed control with chemicals was with inorganic salts, which were used widely in the early 1900s. The work of Bolley in North Dakota, Bonnet in France, and Shultz in Germany, who all used copper salts and later sulfuric acid to control weeds in cereals, is most often cited as the earliest accounts of weed control with chemicals (Zimdahl 1991,1993). Later, agricultural scientists in Europe and the United States observed the herbicidal effects of other metallic salt solutions on weeds in cereal crops. Historical accounts always cite the synthesis of 2,4-D in 1941 by Pokorny (1941) and the subsequent discovery of its plant growth regulating and herbicidal properties as the first account of an organic chemical used to control weeds. However, dinitrophenolic chemicals were used extensively for weed control in the 1930s. Nevertheless, Weed Science and hence planned weed control tactics began after the discovery of the herbicidal properties of 2,4-D. Later, Bucha and Todd (1951) reported the first herbicide that was not a derivative of a phenoxy acid, like 2,4-D. This chemical was monuron—the first of many phenyl urea herbicides that disrupt photosynthesis (see Chapter 10 for more details about these chemicals).

Zimdahl (1991, 1993) also notes that the major players in the history of Weed Science all developed careers during the twentieth century. He includes seven men as prominent figures in the creation of Weed Science in the United States: Wilfred Robbins (1884–1951), University of California, Davis; James Zahney (1884–1975), Kansas State University; Charles Willard (1889–1974), Ohio State University; F.L. "Tim" Timmons (1905–1994), U.S. Department of Agriculture; Erhart "Dutch" Sylvester (1906–1975), Iowa State University; Kenneth Buchholtz (1915–1969), University of Wisconsin; and Alden Crafts (1897–1990), University of California, Davis. Each of these scientists pioneered investigation of organic chemicals for weed control, developed weed control strategies or programs that became models for others in the United States, and were founders of regional and national Weed Science societies.

The above honor roll suggests that the history of Weed Science has been very recent, dominated by the U.S. land grant system of higher education, and highly chemically oriented. During this same time period, however, other scientists—botanists and plant scientists—also had begun to study weeds as unique and interesting organisms. A list of these scientists would include: Herbert Baker, botanist and plant evolutionist, University of California, Berkeley; Jack Harlan, botanist, evolutionist, ecologist, University of Illinois; John Harper, ecologist and plant population biologist, University of Wales; and Edward Salisbury, botanist and natural historian, Royal Botanic Gardens, Kew. Although not known as Weed Scientists, these men devoted much of their careers to the study of weeds, including their biology, life cycles, and evolution. As the result of such pioneering efforts, a new era of understanding and activity in Weed Science is beginning to emerge—management based on biological principles and ecological concepts.

Recent Directions in Weed Science

In 1992, the Weed Science Society of America (WSSA) developed a list of research needs for its discipline (Hess, 1994). The Society concluded that Weed Science research should focus on:

- increasing knowledge related to the biological and economic impact of weeds across a wide range of environments,
- improving understanding of the biology, ecology, and genetics of weeds to optimize their management,
- biological agents and natural products to optimize their performance in the field,

- weed populations resistant to herbicides so the occurrence of such populations can be minimized,
- new technology for the application of herbicides to optimize their performance and minimize their environmental impact, especially in surface and ground water, and
- better methods to detect residues of herbicides in water, soil and vegetation.

The research identified by WSSA points out the need for greater understanding of weed biology as well as the environmental and biological impacts of weed control measures. To enhance such understanding is the objective of this text.

Disciplinary Composition and Activities of Weed Science

Weed Science, as it exists today, is comprised of six fundamental scientific disciplines that are organized in Figure 1.15 to show the predominant areas of activity—*weed control technology, weed biology* and the *societal values and issues affecting weed control.* Each fundamental discipline depicted in Figure 1.15 uses the scientific approach, as do weed scientists. But Weed Science, like many other applied disciplines, has rarely developed its own theories. Rather, weed scientists rely upon the fundamental disciplines (Figure 1.15) to generate theories that also apply to weeds and crops. The result has been a body of empirically generated information that is used to develop, improve, modify, and sometimes justify the technology used to control weeds.

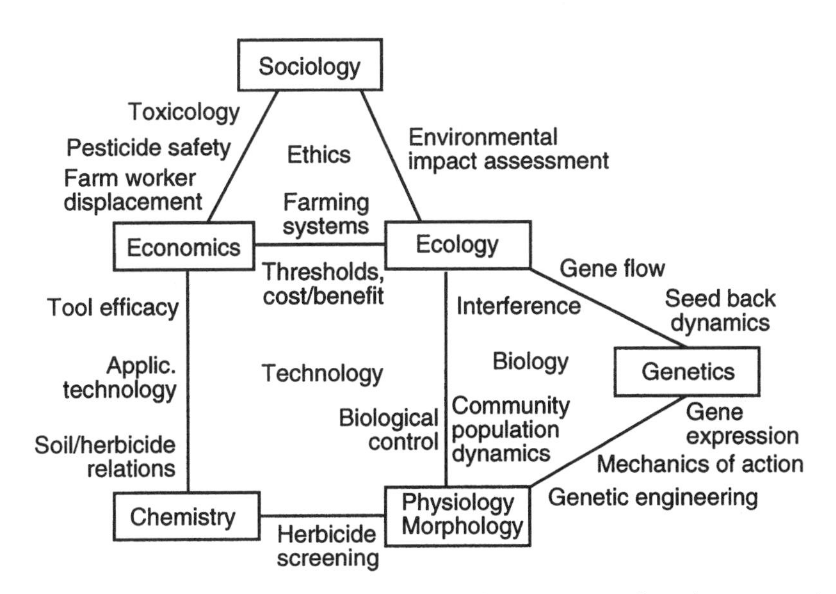

FIGURE 1.15 Interrelationships of six fundamental disciplines in weed science. Major areas of research activity are weed control technology, weed biology, and ethics of weed control. (From Radosevich and Ghersa, 1992.)

WEED CONTROL TECHNOLOGY

The technological goal of Weed Science is to develop efficient methods of weed control in crops, forest plantations, and some rangeland and non-crop situations. Information from the disciplines of plant physiology/morphology, organic chemistry, economics, and ecology (Figure 1.15) is used to develop or improve tools to control weeds. The search for cost-effective ways to control weeds often has focused on tillage and herbicides (Chapters 8, 9, and 10) as a means to reduce labor requirements, reduce production costs, or increase crop yields. Several reasons to control weeds are listed below. The question of whether to control weeds is more complicated and is addressed in Chapters 5, 8, and 11.

Improve crop production. The threat of weeds to continued crop productivity accounts for most of the human effort devoted to weed control. It has been estimated that 10 to 15 percent of the total market value of farm products is lost because of weeds. This loss amounts to about four billion dollars per year in the United States (Bridges and Anderson, 1992; Bridges, 1994). Direct losses to forests and rangeland are more difficult to estimate than agricultural losses. Walstad and Kuch (1987) believe that a nearly 30 percent reduction in forest productivity will result over the next several decades because of weed occupation during the early stages of forest plantation formation.

Enhance product quality. Weeds have a detrimental effect on crop quality as well as quantity, especially crops that must meet size, color, nutrient content, or contamination-free standards. For example, yields of alfalfa hay in California often are highest during the first cutting when annual weeds are present. However, hay quality is often low when weeds are present in the crop. Protein contents below 10 percent often are found when the hay contains large amounts of weeds, but when the weeds are controlled and few weeds are present with the alfalfa, the protein content of the hay can exceed 20 percent. Such increases in grade or quality often mean increased revenue for growers, since a premium price is usually paid for commodities of high quality.

In some cropping systems, the crop seed and weed seed are so similar in weight and shape that separation at harvest is difficult. Examples are alfalfa and dodder (*Cuscuta* spp.) seed, soybean seed and nightshade (*Solanum* spp.) fruits, and pea seed that are mixed with the immature flowers of Canada thistle (*Cirsium arvensis*). If the weed material is not removed from these crops by screening, lower price for the commodity will result. For seed crops, the presence of a few noxious weed seeds, even less than one percent, usually makes the commodity unmarketable.

Reduce costs of production. Weed control is a major reason for many cultural practices associated with crop production. For example, weeds are killed during plowing and cultivation to prepare seedbeds for planting. A report by the United States Department of Agriculture indicated that approximately 42 percent of all pest control costs were for weeds (Ross and Lembi, 1985). Over 65 percent of the total pesticide sales in the United States in 1984 was for herbicides

(Zindahl, 1993). There is no doubt that weed control is a costly endeavor in the production of most crops.

Weeds also interfere with harvesting operations, often making harvesting more expensive and less efficient. For example, weeds sometimes get wrapped around rollers or cylinders of mechanical harvesters, causing equipment breakdowns and longer harvest times. Up to 50 percent loss in efficiency and 20 percent loss of yield can result from the presence of weeds at harvest time.

Reduce other pests. Some weed species act as alternate hosts or harbor insects, pathogens, nematodes, or rodents that are crop pests. Numerous specific examples exist of various pest organisms that benefit from the presence of weeds. Aphids and cabbage root maggots live on wild mustard, later attacking cabbage and other cole crops. Nightshades are hosts of the Colorado potato beetle. Disease organisms, such as maize dwarf mosaic and maize chlorotic dwarf virus, use Johnsongrass rhizomes to overwinter. Black stem rust uses barberry, quackgrass, and wild oat as hosts prior to infesting cereal crops. Rodent damage to orchards can be prevented by weeding around trees before winter.

It also is possible for weeds to aid in the prevalence or spread of certain beneficial organisms that are used to control other pests. In such cases, the weeds act as an alternate source of food or cover for the beneficial organisms, allowing them to survive when the preferred host is not available.

Improve animal health. Some weeds are poisonous to animals. However, plants toxic to one species of animal may be harmless to others. For example, larkspur (*Delphinium* spp.) will kill cattle if eaten in sufficient quantity, but sheep and horses are relatively unaffected by this rangeland weed. In contrast, fiddleneck (*Amsinckia* spp.) is highly toxic to horses, while other livestock are relatively tolerant of it. It is estimated that up to 10 percent of range-grazing livestock may become afflicted by poisonous plants at some time during each growing season.

In addition to direct poisoning, animals may experience other discomforts from association with certain weed species. Some plants [e.g., St. Johnswort (*Hypericum perforatum*), buckwheat (*Eriogonum longifolium*), and spring parsley (*Alchemilla arvensis*)] contain chemicals that make animals abnormally sensitive to the sun—a phenomenon called *photosensitization*. Other plants contain teratogenic materials that result in fetal malformations. For example, malformed lambs can result if false hellebore (*Veratrum californicum*) is ingested by sheep around the fourteenth day of gestation. Bracken fern (*Pteridium aquilinum*) causes a disease of cattle called "red water" because of the blood-colored urine that is its symptom. This weed causes cancer of the bladder if eaten in sufficient quantities.

Other forms of animal discomfort sometimes result from weeds. The sheep shown in Figure 1.16 suffers from the "burs" of burclover, which became lodged around the animal's eyes, mouth, and throat while it was grazing. Foxtail and downy brome caryopses lodge in the feet of dogs, often resulting in infections.

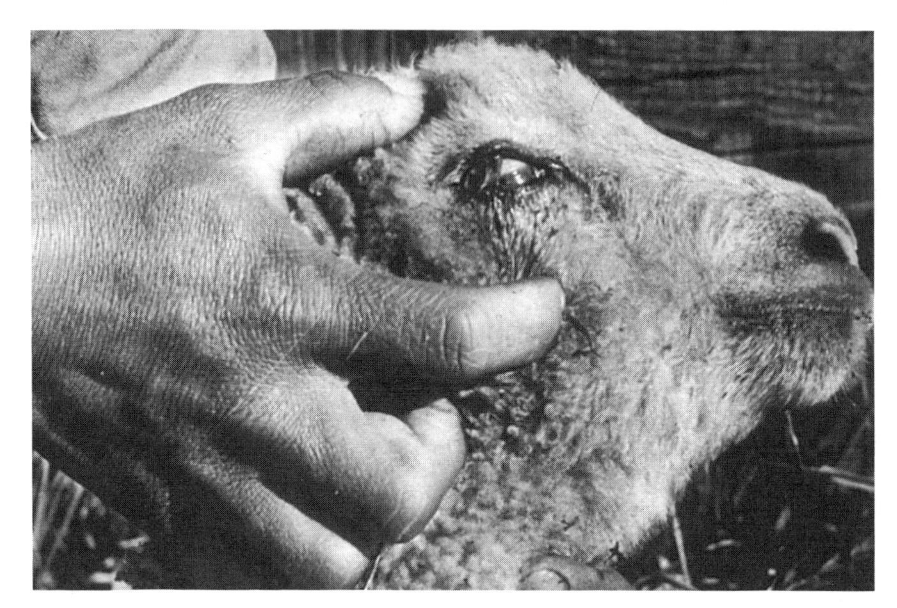

FIGURE 1.16 A sheep with the "burs" of burclover around its eyes and mouth. (Photograph by W.B. McHenry, University of California, Davis.)

Enhance human activities. Weeds affect a number of human activities that are difficult to assess in monetary terms. The presence of weeds can reduce real estate values because of the unkempt and unsightly appearance of the property. Dense moisture-holding weed growth furthers the deterioration of wooden and metal structures and machinery, further reducing property value. Access and enjoyment of recreation areas also are reduced by the presence of weeds.

Weeds adversely affect transportation. Some rivers and lakes in the tropics and subtropics are clogged by aquatic weeds, making travel on them nearly impossible (Figure 1.17). Ross and Lembi (1985) provide an interesting example of how weeds influence transportation costs. Citing data from Shuttleworth (1973), they indicate that 487,000 tons of wild oat seed were inadvertently transported from Canada to the United States along with 16 million tons of grain. The transportation costs for the wild oat were estimated at 2 million dollars.

Weeds are kept free from highway intersections to prevent accidents. Airports and railways also keep signs and lights free of weeds so that maximum visibility can be maintained. Powerline rights-of-way are kept free of tall-growing vegetation to prevent power outages that might result if trees contacted powerlines during storms and to increase access to downed powerlines.

Reduce risks to human health. Toxicants or irritants produced by weeds can cause serious health problems for some people. These discomforts or illnesses include hay fever, dermatitis, and direct poisoning (see Table 1.4). Hay fever afflicts millions of people each year. It is caused by an adverse effect of

FIGURE 1.17 A waterway made unnavigable by aquatic weeds. (Photograph by W.B. McHenry, University of California, Davis.)

proteins associated with the pollen of certain plants on the respiratory system of susceptible people. Ragweed (*Ambrosia* spp.) is best known for causing hay fever. However, pollen from many other broadleaf plants, grasses, trees, and shrubs causes similar allergic reactions. Each year, many people are troubled by poison ivy (*Rhus radicans*), poison oak (*Rhus diversiloba*), and poison sumac (*Rhus vernix*). These plants produce and store a toxic substance called urushiol that causes intense itching and rash upon contact with the skin. Many plants contain toxic substances that, when ingested, cause sickness or death to humans. Toxic substances in weeds are alkaloids, glycosides, oxalates, resins and resinoids, volatile oils, acrid juices, phytotoxins (toxalbumens), and minerals. There are few poisons, including synthetic substances and minerals, that can produce illnesses that match the strength and violence of illnesses caused by some plant-produced toxins.

WEED BIOLOGY

The technology of Weed Science is closely connected to biological research (Figure 1.15). The linkages among the disciplines of ecology, genetics, and plant physiology/morphology delineate this activity. By integrating weed and crop biology with technology, weed scientists have developed better tools to control weeds. For example, weed identity and stage of physiological development have been used to improve specific herbicide uses and application methods (Chapter 9). Knowledge about plant anatomy, morphology, physiology, and biochemistry also enhanced the weed scientist's understanding about plant structure and function and, thus, led to improved herbicide technology. Research on plant hormone regulation, photosynthesis, respiration, water and carbohydrate movement, cell division, and metabolism facilitated development of herbicides and the understanding of how those chemicals inhibited plant function (Chapter 10). Understanding soil physical properties led to improved efficiency and selectivity of soil-applied herbicides and now allows insight into the mitigation of soil and water contamination by such chemicals.

Many agricultural and forest systems now focus on monospecific crops grown in the absence of weeds and other pests. Therapeutic procedures are often followed to control unwanted vegetation in these systems. Justification for such control is based on the knowledge that, under current practices, weed populations are very difficult to prevent and when established, they are stable because of persistent reservoirs of seed or buds in the soil. However, another perspective about weeds considers that weeds have specific "regeneration niches" which mimic those of crops or fit the environments humans create to grow crops, forest plantations, pastures, and so on. Perhaps establishment and growth of weeds could be prevented or reduced by manipulating those same environmental or biological variables. This ecological approach to weed control has begun to receive wider attention in recent years. The fundamental principles of plant ecology that are pertinent to weed biology are discussed in the following chapter. In addition, weed management in relation to these principles will be explored in Chapters 3 through 7.

WEED SCIENCE IN A SOCIAL CONTEXT

Earlier sections of this chapter introduced the concepts of weeds and weed control. Later chapters consider some important ecological principles that influence our understanding about weeds and weed management. However, the interrelationships of weeds and humans, and their possible benefits, costs, risks and opportunities, extend much farther than local fields, farms, or forest plantations. For example, consumers of agricultural and forest products and members of the public concerned about environmental well-being also are influenced by weeds and now influence decisions about them. The aims and expectations of the human community may differ substantially from those of the individual farmer or forester, or there may be conflicts among the various interest groups within the larger human community. This linkage of the social concerns about weeds and the science and technologies developed to manipulate vegetation is also shown in Figure 1.15. The linkage of economics, ecology, and sociology, and some specific examples of how it has affected weed science are explored in Chapter 11.

Answers to such questions as the long-term effects of herbicides on the induction of cancer or birth defects in humans, interactions of herbicides with other chemicals in the environment, and chronic impacts of low-level chemical residues in surface and ground water are now being asked of weed scientists. We are also being challenged about the effectiveness of our tools, and their long- and short-term impacts on biological and genetic diversity, soil productivity, and the sustainability of a production system that relies so much on external inputs of money, energy, and chemicals. There is no doubt that, over the last fifty years, society's emphasis has changed so that it no longer concentrates primarily on productivity and economic efficiency but also gives attention to environmental safety and human health.

One mark of a mature profession is an awareness of its history and, as Zimdahl (1991) points out, the history of Weed Science does not differ substantially from many other pest management and production-oriented agricultural disciplines. Unquestionably, weed scientists and other applied scientists have been willing to tackle the hard problem of local and possibly world-wide food shortage, which explains the predominant emphasis on productivity in these disciplines. These scientists also demonstrate a sincere willingness to develop technological improvements that increase production efficiency, reduce time, or decrease costs. However, nagging questions still exist over fundamental values that were simply incomprehensible, and therefore "unmeasureable," to many weed scientists even a decade ago. These same questions also were considered irrelevant by practitioners of weed control. In many cases, questions about value, benefits of weed control, or risks to environmental well-being or human health were simply taken as "givens," as "societal goods," and were not even thought of being asked or re-examined until they could no longer be ignored. Agricultural and natural resource scientists, who "knew" they were helping to produce more food, fiber, or wood for a hungry world, were simply not ready for the criticism heaped upon them for pursuing such an admirable and universally accepted goal. These are some of the issues and value questions we consider in Chapter 11.

SUMMARY

Weeds are a category of vegetation that exists because of the human ability to select among plant species. In most cases, the value of a weed is determined by the perception of its viewer. Weeds have been described and defined in both anthropomorphic and biological terms. They also may describe a condition of the land or environment and they affect almost everyone at some time or place. Some of the negative aspects of weeds are lowered crop yields, animal discomfort and death, poor product quality, increased costs of production and harvest, higher incidence of other pests, and reduced human health and activities.

Weeds have been classified in numerous ways. Some methods used to classify weeds are by taxonomic characteristics, life history (annuals, biennials, perennials, etc.), physiological differences, habitat, degree of undesirability, and evolutionary tactic. Weeds are distributed widely throughout the world, inhabiting most agricultural, forest, and rangeland systems. However, weeds account for less than 0.1 percent of the worldwide taxa. Many environmental, biological, and human factors influence the distribution of weed species, although humans are the main factor in the continued evolution of weeds.

Weed Science is an integrative, applied, scientific discipline that has been extremely successful in developing tools and tactics to suppress weeds. Although weeds have probably been with human societies since the beginning of agriculture, the history of weed science is largely the history of chemical weed control. There are three predominant areas of activity that comprise the discipline today: weed control technology, Weed Biology, and the social context that surrounds weeds and the tools and information used to manipulate vegetation. Weed control technology is the information and tools necessary for practitioners to suppress weeds in crops, forest plantations, and other situations where some types of vegetation are unwanted. Reasons to control weeds include improved crop production, enhanced product quality, reduced costs of production, reductions in other pests—insects and diseases—improved animal health, enhanced human activities, and reduced risks to human health. Weed biology is the study and use of ecological, physiological, and genetic information about weeds and their associated plants, crops, to improve weed management. The issues and concerns about weed manipulation and the findings of weed biology also must be considered in a broader social context because their consequences often extend well beyond local fields, farms, or forest plantations.

Chapter

2

Principles of Weed Ecology

A fundamental goal of plant ecology is to explore the underlying order of vegetation. Some ecologists are most concerned with the overall relationship of vegetation to environment, while others study the biology of certain plant species in relation to local conditions. Applied plant ecologists seek to use basic information to address vegetation management problems. Although the approaches differ among these scientists, of common interest is the manner in which plants adapt to and exist in their respective environments. It is through ecological principles and concepts that land managers can begin to understand the nature of weediness. Once this foundation is established, it is possible to explore the relationships and interactions that exist among environment, weeds, and crops in agricultural, forest, and rangeland systems. In the process, less costly, more environmentally sound weed suppression and improved profitability may result.

ENVIRONMENT

Central to plant ecology is the recognition that plants exist in and therefore respond to a wide array of environments. The *environment* is the summation of all living (biotic) and nonliving (abiotic) factors that can affect the development, growth, or distribution of plants. It is often divided into two components, the macroenvironment and the microenvironment. The *macroenvironment* is the broad-scale regional environment that includes many aspects of soil and climate, such as overall light intensity, rainfall, humidity, wind, and temperature. The *microenvironment* occurs on a smaller scale and is that aspect of the macroenvironment that is influenced by the presence of objects (rocks, trees, etc.), chemicals (organic matter, nutrients), and topography. Although both the macroenvironment and the microenvironment can be measured and therefore expressed in similar terms, it is the microenvironment to which individual plants respond to form the mosaic of vegetation over a local or regional landscape. The microenvironment is emphasized in this book, since weed species have generally adapted to microenvironmental factors.

SCALE

Hierarchy theory (O'Neil, 1986), as an ecological concept, was developed within the context of general systems theory (GST). GST is a holistic scientific theory and philosophy of the hierarchical order of nature, which is thought to consist of open systems with increasing complexity. Biological systems have two subsets within GST: (1) living systems, such as individual plants or animals, and (2) ecological systems. In ecological systems, combinations of organisms and environments are arranged, or arrange themselves, in a nested hierarchy of subsystems (levels) according to differences in process rates. At high levels, complexity is high and process rates (lifespans, turnover rates, rates of activity) are slow, while at low levels, process rates are fast and complexity is generally lower. Temporal and spatial scale are determined by the differences in process rates among levels.

Scale and Ecological Systems

A common hierarchy proposed by ecologists and depicted in Figure 2.1(a) includes *biome, ecosystem, community, guild, species, population*, and *organism*, although the hierarchy could be extended downward to include tissues, cells, and enzymes. The components within any particular level are always linked and perturbations are characterized by feedbacks both within and among levels. Generally, even slight changes in a higher level result in substantial impacts at lower levels, but impacts over large areas or long times are necessary for lower levels to influence higher ones. Most plant ecologists are concerned with the levels from ecosystems to organisms because at these levels interactions can be recognized easily. The everyday activities of agriculture, forestry, and resource management also operate within these levels of organization.

A *population* is a group of organisms within a species that co-occur in time and space. A species is usually composed of several to many populations. The term *community* refers to all the populations that occupy a particular site and is a concept used primarily to examine or explain the biological interrelationships of organisms. Although "all" populations includes soil organisms, insects, birds, plants, and so on, the focus of plant ecologists is usually on the principal

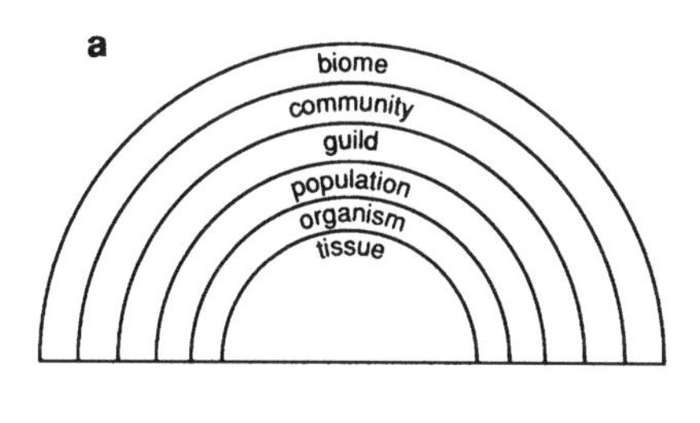

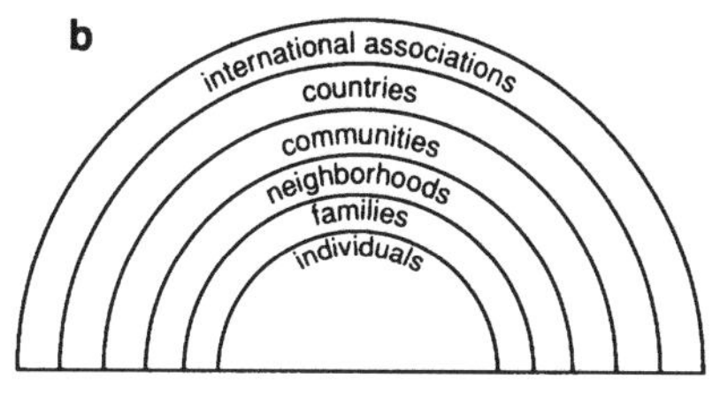

FIGURE 2.1 Hierarchical scale in (a) ecological systems and (b) human social systems arranged according to size and complexity. (Ghersa et al. 1994a.)

higher plants, for example, a corn field and its associated weeds or a stand of Douglas-fir and its associated plants. The basic attributes of a plant community, as described by Barbour et al. (1987) are:

- relatively consistent floristic composition
- relatively uniform physiognomy (structure, height, cover, etc.)
- characteristic distribution in a particular type of environment or habitat.

Specific traits that further characterize plant communities are listed in Table 2.1.

The view that plants occur in discrete communities analogous to organisms was promoted by Frederic Clements and prevailed in plant ecology during the early part of the twentieth century (Barbour et al. 1987). In the latter part of this century, the continuum theory, which holds that each species has its own unique range in relation to the environment, became more widely accepted (Austin, 1985). As discussed by Barbour et al. (1987), debate continues about this important feature of vegetation, in part because sampling method often determines whether plant associations appear as discrete units or as points along a continuum. Nevertheless, the concept of a community as a discrete ecological unit is useful because it allows ecologists to define accurately the vegetation with which they are concerned.

A complete group of communities and their environments when considered together is an *ecosystem*. Within ecosystems, the different organisms are clas-

TABLE 2.1 Some Characteristics of Plant Communities

Physiognomy
Architecture
Life-forms
Cover, leaf area index (LAI)
Phenology
Species composition
Characteristic species
Accidental and ubiquitous species
Relative importance (cover, density, etc.)
Species patterns
Spatial
Niche breadth and overlap
Species diversity
Richness
Evenness
Diversity (within stands and between stands)
Nutrient cycling
Nutrient demand
Storage capacity
Rate of nutrient return to the soil
Nutrient retention efficiency of the nutrient cycles
Change or development over time
Succession
Stability
Response to climatic change
Evolution
Productivity
Biomass
Annual net productivity
Efficiency of net productivity
Allocation of net production
Creation of, and control over, a microenvironment

Source: Barbour et al., 1987. Reproduced by permission of the Benjamin Cummings Publishing Company, Menlo Park, CA.

sified by their function or energy flow, that is, who eats whom. The four trophic levels that function in an ecosystem and form a "food chain" are:

- *Producers*—green photosynthetic plants (autotrophs) that provide the basic food resource to all other organisms (heterotrophs).
- *Primary consumers*—herbivores that feed directly on green plants.
- *Secondary consumers*—predators that feed on the primary consumers, for example, birds that feed on grasshoppers. Many secondary consumers also act occasionally as herbivores.
- *Decomposers*—bacteria and fungi that contribute to decay and are important for nutrient cycling, breakdown of chemical residues, and soil formation.

Food or trophic chains can be constructed to describe the relationships among organisms in a particular community or ecosystem. For example, the chain in a simple agroecosystem might be:

grass/clover	→	cattle	→	humans
(producer)		(primary consumer)		(secondary consumer)

These chains become complex as more populations are introduced into the ecosystem to create a trophic web, as shown below for a hypothetical alfalfa-weed plant community.

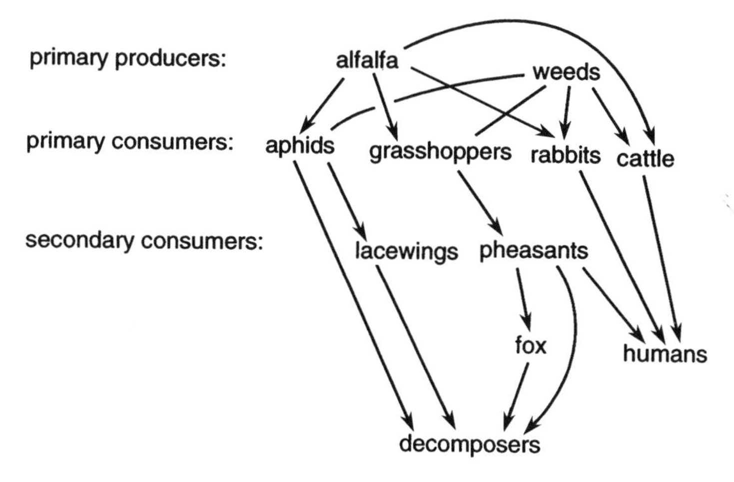

Each arrow in such schemes represents a transfer of food, or energy, from one organism to another. These transfers are important because they define the efficiency of the system in the flow of mass (carbon), energy, and nutrients.

The perceptions of scale are what determine relevant researchable questions in Weed Science. For example, weed control practices can cause other effects besides simple changes in a local, field level weed flora. Management practices that cause changes only in crop or weed composition might eventually influence an entire agricultural or forested landscape or region. Cultivations or herbicide applications certainly reduce weed competition and change vegetation composition, but they also may cause losses in soil fertility through erosion or nutrient leaching, which are larger-scale issues. In addition, weeding stimulates feedback within a community that increases the probability of invasion by new weeds. This feedback may explain why relative and absolute abundance of the weed flora in the United States has increased steadily from 1900 to 1980 (Forcella and Harvey, 1983), despite enormous local efforts to control weeds.

Scale in Agroecosystems

Just as ecological systems can be arranged through general systems theory (GST) into a hierarchical structure (Figure 2.1a), social or human systems also can be arranged similarly according to function and scale (Figure 2.1b). A common hierarchy of human social systems is *individual, family, neighborhood, community, country*, and *global association*. As in ecological systems, levels in human systems can be determined by actual differences in process rates (e.g., adoption of new technology, cultural invasion, education) that define functional boundaries of scale as opposed to arbitrary ones.

Human actions and impacts during the early evolution of agriculture probably operated at a spatial and temporal scale similar to that of the ecosystem (Figure 2.1). That is, individual humans and human populations manipulated individual plants and plant populations, and the reaction times and feedbacks of information were probably similar for both social and ecological processes. As agriculture has evolved, however, the capacity for humans to use energy (Figure 2.2) and acquire and transmit information across greater distances and to future generations has led to a divergence of scale. In other words, human actions now have an impact over larger amounts of space, greater time frames, and longer response times than it is possible for ecological systems to accommodate easily.

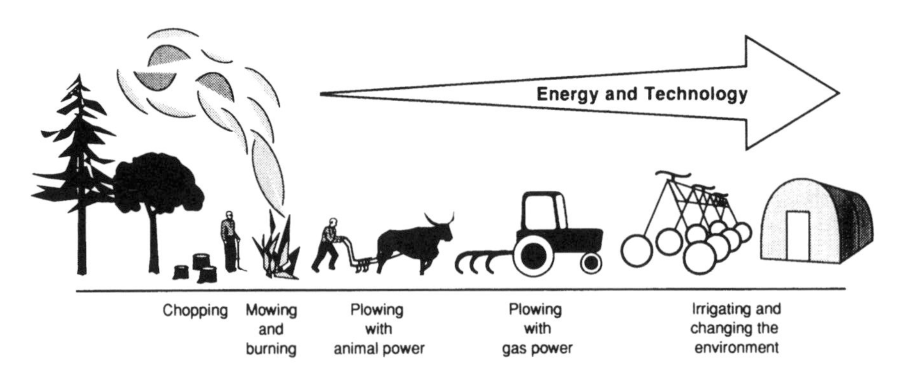

FIGURE 2.2 Evolution of human activities and the agroecosystem. Width of the arrow indicates relative amount of energy and technology. (From Ghersa et al., 1994a.)

Thus, the functional levels of human systems and agro-ecological systems seem to have diverged as modern agriculture has evolved.

A better understanding of the agroecosystem (see Figure 1.14) should assist in the recoupling of human systems to ecosystems they impact. However, the healthy coevolution of human systems and agroecosystems will also require minimizing human impacts. Just replacing herbicides with cultural approaches, breeding weed suppressive or allelopathic intercrops, or using biological control with insects or pathogens may not be enough. It will also require the maximum use of information about human values and biotic interactions when designing new or different weed management strategies. Levins (1986) uses three generalized models of pest management to demonstrate this point (Table 2.2). He calls

TABLE 2.2 Approaches to Pest Management

	Industrial	*Present IPM*	*Ecological Agricultural*
Goal	Eliminate or reduce pest species	Maximize profits	Multiple economic, ecological, and social goals
Target	Single pest	Several pests around a crop and their predators	Fauna and flora of a cultivated area
Signal for intervention	Calendar date or presence of pest	Economic threshold	Multiple criteria
Principal method	Pesticide	Prevention by plant breeding and crop timing, careful monitoring, and multiple intervention	System design to minimize outbreaks and mixed strategies
Diversity	Low	Low to medium	High
Spatial scale	Single farm	Single farm or small region defined by pest	Agrogeographic region
Time scale	Immediate	Single season	Long-term steady-state or oscillatory dynamic
Boundary conditions	Everything as is: crops, cropping system, land tenure, microeconomics, decision rules, social organization	Major crops, land tenure, and decision rules	Societal goals
Research goal	Improved pesticides	More kinds of interventions	Minimize need for intervention

Source: R. Levins, 1986.

these models the *industrial, IPM,* and the *ecological agricultural* approaches to pest management. Of the three models, only the third, which is still hypothetical, requires an understanding of fundamental processes in ecological (Figure 2.1a) and human (Figure 2.1b) systems to manage weeds and other pests.

WEEDS AS A COMPONENT OF COMPLEX ECOLOGICAL AND HUMAN SYSTEMS

As seen above, components of complex systems maintain links that allow the flow of matter, energy, and information. These rates of flow among components determine the behavior of the entire system and the regulation of any component by positive and negative feedback responses. A *positive feedback* response occurs when a process increases in relation to its previous rate. In contrast, *negative feedback* results in a reduced process rate in relation to its previous rate. Such regulation in ecological systems follows cybernetic principles (Checkland, 1981); that is, when negative feedback responses are absent, the population experiences exponential growth (see Figure 2.6). All populations have intrinsic potential for exponential growth when environmental regulation is lacking (Begon and Mortimer, 1986). This growth causes an invasive process often called an infestation, infection, or epidemic, depending on the organism and the point of view of the observer. Considering the cybernetic viewpoint, a weed infestation is a plant population lacking negative feedback control to compensate for the positive response of reproduction and growth. Food webs, nutrient cycling, individual responses to density, and so on are all related to negative regulation. Negative regulation is lost if, for example, in a production system weed seeds are dispersed from year to year by machinery (Ghersa and Roush, 1993). Also, if the soil nutrient cycle is bypassed by fertilization, the negative response to nutrient deficiency is removed. In these situations a plant population outbreak occurs because of the prevalence of positive feedback. If the plant population has human value, we may call it a high-yielding crop; if it is a weed, we call it a problem.

The lack of negative regulation in a system also may occur if a new element is incorporated into the system. When a new population immigrates into a system, it often lacks links to other components. For example, grazers may not immediately recognize a new plant as food. This is why many exotic species in a particular area are weeds (Salisbury, 1961). This systemic view of weeds using cybernetic theory emphasizes understanding and correcting what is allowing the outbreak of a population, rather than concentrating on eliminating the invader. While these ideas are most appropriate for ecological systems, such as weeds and crops, they are applicable to socioeconomic systems, as well. For example, a farmer is usually willing to spend more money for weed control in relation to how densely invaded his crops are.

COMMUNITY DIFFERENTIATION

As mentioned above, questions remain about whether plants occur in discrete associations or communities in which members have similar distribution limits within a particular habitat, or whether their distributions occur in an overlapping fashion along continuous environmental gradients. In either case, it is clear that groups or stands of vegetation exist where there are discontinuities in the environment, such as a change in soil type or disturbance. Thus, regardless of their cause, it is valuable to explore the attributes of these stands of vegetation which we call plant communities.

Most plant communities exhibit both vertical and horizontal differentiation; that is, different species occur at various heights above the ground and also are distributed differently along the ground surface. The vertical distribution of species is usually determined by a gradient of sunlight, with the upper canopy being in full sun and lower canopies occurring in diminished light intensity.

Horizontal distribution is more complex. Whittaker (1975) has identified four ways in which species in a community (also individuals in a population) can be distributed horizontally. These are shown in Figure 2.3. In natural communities, species often appear to be scattered at random (Figure 2.3a) and, indeed, regular spacing of plants (Figure 2.3c) is usually rare. This observation does not apply to the agricultural plant community or forest plantation because at least one species, the crop, is usually planted in rows. Departures from randomness also are known to occur in natural plant communities. In these cases, the species are concentrated into patches (Figure 2.3b and d). Patchy distribution may result from the dispersal pattern of parent plants, gradients in the microenvironment, or species interactions, that is, positive or negative associations of one species with another.

It is generally held by ecologists that each species within a plant community has a unique pattern of distribution. These patterns may be correlated with those of other species, but they are not identical to them. Thus, a plant community is a composite of numerous species distributions, each superimposed upon the other and sometimes confounded by disturbance, such that a myriad of subtle interactions exist.

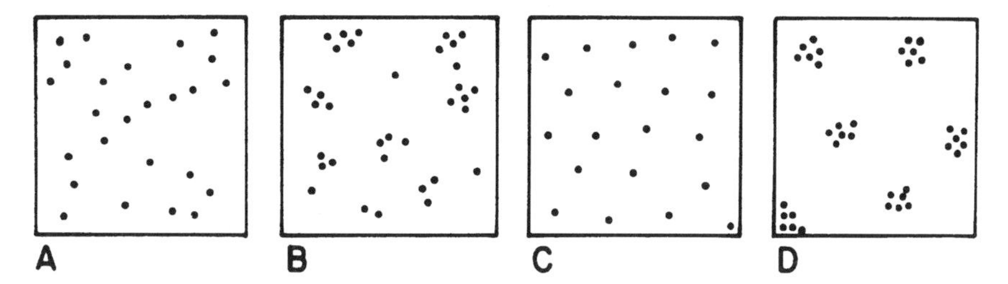

FIGURE 2.3 Four ways in which individuals of a population can be distributed in horizontal space in a community. (a) Random dispersion (note its apparent irregularity). (b) Clumped or contagious distribution. (c) Regular or negatively contagious distribution. (d) Combinations of strong clumping of individuals into colonies, and regular distribution of the colonies. (Reprinted with permission of Macmillian Publishing Co., Inc. from *Communities and Ecosystems* by R.H. Whittaker. Copyright 1975 by Robert Whittaker.)

SUCCESSION

The species composition of a plant community often changes over time, a process called *succession*. If the microenvironment remains relatively constant, the change in species composition becomes very slow or even ceases. The last stage of succession is termed the *climax* and is often idealized as a community having a constant species composition. According to Gause's competitive exclusion principle (see niche differentiation, below), species that are direct competitors cannot coexist permanently in the same niche. Thus, the climax stage must represent a system of interacting, niche-differentiated species that tend to complement rather than directly compete with one another. Other phases of community development also may be evident over time and are termed pioneer, early, or intermediate *seral* stages.

Succession is usually divided into two components, primary and secondary. *Primary succession* is the establishment of plants on land that has never been vegetated before, for example, land that is created as a lake fills with silt or rock weathers to soil. *Secondary succession* is the pattern of change after a radical disturbance so that a new patch in the physical environment is once again available for colonization by plants. Secondary succession is most often of concern to land managers. For example, Figure 2.4 depicts several stages of secondary succession following clear-cut logging. A forest resembling the original should eventually result if the area is not disturbed once again.

FIGURE 2.4 Secondary succession after logging. Various stages of community development are evident depending on the time (years) after canopy removal. (Photograph by S.R. Radosevich.)

Mechanisms of Succession

It is obvious that species replace one another through succession. Connell and Slatyer (1977) described several alternative mechanisms through which species may replace each other during succession (Figure 2.5). After a major perturbation of the environment, "opportunistic" species with broad dispersal powers, rapid growth, and short life spans usually arrive first and occupy the empty space. These species usually do not invade or grow in the presence of living adults of their own or other species. According to Connell and Slatyer, the species that

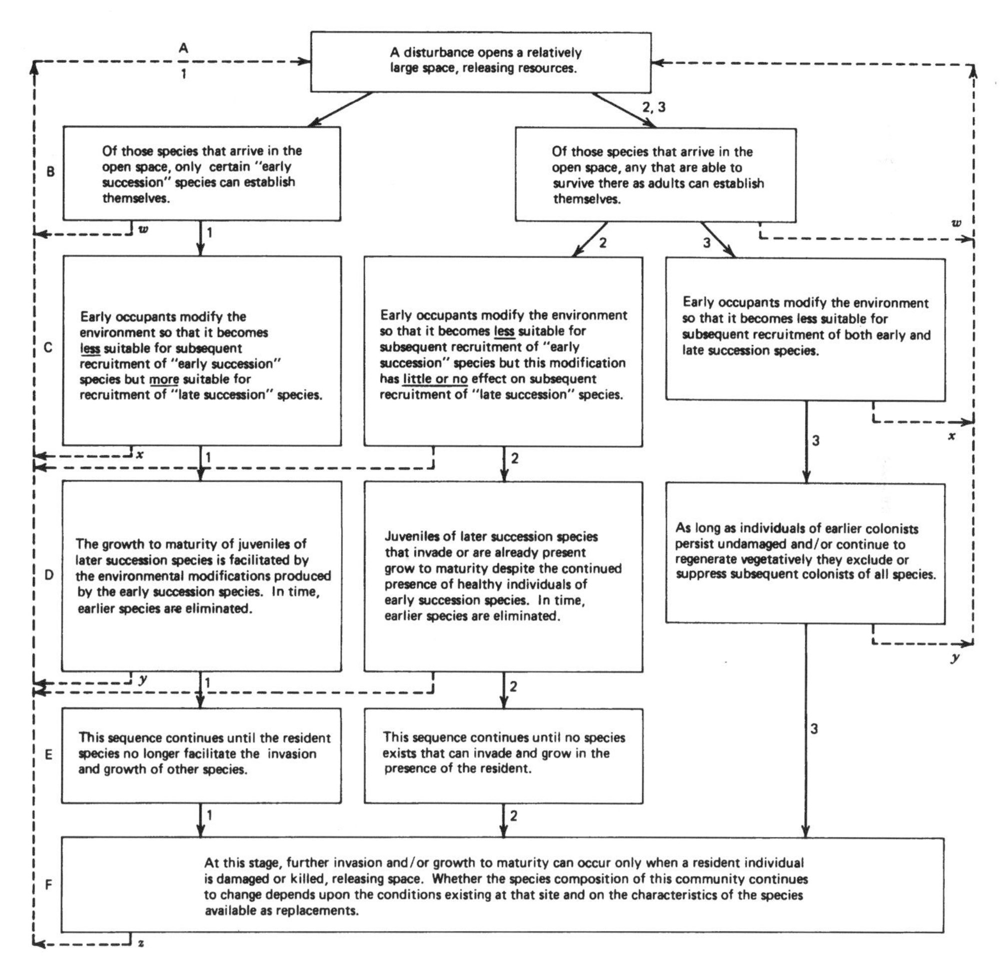

FIGURE 2.5 Three models of the mechanisms producing the sequence of species succession. The dashed lines represent interruptions of the process, in decreasing frequency in the order *w*, *x*, *y*, and *z*. The numbers 1, 2, and 3 refer to facilitation, tolerance, and inhibitory models, respectively, which are discussed in the text. (From Connell and Slatyer, 1977, *American Naturalist* 111: 1119–1144. Copyright by University of Chicago.)

replace these earliest occupants may be determined by any of three mechanisms (Figure 2.5).

In the first situation, the *facilitation* model, the entry and growth of the later species requires the earlier species to "prepare the ground" for them. Only after a suitable change occurs in the microenvironment can later species colonize the area. This is the traditional model of Clements and his followers, but may pertain mainly to certain primary successions. Lack of experimental evidence has caused rejection of this model by most ecologists. The second, *tolerance*, model suggests that a predictable successional sequence occurs because of the existence of species that have evolved different strategies for exploiting environmental resources. Later species are those that are able to tolerate lower levels of resources than earlier ones. Thus, later species are able to invade and grow to maturity in the presence of those that preceded them. This model seems most appropriate for forest vegetation. In the third, *inhibition*, model all species, even the earliest, resist invasion by competitors. The first occupants pre-empt the space and continue to exclude or inhibit later colonists until the former die or are damaged, thus releasing environmental resources. Only then can later colonizers become established and eventually reach maturity. This model also suffers from a paucity of experimental evidence.

An alternative theory of succession, the resource-ratio hypothesis, was proposed by Tilman (1985). This hypothesis holds that succession results from a gradient over time in the relative availabilities of limiting resources in a particular habitat. Assuming that each plant species is a superior competitor for a particular component of the limiting resources, community composition should change whenever the relative availability of two or more limiting resources changes. Thus, in Tilman's theory, competition essentially drives succession. Other ecologists have suggested alternate processes besides competition that drive succession, such as colonization, allelopathy, herbivory, or natural selection at the community level (Barbour et al., 1987; Tilman, 1988). While succession is a major focus of ecological research, there is no clear consensus yet as to which mechanisms are most important.

In the majority of natural communities, and certainly in agricultural ones, succession is often interrupted by disturbance, which starts the process over again. However, if not interrupted, the community eventually reaches a stage in which further change is on a small scale, occurring only as individuals die and are replaced. The pattern of these changes depends on whether individuals are more likely to be replaced by members of their own or another species.

Succession in Agriculture

The cultivated field represents a special example of secondary succession because it is continually being disturbed. Succession on abandoned cropland is called *old-field* succession (see Figure 1.11). Once agricultural operations cease, the systematic replacement of early and intermediate seral stages occurs through time until a climax community similar, but not identical, to the original pristine one appears. If disturbance occurs repeatedly, as in the agricultural system by tillage and in forests by frequent fires, succession may become cyclic such that earlier stages are favored. For example, after logging or a severe forest fire, first herbaceous pioneers then intermediate shrub communities may dominate a site for decades or more, especially if it is frequently disturbed (see Figure 1.10). Similarly, in agriculture, dominance by herbaceous annual species usually delays the establishment of perennial species that normally would succeed the annuals. In these situations, herbaceous perennials appear to have lower competitive ability than annuals in the seedling stage, but greater persistence and stability associated with vegetative reproduction, and therefore have difficulty becoming established as seed. Once established, however, perennial species tend to replace annuals in the community.

Under the disturbed regime typical of agriculture, it is unnecessary for plant species to have different life history characteristics in order to replace earlier residents. Direct control of weeds is common on most farms and forest plantations, which provides an environment of continued disturbance. Thus, no difference in competitive ability needs to occur among weed species, since the earliest stage of succession is constantly being recycled. Replacement of weed species over time may occur at random or due to subtle year-to-year changes in meteorological conditions or management practices. However, the entire weed-crop community can respond to such management manipulations. These responses are usually short-term owing to the transitory nature of most cropping systems, but under some conditions long-term responses are possible. Once they have occurred, neither short- nor long-term responses are easily reversed, and both can have significant impacts on continued weed and crop management. These topics are discussed further in Chapters 3 and 4.

NICHE DIFFERENTIATION

Niche is a term used to describe a species' place within a community, including its place in the space, time, and function of that community. The concept of a niche denotes specialization. As Whittaker (1975) points out in his analogy of a niche to human society, an individual may gain from professional specialization to acquire the resources (income) needed to live. Two or more individuals may gain by following different specialties since they are not in direct competition, and society at large may gain if the specialization of one individual satisfies the needs of another. Thus, considerable evolutionary advantage must underlie the specializations of the plant species within any plant community. Through differential specialization, species avoid at least some degree of direct competition.

In order to understand the importance of niche separation in natural communities, and perhaps in managed ones, we must consider the logistic equation of Volterra (1926) and Lotka (1925). As described by Whittaker (1970), if environmental resources are not limiting, a population may increase geometrically, that is

$$dN/dt = rN \qquad (eq.\ 2.1)$$

in which the rate of growth in numbers of individuals per unit time (dN/dt) equals the number of individuals (N) in the population at a given time, multiplied by r, the intrinsic rate of increase for that population in the absence of crowding or competition effects on growth. If environmental resources are limited, the growth rate of the population is continually lessened by competition as the number of individuals approaches the maximum number the environment can support. This maximum number is the carrying capacity of the environment, K. The logistic curve (Figure 2.6) generated from the following equation is a convenient first approximation for growth rate of a population to a ceiling level set by a limiting environment:

$$dN/dt = rN\,(K - N)/K \qquad (eq.\ 2.2)$$

In the above equation, (K - N)/K specifies that population growth will be reduced as population number, N, approaches carrying capacity, K, and will be zero when N = K; the population is then stabilized at carrying capacity. The logistic equation now may be applied to two competing populations:

$$\frac{dN_1}{dt} = r_1N_1\,\frac{K_1 - N_1 - \alpha N_2}{K_1} \qquad (eq.\ 2.3)$$

$$\frac{dN_2}{dt} = r_2N_2\,\frac{K_2 - N_2 - \beta N_1}{K_2} \qquad (eq.\ 2.4)$$

In these equations, N_1 and N_2 are the populations of species 1 and 2 at a given time, r_1 and r_2 are their intrinsic rates of population increase, and K_1 and K_2 are the environmental resource limits (carrying capacities) for each species in the absence of the other. α and β are competition coefficients that express, through αN_2 and βN_1, the effects of the population level of one species on the population change of the other species. The equations imply that, for most values of α and β, one species increases while the other competitor declines, until at equilibrium the latter is extinct. This idea that two species cannot coexist permanently in the same niche is known as *Gause's competitive exclusion principle* (Gause, 1934).

Species that divide a shared resource among themselves are collectively called a *guild*. According to Gause's principle, if two species in a guild are direct competitors, one species should approach extinction. This suggests that the competitive relationships that nearly always develop between weeds and crop plants in crop production systems might be regulated to some extent by natural (competitive exclusion) processes. In the agricultural field where seeds or propagules are repeatedly introduced, the competitive exclusion principle would be manifest as extreme dominance or suppression of one species by another rather than local extinction. If, however, the species differ in their requirements or specializations,

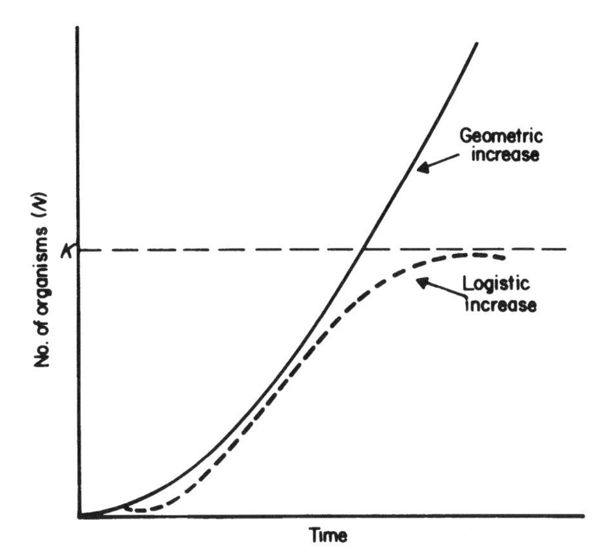

FIGURE 2.6 Geometric (solid line) versus logistic (dashed line) population growth over time. K is the carrying capacity of the environment for a population showing density-dependent logistic growth (dashed line). (From Barbour et al., 1987, by permission of the Benjamin Cummings Publishing Co., Inc., Menlo Park, CA.)

then it is possible for them to coexist. Because of niche separation, many natural systems are typified by a high degree of species diversity, coexistence, and uniform total productivity. In rangeland and forest systems, species diversity and uniform productivity are acceptable when coexistence of several species is the ultimate goal. In addition, because of the range of vegetation types often present in rangelands and forests, some spatial and temporal specializations are evident among particular weed and crop plants that would allow coexistence without significant reductions in productivity of desirable species. In contrast, when productivity of a single species is of concern, most of the environmental potential (resources) should be directed toward the crop; weed suppression, not coexistence, is the desired goal. Although some agricultural crops are superior competitiors compared to many weeds, it is not enough simply to allow them to compete with the hope of eventual weed suppression or even extinction since some loss in crop yield would inevitably occur over the time frame of a typical production season. Furthermore, the niche differences between weeds and agricultural crops usually are not great enough to allow maximum crop productivity to occur without some human intervention for weed control.

PATTERNS OF EVOLUTIONARY DEVELOPMENT

It is generally recognized that organisms are capable of budgeting energy or resources in order to complete their life cycle successfully. This process is called *resource allocation.* Allocation is closely linked to species survival, and the patterns of resource allocation that are retained are generally viewed as adaptations that minimize extinction. In plants, the resources available to a species must be divided among various organs and activities in order to complete the life cycle successfully. The amount of photosynthetic energy allocated to root, shoot, leaf, and reproductive portions and the amount of time (implied resources) spent in dormancy, growth, and maintenance are important attributes that govern plant species success. Figure 1.3a illustrates those major activities performed by annual plants that require resource allocation. Figure 1.3b illustrates those activities necessary for a perennial species. Several points of view are possible concerning the patterns of resource allocation that exist among species; however, these theories all recognize the importance of resource allocation for species survival and plant community development.

r and K Selection

The most widely held theory dealing with patterns of evolutionary development is that of r and K selection. This idea was derived from the logistic equation of population growth (Lotka, 1925; Voltera 1926, see page 56). As shown in Figure 2.6, population growth in an ideal (limitless) environment would be expected to increase geometrically, whereas in real (limited) environments growth declines as the population approaches K, or carrying capacity.

The theory of r and K selection, first proposed by MacArthur (1962) and later Pianka (1970, 1994), is that organisms lie on a continuum between two extremes of resource allocation that represent two strategies for survival. In the extreme cases, species may be r-selected or K-selected. Table 2.3 lists various traits associated with each strategy. Extreme K-selected species tend to be long lived, have a prolonged vegetative stage, allocate a small portion of biomass to reproduction, and occupy late stages of succession. The population size is near carrying capacity and is regulated by biotic factors. Extreme r-selection leads to a short-lived plant that occurs in open habitats and early stages of succession. A large portion of biomass is allocated to reproduction, and the population is regulated by physical factors. It should be noted that few plant species, if any, are entirely r-selected or K-selected. Most species represent a compromise between the two strategies. Weeds associated with agricultural lands and highly disturbed sites in forests and rangelands seem to fit most closely the characteristics of r-selection noted in Table 2.3.

C, R, and S Selection

Another theory concerning plant resource allocation and evolutionary pattern was proposed by Grime (1979), although this view may be regarded as an extension of the more widely acknowledged r and K continuum. Grime proposed that there are two basic external factors that limit the amount of plant material in an environment: stress and disturbance. He defined *stress* as

TABLE 2.3 Traits of r and K Selection

Trait	*r Selection*	*K Selection*
Climate	Variable and/or unpredictable; uncertain	Fairly constant and/or predictable; more certain
Mortality	Often catastrophic; density-independent	Density-dependent
Survivorship	Mortality at early age	Continuous mortality through life span or more as age increases
Population size	Variable in time; not in equilibrium; usually well below carrying capacity of the habitat; recolonization each year	Fairly constant in time; in equilibrium; at or near carrying capacity of the habitat; no recolonization necessary
Intraspecific and interspecific competition	Variable; often lax	Usually keen
Life span	Short, usually less than one year	Long, usually more than one year
Selection favors	Rapid development; early reproduction; small body size; single reproduction period in life span	Slower development; greater competitive ability; delayed reproduction; larger body size; repeated reproduction periods in life span
Overall result	Productivity	Efficiency

Source: Pianka, 1994.

external factors that limit production, such as reduced or limiting light intensity, water availability, nutrients, or suboptimal temperature. *Disturbance* is the partial or total disruption of plant biomass, for example, by mowing, tillage, grazing, or fire. As with the r and K continuum, the spectrum of these two factors can vary widely, but if only the extremes of high and low stress and disturbance are considered, four possible combinations occur (see Table 1.5). Of these four combinations, only three possible evolutionary strategies are apparent: *ruderals, stress tolerators*, and *competitors*. The fourth possible combination, high stress and high disturbance (Table 1.5), creates an environment unsuitable for plant survival. Plants that fall into each of these strategies can be classified according to their common adaptations (Table 2.4, Figure 1.6).

Grime prefers to arrange the three evolutionary strategies into a triangular model (Figure 2.7) to describe the various equilibria between stress (I_s), disturbance (I_d), and competition (I_c). In this model C, R, and S represent the three extremes of specialization. Since few species have all the characteristics listed in

TABLE 2.4 Some Characteristics of Competitive, Stress-tolerant, and Ruderal Plants

	Competitive	*Stress-tolerant*	*Ruderal*
Morphology			
1. Life forms	Herbs, shrubs, and trees	Lichens, herbs, shrubs, and trees	Herbs
2. Morphology of shoot	High dense canopy of leaves; extensive lateral spread above and below ground	Extremely wide range of growth forms	Small stature, limited lateral spread
3. Leaf forms	Robust, often mesomorphic	Often small or leathery, or needlelike	Various, often mesomorphic
Life History			
4. Longevity of established phase	Long or relatively short	Long to very long	Very short
5. Longevity of leaves and roots	Relatively short	Long	Short
6. Leaf phenology	Well-defined peaks of leaf production coinciding with period(s) of maximum potential productivity	Evergreens, with various patterns of leaf production	Short phase of leaf production in period of high potential productivity
7. Phenology of flowering	Flowers produced after (or more rarely, before) periods of maximum potential productivity	No general relationship between time of flowering and season	Flowers produced early in the life history
8. Frequency of flowering	Established plants usually flower each year	Intermittent flowering over a long life history	High frequency of flowering
9. Proportion of annual production devoted to seeds	Small	Small	Large
10. Perennation	Dormant buds and seeds	Stress-tolerant leaves and roots	Dormant seeds
11. Regenerative strategies*	V,S,W,B_s	V,B_r	S,W,B_s

TABLE 2.4 Continued

	Competitive	*Stress-tolerant*	*Ruderal*
Physiology			
12. Maximum potential relative growth rate	Rapid	Slow	Rapid
13. Response to stress	Rapid morphogenetic responses (root-shoot ratio, leaf area, root surface area) maximizing vegetative growth	Morphogenetic responses slow and small in magnitude	Rapid curtailment of vegetative growth, diversion of resources into flowering
14. Photosynthesis and uptake of mineral nutrients	Strongly seasonal, coinciding with long continuous period of vegetative growth	Opportunistic, often uncoupled from vegetative growth	Opportunistic, coinciding with vegetative growth
15. Acclimatization of photosynthesis, mineral nutrition and tissue hardiness to seasonal change in temperature, light, and moisture supply	Weakly developed	Strongly developed	Weakly developed
16. Storage of photosynthate and mineral nutrients	Most are rapidly incorporated into vegetative structure with some storage for growth the following season	Storage systems in leaves, stems and/or roots	Confined to seeds

Source: Grime, 1979.

* Key to regenerative strategies (11): V, Vegetative expansion; S, seasonal regeneration in vegetation gaps; W, numerous small wind-dispersed seeds or spores; B, persistent seed (s) or seedling (r) bank.

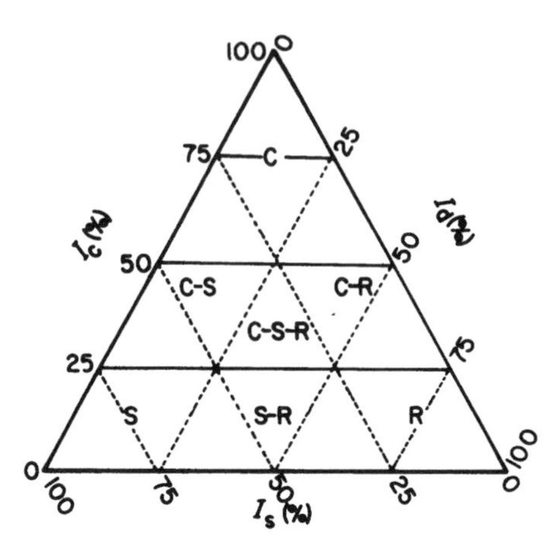

FIGURE 2.7 Model describing the various equilibria between competition, stress, and disturbance in vegetation and the location of primary and secondary strateges. I_c, relative importance of competition; I_s, relative importance of stress; I_d, relative importance of disturbance. C, competitive strategy; S, stress-tolerant strategy; R, ruderal strategy; secondary (combination) strategies are discussed in the text. (From Grime, 1977, *American Naturalist* 111: 1169–1194. Copyright 1977 by University of Chicago.)

Table 2.4, Grime "maps" the species according to certain traits using triangular ordination. Although the indices for stress, disturbance, and competition are difficult to establish quantitatively, this procedure provides a tool to categorize plants according to life history and successional stage.

In terms of evolutionary strategy, many weeds possess characteristics common to both competitors and ruderals (Table 2.4). From Figures 2.7 and 1.7, it appears that many herbaceous annuals, biennials and certain herbaceous perennials follow a pattern of *competitive-ruderals*. Trees and shrubs, however, most closely follow the pattern of *stress-tolerant competitors* (see Chapter 1). Although Grime describes many other patterns of vegetation in relation to both life form and evolutionary strategy, it seems that these two classes warrant further investigation in order to characterize the nature of weediness. This aspect of weed evolution will be explored later in Chapter 3.

PLANT DEMOGRAPHY AND POPULATION DYNAMICS

Stages in the life history of plants provide the opportunity to assess how changes in population size or structure occur over time. These basic changes in plant life history are shown in Figure 2.8. Beginning with seeds, the population of seeds in the soil is generally referred to as a *seed bank*, or reservoir. Some seeds in this population germinate, graduate to the next stage and become *seedlings*, while others remain dormant in the reservoir or die. Seedlings that germinate at nearly the same time are a *cohort*, although agriculturists often refer to the phenomena as "flushes" of germination. *Recruitment* is the transi-

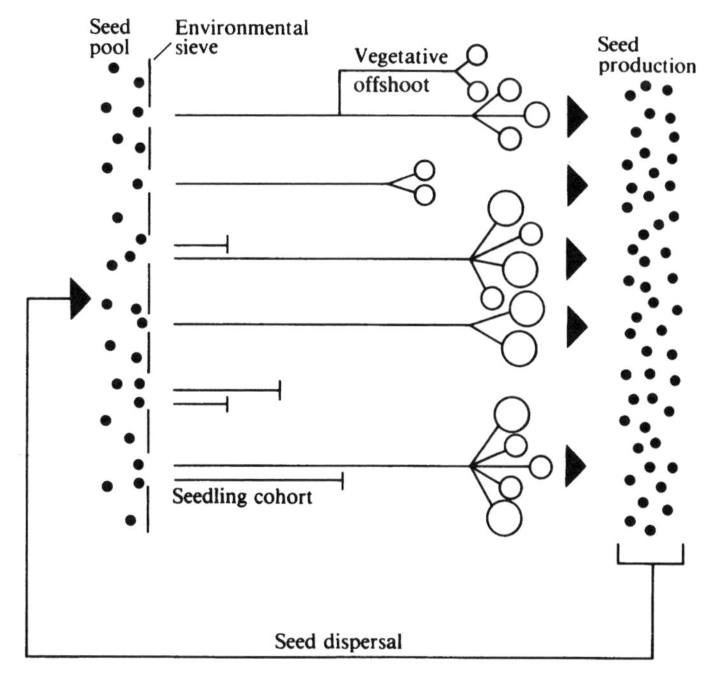

FIGURE 2.8 An idealized plant life history. (Adapted from Harper and White, 1971, in Silvertown, 1987).

tion from juvenile stages, seeds and seedlings, to adult form, in which independent existence and reproduction are possible. Seeds are a primary method of recruitment but vegetative reproduction also occurs in many plants. These vegetative offshoots, called *ramets* or *clones*, may remain attached to the "mother" plant or be separated from it. *Genets*, in contrast, are genetically distinct individuals that arise from a seed.

Plant demography is the statistical study of population changes and their causes throughout the life cycle (e.g., Figures 1.3 and 2.8). There are four basic demographic processes that determine how a population of plants changes over time:

- birth (B)
- death (D)
- immigration (I)
- emigration (E)

Population ecologists describe how these processes change the size of a plant population (N) between one time interval (t) and another (t + 1) by using the following difference equation:

$$N_{t+1} = N_t + B - D + I - E \qquad \text{(eq. 2.5)}$$

All experiments or analyses about the population dynamics of plants, weeds, crops, or natural systems ultimately come back to the above simple equation (Chapter 4).

Selection

Significant evolutionary change in a population occurs when three criteria are met: (a) there is phenotypic variation, (b) some of this variation is heritable, and (c) selection acts differentially upon the range of genotypes (Silvertown and Lovett Doust, 1993). This potential for evolutionary change is easily demonstrated in the grocery store or plant nursery. The wide array of cole crops (crops in the mustard family, Brassicaceae)—cabbage, broccoli, cauliflower, brussel sprouts, kohlrabi—is a result of artificial selection of the same ancestral species of wild plant, wild mustard, *Brassica campestris* (Silvertown and Lovett Doust, 1993). Similarly, all of the strains and varieties of roses originated from a common wild plant. Natural selection can produce results just as dramatic as artificial selection; the process is just not as rapid. Natural selection occurs when one phenotype leaves more descendants than others because of its superior ability to survive or produce offspring in a particular environment. When it is possible to analyze the demography (B, D, I, and E) of individual phenotypes, it becomes possible to determine which phenotypes are likely to leave the most offspring, and thus to determine the direction of natural selection.

Fitness

Because survival and reproduction are both demographic processes, natural selection is also a demographic process. *Fitness* is a single value of relative evolutionary success that combines both survival and reproduction. It is not fixed. Rather, fitness is determined within a particular environment or suite of ecological conditions and is relative to the success of other phenotypes that also exist in the same population. Fitness is an important factor in determining the ecological success of many, if not most, weed species.

SUMMARY

Plant ecology is the study of interrelationships between vegetation and its environment. It is through such study that new information about weeds and associated plants, crops, is generated. Furthermore, it is through application of plant ecological principles that land managers can begin to understand the nature of weediness and develop less costly, environmentally sound suppression tactics. Some general principles of ecology that are relevant to weed biology are environment, scale and hierarchical structure, community and niche differentiation, succession, evolutionary development, and plant demography. Environment is the summation of all living and nonliving factors that can affect the development or distribution of plants. The concept of scale, derived from hierarchy theory, explains how organisms are grouped or ordered in nature. A common ecological hierarchy is biome, ecosystem, community, species, population, and organism. Human social systems also can be ordered in a hierarchical manner, such as country, region, community, neighborhood, family, and individual. Ecological and human systems often interact in agroecosystems and it is these interactions that often determine relevant researchable questions in weed science. Plant communities can be both horizontally and vertically differentiated through space. Communities also change over time through the process of succession. Niche is a term used to describe a species' place in space, time, and function in a plant community. It is generally held that species, through differential specialization, avoid direct competition with one another, at least to some degree. The r and K continuum is a generally accepted theory regarding evolutionary development of most plant and animal species. The C, R, and S strategies are part of an alternate theory which is an extension of the r and K continuum. Within this structure, weed species may follow a combined strategy of either stress-tolerant competitors or competitive ruderals. Plant demography is the study of how plant populations change in size and structure during various stages of their life cycle. It is possible, using demographic principles, to assess how weed populations might change through time or respond to perturbations in their habitat or environment.

PART

II

Weed Biology

Weeds are likely to undergo considerable evolutionary adaptation in response to disturbance and association with neighboring plants. Baker (1974) states that evolutionary success of any organism must be measured in terms of numbers of individuals, reproductive output, area of the world's surface occupied, range of habitats entered, and potential for putting descendants in a position to continue the genetic line. Three characteristics are proposed most often to characterize weedy vegetation: prolific reproduction, widespread dispersal, and rapid growth. In the following chapters we will explore these and other characteristics that make weeds successful organisms in their environments. Central to this endeavor is the necessity to unravel the life histories of the plant (weed) species. Although this is not easy to accomplish, it has been done for some species, and such studies provide insight into the nature of weeds as successful organisms in agricultural, forest, and range environments.

Chapter

3

Genetics and Evolution of Weeds

Darwin's theory of evolution is based on the principles that variation exists within populations, that variation is heritable, and that the phenotype of the individuals in a population change through generations because of natural selection (Griffiths et al., 1993). Population geneticists strive to understand the genetic composition of populations and the forces that determine and change that composition. Unfortunately, information on the population genetics and evolution of weed populations in relation to the selection pressures imposed by agriculture is generally limited. Several classic studies exist in the literature which remain relevant and shed light on the nature of weed evolution. However, it is likely that increased research in the area of population genetics and evolution of weeds could be valuable in designing better strategies for their prevention and management.

It is generally accepted that many weed populations arise as a result of evolution of wild plant colonizers through selection and adaptation to continuous habitat disturbance. Weed populations also may originate as "escapes" from crops or by hybridization between wild and cultivated races of plant species. It is apparent that weeds have evolved and will continue to evolve in response to human activities. To understand the nature of weed evolution, it is important to address the genetic variation in weeds, upon which selection acts, the breeding systems of weeds, which influence the distribution of genetic variation within and among populations, and the selective forces imposed on weed populations by agricultural practices. These topics will be addressed in this chapter.

RELATIONSHIP OF GENETIC AND PHENOTYPIC VARIATION TO WEED MANAGEMENT

There is abundant genetic variation among individuals in most plant populations, and weeds are no exception to this generalization. The response of organisms to selection depends on this heritable variation. Phenotypic variation refers to the appearance of biological traits in an organism, that is, the joint expression of genetic and environmental variation leading to chemical, structural, and behavioral characteristics of the organism. Although response to selection depends on the existence of heritable variation among the individuals being selected, one of the consequences of selection is that variation is reduced among succeeding individuals. Thus, any view of genetic evolution must contain assumptions about how variation is created and maintained in a species. This feature of selection is known as the *paradox of variation* and must be considered as we try to explain how weed species occur in new crops, the presence of new weed species in existing crops, or the development of tolerant races to weed control practices.

Individual Selection

Geneticists distinguish between two viewpoints regarding the maintenance of genetic variation (Lewontin, 1974; Hartl and Clark, 1989). One view considers most heritable variation to be non-adaptive, that is, the role played by individual selection on most traits is to remove deleterious mutations. The typical model of this viewpoint is of a locus (location of a gene on a chromosome), with mutations to deleterious alleles (alternative forms of a gene) being balanced by selection against those alleles. When environmental conditions change, alleles that were deleterious might become advantageous, thereby allowing a population to adapt to the new condition. The second school of thought considers genetic variation to be adaptive and maintained in species by individual selection. The typical genetic model of this view is of a locus with two or more alleles, for which the heterozygotes have higher fitness than the homozygotes (overdominance). When environmental conditions change, a different genetic balance may be attained, but selection would still act to preserve variation.

Many evolutionary geneticists adopt an intermediate position between these two viewpoints, yet the difference between them is important, especially for studies of coevolution involving human activities and weed species. If the first view is adopted, for example, to understand the rate at which herbicide resistance will appear, a delay in the species' response would be expected because the appropriate mutation may not be present. If, on the other hand, the alternative view is adopted, it is reasonable to assume that sufficient genetic variation already exists for weed species to respond quickly to any new condition created by changes in management practices, such as the use of herbicides. There is no way at present to determine which view of genetic evolution is correct. However, an important element of coevolution in agricultural systems is that evolutionary responses to environmental conditions occur rapidly, that is, over a few generations. If evolutionary changes cannot occur on an ecological time scale, which in the agroecosystem is related to shifts in cropping activities and weed control practices, then the coevolution of crop mimics, herbicide-tolerant races, and so on, would not be found or even expected. That these special cases of weed evolution are found suggests that genetic variation is maintained by selection, as believed by proponents of the second school of thought.

BREEDING SYSTEMS OF WEEDS

Sexual Reproduction

Many weeds possess an apparent ability to colonize readily a wide range of disturbed habitats. This observation has led many plant geneticists, evolutionists and demographers to speculate about the breeding systems of weeds in relation to their colonizing abilities. Although several hypotheses have been proposed, a recurring theme has emerged that considers the adaptive value of uniparental reproduction with occasional genetic recombination. This particular hypothesis led to the so-called *Baker's rule* concerning the genetics of weeds and other colonizing species. According to Baker (1974),

> A notable feature of most weeds, especially annuals, is their ability to set seed without the need for pollinator visits, either by autogamy (self-fertilization) or agamospermy. Even when outcrossing does take place, wind or generalized flower visitors are adequate. The advantages of autogamy or agamospermy for a weed include providing for starting a seed-reproducing colony from a single immigrant or regeneration of a population after weed-clearing operations have removed all but a single plant. In addition, they allow rapid build-up of the population by individuals virtually as well adapted as the founder. Where the weed is a perennial, self-compatibility is less certain to be found (and some such weeds are even dioecious), but an extra emphasis upon vegetative reproduction here achieves the same end, i.e. the rapid multiplication of individuals with appropriate genotypes.

Many weed species utilize breeding systems adapted for *inbreeding* to produce stable duplicates of successful genotypes coupled with occasional (environmentally controlled) *outcrossing* for recombination to occupy new or changing niches (microenvironments) (Allard, 1965; Baker, 1974; Barrett, 1982). Young and Evans (1976) provide a striking example of this breeding system in a study of downy brome (*Bromus tectorum*).

Downy brome is an annual grass which, upon introduction, invaded much of the *Artemisia*-dominated grassland of western North America known as the Great Basin (Chapter 1). This weed increases dramatically in number or population size following removal of the native perennial grass by overgrazing. It also is abundant on disturbed sites and is considered to be one of the most widespread weeds in the western United States. According to Young and Evans (1976), the downy brome growing in degraded *Artemisia* communities are predominantly self-pollinated. Any stable habitat in the Great Basin has a population of relatively similar phenotypes, but each plant is probably an individual genotype (Figure 3.1a). These populations can reproduce year after year from seed. Destruction of the *Artemisia-Bromus* communities by fire or tillage causes this system to change. The shrubs and much of the competing vegetation, including a large portion of the downy brome seeds (caryopses), are destroyed. The surviving downy brome seeds germinate in a seedbed where resource availability has been increased. These plants (Figure 3.1b) produce thousands of flowers from many tillers that stay green much longer than unburned or nontilled populations. In response to the increased availability of resources, locules stay more turgid, stigmas are receptive, and anthers are exerted for longer periods of time. Chances for cross-pollination thus increase. Because each plant is essentially an inbred line, the hybridization of these lines results in a great population expression of heterosis (hybrid vigor) the second year after the disturbance (Figure 3.1c). Following the hybrid generation, recombination occurs (Figure 3.1d), but many of the new genotypes have no particular advantages for success. However, owing to the wide expression of genetic variation, the

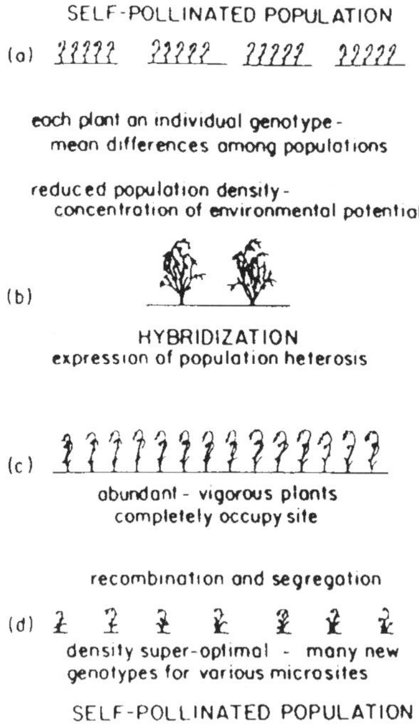

SELF-POLLINATED POPULATION
(a)
each plant an individual genotype -
mean differences among populations
reduced population density -
concentration of environmental potential
(b)
HYBRIDIZATION
expression of population heterosis
(c)
abundant - vigorous plants
completely occupy site
recombination and segregation
(d)
density super-optimal - many new
genotypes for various microsites
SELF-POLLINATED POPULATION

FIGURE 3.1 Model for hybridization in a largely self-pollinated population of downy brome (*Bromus tectorum*). Environmental concentration can be caused by fallow operations for weed control or by wildfires. (From Young and Evans, 1976, by permission of Weed Science Society of America.)

species can occupy all the microenvironments of the site. Successful genotypes (Figure 3.1d) resume self-pollination, once again producing stable duplicates on the site year after year (Young and Evans, 1976).

Although there are examples of successful weed species that are not predominantly self-pollinated, the pattern just described has been observed for numerous annual weed species. Such species are usually conspicuous colonizers, implying a common genetic breeding system based on compromise between the high level of recombination of outbreeders and the stability of self-pollinated species. The genetic variation that results from recombination in a nearly completely autogamous (self-pollinating) colonizing (weed) species helps its establishment in an area being newly colonized, whereas self-fertility is of value in building up the adapted population from its small beginning (Allard, 1965; Baker, 1974; Barrett, 1982).

INFLUENCE OF POLYPLOIDY

Polyploidy is a condition in which an individual possesses an excess of entire sets of chromosomes. Because polyploidy confers a selective advantage in many situations, it has been very important in plant evolution. Since some species are more successful as weeds than others, it is thought that polyploidy might influence the ability of plants to behave as weeds. Heiser and Whitaker (1948), using the weed flora of California that they developed from Robbins et al. (1941) as a base, studied the relationship of chromosome complement to weediness. They examined a total of 175 weed species and found that they occurred in approximately equal proportions as diploids and polyploids. This suggests that polyploidy, in general, confers no particular advantage upon weedy species, since approximately the same proportion of polyploidy was found among all higher plants. Heiser and Whitaker also compared the chromosome numbers and life cycles of weedy species of Gramineae (Poaceae) and Compositae (Asteraceae) to the ploidy levels and life cycles of those two families as a whole (Table 3.1). A striking tendency toward the annual life cycle was observed in weedy species of both families. These comparisons led Heiser and Whitaker to conclude that annuals in general and annual polyploids in particular are more likely to occur as weeds than are other species in those two families that have different life cycles or chromosome numbers. No other studies since that of Heiser and Whitaker have examined systematically the degree of polyploidy among weedy taxa. However, it is frequently assumed that the polyploidy level in weeds must be high because greater amounts of genetic variation and greater ability to colonize new habitats are thought to be associated with polyploidy (Barrett, 1982; Ehrendorfer, 1980 [cited in Hilu, 1993]; Warwick, 1990).

Electrophoretic studies of isozyme variation in plants have provided valuable information about the mating systems and genetic makeup of weed populations. Weeds have a wide range of levels of genetic diversity, from plant species that are genetically uniform to those that are extremely diverse (Barrett, 1982). As a group, however, weeds tend to be less genetically diverse than other

TABLE 3.1 Comparison of the Gramineae (Poaceae) and Compositae (Asteraceae) Plant Families with the Weedy Species of These Families in California

	Weedy Gramineae of California (%)	*Gramineae in General (%)*	*Weedy Compositae of California (%)*	*Compositae in General (%)*
Diploids	33	34	65	67
Polyploids	67	66	35	33
Annuals	64	24	57	35
Perennials	36	76	43	65
Annual—diploid	40	59	67	81
Annual—polyploid	60	41	33	19
Perennial—diploid	21	34	63	56
Perennial—polyploid	79	66	37	44

Source: Heiser and Whitaker, 1948.

groups of plants (Warwick, 1990). A number of weed species have been reported to lack isozyme variation (Barrett, 1982, 1988). A study by Warwick (1990), in which isozyme variation of five agricultural weed species in Canada was surveyed, illustrates this point (Table 3.2). Four of the five species were found to be polyploids and self-pollinating annuals. Electrophoretic analysis

TABLE 3.2 Isozyme Variation in Five Weeds of Corn and Soybean Monocultures from Eastern Ontario, Canada

	Abutilon theophrasti	*Panicum milaceum*	*Setaria faberi*	*Sorghum halepense*	*Datura stramonium*
Chromosome no.	2n = 24	2n = 36	2n = 36	2n = 40	2n = 24
No. of populations	39	39	8	13	9
No. of enzymes	16	11	14	14	12
No. of loci	27	19	24	21	22
No. (%) of loci monomorphic	25 (93%)	18 (95%)	21 (88%)	18 (86%)	22 (100%)
No. (%) of loci polymorphic	2 (7%)	1 (5%)	3 (12%)	3 (14%)	0
No. (%) of duplicated loci with enzyme multiplicity	14 (52%)	8 (42%)	13 (54%)	3 (14%)	2 (9%)
No. of multilocus genotypes	4	2	9	10	1

Source: Warwick, 1990.

revealed a striking lack of genetic polymorphism at isozyme loci, with each species usually composed of only one to several multilocus genotypes (Table 3.2). In the polyploid species, fixed heterozygosity was apparent for 14 to 54 percent of the loci screened. It has been suggested that this type of genetic variation may increase biochemical versatility and thus allow individuals of polyploid weeds to extend the range of environments that they can colonize successfully (Barrett, 1988).

Polyploids are believed to grow faster, occupy larger areas, and invade more disturbed sites than diploid plants, which is an advantage if genetic diversity within a species is restricted, as Table 3.2 suggests. This observation suggests that polyploidy might be important in weediness in some situations. Furthermore, many weed species occur in the grass and sunflower families (Table 1.3). Perhaps the rapid reproductive and growth rates of polyploid annuals account for the large number of weeds found in these two families. However, if polyploidy does confer some advantage for weediness, it is not readily apparent in the total weed flora.

Vegetative Reproduction

A common characteristic of many weeds of agricultural and forested lands is the ability to reproduce vegetatively. This trait is most common in, but is not restricted to, weeds with a perennial life cycle. Perennial plants may be either herbaceous or woody in growth habit. Although perennials are defined as plants that live for longer than two growing seasons, it usually is difficult to determine actual age of these plants, since different parts of the plant may be very different in age. Following the terminology of Harper (1977), a single unit of clonal growth is called a *ramet*, and genetically distinct individuals are called *genets*. Thus most perennial weeds can be either herbaceous or woody and can reproduce sexually, giving seeds (genets), or vegetatively, producing ramets.

TYPES OF VEGETATIVE REPRODUCTION

Abrahamson (1980) describes several methods of vegetative reproduction. There are numerous examples of common perennial weeds that have one or more of these methods of propagation (Figure 1.4). The most common forms of vegetative propagation found in weed species are described below.

Stolons and runners. These are long, slender stems that grow along the soil surface and produce adventitious roots and new shoots. Several examples are bermudagrass (*Cynodon dactylon*), a perennial grass; large crabgrass (*Digitaria sanguinalis*), an annual grass; and chinquapin (*Castanopsis sempervirens*), a shrub.

Rhizomes. Rhizomes are underground stems that produce adventitious roots and shoots, for example, Johnsongrass (*Sorghum halepense*) and quackgrass

(*Agropyron repens*), perennial grasses; purple nutsedge (*Cyperus rotundus*) and yellow nutsedge (*C. esculentus*), perennial sedges; goldenrod (*Solidago* spp.), a perennial herb; and bearmat (*Chamaebatia foliolosa*), a shrub.

Tubers. Enlarged terminal portions of rhizomes are called tubers. These possess extensive storage tissue and axillary buds. Examples are *Cyperus rotundus* and *C. esculentus*, perennial sedges, and Jerusalem artichoke (*Helianthus tuberosus*), a perennial herb.

Bulbs. Bulbs are underground modified buds consisting of a stem and fleshy scale leaves. Food storage is in the leaves. An example is wild garlic (*Allium vineale*), a perennial herbaceous lily.

Corms. Corms are enlarged, vertical underground stems covered with one or more layers of leaf bases. Food storage is in the stem. An example is bulbous buttercup (*Ranunculus bulbosus*), a perennial herb.

Roots. Many species produce long horizontal roots that give rise to shoots. Examples include Canada thistle (*Cirsium arvense*), field bindweed (*Convolvulus arvensis*), perennial herbs, and many others.

Stems. Some species produce adventitious roots and new shoots near the tips of branches, as *Rubus* spp. (a perennial vine) does, whereas others produce sprouts from a stem base or stump. Examples of the latter are dandelion (*Taraxacum officinale*), a perennial herb; *Rubus* spp., a vine; and bigleaf maple (*Acer macrophyllum*) and tanbark oak (*Lithocarpus densiflora*), which are trees. A special structure, the underground burl or lignotuber, is common in many shrubs that sprout prolifically after fire. This type of stem is an enlarged woody storage structure covered with adventitious buds. Many species in the genera *Arctostaphylos* and *Ceanothus* possess this structure.

Fragmentation. Spread and establishment of a ramet can occur by fragmentation of various plant parts, such as excised leaves or stems. This method of propagation may occur with segments of most underground portions of the plant. It also is evident with stems or leaves of, for example, purslane (*Portulaca oleracea*), an annual herb.

THE BUD RESERVE

A notable feature of most of the forms of vegetative propagation listed above, especially those occurring below ground, is the large number of vegetative propagules that can be produced. This phenomenon can be likened in many ways to the seed reserve in soil (Chapter 4), characteristic of annual weeds. Keeley (personal communication) presents the example of purple nutsedge (*Cyperus rotundus*), a perennial sedge in cotton-growing areas of the San

Joaquin Valley, California. Keeley and his associates indicate that purple nutsedge may reproduce vegetatively from rhizomes, tubers, or basal bulbs, though the tuber is the main reproductive and storage organ. Based on their studies, it was estimated that 10 to 15 million tubers per hectare could be produced within the top 15 cm of soil per year under optimal conditions. Low soil moisture impeded tuber production, whereas high soil temperatures and high light stimulated shoot development from dormant tubers.

Quackgrass (*Agropyron repens*) is another herbaceous perennial weed that has been studied often. Quackgrass can reproduce vegetatively from each node on a rhizome, with the reserve of dormant buds being up to 1200 buds per plant (Westra and Wyse, 1980). Common to most forms of vegetative reproduction is that buds are held in a dormant state until some form of separation from the parent plant occurs. Thus buds on rhizomes and tubers, for example, remain dormant until some damage to the shoots or underground reproductive systems occurs, after which abundant shoots are produced from fragments. Similarly, many stump and burl sprouting shrub species do not produce new shoots until the existing canopy has been removed. Following fire or some other disturbance that kills the top of the shrub, a profusion of sprouts is produced.

An important distinction should be made between the seed bank (Chapter 4) and the vegetative reserve of buds. Buds represent clones from a single plant that is successful in its microenvironment, whereas the population of buried seeds represents a reserve of untested genotypes. Thus, vegetative reproduction should be most advantageous when environmental conditions are relatively stable (unchanging) and the chance of disturbance is infrequent or predictable. Perennial herbaceous weeds are usually disfavored by frequent cultivations that can alter the microclimate of a site and also prevent establishment of new plants by vegetative means. They can be favored, however, by periodic (annual) tillage such as occurs in orchards or vineyards. The ability to propagate vegetatively is also of value during the early and middle stages of forest establishment, during which early site capture following disturbance is essential.

OCCURRENCE OF VEGETATIVE PROPAGATION

As already described, vegetative reproduction is probably most advantageous to weedy species in disturbed but relatively stable environments. These conditions are met in a variety of situations. Vegetative reproduction is especially well developed in aquatic habitats (Figure 1.5). In many cases, production of vegetative organs and fragments constitutes the major reproductive and dispersal efforts of aquatic weeds. In fact, for many aquatic weeds, seed production is markedly reduced or nonexistent. Notable examples of this lack of sexual propagation include the successful spread of elodea (*Elodea canadensis*) even though only one sex of elodea was introduced into Europe, and the sterility of water hyacinth (*Eichornia crassipes*) in many areas where it was introduced and is now a troublesome weed. It is apparent that vegetative reproduction is a dominant feature of aquatic weeds.

Vegetative reproduction is also common in plant communities influenced by fire. These areas are dominated most commonly by shrubs or trees, although fire-adapted grasslands also occur. Post-fire vegetative regeneration appears to be beneficial in that it allows reestablishment much faster than can occur by seed. Examples include chaparral (shrub)-dominated ecosystems of Mediterranean-type regions, temperate deciduous and coniferous forests, and certain perennial grasslands.

Finally, a habitat often occupied by herbaceous perennial weeds is arable land that receives some periodic but usually not extensive tillage. This habitat is most likely to occur in association with perennial crops. Bermudagrass (*Cynodon dactylon*) in asparagus and field bindweed (*Convolvulus arvensis*) in alfalfa are examples of perennial weed/crop associations in agricultural systems. Herbaceous perennial weeds also are found near trees in orchards where cultivation cannot be accomplished easily or in row crops where herbicides are used extensively for annual weed control and tillage is therefore infrequent.

ADVANTAGES OF VEGETATIVE REPRODUCTION

Several authors over many years have indicated that many widespread weed species have vegetative reproduction as an alternative to seed production (Salisbury, 1942a; Clausen et al., 1940; Abrahamson, 1980; Hegazy, 1994). They argue that such species have a considerable range of tolerance to environmental conditions and can reproduce efficiently under suboptimal or extreme conditions. A study by White (1979) concerning the physiological adaptations of Canada thistle (*Cirsium arvense*) ecotypes to different environmental conditions demonstrates this point. She collected rhizomes of two ecotypes, a northern ecotype originally from Yellowstone County, Montana, and a southern ecotype from an agricultural field near Hollister, California. Seed production of each was similar, but germination of the southern ecotype was much less than the northern one. The amount of rhizome production was also similar, but the southern ecotype was better adapted than the northern ecotype for ramet production from rhizomes at high temperatures (37°C). At lower temperatures (17°C) the reverse was true; the northern ecotype was more prolific. The southern ecotype represents the southern range extreme of Canada thistle, and it appears to reproduce itself solely by vegetative means in the suboptimal environment to which it has adapted. Spread of the southern ecotype occurs only in conjunction with mechanical disturbance of the rhizome system, in sharp contrast to the northern ecotype. By maintaining efficient seed and vegetative reproduction, the northern ecotype is capable of colonization in new locations by wind dissemination as well as continued occupation of the habitats already colonized.

Clearly, vegetative and sexual reproduction in flowering plants, especially weeds, are both advantageous under optimal environments. If they were not, evolutionary processes would eliminate the mode of reproduction of least value to the species. However, most perennial weeds utilize both mechanisms, indicating that in terms of energy balance the return per investment may be approximately equivalent for both reproductive processes (Abrahamson, 1980). Where both vegetative and sexual reproduction occur simultaneously, the vegetative offspring develop quickly to maintain the local population. Sexual reproduction, on the other hand, provides for many diverse propagules that can colonize other microenvironments or sites.

WEEDS AS STRATEGISTS

As already discussed in Chapter 2, many weeds possess characteristics common to both competitors and ruderals (Table 2.4) and may exist predominately in two combined strategies, competitive ruderals and stress-tolerant competitors (Figures 1.6, 1.7 and 2.7; Grime, 1979). It also is tempting to consider the pattern that Grime calls C-R-S strategists as appropriate for weeds. However, Grime notes that such environments are usually subject to pronounced seasonal and temporal variation in disturbance, stress, and competition. For example, in temperate zones in certain unfertilized pastures that experience constant grazing pressure, the vegetation is usually composed of species mixtures that develop characteristics intermediate to competitors, ruderals, and stress tolerators. We should keep in mind that weeds are highly specialized and successful organisms of productive environments. For this reason, it seems unlikely that many weed species would adapt in such a broad manner as the C-R-S category suggests.

Competitive Ruderals

Plant species with the adaptations of competitive ruderals should be found on productive sites where dominance by true competitors is diminished by some disturbance (Table 1.5, Figure 2.7). Only occasional disturbance is expected in this case since very frequent or severe disturbance would favor strictly ruderal vegetation. Conditions favorable for competitive ruderals might result when damage to vegetation occurs once or twice annually or during the life cycle, but does not affect or eliminate all the individuals from the community. Grime (1979) states that examples of this habitat include fertile meadows and grasslands subject to seasonal damage (e.g., grazing), flood plains, eroded areas, and margins of lakes and ditches. Arable land also is included in this habitat type.

Plants having the competitive ruderal strategy would be expected to have a rapid early growth rate, and the onset of competition between individuals should occur before the initiation of flowering. Annual herbs that Grime places in this category—common ragweed (*Ambrosia artemisiifolia*), Pennsylvania smartweed (*Polygonum pensylvanicum*), and velvetleaf (*Abutilon theophrasti*)—are characterized by a relatively long vegetative phase (Figure 1.7). Grasses, such as ryegrass (*Lolium multiflorum*) and others, also are capable of rapid and large dry matter production (Figures 1.7 and 2.7). Optimization of resource capture and seed production appears to be an important criterion for competitive ruderal species.

Many crops, such as barley, rye, corn, and sunflower, are annuals with rapid early growth rates and the capacity to produce a high leaf area index. Grime believes these species also should be classified as competitive ruderals.

Table 1.3 lists the annual weed species determined by Holm et al. (1977) to be of worldwide importance. Each species is found primarily on productive arable land. These weed species generally are characterized by high plasticity in vegetative growth, rapid early growth rates, and a prolonged vegetative phase prior to and during reproduction. Holm et al. (1977) repeatedly emphasizes the ability of these weeds to compete against crop species. Most of the weeds allocate a large proportion of resources to seed production. Many of the annual species listed by Holm as the world's worst weeds appear to fit the category of competitive ruderals.

It would be enlightening if a comparative analysis of relative growth rates were conducted using the species in Table 1.3, similar to that accomplished by Grime and Hunt (1975), in which species of various habitats were compared on the basis of early biomass accumulation. If this analysis were conducted under optimal and suboptimal conditions for growth and using the techniques of plant growth analysis (Chapter 6), it would likely demonstrate the correlation of success or noxiousness with early biomass production for weed species of arable lands. This result would clearly demonstrate the evolutionary significance of rapid early biomass production. It also might demonstrate the management value, from an evolutionary viewpoint, of maintaining a vigorous "competitive" crop such that early biomass accumulation by a weed is of reduced competitive advantage. The importance of early weed control for optimum crop production has been recognized for many years by agronomists. Perhaps the basis for this view is the morphological and physiological adaptations of common evolutionary origin among weed species and crops that cause early onset of competition in weedy agricultural fields.

Although many weed species classified as competitive ruderals are annuals, others are not. The notable exceptions are the herbaceous perennial species (Figures 1.7 and 2.7), such as Canada thistle (*Cirsium arvense*), quackgrass (*Agropyron repens*), coltsfoot (*Tussilago farfara*), and Johnsongrass (*Sorghum halepense*). Such species tend to be strongly rhizomatous or stoloniferous and have a high capacity for vegetative growth. Many also maintain high seed production. Their competitive behavior is often noted; however, they can be displaced as seedlings by more competitive annual species, especially under a frequently tilled regime. However, tillage can promote proliferation of growth from vegetative fragments once herbaceous perennials become established. Thus the establishment and spread of these species is enhanced by occasional disturbance.

Most herbaceous weeds common to arable land have adapted to the combined strategy of competitive ruderals. As ruderals, these species appear to require the soil disturbance associated with agriculture for establishment and growth. Since it usually is not possible to maintain a completely disturbed environment and still grow a crop, plants of this strategy necessarily developed competitive characteristics as well. Perhaps the arable weeds developed initially as ruderals in their "natural" habitats, and with the advent of agriculture, a relatively recent event on the evolutionary scale of time, also developed characteristics that allowed success in more competitive or less disturbed environments.

Stress-Tolerant Competitors

In order to understand the role of the stress-tolerant competitors as weeds we must consider once again the process of succession, particularly in productive environments. As we have seen already, this process involves initial colonization followed by progressive microenvironmental modification and replacement of the vegetation over time (Chapter 2). The role of stress-tolerant competitors, primarily shrubs and trees, is particularly evident within the time frames common to forest succession. In this case shrubs often dominate an early to intermediate seral stage (Figures 1.9 and 1.10) for a prolonged period of time.

The shrubs and some trees that appear early in forest succession on productive habitats are similar in many ways to competitive-ruderal herbs. The common characteristics include rapid dry matter production, at least in comparison to other trees and shrubs, rapid stem extension and leaf production through most of the growing season, and rapid phenotypic responses of leaf or shoot morphology to shade (Figure 1.7). Grime (1979) indicates that such features are particularly conspicuous among deciduous trees (*Ailanthus, Betula, Populus*) and shrubs that occur in the early phases of natural reforestation in disturbed woodlots of eastern North America (Figure 1.6). Species that assume a similar role in the forested areas of northwestern North America occur in the genera *Arctostaphylos, Ceanothus, Rubus, Alnus*, and *Acer*. These species often are deterrents to forest regeneration efforts in the western United States.

Most of the woody species that initially colonize an area following disturbance by fire or logging do so from long-term seed reserves in the soil. Initial seedling growth is usually slow, but once the plant is established, extensive and rapid vegetative growth results. In addition, many species, including greenleaf manzanita (*Arctostaphylos patula*), bigleaf maple (*Acer macrophyllum*), and tanbark oak (*Lithocarpus densiflora*) in western North America, sprout readily from root crowns or stumps if another disturbance occurs to remove top growth. In that event, total canopy coverage can approach the predisturbance levels in only a few years. Maximum photosynthate production and usually vegetative growth are coincidental with periods of low moisture stress. These species are notably shade intolerant, however, and the ultimate dominance of more shade tolerant trees is assured.

It appears that many shrub and tree species that are considered weeds on disturbed forest lands follow the combined strategy of stress-tolerant competitors. These species usually dominate the vegetation of early and intermediate seral communities following a major disturbance to the forest. They are long-lived, often existing for decades, and tend to allocate a significant amount of their resources to stems and branches for canopy (foliage) support. However, as competitors these species also possess rapid early rates of vegetative growth, especially if sprouting occurs following top removal. Because of the competitive ability of these weeds, succession may proceed beyond the intermediate stages only as more stress-tolerant species gradually outlive them (Figure 2.5).

INFLUENCE OF HUMANS ON WEED EVOLUTION

Weeds, Domesticates, and Wild Plants

Plant species react in different ways when their habitats are disturbed by humans. Some species flourish because of the disturbance, whereas others migrate or die and are replaced. De Wet and Harlan (1975) describe three classes of vegetation based on the degree of association with human-caused disturbance. According to de Wet and Harlan, *wild plants* grow naturally outside of human-disturbed habitats. They are aggressive colonizers, and when the habitat is not frequently disturbed, successive waves of different species invade until dynamic, but eventually stable, population balances are achieved. When the habitat is continuously disturbed by humans, a much different set of species becomes established. De Wet and Harlan describe these species as *weeds* and *domesticates* (crops). Neither class of vegetation can compete successfully with "wild" species for "wild habitats." Weeds may invade newly disturbed habitats, but they are usually replaced by wild colonizers if the habitat is not disturbed further.

De Wet and Harlan (1975) indicate that domesticates can be weedy and plants generally considered to be weeds are sometimes grown as crops. However, it is unusual for weeds to require artificial propagation as do crops. Thus, weeds are able to establish new populations within the human-disturbed habitat without further assistance from humans. Domesticates, in contrast, require cultivation and the continual help of humans in order to propagate.

De Wet and Harlan believe that weeds have evolved in response to human-caused disturbances in three principal ways: (1) from wild colonizers through adaptation to and selection for continuous habitat disturbances, (2) as derivatives of hybridization between wild and cultivated races of domesticated species, and (3) from abandoned domesticates by selection toward a less intimate association with humans. They propose that most weeds have evolved directly from wild species that invaded human-disturbed habitats. As evidence, many weed species—for example, dandelion (*Taraxacum officinale*), henbit (*Lamium amplexicaule*), and crabgrass (*Digitaria sanguinalis*)—have been distributed far beyond their "natural" range and these weeds have wild races in their native Old World habitats. In addition, most domestic plant species also have weed and wild races, supporting the second avenue of weed evolution. De Wet and Harlan indicate that hybrids between wild and cultivated forms rarely invade successfully the natural habitat of the species but are common in disturbed habitats associated with cultivation. Domesticated races (crops) also can revert to a weedy growth habit when they are no longer cultivated. However, eventual replacement by other races or species less dependent upon cultivation than crops is likely. In cases of reversion to weedy forms from a domesticate, seed dispersal mechanisms of the weed race are often similar to those found in wild types. Cultivated forms of cereals, for example, are characterized by spikelets that persist on the inflorescence at maturity, whereas wild-type cereals disperse seed by means of an abscission layer that forms between the rachis and the spikelet. Weed races of cereals (e.g., wild oat) are similar to the wild type regarding the method of seed dissemination.

It is certain that weeds have evolved and will continue to evolve in response to agricultural activities. This fact must be recognized in the development of management practices for weed control, since the reliance on a single method or tool will most likely result in weed species or races that have become tolerant to it.

Crop Mimics

The selective forces created by agricultural practices often result in the evolution of agricultural races, agroecotypes, of weeds (Barrett, 1983). Some of these agroecotypes are intimately associated with a specific crop. The more closely a weed species resembles the crop, generally the more difficult it is to control without crop damage. Some crop-weed associations are so intimate that their response to selection pressure is mimetic forms of weeds which often resemble the crop in morphology or behavior, thereby avoiding control tactics. Barrett (1983) indicates that although mimicry has been studied extensively for over a century in animals and insects, there are few studies of this type of coevolutionary phenomenon for weeds. Rather, most accounts of mimicry in weeds that occur in the literature exist as reports that document the form of resemblance between a crop and weed species. Weins (1978) suggests that the following terms are central to the concept of mimicry:

- *model*—the animate or inanimate object or function that is being imitated;
- *mimic*—the imitating organism; and
- *operator*—the organism that is unable to discriminate effectively between the model and the mimic.

In the case of weed mimicry, the operators or selective agents are usually mechanical devices designed and operated by humans.

As already indicated, some weeds and crops grow in conjunction with one another and resemble each other in form or function. For example, barnyardgrass (*Echinochloa crus-galli*) occurs wherever rice is grown and is a weed of major importance in that crop (Figure 3.2a). Wild oat (*Avena fatua*) is considered to be a nearly cosmopolitan weed wherever cereal crops occur (Figure 3.2b). In

FIGURE 3.2 Barnyardgrass (*Echinochloa crus-galli*) occurring in rice grown in California (top), and wild oat (*Avena fatua*) occurring in wheat in North Dakota (bottom). (Photographs by E.J. Roncoroni and L.W. Mitich, respectively.)

Europe, large-seeded false flax (*Camelina sativa*) has evolved several ecotypes that are closely associated with flax production. In such cases, these crops and associated, or satellite, weeds may have evolved together under similar cultural practices or environments so that growth and reproduction of the weed matches the life cycle of the crop.

One of the finest examples of crop mimicry involves weeds associated with flax. Baker (1974) suggests that this may be because flax is one of the oldest plants grown as a crop in Eurasia. In open-grown situations the weed, *Camelina sativa* var. *sativa*, is a generalized annual plant with a wide-branching growth habit. When found with flax, however, this weed takes on a taller, less branched form which resembles a flax plant. In some areas where flax cultivation is very intensive, *C. sativa* var. *sativa* is replaced by another ecotype, *C. sativa* var. *linicola*, which is even more specialized. In the variety *linicola*, the life cycles of the weed and flax are so closely aligned that the weed is always harvested with the crop. As in flax, the fruits resist shattering at harvest and the seeds so closely resemble flax seeds that they cannot be separated readily by winnowing. Thus seeds of both plants are often sown together. In his review of this subject, Baker (1974) indicates that *C. sativa* var. *linicola* undoubtedly evolved from var. *sativa*, and that crop mimicry in this case seems to be fixed genetically. Perhaps satellite weeds of crops and crop mimics once possessed a more general habit of growth or development which became more croplike as selection for specialization increased.

Another striking example of crop mimicry by a weed species is barnyardgrass in rice. Barrett (1983) indicates that although mimetic forms of barnyardgrass certainly originated under "primitive" agricultural systems, they also are present and presumably evolved under modern mechanized rice culture, as well. In California, *E. crus-galli* var. *oryzicola* has replaced *E. crus-galli* var. *crus-galli* as the major weed in rice fields. According to Barrett and Seaman (1980) two distinct races of *E. crus-galli* var. *oryzicola* were introduced into California as rice seed contaminates in 1912–1915 when rice culture was just beginning in the state. These two races now behave as distinct biological species and differ in chromosome number, morphology, flowering, and distribution (Table 3.3). Despite the occurrence of populations containing millions of individuals of *E. crus-galli* var. *oryzicola* in California rice fields each year, as well as their continued spread to uninfested rice producing areas, few populations are found outside the rice agroecosystem (Figure 3.3). Barrett indicates that this restricted habitat preference is typical of crop mimicry and that the behavior and spread of *E. crus-galli* var. *oryzicola* contrasts significantly with *E. crus-galli* var. *crus-galli*. Although these two varieties are both pernicious weeds of worldwide importance, they exhibit very different adaptive strategies. *E. crus-galli* var. *crus-galli* exhibits many of the traits of Baker's general purpose genotype, whereas the rice mimic var. *oryzicola* is a specialized biotic ecotype with limited ecological amplitude (Barrett, 1983). The analysis of the contrasting life histories of these two barnyardgrass varieties would be helpful in understand-

TABLE 3.3 Barnyardgrasses in California Rice Fields

Taxon (frequently used synonyms in parentheses)	*Ploidy*	*Origin*	*Rice Weed Status*	*Distribution in California*	*Microhabitat in Rice Agroecosystem*
E. crus-galli (L.) Beauv. Var. *crus-galli Barrett 1201*, TRT	6x	Eurasia	worldwide	widespread	rice field edges, levees, and shallow water areas in fields
E. crus-galli (L.) Beauv. Var. *oryzicola* (Vasing) Ohwi a) early flowering form (*E. oryzoides* [Ard.] Fritsch) *Barrett 1202*, TRT	6x	Eurasia	Europe, Asia, Australia, Argentina	generally restricted to Central Valley	rice fields
b) late flowering form (*E. phyllopogon* [Stapf.] Koss) *Barrett 1203*, TRT	4x	Asia	Europe, S.E. Asia, India, Russia, China, Japan	restricted to rice-growing regions of Central Valley	rice fields
E. muricata (P. Beauv.) Fern. (*E. pungens*) *Barrett 1204*, TRT	4x	North America	U.S. and Australia	scattered localities in central and northern California	rice field edges, levees, and drainage ditches
E. colona (L.) Link *Barrett 1294*, TRT	6x	Asia	worldwide (mostly in upland rice)	scattered localities mostly in southern California	rice field edges and levees

Source: In Barrett, 1983; after Barrett and Seaman, 1980.

ing the selective forces operating in production systems that cause shifts in the spectrum of weed species over time.

Shifts in Weed Species Composition

Over a short period of time plant communities often are perceived as being stable (Williamson, 1991), their composition of species appearing relatively constant from season to season and year to year. Nevertheless, evolutionary change is inevitable in all plant communities, and shifts in species composition

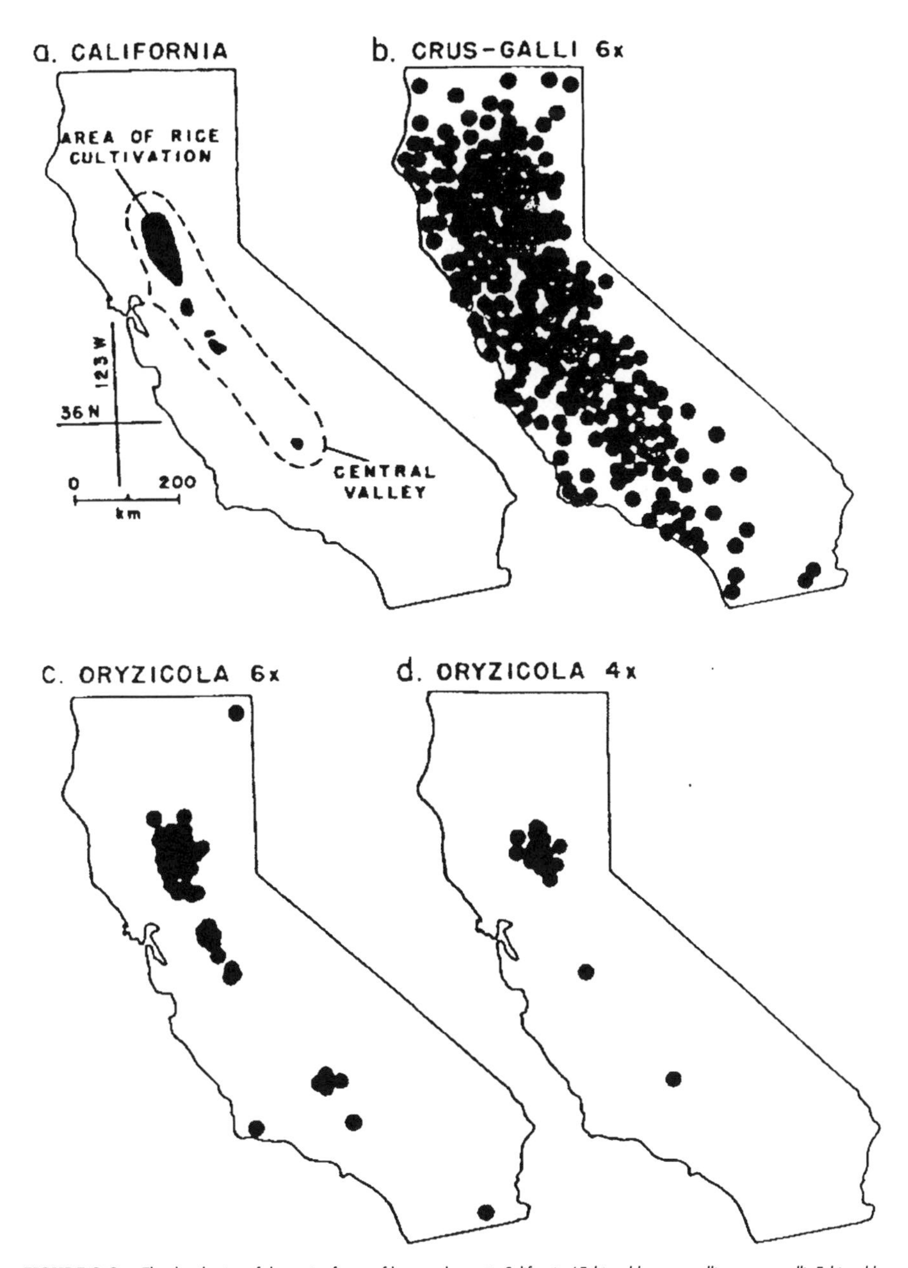

FIGURE 3.3 The distribution of the major forms of barnyardgrass in California (*Echinochloa crus-galli* var. *crus-galli, Echinochloa crus-galli* var. *oryzicola* [4x, 6x races]) in relation to the areas of rice cultivation (Barrett, 1983).

over time and as the result of human disturbance should always be expected. Thus, a discussion of plant succession is particularly relevant at this point. As seen in Chapter 2, plant succession can be either primary (establishment of plants on land never vegetated before) or secondary. Secondary succession is the pattern of change after a radical disturbance so that a new patch of the physical environment is available once again for colonization by plants (Figure 2.4). Secondary succession is most applicable to human-disturbed systems, such as farming and plantation forestry. In addition, a "farming cycle" of repeated disturbance may be superimposed onto the plant colonization process of secondary succession in certain agricultural systems (Soriano, 1971), where, for example, each year begins with tillage and sowing, followed by crop production, and finishes with a harvested field. Since weed communities must respond to several patterns of change in the physical environment during secondary succession—for example, (1) long-term environmental impacts from vegetation disturbance such as soil erosion or climate change, (2) seasonal and annual variation, and (3) finer scale effects resulting from the farming cycle—their stability in species composition should be difficult to observe. However, weed communities often are quite predictable for particular geographic areas and cropping systems (Salisbury, 1961; Leon and Suero, 1962; Crawley, 1987). This apparent contradiction can be explained if weed species coevolve with their cropping systems, which allows adaptation to intensely and regularly disturbed environments (Harper, 1957; Ghersa et al., 1994a).

Plant species in weed communities follow the general principles of secondary succession that are outlined in Figure 2.5 (Connell and Slatyer, 1977; Vitousek and Walker, 1987). Genetic diversity also will change through time in such communities, with or without human disturbance. Land exposed by regular disturbance has periods with very low soil cover from plant biomass but high levels of resource availability. These conditions make the crop-weed community susceptible to invasion by other plants (Crawley, 1987). Each invasion of the community creates a new scenario of community instability that Williamson (1991) calls "press" perturbation, a situation in which the structure and functional properties of the community are modified by new species rather than the elimination of ones already present. This continual state of disequilibrium of weeds in arable land was demonstrated clearly by Forcella and Harvey (1983) in their study of alien arable weeds in the northwestern United States. In that study, both total species richness and species uniformity throughout the area were found to have increased since the turn of the century. A similar response of weeds to continual environmental modification was found in the rolling pampas of Argentina by Suarez et. al. (1995).

Because successional changes in the weed flora occur at the same time that cultural practices are varied, it is difficult to isolate which factors actually cause changes in the relative abundance of weed species. This is especially true for weed compositional shifts that might occur over an entire region, as discussed above. On the other hand, on a smaller scale such as a patch or field, shifts in species composition may be easily attributed to a particular change in the suite of cultural practices used to grow crops. Unfortunately, weed researchers often have overlooked the implications of ecological processes such as colonization and succession when developing weed control tactics. This oversight, although understandable, limits our predictive ability about future weed invasions, species losses, or evolution of herbicide-resistant weed populations. However, we do know that weeds follow the general rules of plant biology, pertaining to genetic diversity, breeding systems, and secondary succession. For this reason, it should be possible not only to predict shifts in weed communities but also to design cultural practices or generate microenvironments that are at least temporarily unsatisfactory to certain weeds.

Impacts of Weed Control on Weeds

The spatial distribution of weeds within a particular cropping system is influenced by a wide range of cultural practices. Furthermore, a change in the way a crop is grown also may influence the weed species associated with that crop. Haas and Streibig (1982) provide striking evidence about how shifts in various agronomic practices from 1911 until the present have affected the weed flora of Denmark. By analyzing the distribution of weed life forms in relation to the associated crops, Haas and Streibig documented a general rise in the abundance of annual weed species and a concomitant decrease in herbaceous perennials over the 60 years of their study. They believe that this change in life form of the Danish weed flora has been enhanced by the greater soil disturbance that occurs from tractor-drawn, rather than horse-drawn, tillage implements. Frequent tillage was considered to favor annual weeds, whereas infrequent tillage practices, which are less common today, allowed biennial or perennial species to predominate.

Haas and Streibig (1982) also analyzed the composition of the weed flora that occurred in several crops grown in Denmark. They found that the life forms of both weeds and crops usually were similar and correlated. Furthermore, they observed that certain weed species are more prevalent in certain crops now than they were in the same crops 60 years ago. Other species examined are less common now or else unchanged in frequency. Table 1.6 summarizes the occurrence of weed species in Danish spring cereals since 1918. It is apparent from this table that no single factor can account for the changes in weed species distribution that have occurred over the last five decades. However, there is a general tendency toward increased nitrophily (plants with a high nitrogen requirement) in those species that have remained unchanged or have increased. *Rumex acetosella* (sheep sorrel) has actually been disfavored by soil amendments (lime), since it is a calcifuge (plants that occur in soils with low calcium and pH) species and liming is now a common practice in Denmark.

Most species in Table 1.6 that have increased over the years also tolerate or are favored by the soil compaction associated with combine harvesting. *Matricaria matricarioides* (pineappleweed) is noted for its ability to tolerate compacted ground. Haas and Streibig also showed that dicot species that are increasing in frequency in Danish cereal fields are more tolerant to certain herbicides (MCPA and dichlorprop) than species that are decreasing in frequency. The fact that cereal production has increased markedly in Denmark and many other countries in western Europe also was noted. Thus a general decrease in crop rotation occurred during the last 30 years of their study because cereal crops are now planted with greater frequency. Since each crop has its own unique history in terms of tillage practices, fertility regimes, and harvesting methods, the shift toward cereal monocultures also has contributed to some of the species changes shown in Table 1.6. These and other data demonstrate that the weed species composition of many crops is quite dynamic and responsive to changes in cultural practices over time. Although it is often difficult to determine a single factor or group of factors responsible for shifts in weed composition, these shifts do occur and must be considered for weed/crop management.

Populations within a weed species also may be influenced by cultural practices or harvest methods. For example, Schoner et al. (1978) have observed the predominant occurrence of prostrate forms of yellow foxtail (*Setaria lutescens*) in alfalfa fields of California, in contrast to other alfalfa-growing areas in other regions of the United States (Figure 3.4). The prostrate form of this weed species apparently is favored by the frequent cutting cycle (21–28 days) common for alfalfa production in California. Studies by Price et al. (1980) suggest that wild oat (*Avena fatua*) can be manipulated genetically by cultivation practices. By using gel-electrophoretic techniques, those workers determined that distinct wild oat populations exist under rangeland and cultivated (cereal) conditions. Though some overlap of wild oat populations exists among different environments, these data suggest that cultivation commonly associated with cereal production can result in subtle changes in the composition of the weed community from that of a rangeland environment.

SPECIES SHIFTS ASSOCIATED WITH HERBICIDES

Although changes in weed species are continually taking place, the extreme efficacy of herbicides increases selective pressure to very high levels, sometimes producing local extinctions. Many authors (Ashton and Crafts, 1981; Ross and Lembi, 1985; Devine et al., 1993) have reviewed numerous publications that demonstrate differential susceptibility of plant species to herbicides. Because of interspecific selectivity, continued use of a particular herbicide often causes a shift within a weed community from susceptible to more tolerant species. For example, a species shift that favors grasses is readily observed from applications of 2,4-D for broadleaf weed control in cereals. The selective activity of herbicide use in crops is expected and in fact is encouraged, since it is the basis for effective chemical weed control in crops. However, applications of 2,4-D to cereals, when made annually for a number of years to control broadleaf weeds, also have favored the occurrence of grass weeds in the cereal cropping system (Fryer and Chancellor, 1979; Hay, 1968). Figure 3.5 illustrates the changes in the weed flora that Fryer and Chancellor (1979) believe

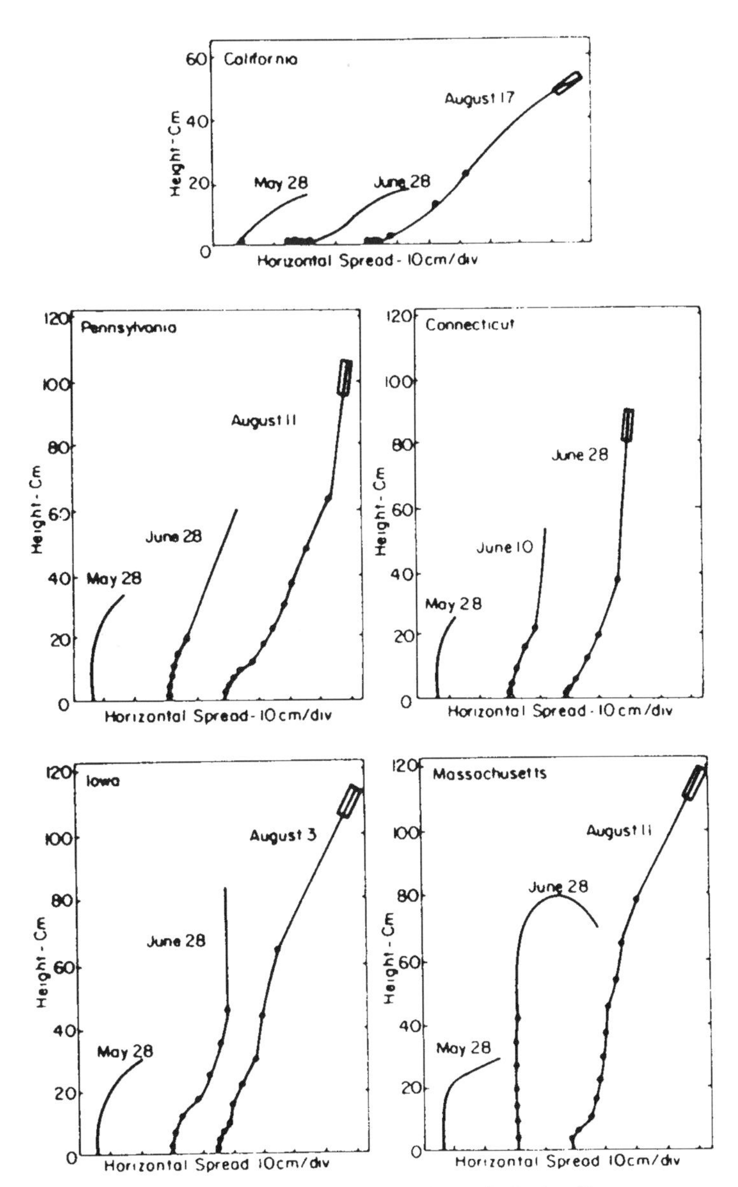

FIGURE 3.4 Diagrammatic comparisons of the mean growth pattern of culms of yellow foxtail biotypes. Drawings represent culm orientation at seeding, immature, and mature stages of growth of each biotype. Each drawing was constructed from measurements of extended culm length from base to tip, culm vertical height from soil surface to culm tip, number of nodes, vertical height of each node from the soil surface, internode length, and distance of each node from the culm base. Blackened circles represent the position of nodes. Open rectangles represent mature panicles. (From Schoner et al., 1978.)

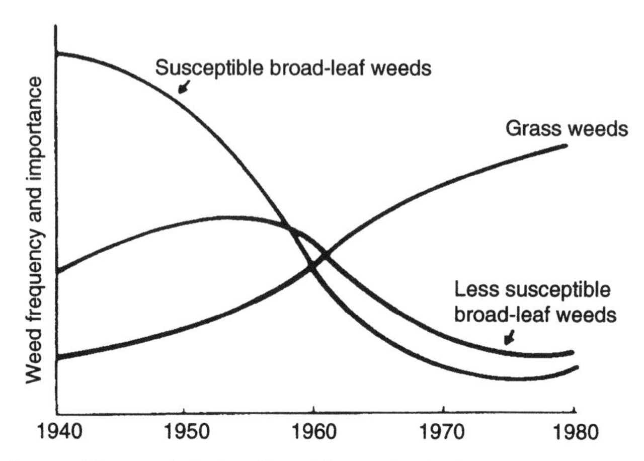

FIGURE 3.5 Conjectured changes in the British arable weed flora over four decades. (From Fryer and Chancellor, 1970; reproduced by permission of the British Crop Protection Council.)

have occurred over the last four decades in Great Britain. As a result of routine spraying of 2,4-D and related herbicides, many susceptible broadleaf weed species have declined in frequency, whereas grass weeds, such as wild oat (*Avena* spp.) and blackgrass (*Alopecurus myosuroides*) have increased in importance (Fryer, 1982). The observations of Fryer and his associates agree with those of Bachthaler (1967) in his study of the German weed flora. Over the 17-year period of Bachthaler's study (1948–1955 and 1958–1965), he observed a reduction in weed species that were easily controlled by herbicides but an increase in the predominance of four grass weeds.

Another example of a shift in weed species composition from herbicide use occurs when both crop and weed species are closely related taxonomically. Plants frequently respond similarly to herbicides according to taxonomic families, probably because similar taxonomic characteristics reflect similar physiological or morphological traits that also affect herbicide performance. Trifluralin, for example, does not readily control plants in the family Solanaceae, and for that reason it is used for weed, especially grass, suppression in such crops as tomato and potato that are in that family. Repeated use of trifluralin in tolerant crops on an annual basis frequently causes a shift in weed species composition toward nightshades (*Solanum* spp.). Observations of species shifts are not restricted to a few herbicides such as 2,4-D and trifluralin, but are of rather widespread occurrence and involve many herbicides and many weed species. Other examples include the increased occurrence of pigweeds (*Amaranthus* spp.) from use of napropamide, mustards (*Brassica* spp.) from use of benefin, and common groundsel (*Senecio vulgaris*) from use of diuron and terbacil.

Shifts in weed species composition within a crop/weed community are relatively common following repeated annual herbicide application. Such events were predicted in the mid-1950s by Harper (1957), who observed two phenomena: (1) the selective control of weeds that occurred with herbicides and (2) the occurrence of pest resistance to other pesticides and antibiotics. He predicted that plant species tolerant to herbicides eventually would replace susceptible ones in situations of repeated herbicide use. Also suggested by Harper was a cost to fitness associated with herbicide tolerance, implying that tolerant species would be less competitive than those they replaced.

Because of the apparent widespread adaptability of weed species to herbicides, more than a single chemical tool for each cropping system is needed. The phenomenon of herbicide tolerance also demonstrates the need for integrating chemical and mechanical weed control measures within a cropping system, or for using crop rotations. Such a variety of control measures will help to ensure that herbicide-tolerant weed species do not build up in weed/crop communities.

HERBICIDE RESISTANCE

There are no uniform and consistent definitions of "tolerance" and "resistance"; the terms often are used interchangeably. LeBaron and Gressel (1982) refer to *tolerance* as the normal variability in response to herbicides that exists among plant species and that can build up quickly in a population. Thus one species may be more tolerant than another to a given herbicide, and may increase in the population following use of that herbicide, even though both may be controlled at some level of exposure to it. *Resistance* involves the altered response to a herbicide by a formerly susceptible weed species to the extent that some individuals in that species are no longer susceptible. Therefore, herbicide resistance may be viewed as the extreme level of herbicide tolerance. While the word tolerance is often used loosely, particularly when the mechanism of a plant's response is unknown, resistance usually implies an evolved response to selection by a herbicide.

Though the occurrence of pesticide resistance in other organisms is widespread, there are still relatively few examples of herbicides to which formerly susceptible weed species have evolved resistance. Furthermore, herbicide resistance has a more recent history than either insecticide or fungicide resistance. The lower number of cases of herbicide resistance in plants as compared to cases of insecticide and fungicide resistance is probably due to a combination of factors. These factors include the relatively low persistence of many herbicides, lower fitness of some resistant weed strains compared to susceptible ones, the ability of herbicide-thinned stands of susceptible weeds to continue to produce seed, and a large soil reservoir of susceptible weed seeds. All these factors tend to maintain a large susceptible gene pool even under a regimen of repeated annual herbicide applications.

Nevertheless, over the last twenty-five years it has been well documented that repeated applications of herbicides with similar modes of action onto the same field site impose sufficient selection pressure to increase resistance within species that formerly had been susceptible. Herbicide resistance has become well known in scientific and agricultural communities around the world since the discovery of triazine resistance in common groundsel (*Senecio vulgaris*) in

1970. Cataloging the cases of resistance has continued since that time (Caseley et al., 1991; LeBaron and Gressel, 1992; Powles and Holtum, 1994). In the 1980s, it was noted that the rate of increase or perhaps of detection of herbicide resistance cases had equaled the rate of increase of cases of resistance in arthropods and pathogens (Holt and LeBaron, 1990).

Table 3.4 summarizes the known cases of herbicide resistance through 1993, which are presented in detail by Holt and LeBaron (1990) and LeBaron (1991). At least 57 weed species, including both dicots and monocots, have been found with resistance to the triazine herbicides. In addition, at least 64 species have been reported to have biotypes resistant to one or more of 14 other herbicides or herbicide families (Table 3.4). The known or suspected mechanism of resistance to the herbicides for which resistant weed biotypes are most numerous or widespread is shown in Table 3.5. A common feature of these herbicides is their specificity for a single target site, which is often determined by a single or very few genes. Where the genetics of evolved resistance have been studied, results indicate that one or very few genes are probably involved in conferring resistance, as well. These randomly occurring, single-gene mutations that confer resistance may be selected for easily in situations where the herbicide is used repeatedly.

TABLE 3.4 Occurrence and Distribution of Herbicide-Resistant Weed Biotypes Selected under Agricultural Conditions[a]

Herbicide or Class	*Year First Detected*	*Number of Species with Resistant Biotypes*	*Number of Known Sites[a]*
ACCase inhibitors	1982	8	>1000
ALS inhibitors	1986	8	>1000
Amides	1986	2	2
Aminotriazoles	1986	2	2
Arsenicals	1984	1	2
Benzonitriles	1988	1	1
Bipyridiliums	1976	18	>50
Carbamates	1988	2	70
Dinitroanilines	1973	5	>20
Phenoxyacetic acids	1962	6	5
Picloram	1988	1	1
Pyridazinones	1978	1	3
Substituted ureas	1983	7	>50
Triazines	1968	57	>1000
Uracils	1988	2	1

Source: Holt et al., 1993.

[a]A known site is a site where a resistant biotype is thought to have evolved.

TABLE 3.5 Herbicides or Classes of Herbicides for which Resistant Weed Biotypes Are Most Numerous or Widespread, and the Known or Suspected Mechanism of Resistance to Each

Herbicide or Class	*Mechanism of Resistance*
Aryloxyphenoxypropionics	Enhanced detoxification
Bipyridiliums	Enhanced detoxification and/or sequestration
Dinitroanilines	Hyperstabilized microtubules
Substituted ureas	Enhanced detoxification
Sulfonylureas	Modified ALS enzyme binding site
Triazines	Modified photosystem II protein binding site

Source: Adapted from Holt, 1992.

With the rapid increase in occurrence of resistance to triazines and other classes of herbicides (Table 3.4), the phenomenon of resistance has become more than a scientific curiosity. Some manufacturers of agricultural chemicals now address this problem through changes in their label recommendations for use of herbicides that have high risk for selecting resistance. Another offshoot of herbicide resistance is the potential for transferring this trait to crop species. Recent advances toward the development of herbicide-resistant cultivars have led to both optimism and concern about their use in agriculture and forestry.

IMPLICATIONS OF HERBICIDE RESISTANCE EVOLUTION TO WEED MANAGEMENT

The evolution of herbicide resistance is simply another item in a growing list of constraints that potentially limit pesticide uses in agriculture and other production systems. A critical step needed to address the resistance problem is to consider how herbicide resistance fits into the broader function of weed/crop ecosystems. In other words, if the approach is only to replace the herbicide that has selected for resistance with a new one, even with a different mode of action, then the problem is being perpetuated and its cause ignored. A more effective approach is to examine the biological and ecological processes that govern the potential evolution and spread of resistance in cropping systems. For example, triazine resistance is due to a mutation in a single-protein binding site associated with photosystem II of photosynthesis (Pfister et al., 1979). The function of this protein was linked to differences between susceptible and resistant biotypes in chloroplast and whole-plant performance (Holt et al., 1981, 1983), which also influenced the fitness of the two biotypes (Conard and Radosevich, 1979; Holt and Radosevich, 1983; Holt, 1988). It is often suggested that selection for particular traits that accommodate an environment involves a biological cost to the organism (Dobzhansky, 1970; Gressel and Segel, 1978). In this case, plant fitness is influenced by complex interactions between the maternally inherited single gene that codes for the triazine binding protein (Darr et al., 1981; Souza

Machado et al., 1978; Warwick and Black, 1980) and nuclear inheritance of other plant performance traits (Stowe and Holt, 1988; McCloskey and Holt, 1990, 1991). These genetic, physiological and ecological processes ultimately interact to influence the population dynamics of triazine resistant and susceptible weed populations.

An important tool for management of resistance is the use of models, because they provide a logical structure with which to conceptualize and organize relevant biological processes and to simulate weed responses to management (selection) options. Gressel and Segel (1978, 1982) developed the first simulation model for investigating the dynamics of herbicide resistance. Their model has been used extensively to evaluate potential evolution of resistance. Because the model simplified several life-history traits, it generated pessimistic predictions about the evolution and potential management of herbicide resistance; that is, removal of the herbicide was the only viable management option. More recent work (Maxwell et al., 1990; Putwain and Mortimer, 1989; Gressel and Segel, 1990) has increased the complexity and comprehensiveness of resistance models that now include the influence of other factors like fitness and gene flow (Chapter 2), which appear to regulate the dynamics of resistance.

Fitness. Harper (1956) recognized fitness as an important factor influencing the evolution of herbicide resistance long before herbicide resistance evolved. Fitness describes the potential evolutionary success of a phenotype, based on its survival and reproductive success. All herbicide resistance models now strongly weight the influence of fitness because the increase in fitness of the resistant phenotype compared to the susceptible during repeated herbicide use selects for the evolution and spread of resistance. Because of the experience with triazine herbicides, in which resistance was accompanied by reduced fitness when the herbicide was not used, managing fitness provided some hope for reducing resistance after it had occurred in a weed population. When the resistant phenotype is less fit than the susceptible, removal of the herbicide leaves natural selection to restore susceptibility. There are now numerous reports of weeds resistant to herbicides other than triazines (Table 3.4), where the resistant phenotype may be as fit as the susceptible one (Holt et al., 1993; Jasieniuk et al., 1996; Warwick and Black, 1994). However, fitness in these reports typically has been estimated only by measuring relative competitive ability during vegetative growth stages or seed output during only a portion of the life cycle. The more recent demographic models of resistance emphasize that an adequate assessment of relative fitness must also include other stages in a plant's life history, for example, germination, establishment, survival, and fecundity, as well as plant growth. It is quite likely that many potential avenues for managing these new forms of resistance will become apparent as the role of differential fitness between the phenotypes becomes better understood.

Gene flow. Gene flow is the result of processes such as pollen and seed movement that influence the maintenance of a genotype in a population. Gene flow alters allele frequency in populations through (1) immigration of external genes into a population, (2) breeding systems and genetic drift that limit the flow of genes within a population, and (3) the presence of a seed bank and dormancy that slow the loss of genes from a population (Levin and Kerster, 1974; Maxwell et al., 1990). The most important implication of these processes for herbicide resistance is the recognition that the susceptible genes can play a role in management of resistance (Roush et al., 1990). Model simulations suggest that gene flow can significantly decrease the proportion of resistance in a population. For example, gene flow in cross-pollinating species with large, nearby sources of susceptible genes would reduce dramatically the proportion of resistance once herbicide use had been stopped. However, resistance could also be reduced in outcrossing species by small susceptible sources that are near the resistant treated population (Roush et al., 1990; Maxwell et al., 1990). Gene flow alone would probably not reduce effectively the proportion of resistance in predominately self-pollinating species, unless the susceptible sources were quite large and near the treated population so that seed dispersal into the resistant area would occur. Nevertheless, such simulations suggest that management strategies that increase flow of susceptible genes through pollen or seed dispersal would be a fruitful area of research in herbicide resistance management.

Maxwell et al. (1990) evaluated several management scenarios using their herbicide resistance model. Most of the scenarios examined the influence of the susceptible biotype on resistance evolution and the potential role of a susceptible source on the prevention and recovery from resistance. One scenario evaluated was changes in "herbicide effective kill" (Figure 3.6), which increase the

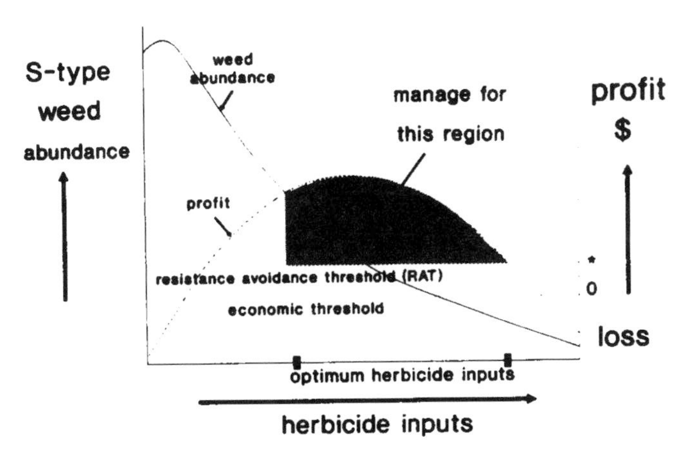

FIGURE 3.6 Hypothetical relationship between an economic threshold and a resistance avoidance threshold (RAT). The graph indicates the response of susceptible weed abundance to increasing herbicide inputs (solid line), the response of profit to herbicide input (dashed line), the economic threshold, and a threshold of S-biotype abundance for avoiding resistance (dotted lines). The shaded region corresponds to the levels of herbicide input that optimize both the profit and the avoidance of resistance. The RAT is illustrated at a level greater than the economic threshold; however, RAT conceivably could be equal to or less than the economic threshold (Roush et al., 1990).

probability that some susceptible individuals will remain in a treated field. In this simulation, the reduced herbicide efficacy (achieved by using lower herbicide rates or strip treatments over crop rows) resulted in sufficient gene flow of susceptible alleles to effectively slow evolution and reduce the proportion of resistance in the treated population. This recognition of the value of herbicide-susceptible plants also emphasizes the need for economic thresholds for weed competition in cropping systems (Figure 3.6 and Chapter 5).

POSSIBLE PREVENTION AND MANAGEMENT OF HERBICIDE RESISTANCE

Ghersa et al. (1994c), Maxwell et al. (1990), Putwain and Mortimer (1995), and Roush et al. (1990) present ways to prevent and manage herbicide-resistant genes. These authors consider selection by the herbicide, gene flow among sympatric populations, and seed flow as predominant factors for resistance management. The case of diclofop-methyl resistance in Italian ryegrass (*Lolium multiflorum*), an important annual weed in wheat-growing areas of western Washington and Oregon, USA, serves as an example of how herbicide resistance might be prevented or managed.

Diclofop-methyl resistance in Italian ryegrass: an example. Flower production and pollen dispersal patterns of diclofop-methyl-susceptible and resistant plants were examined to determine if evolution of resistance could be controlled by cross-fertilization between biotypes. The susceptible and resistant biotypes were found to differ in timing and abundance of both ovule production and pollen release such that pollen from the susceptible plants had a much greater chance of fertilizing the resistant plant population than vice-versa. Susceptible Italian ryegrass, growing with or without wheat competition, produced more than 60 percent of its seeds before any pollen from resistant plants was released. In contrast, throughout the course of resistant plant seed production, pollen from susceptible plants composed at least 30 percent of the total pollen load. These phenological differences suggest that gene flow can be used to reduce evolution of diclofop-methyl resistance in Italian ryegrass populations within wheat-ryegrass cropping systems. Since Italian ryegrass is also an important crop in the area where diclofop-methyl resistance occurs, cessation of herbicide application and sowing or reinvasion of a susceptible Italian ryegrass genotype into a wheat field infested with resistant Italian ryegrass can be expected to reduce the evolution of diclofop-methyl resistance by at least 6 percent per year (Ghersa et al., 1994c).

Intensive use of diclofop-methyl to eliminate Italian ryegrass seedlings from winter cereals has led not only to evolution of herbicide resistant ryegrass populations (Cussans, 1978; Gronwald et al., 1989) but also a more dormant seed population. Ghersa et al. (1994c) conducted field and greenhouse experiments to compare seedling emergence patterns of diclofop-methyl resistant and susceptible Italian ryegrass. Before current year seed shed, the seed bank in the experimental plots was found to contain three times as many viable resistant seeds as susceptible seeds. They also found that Italian ryegrass seeds from resistant plants had a greater degree of dormancy than seeds from susceptible plants. These results support the hypothesis that the timing of field tillages and herbicide use is a selective force that alters seed dormancy and thus changes seedling emergence patterns as herbicide-resistant Italian ryegrass weed populations evolve.

Linkages between seed dormancy and herbicide resistance have ecological and agronomical implications. Considering this selective link between herbicide resistance and seed bank persistence traits in Italian ryegrass, the use of crop rotations to reduce a monoculture's selective pressure for evolution of herbicide resistance in Italian ryegrass (as proposed by Gressel and Segel, 1978 and Gressel, 1991) may sometimes be counterproductive. During the growing cycle of the alternate crop, relatively little Italian ryegrass seed would be produced to replenish the soil seed bank, and therefore the growth of a new Italian ryegrass population after rotating back to the cereal crop would depend on the viability of seeds deposited in the soil during the previous cereal cropping cycle. Because of the lower persistence in the soil seed bank of the susceptible biotype relative to that of the resistant biotype, the new weed population would be characterized by greater numbers of resistant plants and fewer susceptible plants.

Ghersa et al. (1994c) also observed that germination of herbicide-resistant Italian ryegrass usually occurs during the winter wheat sowing season in the fall when maximum soil temperature falls below 35°C. This selected trait in the weed could be exploited to improve wheat yield in crops infested with herbicide-resistant Italian ryegrass. By sowing wheat earlier when soil temperature is warmer, emergence of the resistant ryegrass would be delayed compared to the crop and the competitive effect of the weed would be minimized. In the many fields infested with resistant Italian ryegrass, this strategy should reduce significantly the ryegrass population. As early sowing continues and herbicide use is halted (removing the selective pressure), the new cultural practices should select for an increasing proportion of early germinating non-resistant ryegrass in the population through seed and gene flow. A periodic return to later sowing and use of diclofop-methyl herbicide would allow selection for populations that germinate in a predictable narrow range of environmental conditions. This cycle should slow the evolution of highly resistant populations of Italian ryegrass at a regional scale, because it should improve seed and gene flow between herbicide resistant and susceptible biotypes (Roush et al., 1990).

Value of genetic and demographic studies. With the exception of diclofop-methyl resistance in Italian ryegrass, there have been very few comparative genetic and demographic studies conducted on herbicide resistant and susceptible populations. Such studies are critical for resistance management, since theoretical (Maxwell et al., 1990; Putwain and Mortimer, 1995) and experimental approaches suggest that manipulation of life history and genetics could reduce the intensity of selection and hence constrain the rate of evolution of herbicide-resistant weed populations. Such an approach seems feasible to prevent and manage herbicide resistance because selection by agricultural practices often influences other biological traits that can be manipulated agronomically to reduce or reverse the impacts of herbicide selection.

Recently, O'Donovan et al. (1995) reported that wild oat populations resistant to triallate and difenzoquat have the capacity for rapid germination over a wider range of temperatures than the susceptible populations. Because they could not detect major differences in growth, development, and competitiveness of the susceptible and resistant populations, it was concluded that little or no decline in the resistant populations would occur, even after use of triallate and difenzoquat ceased. However, in this case, it may be possible to exploit the low level of dormancy characteristic of the resistant populations by using cultural management approaches such as precisely scheduled tillage.

Herbicide Resistant Crops

In recent years, considerable research has been directed toward the introduction of herbicide tolerance or resistance into normally susceptible crop species (Schulz et. al., 1990). The procedure used to develop, analyze, and deploy such products is outlined in Figure 3.7 (Strauss et al., 1995). The procedure includes identification, insertion, and transfer of appropriate genetic material into plant cells, plant regeneration and propagation, testing, and containment. Interest in developing herbicide tolerant crop cultivars (HTC) has been spurred by several factors, such as, (1) a reduction in rate of discovery of new herbicides, (2) increased costs to develop new herbicides, and (3) new tools in biotechnology that greatly increase the feasibility of developing HTCs. Harrison (1992) and Dyer et al. (1993) indicate that several potential benefits could result from the development of crops tolerant to herbicides, including:

- an increased margin of safety with which herbicides can be used and subsequent reductions in crop loss due to herbicide injury,
- reduced risk of crop damage from residual herbicides used in previous rotational crops, and
- introduction of new herbicides for use on formerly susceptible crops.

The risks or concerns with development of herbicide tolerant and resistant crops were also identified. According to Harrison (1992), these include

- potential for increased use of herbicides,
- abandonment of alternative weed control practices other than herbicides,
- concern that HTCs may become weed problems or resistance may be transferred [through gene flow] to other species,
- accelerated selection for resistant weed populations from use of higher rates of herbicides,
- misuse of herbicides, leading to ground water contamination or other environmental problems, and

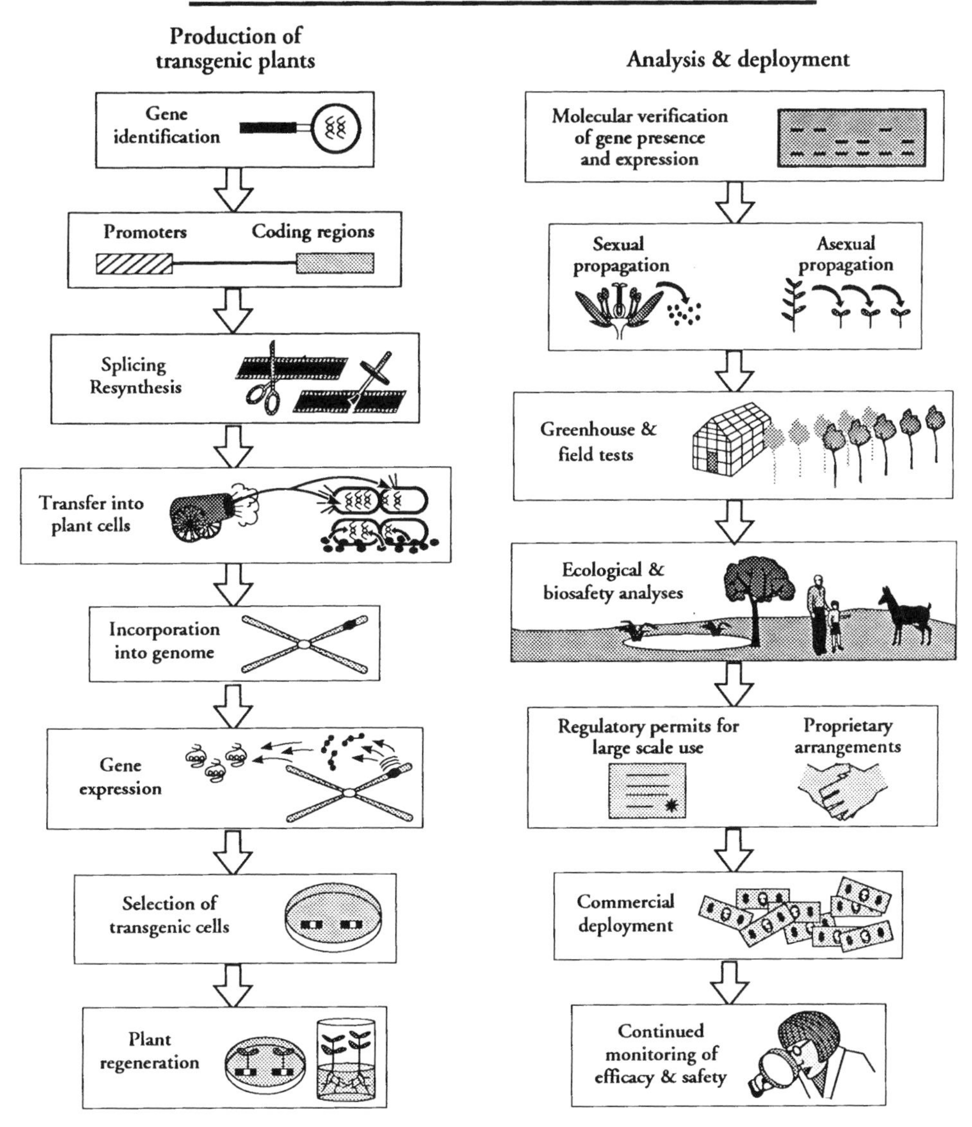

FIGURE 3.7 A summary of the steps required to produce transgenic plants and bring them to the marketplace (Strauss et al., 1995).

- general public concern about the release of genetically engineered organisms into the environment.

Burnside (1992) also notes that the significance of these benefits and concerns depends very much on the viewpoint of various researchers and citizens. This is a topic explored further in Chapter 11. The controversy underscores the necessity to understand the biological and ecological factors that influence herbicide resistance and the economic and social values that underpin modern weed control (Chapter 11).

SUMMARY

It is apparent that weeds have evolved and will continue to evolve in response to human activities. The response of organisms to selection depends on heritable variation and there appears to be sufficient genetic variation among weed species for this to occur. Both natural and artificial selection by humans are important processes through which weed evolution occurs. Weeds also have the ability to colonize readily a wide range of disturbed habitats, most notably through a breeding system of self-pollination and environmentally controlled outcrossing. The occurrence of polyploidy and vegetative reproduction in weed species also may increase the ability of weeds to occupy successfully a wide range of habitats. Weeds are thought to occur predominantly in two combined evolutionary strategies, competitive-ruderal and stress-tolerant competitors.

Plant species react in different ways when their habitats are disturbed. Three classes of vegetation based on its association with human-caused disturbances are wild plants, weeds, and domesticates. The selective forces created by agricultural practices often result in agricultural races of weeds, called crop mimics. Shifts in weed species composition also are likely to occur as a result of the selective forces of cultural practices used to grow crops. These shifts in species composition may be among species in a crop-weed community or within a species. Herbicide resistance is the most notable modern-day shift in weed species and population composition. Both genetic and demographic studies can be useful in developing strategies or tactics to prevent and manage herbicide resistance.

Chapter

4

Weed Demography and Population Dynamics

Population ecology is the branch of ecology that deals with the impact of the environment on a population, or group of individuals of a particular species occurring within a defined geographical area. In the case of weeds, populations are often selected by the tools or tactics designed to suppress them (Harper, 1956). Demography is the study of the numerical changes in a population through various stages of development. Analysis of these numbers can suggest reasons for changes in population size or species composition over time. The consequences of various management practices on a weed flora also can be determined through demographic study.

PRINCIPLES OF PLANT DEMOGRAPHICS

The life cycle is the fundamental descriptive unit of an organism. Specifically, an individual plant is the result of a series of processes that started at fertilization and continued through embryonic growth, seed germination, seedling establishment, development into adulthood, and finally senescence and death (Figures 1.3 and 2.8). Each of these processes occurs at a measurable, requisite rate. These *vital rates*, which form the basis of plant demography, describe a plant's development through its life cycle. The response of vital rates to the environment determines population dynamics in ecological time and evolution of life histories in evolutionary time.

By focusing on the vital rates of the life cycle, demography addresses both the dynamics and the structure of plant populations. The potential for exponential growth of groups of individuals is one fundamental fact of population dynamics (Figure 2.6). Another is that populations do not continue unrestricted growth for long (Figure 2.6), because the eventual resource limitation and biotic interactions curtail population growth until an equilibrium between birth rate and death rate is attained (Caswell, 1989).

Natality, Mortality, Immigration, and Emigration

Plants differ substantially among species in life form and timing of stages of development, but certain basic population processes are common to all (Chapter 2). For example, a large number of plants inhabiting a field would not remain static over time. If the number of plants increases, either there has been an influx of individuals from somewhere else, new plants have been created (born), or both events have occurred. These are two of the most basic processes that affect plant population size: *immigration* and *birth*. Alternatively, if the number of plants in the hypothetical field declines, some of them must have died since it is difficult to explain how plants could simply leave a field. The processes that reduce plant numbers are *death* and *emigration*. Dispersal is an example of how plants emigrate. All four of the basic processes can occur simultaneously in a population (see below). If the population declines, then death and emigration together outweighed birth and immigration, and vice versa, if the population increases.

In Chapter 2, birth, death, immigration, and emigration were combined in a simple algebraic equation to describe the change in numbers of a population between two points in time.

$$N_{t+1} = N_t + \mathrm{B} - \mathrm{D} + \mathrm{I} - \mathrm{E} \qquad \textit{(eq. 2.5)}$$

However, if the population is so large that absolute numbers cannot be used, then the equation is constructed in terms of density, so N_t becomes, for example, the number of plants per square meter at time t. Equation 2.5 demonstrates

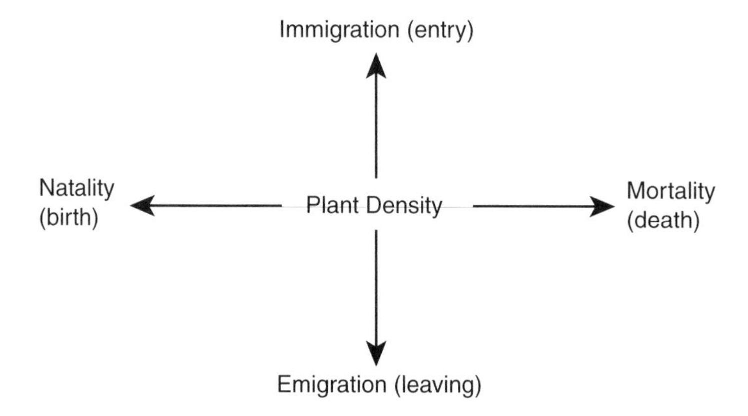

that understanding demography requires measurement of the four basic processes and accounting for their values. Unfortunately, this task is rarely simple or straightforward because every plant species passes through a series of stages in its life cycle (seed, seedling, adult plant, and so on) and each stage must be identified individually. Therefore, equation 2.5 represents a general, ideal model upon which more realistic descriptions are built.

The complexities of weed populations are best described using diagrammatic life table models (to be discussed later). This demographic approach to the study of weed population dynamics was first introduced by Sagar and Mortimer (1976) and has been used widely by weed ecologists for the past twenty years. Complete reviews are found in Mortimer (1983), Cussans (1987), and Cousens and Mortimer (1995).

Life Tables

In order to study the demography of any plant, its life cycle must be divided into fractions or components. For example, the plant's life cycle could be divided into an active fraction (growing plant) and a passive one (dormant seed and vegetative plant parts such as stolons, bulbs, and rhizomes). It might also be divided into the sporophyte (plants as we see them) and the gametophyte (a phase that is very reduced in higher plants and includes reproductive structures in the flower). The number of fractions included in a study and the level of detail necessary largely depend on the purpose of the project or study. The aims of some projects may only require simple and general descriptions in which a few components are considered in terms of a similar level of detail. In other cases, when deeper understanding is required, more fractions would be included and those determined to be critical would be studied in great detail. An approach followed by many plant demographers is to start with a simple general model to find critical components to be studied later in greater detail.

A simple diagrammatic life table of an idealized higher plant is shown in Figure 4.1. The number of individuals (N) at the start of each developmental stage (seed, seedlings, adults) is given inside the rectangular boxes. The N_{t+1} adults alive at time $t + 1$ (i.e., the next generation) come from two sources: (1) survivors of the N_t adults alive at time t and (2) those coming from birth, which in Figure 4.1 is a multi-stage process involving seed production, germination, and the survival and growth of seedlings. In the first source, the probability of survival (the proportion of them that survive) is placed inside the triangle and noted by p in Figure 4.1. For instance, if $N_t = 100$ plants and p, the survival rate, is 0.9, then there are 100×0.9 or 90 survivors contributing to N_{t+1} at time $t + 1$ (10 individuals have died; the morality rate, $1 - p$, between t and $t + 1$ is 0.1).

For the other source of plants in Figure 4.1, birth, the average number of seed produced per adult (the average fecundity of the plant population) is noted by F in Figure 4.1 and placed in a diamond. The total number of seed produced

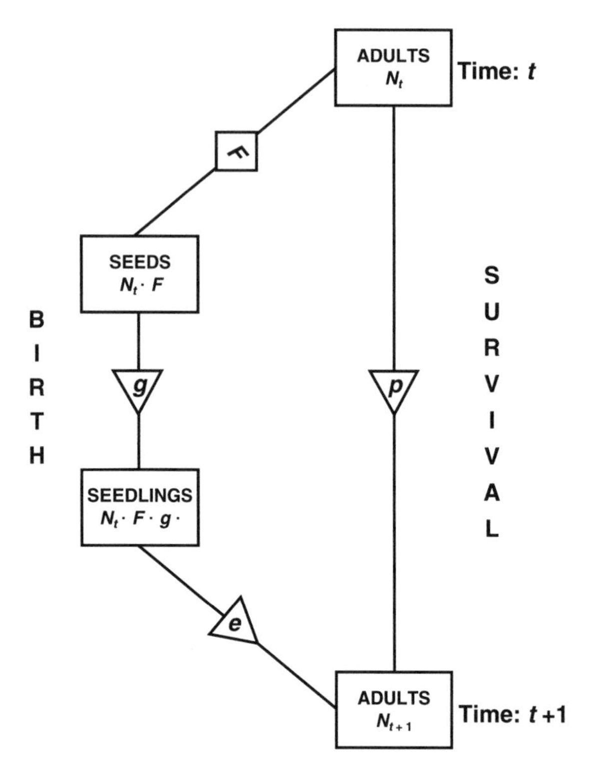

FIGURE 4.1 A diagrammatic life table for an idealized higher plant. *F*: number of seed per plant; *g*: chance of a seed germinating ($0 \leq g \leq 1$); *e*: chance of a seedling establishing itself as an adult ($0 \leq e \leq 1$); *p*: chance of an adult surviving ($0 \leq p \leq 1$). (From Begon and Mortimer, 1986.)

is, therefore, $N_t \times F$. The proportion of these seed that germinate is denoted by *g*. Multiplying $N_t \times F$ by *g* gives the number of seed that germinated successfully. The final step of the process is the establishment of seedlings, as independently photosynthesizing adults. The probability of surviving this stage of plant development is denoted by *e* in Figure 4.1 and the total number of births is, therefore, $N_t \times F \times g \times e$. The number of the population at time t + 1 is the sum of this calculation and $N_t \times p$.

It is now possible to substitute the terms of the life table (Figure 4.1) into equation 2.5, giving a basic equation for population growth of this hypothetical species:

$$N_{t+1} = \overbrace{N_t - N_t}^{\text{survival}} \underbrace{(1-p)}_{\text{death}} + \underbrace{N_t \times F \times g \times e}_{\text{birth}} \qquad \textit{(eq. 4.1)}$$

In this example, immigration and emigration were ignored, thus this description of how a plant population may change over time is incomplete. Moreover, death was calculated as the product of N_t and the mortality rate $1 - p$, because survival and mortality are opposite processes whose sum is one.

Modular Growth

A major distinction between species of the plant and animal kingdoms is how they grow and develop, which provides a pattern to the organization and differentiation of tissues. Most animals grow in a unitary and linear manner, while the growth of plants is modular (Harper and Bell, 1979; Harper, 1981). In animals, development from the zygote to the adult involves an irreversible process of tissue differentiation leading to organ development. In contrast, growth and differentiation in plants are usually initiated in *meristems* at the apices of shoots and roots (Esau, 1965). Cell division occurs in these meristems, which results in root and shoot elongation and the creation of more meristems. Thus, growth from meristems leads to a repetitive modular structure in the plant body, a *phytomer*. Botanically, a module is an axis with an apical meristem at its distal end. The axis is subdivided by nodes at which leaves, axillary meristems, and vegetative outgrowths may occur. A meristem may further differentiate into a terminal flower, at which time extension growth of the axis ceases.

Three demographic consequences arise from the modular construction of higher terrestrial plants. First, the addition of modules generates a colony of repeating units arranged in a branched structural form. The exact architecture of the plant depends on (1) whether modules vary in form, (2) their rate of production, and (3) their position relative to one another. The way the modules are structured influences the size and shape of the organ and, therefore, interactions among static individuals, which has demographic implications (Horn, 1971). Second, phytomers are relatively autonomous; therefore, herbivory or other physical damage may harm the plant but rarely kill it. The relatively autonomous meristem system also allows reiteration of many parts of the individual. In most plants, removal of vegetative branches will often lead to the replacement of the branch, but in unitary organisms, even though tissue regeneration does occur, removal of a whole organ can cause death. The third consequence of modularity is the opportunity for natural cloning of plants (Harper, 1984), which is only possible when the meristems at the nodes retain the ability to produce new shoots and roots. Fragmentation of an individual into independent clones may arise through physical agents, such as tillage, trampling, and grazing of herbivores, or it may be determined genetically. Cloning is an important characteristic for the persistence and dispersal of many perennial weeds (Holzner and Numata, 1982).

Harper (1977) indicates that the fundamental equation of population biology (eq. 2.5) applies not only to genets (the plant as a whole), but also at the lower level of ramets (i.e., modular units of the clone, Figure 1.4). Demographic approaches to modular dynamics can use the same techniques as for populations of unitary organisms (Harper and Bell, 1979).

Models of Plant Population Dynamics

Models of how a population behaves are needed to understand how the fundamental demographic processes of birth, death, immigration, and emigration influence the stability or change in population size. We have already seen the simplest form of a demographic model (eq. 2.5), which Silvertown and Lovett Doust (1993) suggest may approach being an algebraic truism. However, much more complex equations for population change arise from attempts to account for the way in which birth rates, death rates, and migration rates change along with population density and age structure and with the effects of competitors, predators, pathogens, and mutualists (Silvertown and Lovett Doust, 1993). Although detailed mathematical treatment is not possible in this text, some of these modeling approaches are described below.

MODELS BASED ON DIFFERENCE EQUATIONS

Two life cycle models are needed to examine the population dynamics of any higher plant species: (1) for species in which vegetative multiplication does not occur (Figure 4.2), such as most annual and biennial plants; and (2) for species in which vegetative multiplication does occur (Figure 4.3), as in perennial plant species.

When vegetative reproduction does not occur. Two routes are shown in Figure 4.2 by which a population of individuals of size A_1 in generation G_t may occur in the succeeding generation G_{t+1}. The population A_2 in generation G_{t+1} may come from genet reproduction (Kays and Harper, 1974), and sometimes from survivors of the previous generation. Route I (Figure 4.2) represents genet reproduction and plant establishment from seed which is divided into six intermediate phases. B is the total number of viable seed produced by the population A_1; C is the total number of viable seed falling onto the soil surface, to which are added any seed arriving by invasion (G); D_1 is the total number of viable seed that are present in the surface seed bank. This seed bank may lose individuals to the buried seed bank (D_2), which includes the "carry over" seed from previous generations. [There also may be inputs to the surface seed bank from invading seed or to the buried seed bank by sowing crop seed contaminated with weed propagules]. E is the number of seedlings germinated from either seed bank and F is the number of plants established.

Seven interphases are also recognized in Figure 4.2, including *a* through *g*, where, for example, *d* is the probability of a seed giving rise to a seedling within the time span of a generation. The invasion interphase *g* is further subdivided to distinguish contributions to the seed rain g_1, to the surface seed bank g_2, and to the buried seed bank g_3.

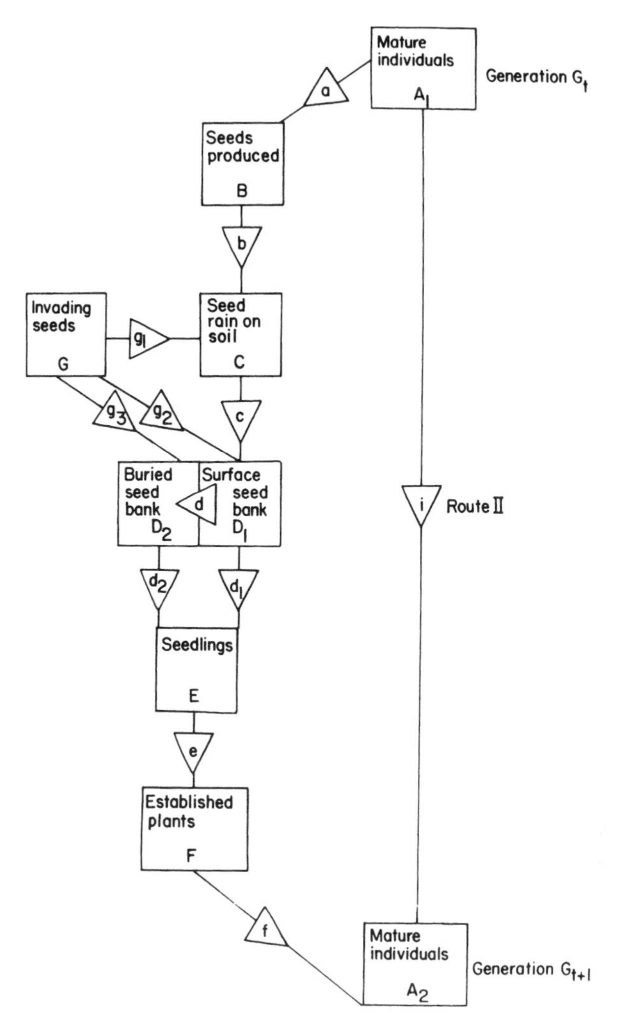

FIGURE 4.2 The generalized life table for a higher plant species that does not have ramet production. Symbols are described in the text. (From Sagar and Mortimer, 1976.)

Route II (Figure 4.2) is found in all species except ephemerals and annuals. This route indicates an interphase probability for the fraction of the population A_1 that survives to generation G_{t+1}. For a biennial species, the interphase (i) may theoretically carry a value of 0.5, for half the plants in the population A_1 would flower and die and half would remain vegetative and survive into population A_2. However, Figure 4.2 requires some slight modification for biennial species because of the overlap of generations and is inappropriate for species that have mixed populations of genets and ramets (Kays and Harper, 1974).

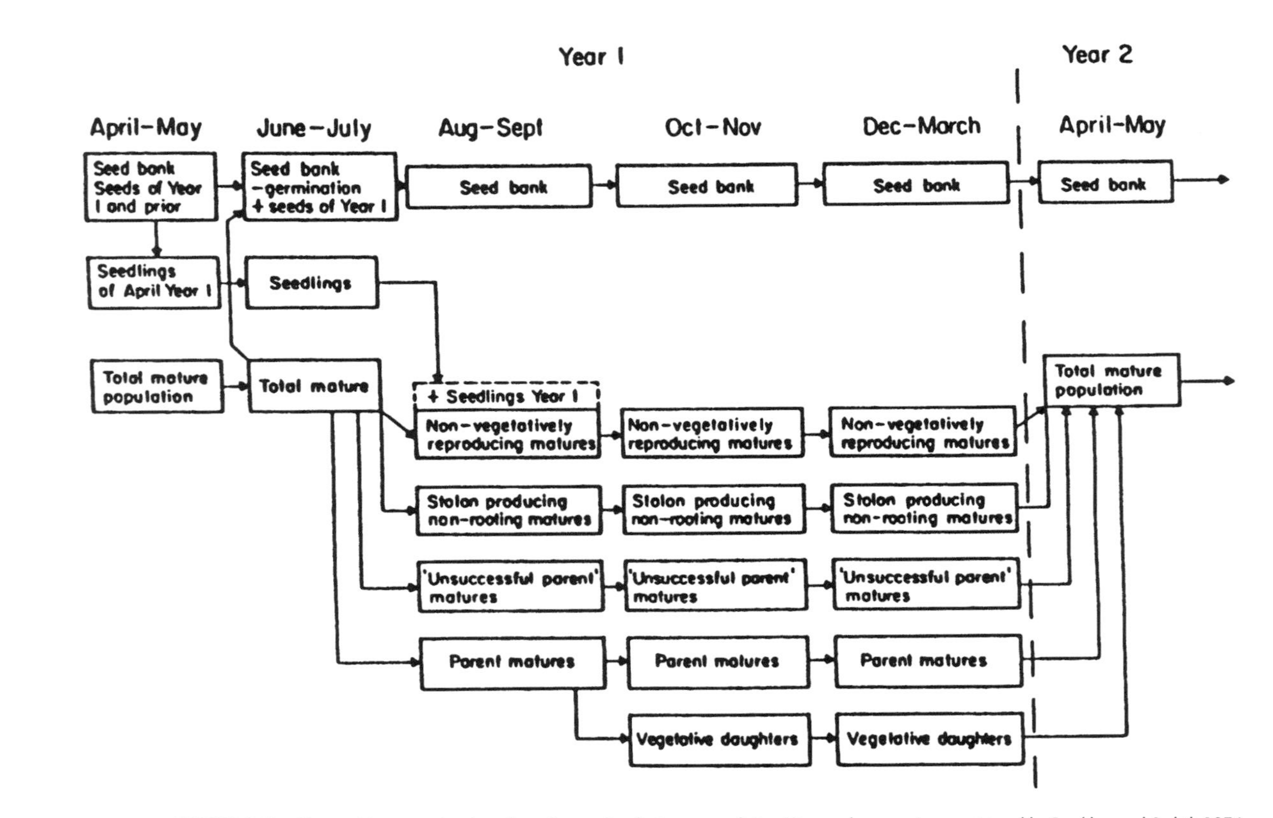

FIGURE 4.3 The transitions occurring throughout the year in a buttercup population (*Ranunculus repens*), as envisioned by Sarukhan and Gadgil, 1974.

When vegetative reproduction occurs. Sarukhan and Gadgil (1974) used transitions depicted in Figure 4.3 to describe the population dynamics *of Ranunculus repens* in Great Britain. This species reproduces sexually by seed and asexually by vegetative propagation, though recruitment by these means occurs at different times during the year. Seed germinate in late spring and early summer while new "daughter" plants become established in late summer as separate adult plants from shoots borne at nodes along creeping stolons. In essence, this complex flow diagram (Figure 4.3) is an age-state classification in which the fluxes from one stage to another are precisely defined chronologically. This approach makes an additional distinction in that asexually produced vegetative daughters are classified separately from sexually produced seedlings, at least during the first year of life. This demographic approach has also been used to describe the dynamics of many other perennial weed species (Sagar and Mortimer, 1976).

Age specific models. Mortality and fecundity are often age specific. In order to solve complex age and stage problems that may arise when describing plant populations, it is necessary to increase the complexity of the general model (e.g., Figure 4.1) into a diagrammatic life table like Figure 4.4. Here the population is divided into four age groups: a_0, a_1, a_2, and a_3; a_0 represents the youngest adults and a_3 the oldest. In a single time-step, t_1 to t_2, individuals from group a_0, a_1, and a_2 pass to the next respective age group; each age group contributes new individuals to a_0 (through birth); and the individuals in a_3 die. This model clearly rests on the assumption that the population consists of discrete age groupings and has discrete survivorship and birth statistics, in contrast to the reality of a continuously aging population.

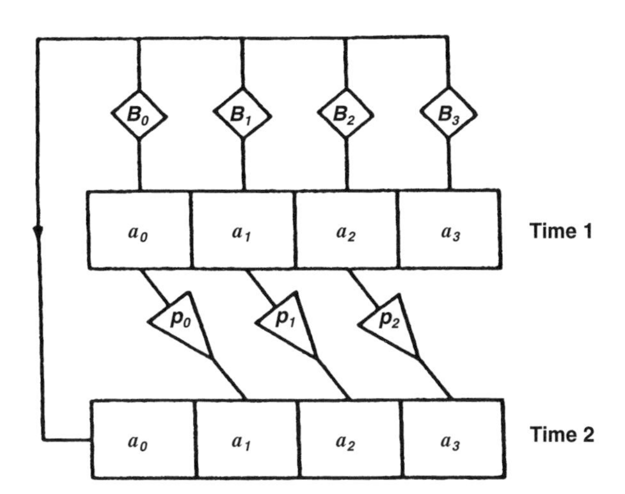

FIGURE 4.4 The diagrammatic life table for a population with overlapping generations: a = numbers in different age groups, B = age specific fecundities, and p = age specific survivorships. (From Begon and Mortimer, 1986.)

It is now possible to write a series of algebraic equations to express the changes that might occur in Figure 4.4:

$$t_2a_0 = (t_1a_0 \times B_0) + (t_1a_1 \times B_1) + (t_1a_2 \times B_2) + (t_1a_3 \times B_3) \qquad \text{(eq. 4.2)}$$
$$t_2a_1 = (t_1a_0 \times p_0) \qquad \text{(eq. 4.3)}$$
$$t_2a_2 = (t_1a_1 \times p_1) \qquad \text{(eq. 4.4)}$$
$$t_2a_3 = (t_1a_2 \times p_2) \qquad \text{(eq. 4.5)}$$

where the numbers in the age groups are subscripted t_1 or t_2 to identify the time period to which they refer. There are four equations in this model because there are four age groups, and they specifically state how the numbers in age groups are determined over the time step t_1 to t_2. An example of how to use this type of life table is given in Appendix 1.

TRANSITION MATRICES

Another way to describe the behavior of populations with overlapping generations that have individuals that fall into different age or size classes, that is, have different rates of reproduction and death depending upon age or size, is with *matrix models*. These models are generally simpler to use and more realistic than those using difference equations.

The matrix model was introduced to population biology by P.H. Leslie in 1943 and is often called the Leslie model. In general form, for n age groups, it is written as:

$$\begin{bmatrix} B_0 & B_1 & B_2 \ldots\ldots B_{n-1} & B_n \\ p_0 & 0 & 0 \ldots\ldots 0 & 0 \\ 0 & p_1 & 0 \ldots\ldots 0 & 0 \\ 0 & 0 & p_2 \ldots\ldots 0 & 0 \\ \cdot & \cdot & \cdot \quad\quad \cdot & \cdot \\ \cdot & \cdot & \cdot \quad\quad \cdot & \cdot \\ 0 & 0 & 0 \ldots\ldots p_{n-1} & 0 \end{bmatrix} \times \begin{bmatrix} t_1a_0 \\ t_1a_1 \\ t_1a_2 \\ t_1a_3 \\ \cdot \\ \cdot \\ t_1a_n \end{bmatrix} = \begin{bmatrix} t_2a_0 \\ t_2a_1 \\ t_2a_2 \\ t_2a_3 \\ \cdot \\ \cdot \\ t_2a_n \end{bmatrix}$$

which may alternatively be written as: $\mathbf{T} \times t_1\mathbf{A} = t_2\mathbf{A}$. **T** is called a *transition matrix*, which when multiplied by the vector of ages (**A**) at t_1 gives the age distribution at t_2. For further information, references include Searle (1966) and Caswell (1989). An example of this approach is given in Appendix 2

Simulation of leafy spurge population growth and effects of management. Maxwell et al. (1988) used a matrix modeling approach to simulate population changes of leafy spruge (*Euphorbia esula*) and determine the consequences of several management tactics used on that species. Leafy spurge is a herbaceous perennial weed of range and pasture lands in the northern Great Plains of the United States and southern Provinces of Canada. The species grows under a wide range of habitats, most commonly open grasslands. There are no herbicides, applied as a one-time treatment, or biological controls that provide effective, acceptable, long-term control of the species. Maxwell et al. (1988), following Watson (1985), divided the life history of leafy spurge into five stages: seeds, buds, seedlings, vegetative shoots, and flowering shoots. By identifying these stages (state variables), the processes of population development were determined (Figure 4.5).

It was found that three important transition parameters—basal buds to vegetative shoot (G2), the number of basal buds that flowering shoots produced (V5), and the number of basal buds that vegetative shoots produced (V4)—were sensitive to their own density. When the three density-dependent functions were included in the model simultaneously, initial exponential growth resulted, followed by growth decline and eventual stabilization of the simulated population (Figure 4.6). Maxwell et al. (1988) then subjected the simulated populations to several management tactics: a single application of picloram (Figure 4.6, top) and a foliage-feeding herbivore that removes 40, 50, and 60 percent of the stems (Figure 4.6, bottom). These simulations were compared to effects of actual picloram or sheep grazing treatments on leafy spurge. The accuracy of the simulation was striking in predicting the outcome of the management treatments. The model also indicated what stages in the life cycle of this weed species were most sensitive to manipulation.

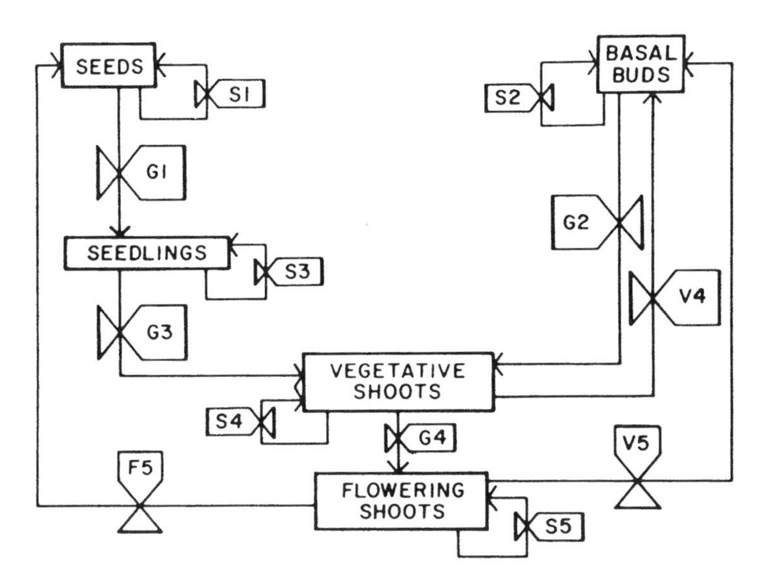

FIGURE 4.5 Diagrammatic model of a leafy spurge population: the boxes represent stages in the lifecycle, arrows indicate processes, valve symbols represent the rate at which a process occurs over a specified iteration time. (From Maxwell et. al, 1988.)

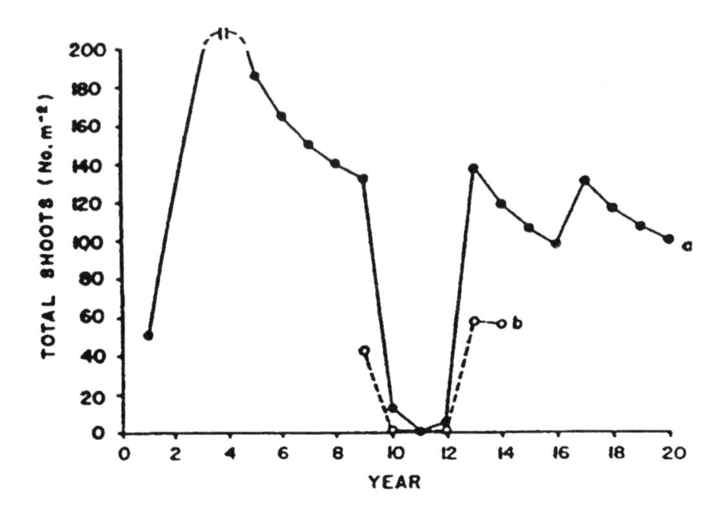

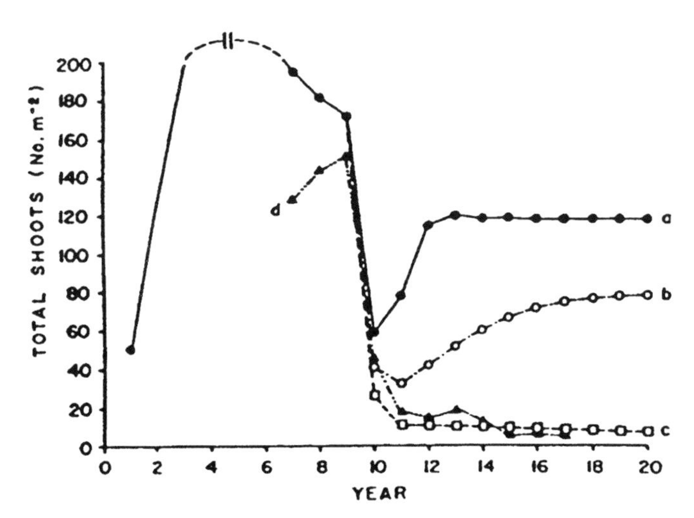

FIGURE 4.6 *Top:* Leafy spurge population simulation with density-dependent functions and a single application of picloram simulated at year 10 (a), and (b) observed effects of picloram application. *Bottom:* Leafy spurge population simulation with density-dependent functions simulating the introduction of a foliage-feeding herbivore at year 10 that removes (a) 40, (b) 50, and (c) 60 percent of the stems. Also shown are (d) observed effects from sheep feeding on leafy spurge. (From Maxwell et al., 1988.)

WEED SEED DYNAMICS

It is sometimes convenient to consider separately the two demographic processes related to plant movement, immigration and emigration, although both processes can be combined under the general term, *dispersal*. Most propagules of weeds are produced on site from a preceding generation and remain there to serve as the primary source of a new population, allowing for the entry of a few immigrants from elsewhere. Thus, immigration is a process of propagule input to an area already inhabited by a species. However, some propagules always leave the site where they were produced (emigration), starting new "colonies" of weeds perhaps in areas previously unoccupied by that species.

Dispersal means scattering or dissemination. Theoretically, if it is to be successful, dispersal should place a seed in a location that allows a greater likelihood of survival than its location with the parent plant. Weed seed disperse in *space* and through *time*. Dispersal in space involves the physical movement of seed from one place to another. Harper (1977) indicates that the amount of seed falling on a given unit of area is a function of several factors:

- height and distance of the seed source,
- concentration of seed at the source,
- dispersability of the seed (appendages, seed weight, etc.), and
- activity of dispersing agents.

Dispersal in time refers to the ability of seed of many species to remain in a dormant condition for some period of time. Thus, the success of a species is enhanced by dormancy if, at some point in the future, the seedling will be in a microenvironment more favorable for survival than if germination were to proceed immediately.

Seed Dispersal through Space

The opportunity for biological invasion begins with dispersal, and many weed species possess well adapted appendages to assist in long-distance movement of their seed (Figure 4.7). Because such appendages enhance the ability to move, they markedly increase the likelihood of seed and seedling survival by removing the individual from sources of parental-associated mortality (Figure 4.8). Most theories about colonization have been developed from studies of natural ecosystems (Crawley, 1987; diCastri, 1989; Mack, 1986). However, colonization of natural ecosystems often differs from that of agroecosystems. In natural ecosystems, for example, areas suitable for occupancy are few and often widely separated. Agricultural land, on the other hand, is exposed to diverse disturbances, periods of high resource availability, and times when plant cover is low (Auld and Coote, 1990; Ghersa and Roush, 1993). These times of disturbance are particularly well suited for establishment of plant colonizers.

On arable and other periodically disturbed sites, weed seed display both horizontal and vertical dispersion, reflecting initial dispersal onto the soil and subsequent movement, often with the assistance of implements used for soil tilling. Thus, most weed seed, even those with special adaptive features for long-distance dispersal, tend to migrate as an advancing front. The greatest concentration of seed generally falls below or only a short distance from the parent plant (on-site

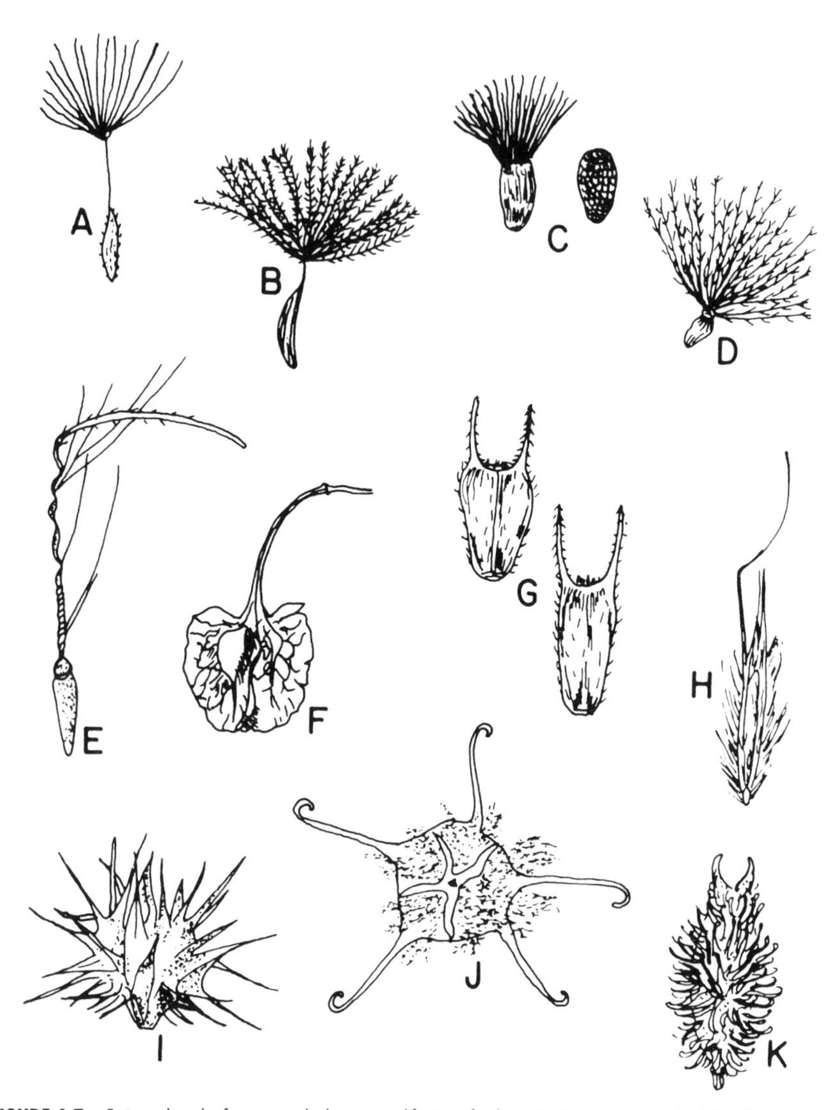

FIGURE 4.7 Fruits and seeds of some weeds showing modifications for dissemination. A, common dandelion (*Taraxacum officinale*); B, meadow salsify (*Tragopogon pratensis*); C, yellow starthistle (*Centaurea solstitialis*); D, Canada thistle (*Cirsium arvense*); E, red-stem filaree (*Erodium cicutarium*); F, curly dock (*Rumex crispus*); G, beggar-ticks (*Bidens frondosa*); H, wild oat (*Avena fatua*); I, sandbur (*Cenchrus pauciflorus*); J, five-hooked bassia (*Bassia hyssopifolia*); K, cocklebur (*Xanthium canadense*). (Compiled from Robbins et al., 1941, *Weeds of California*.)

production) and decreases with increasing distance away from it. This absence of seed and seedlings under parents is a function of initial seed dispersal and seedling mortality. As shown in Figure 4.8, the product of the two factors, dispersal and mortality, is a seedling recruitment surface that indicates the optimum distance between neighboring plants of the same species (Cook, 1980). The result of this

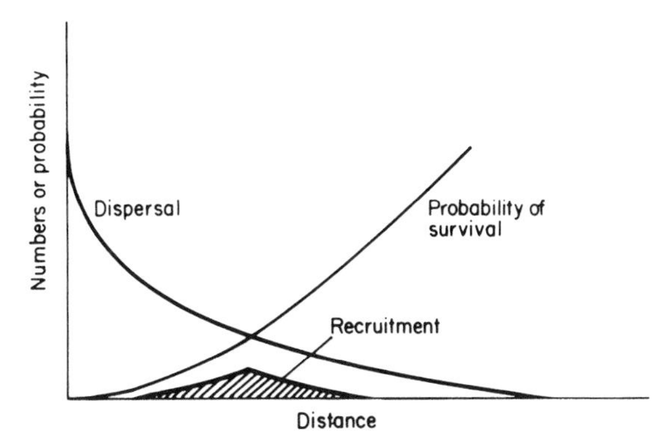

FIGURE 4.8 The recruitment of new genotypes as a function of the number of dispersed seeds and the probability of juvenile survival. (From Cook, 1980, in O.T. Solbrig, ed., *Demography and Evolution in Plant Populations*; reproduced by permission of *Botanical Monographs*, Blackwell Publications, Oxford, U.K.)

interaction is a creeping infestation across a given area. The recruitment effect is also noticeable with seed adapted for wind dispersal when it occurs against the prevailing wind.

Thus, the amount of widely dispersed seed per unit of area is low relative to the total amount of seed produced. Furthermore, plants from seed that have been widely dispersed tend to colonize as isolated individuals and, after high densities are reached, they begin to spread as fronts.

AGENTS OF SEED DISPERSAL

Wind, water, animals, and humans are the usual agents by which seed are dispersed spatially.

Wind. Seed dispersed by wind can have several distinct forms. They can be dusts (such as orchid seed or fungal spores), winged, or plumed. Seed may be adapted for gliding, such as most coniferous seed, or for rotating, such as seed of maple. Of particular interest are the plumed seed characteristic of many species in the family Asteraceae (Figure 4.7A–D), for example, dandelion (*Taraxacum officinale*), meadow salsify (*Tragopogon pratensis*), yellow starthistle (*Centaurea solstitialis*), and Canada thistle (*Cirsium arvense*). Seed of the plumed type are relatively heavy compared to propagules that occur as dusts or spores and ordinarily would not float in air. However, these species have a specialized feather-like structure, the *pappus*, attached to the seed coat, which allows dispersal by wind. Harper (1977) indicates that in Asteraceae the influence of a pappus on dispersal velocity of a seed is best correlated with the ratio of pappus diameter to achene diameter rather than with the ratio of achene weight to pappus weight (Figure 4.9a–b). Weed species in Figure 4.9(a) that produce achenes with a high ratio of pappus to achene diameter have slower terminal velocities, stay in the air longer, and therefore disperse farther than

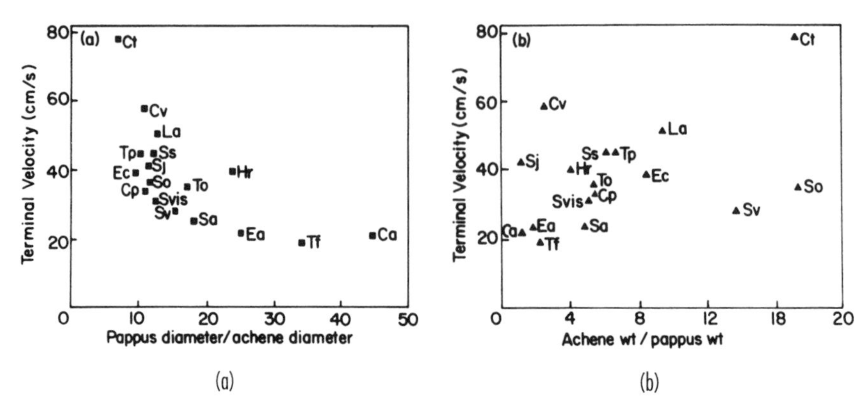

FIGURE 4.9 (a) The relationship between the ratio of pappus diameter to achene diameter and the terminal velocity of the achene-pappus units of selected Compositae. (b) The relationship between the ratio of achene weight to pappus weight and the terminal velocity of the achene-pappus units of selected Compositae. Ca, *Cirsium arvense* (Canada thistle); Cp, *C. palustre*; Ct, *Carduus tenuiflorus* (slenderflower thistle); Cv, *Carlina vulgaris*; Ea, *Erigeron acer*; Ec, *Eupatorium cannabinum*; Hr, *Hypochaeris radicata* (spotted catsear); La, *Leontodon autumnalis* (fall hawkbit); Sa, *Sonchus arvensis* (perennial sowthistle); Sj, *Senecio jacobaea* (tansy ragwort); So, *Sonchus oleraceus* (annual sowthistle); Ss, *Senecio squalidus*; Sv, *S. vulgaris* (common groundsel); Svis, *S. viscosus* (sticky groundsel); To, *Taraxacum officinale* (dandelion); Tf, *Tussilago farfara* (coltsfoot); Tp, *Tragopogon porrifolius* (common salsify). (From Sheldon and Burrows, 1973; reproduced by permission of the trustees of *New Phytologist*.)

species with low ratios. Species such as Canada thistle (Figure 4.9) that are well adapted for dispersal in wind may have small seed, a large pappus, or both.

For most systems of wind dispersal, increasing the height of release brings an immediate reward in enhanced dispersal (Harper, 1977). For example, the flower stalk of dandelion exhibits very plastic growth and elongates significantly, especially after flowering. This is interpreted as an effective way to augment the role of the pappus in achene dispersal. The interesting thing here is not only that greater height increases dispersal distance, but also that many weeds apparently have evolved mechanisms to place their seed structures higher.

Another effective type of wind dispersal is the rolling action of tumbleweeds, such as Russian thistle (*Salsola iberica*). Stallings et al. (1995) studied the seed-scattering ability and distribution patterns of that species in winter wheat stubble fields of eastern Washington. Some plants moved over 4000 meters during the six-week study period, always in the direction of the prevailing wind. When compared to stationary plants, wind-blown plants dispersed up to fifty percent more seed. The total amount of seed produced averaged 60,000 seed per plant, thus the tumbling action of species such as Russian thistle is an effective dispersal mechanism.

Water. Many kinds of weed seed, even those without special modifications, are readily dispersed by water. Irrigation is a very important factor in the spread of weeds throughout many areas of the western United States. Eddington and Robbins (1920) performed the earliest studies of seed dispersal in irrigation water. They found a total of 81 different species of weeds in 156 catches in irrigation ditches in Colorado. Some of the most common species and the amount of seed caught are listed in Table 4.1. It is apparent that large numbers of seed can be dispersed in this way.

Weed seed differ in their ability to float in water, although this depends somewhat on the water conditions and manner in which the seed alight upon it (Robbins et al., 1942). By dropping 100 seed of 57 different species into water, Eddington and Robbins (1920) found that most species floated very well. In their study, only two species—stinkgrass (*Eragrostis cilianensis*) and haresear mustard (*Conringia orientalis*)—floated poorly. There are also various adaptations of fruit and seed that aid water dissemination. For example, the fruit of curly dock (*Rumex crispus*) shown in Figure 4.7F, and arrowhead (*Sagittaria* spp.) have corky "wings" that make them buoyant. Usually weed seed screens are placed in irrigation ditches or water sources to reduce this form of weed seed dissemination.

Animals. Well known are the various forms of hooks and barbs that occur on the outer covering of many fruits and seed of weed species (Figure 4.7E and G–K). Such appendages are particularly well developed in families such as the Asteraceae, Boraginaceae, and in some Poaceae (King, 1966). Many of these "armed" seed and fruit attach to the fur of animals and are thus dispersed over a potentially large area. Some small seed—such as those of crabgrass (*Digitaria sanguinalis*), St. Johnswort (*Hypericum perforatum*), and bermudagrass (*Cynodon dactylon*)—simply lodge temporarily in hair of pasturing animals. Crafts (1975) notes that distribution of medusahead (*Taeniatherum asperum*) and St. Johnswort throughout much of the grazing areas of the Sacramento Valley and Sierra Nevada mountains of California followed cattle and sheep trails.

TABLE 4.1 Estimated Number of Seeds Passing a Point on a 12-ft Irrigation Ditch

Species	*Common Name*	*Seed Number*[a]
Amaranthus retroflexus	redroot pigweed	1,280,448
Chenopodium album	common lambsquarters	1,302,912
Helianthus petiolaris	prairie sunflower	44,928
Polygonum convolvulus	wild buckwheat	157,248
Rumex crispus	curly dock	89,856
Setaria viridis	green foxtail	44,928
Taraxacum officinale	dandelion	10,355,904

Source: Eddington and Robbins (1920).

[a]Based on actual seed catches on a 4.14 in^2 trap for 15 minutes.

A less obvious but equally well known method of seed dispersal by animals takes place in incompletely digested remains of fruit that has passed through the digestive tract. If the animal eats and digests the seed, a loss of dispersal results, but if it eats the fruit and passes the seed in feces, a possible gain to dispersal occurs. Harper (1977) and King (1966) present numerous examples of seed dispersal by birds, rodents, and large ruminants. In addition to ingesting seed, the animal simply may move the seed passively from one area to another, or collect and store the seed. In this case, dense seedling stands may emerge if a seed cache is buried (Harper, 1977).

Humans. The role of humans in dispersal of weed seed is especially well developed in agricultural situations. However, dispersal of seed is a quality selected against by humans in their crop breeding and harvesting operations, since only that portion of a seed crop that has not fallen to the ground can be harvested. For example, Figure 4.10 shows electron micrographs of two varieties of rice. The caryopses on the right dislodged and dispersed readily, whereas those on the left can be removed only by physically stripping the panicles. The rice depicted on the left in Figure 4.10 is a commercial rice variety in California, whereas the one shown on the right was never developed as a crop because of its seed shattering characteristics. Thus, morphological adaptations that allow seed shatter and dispersal may help ensure species success of weeds, but are undesirable traits in crops where seed characteristics that aid collection are more desirable.

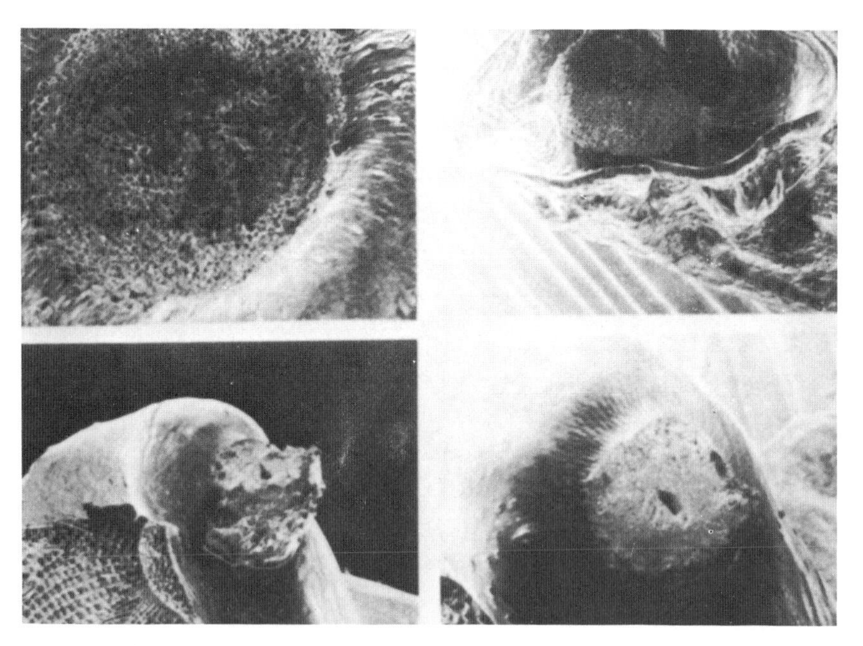

FIGURE 4.10 Electron micrographs of caryopses of two varieties of rice. Micrographs by F.D. Hess and D.E. Bayer, University of California, Davis. Figure described in the text.

Many weed species that grow in close association with certain crops, such as mimics like those discussed in Chapter 3, have some proportion of the seed that shatters and falls. In this way the site continues to be occupied by succeeding generations. However, some seed also remains with the parent plant and are harvested with the crop. This combination of dispersal mechanisms tends to assure and maintain the crop/weed association since the weed seed is usually replanted with the crop.

Barnyardgrass (*Echinochloa crus-galli*) is known to consist of many different populations based on panicle morphology and phenology (see pages 83–87 in Chapter 3). Three distinct varieties of this grass have been identified in the rice-growing region of California (Barrett and Seaman, 1980). These are *E. crus-galli* var. *crus-galli, E. crus-galli* var. *oryzicola*, and *E. crus-galli* var. *phyllopogon* (Figure 4.11a, b, and c, respectively). *E. crus-galli* var. *crus-galli* and var. *oryzicola* mature early and the caryopses usually have shattered from the panicle by

FIGURE 4.11 Three varieties of barnyardgrass (*Echinochloa crus-galli*) associated with rice in California: (a) *E. crus-galli* var. *crus-galli*, (b) *E. crus-galli* var. *oryzicola*, and (c) *E. crus-galli* var. *phyllopogon*. Photograph provided by E.J. Roncoroni and D.E. Bayer, University of California, Davis.

(*continued*)

FIGURE 4.11 Continued.

rice harvest. *E. crus-galli* var. *phyllopogon* apparently germinates later or has a longer life cycle than the other two varieties of barnyardgrass, so that it rarely has shattered completely by the time rice is harvested. Dispersal of *E. crus-galli* var. *phyllopogon*, therefore, occurs as the panicles are stripped by the harvester. Because the life cycles of the three varieties overlap to some degree, some seed of each variety occur with the harvested crop. However, most of the seed of *E. crus-galli* var. *crus-galli* and var. *oryzicola* remain in the field, whereas the largest proportion of *E. crus-galli* var. *phyllopogon* remains with the crop seed. It is also interesting that some cultural practices in rice production, such as high water levels and certain herbicides, are believed to favor the occurrence of *E. crus-galli* var. *phyllopogon* over other barnyardgrass varieties.

Often elaborate attempts are made to break the weed seed/crop association by various "seed cleaning" techniques. These methods often take advantage of differential seed coat morphologies between crop and weed. Examples are using weed seed screens to remove barnyardgrass from rice and using various winnowing procedures in cereals. Dodder (*Cuscuta* spp.) seed is removed from alfalfa by mixing iron filings with the seed and exposing them to electromagnets. The filings attach to the rough, reticulate seed coat of the dodder and are thus separated from the smooth alfalfa seed with the magnet. Harper (1977) indicates that corncockle (*Agrostemma githago*), which produces large tuberculate seed, can be removed readily from grain by screening and is no longer considered to be a serious problem for that reason.

Another way in which humans have attempted to control the unwanted dispersal of weed seed is through governmental regulation (see Chapter 1). Most states and countries maintain seed laws that specify maximum percentages of weed seed contamination allowed in an agricultural crop used for commerce. However, even very little seed dispersed and planted with a crop potentially can infest previously unoccupied fields. For example, as little as 0.25 percent dodder infestation in alfalfa seed could result in as many as 40,000 dodder seedlings per acre after sowing.

Weed Seed Banks

All viable seed and spores (extending the concept widely) present on and in the soil constitute the soil seed bank. Harper (1977) visualized the soil as a seed bank or reservoir in which both deposits and withdrawals are made (Figure 4.12). Deposits occur by seed rain from seed production and dispersal, whereas withdrawals occur by germination, senescence and death, and predation. Storage results from the vertical distribution of seed through the soil profile, with most weed seed occurring at shallow depths. Soil seed banks have become a recognized and indispensable part of plant population ecology to such a degree that substantial amounts of information are now available about seed bank processes (e.g., Leck et al., 1989). Moreover, weed scientists have recog-

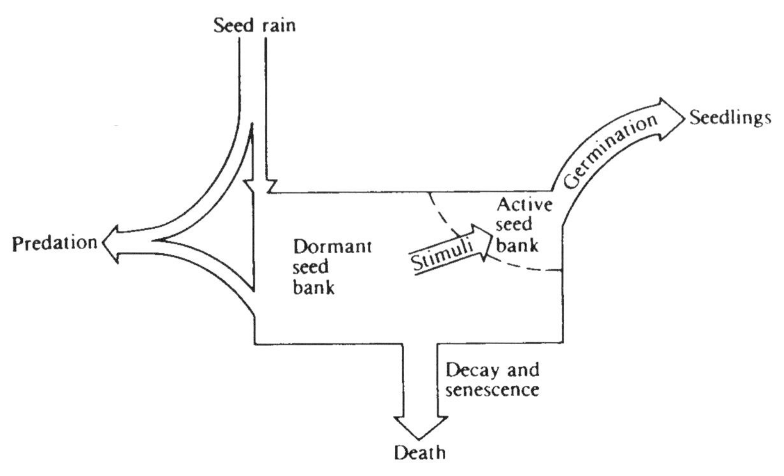

FIGURE 4.12 A model of the dynamics of the seed pool (Harper, 1977).

nized that more information about the dynamics of seed banks may allow improved weed management strategies (Aldrich, 1984; Altieri and Liebman, 1987; Ghersa et al., 1994a; Cavers et al., 1995).

ENTRY OF SEED INTO SOIL

Seed enter the soil from several sources, but most commonly from weeds that are allowed to mature on an already occupied site. Many assessments of weed seed production are available, especially in agriculture (Stevens, 1954, 1957; Holm et al., 1977; Cavers et al., 1995). In general, the amount of seed produced by agricultural weeds is astonishingly high (Table 4.2), but it can also vary markedly due to the high growth plasticity of most weed species. Thus, the actual amount of seed produced per individual plant can vary from nothing to millions, depending on its growing conditions. The importance of high seed production for annual pioneer species is obvious and was discussed in Chapters 1 and 3. Holzner et al. (1982) indicate that, generally, agricultural weeds are species that produce large amounts of small seed. However, their seed production is far exceeded by that of species that colonize habitats for only a short time (ephemerals), such as *Epilobium angustifolium, Verbascum* spp., or *Solidago* spp. in forest clearings, and especially parasitic plants. If weeds are allowed to mature, a substantial amount of seed will continue to enter the soil and become a source for future site occupation by the species. For instance, Wilson (1988) cites an example of weed seed densities in the soil increasing in a single year by 370 percent for grasses and 126 percent for broadleaved weeds when only cultivations were used to control weeds, and presumably the surviving plants matured and dispersed their propagules.

LONGEVITY OF SEED IN SOIL

Weeds vary considerably with respect to the longevity of their seed (Table 4.2), depending upon species, depth of seed burial, soil type, and level of disturbance. Many weed species are noted for the especially long-lived nature of their seed. Information in this area has been collected from two sources: (1) long-term burial studies and (2) seed collections from soils with a history of no disturbance. Burial studies, such as those initiated by Beal (1911) and Duvel (1903) generally agree with the later observations of Lewis (1973), who

TABLE 4.2 Seed Production and Length of Seed Survival in Soil

Weed Species	*No. of Seed Produced per Plant (Stevens, 1954, 1957)*	*Length of Seed Survival in Undisturbed Soil (years)*	
		(Toole and Brown, 1946)	*(Kivilaan and Bandurski, 1981)*
Ball mustard *Neslia paniculata* (L.) Desu.	490	10	
Canada thistle *Cirsium arvense* (L.) Scop.	680/stem	21	
Common lambsquarters *Chenopodium album* L.	72,450	39	
Common mullein *Verbascum thapsus* L.	223,200	39	100
Common purslane *Portulaca oleracea* L.	52,300	30	40
Common ragweed *Ambrosia artemisiifolia* L.	3,380	39	40
Common sunflower *Helianthus annuus* L.	7,200	1	
Curly dock *Rumex crispus* L.	29,500	39	80
Curlycup gumweed *Grindelia squarrosa* (Pursh) Dunal	29,700	10	
Dandelion *Taraxacum officinale* Weber in Wiggers	15,000	6	
Devils beggarticks *Bidens frondosa* L.	7,000	10	
Great burdock *Arctium lappa* L.	31,600	21	
Green foxtail *Setaria viridis* (L.) Beauv.	34,000	39	
Pennsylvania smartweed *Polygonum pensylvanicum* L.	19,300	30	
Prickly lettuce *Lactuca serriola* L.	27,900	9	
Redroot pigweed *Amaranthus retroflexus* L.	117,400	10	40
Shepherdspurse *Capsella bursa-pastoris* (L.) Medic	38,500	16	35
Wild oats *Avena fatua* L.	250	1	
Yellow foxtail *Setaria glauca* (L.) Beauv.	6,420	30	30

Source: Wilson, 1988.

showed that seed of grass and crop species succumb early, whereas seed of legumes and weeds remain viable for a long time (Table 4.3). In other studies summarized by Cook (1980), the species composition of seed has been determined in soils that have not been disturbed for a very long time. In all cases, viable seed of weed and pioneer species were found beneath vegetation of substantial age, indicating that they had not been deposited recently.

TABLE 4.3 Percent Viability of Grasses, Legumes, and Weeds after Storage at Three Depths in Mineral Soil[a]

	Viability (%)									
		Stored 1 Year			Stored 4 Years			Stored 20 Years		
Species	*Initial*	*13cm*	*26cm*	*39cm*	*13cm*	*26cm*	*39cm*	*13cm*	*26cm*	*39cm*
Gramineae										
Alopecurus myosuroides Huds. blackgrass	44	31	28	5	22	10	53	0	0	0
Holcus lanatus L. velvetgrass	88	68	65	46	17	19	5	0	0	0
Bromus mollis L. soft chess	88	3	11	2	0	0	0	0	0	0
Agrostis tenuis Sibth. common bent	72	92	45	36	28	21	3	0	0	0
Festuca rubra L. red fescue	87	17	14	23	0	0	0	0	0	0
Lolium perenne L. perennial ryegrass (early)	89	23	30	26	4	T	T	0	0	0
Lolium perenne L. perennial ryegrass (late)	95	86	53	34	22	10	T	0	0	0
Dactylis glomerata L. orchardgrass (early)	86	48	27	27	0	1	0	0	0	0
Dactylis glomerata L. orchardgrass (late)	94	22	50	49	1	2	11	0	0	0
Phleum pratense L. timothy (hexaploid)	97	85	96	89	81	32	27	0	0	0
Phleum pratense L. timothy (diploid)	89	86	87	61	72	51	39	T	0	0
Festuca pratensis Huds. meadow fescue (early)	89	9	T	1	0	0	0	0	0	0

TABLE 4.3 Continued

	Viability (%)									
		Stored 1 Year			Stored 4 Years			Stored 20 Years		
Species	*Initial*	*13cm*	*26cm*	*39cm*	*13cm*	*26cm*	*39cm*	*13cm*	*26cm*	*39cm*
Gramineae										
Festuca pratensis Huds. meadow fescue (late)	91	4	17	15	T	0	T	0	0	0
Festuca arundinacea Schreb. tall fescue	76	18	21	18	0	0	0	0	0	0
Lolium italicum A. Br. Italian ryegrass	93	29	70	42	3	1	T	0	0	0
Avena fatua L. wild oat	34	19	12	T	3	T	T	0	0	0
Legumes										
Trifolium repens L. white clover	98	36	11	15	10	6	11	1	6	1
Trifolium pratense L. red clover (late)	93	11	6	6	2	2	1	1	8	0
Trifolium pratense L. red clover (early)	55	30	26	27	19	7	5	T	2	T
Trifolum pratense L. red clover (early)	92	14	16	13	19	2	0	0	0	T
Trifolium dubium Sibth. small hop clover	90	13	20	19	9	7	8	1	1	T
Trifolium hybridum L. alsike clover	88	10	8	8	5	7	9	3	2	T
Medicago lupulina L. black medic	84	5	5	3	1	3	1	0	0	T
Medicago sativa L. alfalfa	95	2	3	3	0	0	0	T	1	T
Weeds										
Chenopodium album L. common lambsquarters	88	89	89	88	76	83	92	32	22	15
Geranium dissectum L. cutleaf geranium	99	2	15	T	T	0	0	0	0	0

TABLE 4.3 Continued

		Viability (%)								
		Stored 1 Year			Stored 4 Years			Stored 20 Years		
Species	*Initial*	*13cm*	*26cm*	*39cm*	*13cm*	*26cm*	*39cm*	*13cm*	*26cm*	*39cm*
Weeds										
Geranium molle L. clovefoot geranium	99	0	0	0	0	T	0	0	0	0
Lychnis alba Mill. white cockle	95	92	97	100	83	23	83	1	T	2
Matricaria inodora L. scentless chamomile	91	87	82	82	72	53	55	0	0	0
Plantago lanceolata L. buckhorn plantain	98	97	95	60	76	25	42	0	0	0
Polygonum persicaria L. ladysthumb	95	73	98	79	48	27	33	1	2	2
Ranunculus repens L. creeping buttercup	96	97	90	98	64	62	72	51	55	48
Rumex crispus L. curly dock	63	84	52	98	51	53	82	30	26	0

Source: Adapted from Lewis, 1973.

T = trace, less than 1%.

Perhaps the most interesting reports are those of the longevity of weed seed found in archeological sites. Odum (1965,1974) reports that annual agricultural weeds predominate in ruderal soils beneath ancient human dwellings. Viable common lambsquarters (*Chenopodium album*) and corn spurry (*Spergula arvensis*) were discovered in soil associated with habitations known to be 1700 years old! This information also suggests a very long association of these weed species with humans and their endeavors. Table 4.4 lists a number of weed species with seed that clearly have the capacity to remain viable for long periods of time when buried in soil. Information also has been gathered about the longevity of stored seed versus that of seed occurring in the soil. In most cases, storage life is considerably shorter than seed longevity in the soil.

Basically, seed longevity in the soil depends upon the interaction of many factors, such as the intrinsic dormancy characteristics of the seed populations, the environmental conditions present in the soil that influence dormancy breaking (e.g., light, temperature, water, and gas environment), and biological interactions (e.g., predation and allelopathy) (Fenner, 1994; Simpson, 1990; Fernandez Mendez et al., 1983). The intensity and manner in which these factors interact depend upon seed condition and the location of seed in the soil profile. Seed con-

TABLE 4.4 Seed Longevities in Soil and Decay Rates of Populations

Species	*Common Name*	*Longevity*[a] *(years)*	*Decay Rate*[b] *(g)*
Chenopodium album	Common lambsquarters	1700	0.105
Thlapsi arvense	Field pennycress	30	0.122
Polygonum aviculare	Prostrate knotweed	400	0.156
Viola arvense	Field violet	400	0.161
Fumaria officinalis	Fumitory	600	0.195
Euphorbia helioscopia	Sun spurge	68	0.206
Poa annua	Annual bluegrass	68	0.237
Capsella bursa-pastoris	Shepherdspurse	35	0.244
Stellaria media	Chickweed	600	0.252
Papaver rhoes	Corn poppy	26	0.260
Vicia hirsuta	Tinyvetch	25	0.305
Medicago lupulina	Black medic	26	0.340
Senecio vulgaris	Common groundsel	58	0.340
Spergula arvensis	Corn spurry	1700	0.340
Ranunculus bulbosus	Bulbous buttercup	51	—
R. repens	Creeping buttercup	600	—

Source: Cook, 1980, in O.T. Solbrig (Ed.), *Demography and Evolution in Plant Populations.* Reproduced by permission of Blackwell Scientific Publications Ltd., Oxford.

[a]Longevities from Harrington (1972).

[b]Decay rates from Roberts and Feast (1972).

dition is determined by its genotype, environmental factors during seed development, ripening and after-ripening requirements, seed morphology (size, shape, coat color, presence or absence of hairy coats, fruit covers, coat external rugosities or specific appendages, such as awns, etc.), and seed polymorphism.

DENSITY AND COMPOSITION OF SEED BANKS

The density and composition of seed in soil vary greatly but are closely linked to the history of the land. For example, grassland seed banks generally consist of seed associated with noncropped lands, while croplands contain seed of weeds from cultivated land (Wilson, 1988). When the pattern of seed production, distribution, and storage throughout a successional sequence is studied, the general tendency is for early species to contribute more seed to the seed bank than later ones (Table 4.5). This pattern occurs even though late succes-

TABLE 4.5 Numbers of Seed and the Predominant Species Present in the Seed Pools of Various Vegetation Types

Vegetation Type	*Location*	*Seed m^{-2}*	*Predominant Species in the Soil*	*Source*[a]
Tilled Agricultural Soils				
Arable fields	England	28,700–34,100	Weeds	Brenchley and Warington, 1933
Arable fields	Canada	5000–23,000	Weeds	Budd, Chepil and Doughty, 1954
Arable fields	Minnesota, USA	1000–40,000	Weeds	Robinson, 1949
Arable fields	Honduras	7620	Weeds	Kellman, 1974b
Grassland, Heath and Marsh				
Freshwater marsh	N. Jersey, USA	6405–32,000	Annuals and perennials representative of the surface vegetation	Leck and Graveline, 1979
Salt marsh	Wales	31–566	Sea rush where abundant in vegetation, grasses	Milton, 1939
Calluna heath	Wales	17,500	*Calluna vulgaris*	Chippendale and Milton, 1934
Perennial hay meadow	Wales	38,000	Dicotyledons	Chippendale and Milton, 1934
Meadow steppe (perennial)	USSR	18,875–19,625	Subsidiary species of the vegetation	Golubeva, 1962
Perennial pasture	England	2000–17,000	Annuals and species of the vegetation	Champness and Morris, 1948
Prairie grassland	Kansas, USA	300–800	Subsidiary species of the vegetation, many annuals	Lippert and Hopkins, 1950
Zoysia grassland	Japan	1980	*Zoysia japonica*	Hayashi and Numata, 1971

TABLE 4.5 Continued

Vegetation Type	*Location*	*Seed m^{-2}*	*Predominant Species in the Soil*	*Source*[a]
Grassland, Heath and Marsh				
Miscanthus grassland	Japan	18,780	*Miscanthus sinensis*	Hayashi and Numata, 1971
Annual grassland	California, USA	9000–54,000	Annual grasses	Major and Pyott, 1966
Pasture in cleared forest	Venezuela	1250	Grasses and dicot weeds	Uhl and Clark, 1983
Forests				
Picea abies (100 yr. old)	USSR	1200–5000	All earlier successional spp.	Karpov, 1960
Secondary forest	N. Carolina, USA	1200–13,200	Arable weeds and spp. of early succession	Oosting and Humphries, 1940
Primary subalpine conifer forest	Colorado, USA	3–53	Herbs	Whipple, 1978
Subarctic pine/ birch forest	Canada	0	No viable seeds present	Johnson, 1975
Coniferous forest	Canada	1000	Alder *Alnus rubra*	Kellman, 1970
Primary conifer forest	Canada	206	Shrubs and herbs	Kellman, 1974a
Primary tropical forest	Thailand	40–182	Pioneer trees and shrubs	Cheke et. al., 1979
Primary tropical forest	Venezuela	180–200	Pioneer trees and shrubs	Uhl and Clark, 1983
Primary tropical forest	Costa Rica	742	Pioneer trees and shrubs	Putz, 1983

Source: Silvertown and Lovett Doust, 1993.

[a]All references are cited in the original source, Silverton and Lovett Doust, 1993.

sional species usually are on the site for a much longer time than are pioneers. A significant characteristic of weeds and other pioneer species is the ability to produce a large number of propagules (Cavers and Benoit, 1989). This strategy of high reproductive potential combined with dormancy apparently allows the presence of a large and relatively constant soil seed reserve. In an environment where frequent disturbance is an evolutionary reality, the seed bank must act as a stabilizing factor that serves to assure species survival.

Seed banks of agricultural soils. The studies by Brenchley and Warington (1930,1933) represent early attempts to estimate weed seed abundance in agricultural soil (Table 4.5). The site examined was on the Rothamsted Experiment Station, England, in a field that had been planted continually to wheat for nearly 90 years. The amount of weed seed found was impressive (28,000–34,000 seed/m^2) and represented 47 species. Almost two-thirds of the seed were species of poppy (*Papaver* spp.). In a subsequent study, Brenchley and Warington (1945) found that different fertility levels (manure) often were associated with different species composition in the soil. However, the overall size of the soil seed reserve was found to remain relatively constant despite the use of various cropping systems, since most species could complete their life cycle between annual tillage operations.

Wilson (1988) indicates that the density of agricultural seed banks can range from zero in newly developed soils to between 4,000 and 140,000 seed per m^2 in cropped soil (Table 4.6). He points out that seed densities are influenced by

TABLE 4.6 Weed Seed in Agricultural Soils

Location	Crop History	Soil Depth (cm)	Average No. of Seeds Collected (seed/m^2)	Total No. of Species	Most Frequently Encountered Species		Ref
					Number	% of Total	
England	Vegetables	0–15	4,100	76	9	89	Roberts and Nielsons (1982)
Scotland	Potatoes	0–20	16,000	80	6	78	Warwick (1984)
Colorado	Barley-corn-sugarbeets	0–25	137,700	8	2	86	Schweizer and Zimdahl (1984)
Illinois	Corn-soy-beans-corn	0–18	10,200	25	4	85	Houghton (1973)
Nebraska	Corn-field-beans-sugarbeets	0–15	20,400	19	3	85	Wilson, et al. (1985)
Washington	Potatoes-wheat	0–30	51,000	23	3	90	Kelley and Bruns (1975)

Source: Wilson, 1988.

[a]All references are cited in the original source, Wilson, 1988.

past cropping practices and often vary from field to field. Yet he also notes that seed banks in agricultural fields at different locations often contain the same weed species and share other similarities. Generally, agricultural seed banks are made up of many species, but often only a few of these comprise 70 to 90 percent of the total seed bank (Table 4.6). This set of species may be followed by a smaller subset that comprises 10 to 20 percent of the seed reserve. Wilson indicates that a final set, accounting for only a small proportion of the total seed reserve, consists of species that are remnants of past crops (Wilson, 1988).

Thompson and Grime (1979) measured seasonal variation in the densities of germinable seed in ten contrasting habitats and divided the seed in the various communities into four categories (Figure 4.13). Groups I and II consist of *transient* seeds, which are usually from grasses and forbs and generally do not persist for longer than one year. Groups III and IV identified by Thompson and Grime consist of the *persistent* seed bank, which is represented by species from a wide range of habitats but which persist long enough to become buried in the soil. As long as the cropping pattern in a field remains unchanged, the seed bank changes little from year to year (Figure 4.13). Additionally, if seed of many weed species—for example, *Amaranthus retroflexus, Sorghum halepense*, or *Datura ferox*—are buried in the surface layer of the soil (0 to 1 cm), they may appear as transient seed banks. However, if seed of these same species are buried deeper (e.g., 10 to 15 cm) in the soil, their seed banks will appear to be persistent (Chepil, 1946; Ballaré et al., 1988; Van Esso and Ghersa, 1989). Thus, a seed bank should be described for a particular ecological scenario, using seed demography and considering changes in both seed number and seed condition.

It also should be pointed out that seed banks have spatial and temporal dimensions that are not independent of each other. The spatial dimension is given by the vertical and horizontal distribution patterns of the seed in the soil, while the temporal dimension refers to the life span of the seed in the soil. In agricultural soils, tillage accounts for most of the changes in the vertical distribution of seed, but many factors are involved in the horizontal distribution, as discussed earlier. Soil tillage also influences the rate at which seed viability is lost (Roberts and Feast, 1973; Lueschen and Anderson, 1980; Froud-Williams et al., 1983).

FATE OF SEED IN SOIL

Seed is important both for growth and maintenance of existing plant populations and for the initiation of new populations. However, the relative importance of seed recruitment, especially from seed banks, varies among plant species and among communities (Harper, 1977; Louda, 1989). The dynamics of seed banks are influenced by losses, as well as inputs to the soil. Seed that are lost from the soil seed bank are (1) eaten by rodents, insects, or microbes (predation), (2) killed by senescence and decay, or (3) removed by germination and emergence. Seed predation and senescence/decay are considered here; the processes of dormancy and germination are discussed later in this chapter.

A

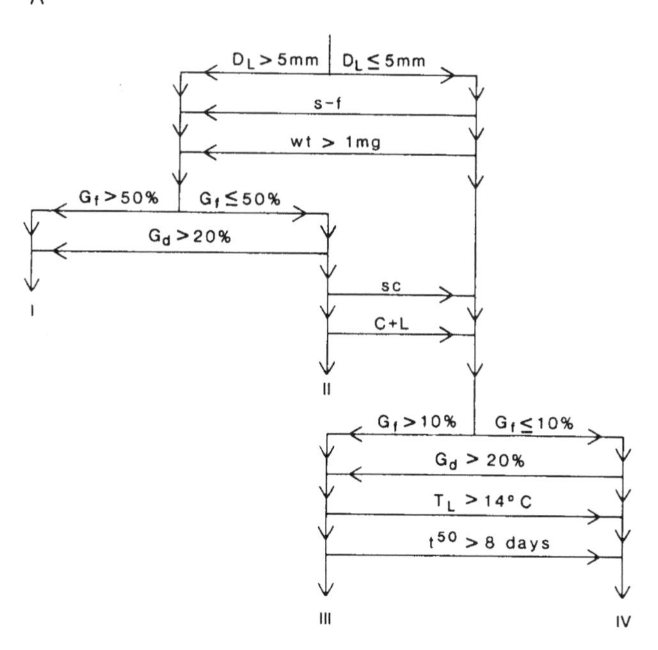

B

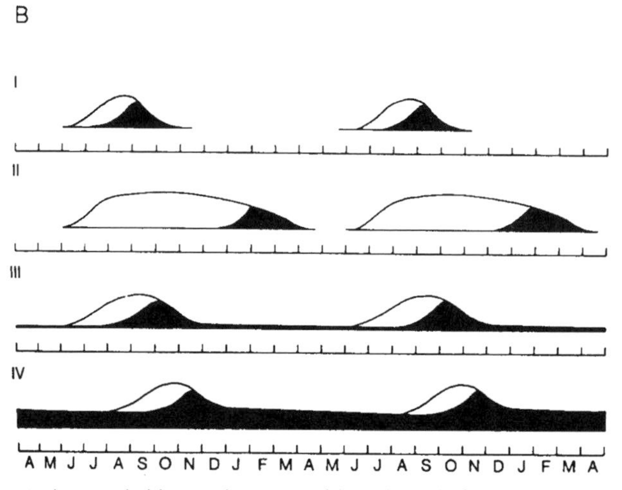

FIGURE 4.13 (A) A key using the laboratory characteristics of the seed to predict four seed bank types developed in the field (after Grime and Hillier, 1981). Guide to symbols: D_L, length of dispersule; s-f, seed not readily detached from fruit; wt, weight of seed; G_f, maximum percentage germination achieved by fresh seed; G_d, maximum percentage germination achieved by seed stored dry at 20°C for one month; SC, seed requires scarification; C + L, seed requires chilling and light to break dormancy ; T_L, lowest temperature at which 50% germination is achieved; t^{50}, time taken for seed stored dry at 20°C for 1 month to reach 50% germination. (B) Scheme describing four types of seed banks of common occurrence in temperate regions (after Thompson and Grime, 1979). Shaded area: seeds capable of germinating immediately after removal to suitable laboratory conditions. Unshaded area: seeds viable but not capable of immediate germination. (I) Annual and perennial grasses of dry or disturbed habitats (e.g., *Hordeum murinum, Lolium perenne,* and *Catapodium rigidum*) capable of immediate germination. (II) Annual and perennial herbs, colonizing vegetation gaps in early spring (e.g., *Impatiens glandulifera, Anthriscus sylvestris,* and *Heracleum sphondylium*). (III) Annual and perennial herbs, mainly germinating in autumn but maintaining a small seed bank (e.g., *Arenaria serpyllifolia, Holcus lanatus,* and *Agrostis tenuis*). (IV) Annual and perennial herbs and shrubs with large, persistent seed banks (e.g., *Stellaria media, Chenopodium rubrum,* and *Calluna vulgaris*). (From J.P. Grime in Allessio Leck et al., 1989.)

Predation. Predators can influence the input and output of seed to the seed bank at virtually every stage of a plant's life cycle. Louda (1989) indicates that seed predators select seed differentially and thereby determine the average value of key characteristics of seed that remain in the soil. For example, by finding and using clumped and large seed, predators reinforce other pressures that select for seed traits characteristic of persistent seed banks, including small seed size and hard seed coats. Additionally, fruit and seed consumers change seed distribution by eating, moving, or caching propagules and sometimes increase germinability and recruitment of seed that escape destruction (Louda, 1989).

Seed predation is believed to influence primarily the dynamics of plant populations that are expanding (Harper, 1977). Louda (1989) suggests that there are two groups of species defined in the literature with predictable periods of expansion and significant predator impacts (Figure 4.14). First, seed predation appears to change density and relative abundance of dominant species that have annual life histories (e.g., grasses of annual grassland, some agricultural crops), or that have high dependencies on seed recruitment for population maintenance and recovery after disturbance (e.g., mangroves, some trees in temperate forests). Second, seed predation influences recruitment, occurrence and distribution of moderately large-seeded plants with fugitive [plants that can "escape" through dispersal or other means and establish elsewhere] life histories (Louda, 1989). Generally, the risk of predator impact increases as the canopy matures, because a larger canopy provides greater cover.

There is little doubt that seed predation, especially of seed on the soil surface, has a marked impact on seed bank density and composition. For example,

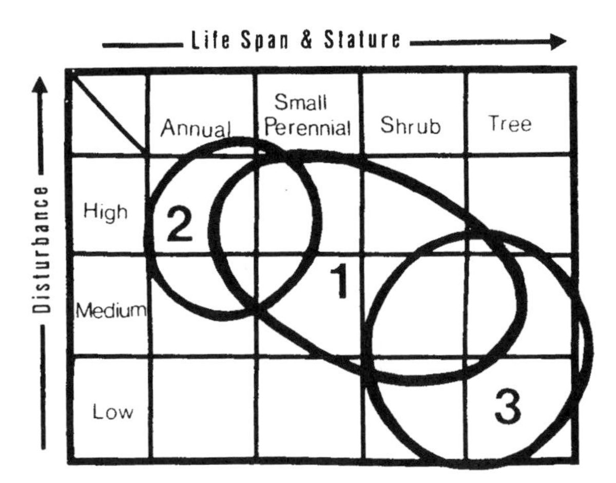

FIGURE 4.14 Predicted relative vulnerabilities of plants with different life histories to seed predators. Group 1 is composed of fugitive and other species whose life history traits can be considered intermediate between group 2, the ruderals and ephemerals specialized for frequent disturbances or harsh environments, and group 3, long-lived perennials adapted for competitive, stable, low-nutrient environments. Increasing populations of group 1 appear most vulnerable to contemporaneous demographic and distributional effects of seed predation. (From Louda, 1989.)

Gashwiler (1967) observed that nearly 70 percent of Douglas fir (*Pseudotsuga douglasii*) seed dispersed into recent clearcuts was eaten, presumably by rodents and birds. Similarly, Ghersa et al. (unpublished) observed that nearly all (99.8 percent) of the current year's seed rain of barnyardgrass (*Echinochloa crus-galli*), redroot pigweed (*Amaranthus retroflexus*), and common lambsquarters (*Chenopodium album*) was eliminated under a canopy of alfalfa. The seed predator in this case was a small field mouse (*Peromyscus* sp.). A several-year rotation of alfalfa is used by farmers in the Pacific Northwestern part of the United States as a means to "clean up" extremely weedy fields. The success of this method is likely due to both the competitiveness of alfalfa and significant seed predation by rodents under the alfalfa canopy.

Senescence and decay. Cook (1980) states that survival of seed stored in soil can be expressed as a negative exponential distribution, whereas shelf storage is best represented by a negative cumulative normal distribution (Figure 4.15). In other words, seed in the soil initially decay faster, then decay of shelf-stored seed increases so that the half-life of both types is about the same. Further decay is lower in the soil but quite rapid on the shelf (Cook, 1980; Roberts, 1972b; Froud-Williams et al., 1983; Ballaré et al., 1988; Van Esso and Ghersa, 1989).

Unquestionably, the environmental conditions surrounding the seed during storage, either in soil or on the shelf, affect seed longevity. Perhaps one reason for these observed differences in longevity is that often seed in soil are maintained in a dormant, yet imbibed condition. Most seed in storage, while also dormant, are air dry. Villiers (1972,1974) points out that dormant but imbibed

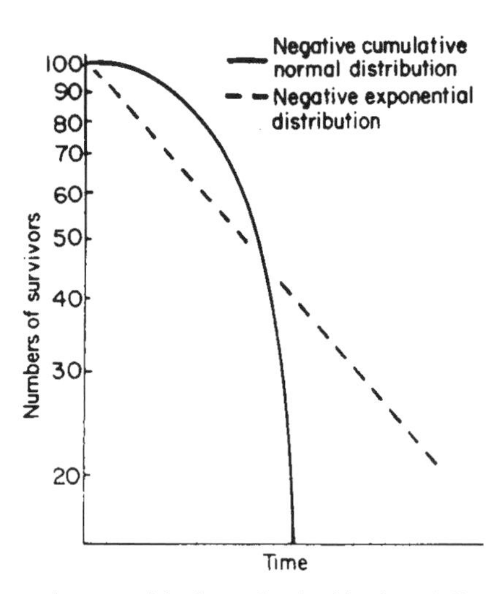

FIGURE 4.15 The negative cumulative normal distribution of seed viability during shelf storage and the negative exponential distribution of survival of seed in soil. (From Cook, 1980, in O.T. Solbrig, ed., *Demography and Evolution in Plant Populations.* Reproduced by permission of Blackwell Scientific Publications Ltd., Oxford, U.K.)

seed should be capable of many metabolic processes and are therefore able to repair damage to membranes and nuclear DNA as it occurs. It appears that the dormancy mechanisms that prevent imbibed seed from germinating play an important role in the longevity of seed of many weed species in soil. Furthermore, the mortality of weed seed in the soil probably is caused by the breakdown of these dormancy mechanisms, resulting in unsuccessful germination and death by predation or by eventual senescence, rather than by viability loss associated with length of storage time as in dry seed.

Wilson (1988) notes that longevity of agricultural weed seed tends to increase with depth of burial and to decline as soil disturbance increases. For example, in Duvel's (1903) experiment, common lambsquarters seed germinated at a rate of 9 percent after burial for 39 years in undisturbed soil. In a later study, common lambsquarters seedling emergence was 9 percent after six years of burial in cultivated soil and 53 percent after six years of burial in undisturbed soil. Similar results have been obtained in numerous other studies using other weed species (Roberts and Feast, 1973; Lueschen and Anderson, 1980; Froud-Williams et al., 1983).

WEED OCCURRENCE IN RELATION TO SEED BANKS

For weeds, the importance of seed banks is primarily in seedling recruitment and subsequent maintenance of high plant densities in crop fields. However, there is no obvious correlation between the half life of a species' seed in the soil and the relative abundance of that species in the field (Roberts and Ricketts, 1979). Effective establishment of weed stands depends less on the presence of propagules than on specific environmental characteristics that are suitable for seed germination and plant growth, because only one successful seedling is necessary to establish a new colony from a whole lifetime of seed production by the parent (Harper, 1977; Cavers et al., 1995).

As noted previously, a significant characteristic of weeds and other pioneer species is production of large numbers of propagules. This combined strategy of high reproductive potential and seed dormancy allows the presence of a large and relatively constant weed seed bank in cultivated land. However, when seed inputs are stopped, the number of viable seed stored in the soil of both cultivated and abandoned fields declines, usually following the negative exponential model depicted in Figure 4.15.

Seed predation, especially at the soil surface, may account for substantial amounts of initial seed loss. For example, when Johnsongrass (*Sorghum halepense*) seeds were buried in the field using plastic mesh fabric bags (5–20 cm), seed viability declined following the curve depicted in Figure 4.15 and about twenty percent were viable after 10 years in the soil. However, 80 to 99 percent of the seed buried in the same soil without plastic mesh bags for protection were lost during the first year (Ghersa, unpublished data). Rapid seed bank decline also has been observed in many other agricultural and rangeland sites (Chepil, 1946; Sarukhan and Gadgil, 1974; Scopel et al., 1988; Ballaré et al., 1988). Nevertheless, many references indicate that seed of various weed species can remain viable for many years in soil. From these data, predictions have been made of long time periods to completely deplete a soil seed bank of weeds (Holzner and Numata, 1982; Aldrich, 1984; Leck et al., 1989).

In order to understand the dynamics of the weed seed bank of a specific field, it is necessary to know the original location of the weed seed and the dispersal mechanisms of those dispersing plant species. Sources of seed loss and locations of their accumulation during the current and subsequent production seasons also are necessary to know. It may be that weed seed banks serve best for short-term dispersal, providing their greatest adaptive importance during their first one or two growing seasons. For example, if a seed bank provides enough seed to allow several cohorts of weed seedlings in a cropping season, some individuals should escape the hazards of tillage or other weed control operations. The large number of seed produced by such survivors would be sufficient, due to the reproductive plasticity of most weeds (Chapter 5), to maintain the soil seed reservoir (Ghersa and Roush, 1993). It has been suggested that few crop species are weedy because they generally are unimportant long-term components of the soil seed bank (Wilson, 1988; Cavers and Benoit 1989). This may be the case, but it is also likely that crop seed are incapable of producing multiple seedling cohorts and thus continually replenishing the seed bank, regardless of their persistence in soil.

Long-term seed survival is most helpful in explaining how specific weed populations can remain throughout an entire successional process and how genetic diversity of weeds is maintained in such a highly selective environment as an agricultural field (Putwain and Mortimer, 1995). To explain the high densities of weeds that are consistently found in many agricultural crops, the adaptive strategies that allow some weed individuals to escape control tactics are very important, especially those factors that influence the occurrence of waves of seedling germination.

Dormancy: Dispersal Through Time

Dormancy is the temporary failure of viable seed to germinate under external environmental conditions that later evoke germination when the restrictive state has been terminated or released. Dormancy is effectively dispersal through time and is especially critical for annual plants, in contrast to perennials, because the seed of annuals represent the only link between generations of those species.

DESCRIPTIONS OF SEED DORMANCY

The extent of both ecological and physiological information on seed dormancy is vast (Roberts, 1972a, b; Taylorson and Hendricks, 1977; Baskin and Baskin, 1989), and it is sometimes confusing, largely because of discrepancies in terminology. Baskin and Baskin (1989) suggest that there are five general types of dormancy exhibited by seed at maturity (Table 4.7). These are distinguished on the basis of

- permeability or impermeability of the seed coat to water (physical dormancy),
- whether the embryo is fully developed or underdeveloped, that is, there is incomplete development of the embryo at seed maturity, and
- whether the seed is physiologically dormant or nondormant.

Potentially, seed of all three types enter seed banks, but most seed found in seed banks in temperate regions have physiological dormancy, with physical dormancy being second in importance.

Physiological dormancy. According to Baskin and Baskin (1989), as seed with physiological dormancy afterripen (time from seed maturation to germination), they pass through a series of states known as conditional dormancy before finally becoming nondormant. In the transition from dormancy to nondormancy, seed first gain the ability to germinate over a narrow range of environmental conditions. As afterripening continues, seed become nondormant and can germinate over the widest range of environmental conditions possible for the species (Figure 4.16). However, if environmental conditions (e.g., darkness) prevent germination of nondormant seed, subsequent changes in environmental conditions (e.g., low or high temperatures) cause them to enter secondary dormancy. As seed enter secondary dormancy, the range of conditions over which they can germinate decreases until finally they cannot germinate under

TABLE 4.7 Types, Causes, and Characteristics of Seed Dormancy

Type	*Cause(s) of Dormancy*	*Characteristics of Embryo*
Physiological	Physiological inhibiting mechanism of germination in the embryo	Fully developed; dormant
Physical	Seed coat impermeable to water	Fully developed; nondormant
Combinational	Impermeable seed coat; physiological inhibiting mechanism of germination in the embryo	Fully developed; dormant
Morphological	Underdeveloped embryo	Underdeveloped; nondormant
Morphophysiological	Underdeveloped embryo; physiological inhibiting mechanism of germination in the embryo	Underdeveloped; dormant

Source: Baskin and Baskin, 1989.

any set of environmental conditions (Figure 4.16). Thus, seed exhibit a continuum of changes as they pass from dormancy to nondormancy and from nondormancy to dormancy (Baskin and Baskin 1989).

Many seed in buried seed banks exhibit annual dormancy/nondormancy cycles as just discussed. These cycles, such as those of obligate winter annuals and spring-germinating summer annuals are shown in Figure 4.17. From such information on changes in dormancy state (e.g., Figure 4.17), it is possible to predict when exhumed seed, as from tillage, will actually germinate. The primary reason that nondormant seed do not germinate while buried is that most of them have a light requirement for germination. This too can be used to manage weed populations, as will be seen later.

Physical dormancy. The exclusion of necessary environmental factors by certain morphological characteristics, especially of the seed coat, accounts for dormancy of a physical nature (Table 4.7). Of particular interest is *somatic polymorphism*—the production of seed of differing morphologies and/or behaviors on the same parental plant. This process is a consequence of divergent cellular differentiation and represents different outcomes of the plant's allocation to seed output. According to Salisbury (1942a) and Harper (1964b, 1977) somatic polymorphism is rather widespread among seed populations of weeds, especially in families of Asteraceae, Brassicaceae, Chenopodiaceae, and Poaceae. Such polymorphism among seed is generally viewed as a mechanism

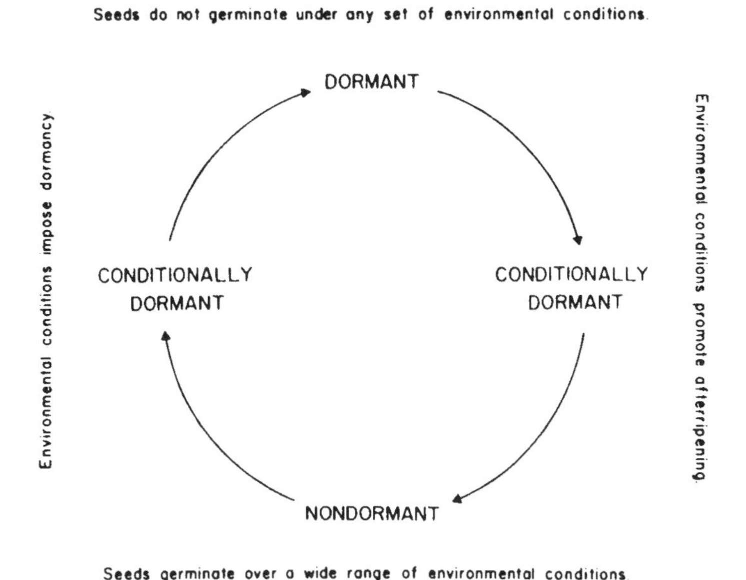

FIGURE 4.16 Changes in the dormancy states of seed with physiological dormancy; the seeds are dormant at maturity and go through all the possible stages of the dormancy cycle. (From Baskin and Baskin, 1989.)

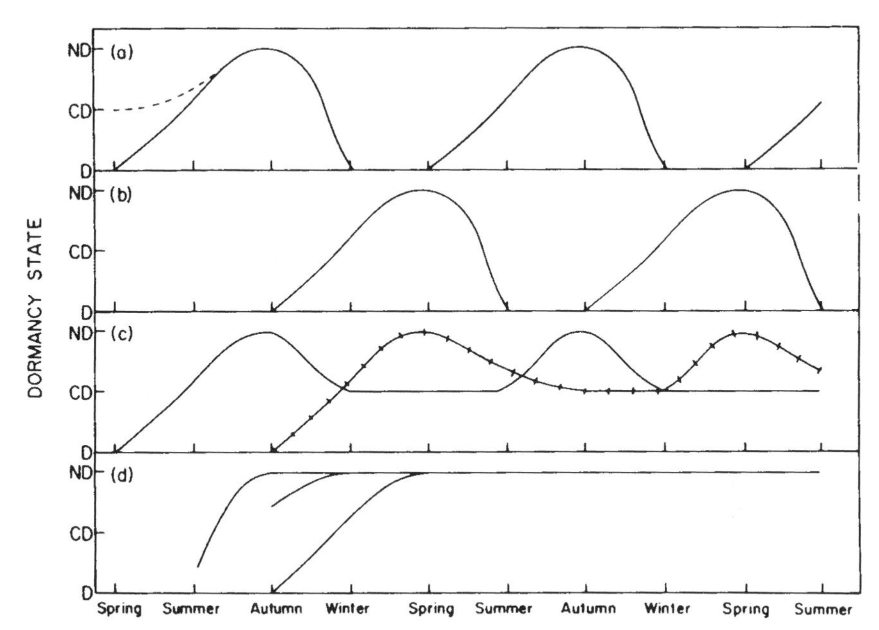

FIGURE 4.17 Patterns of changes in seeds with physiological dormancy. (a) Obligate winter annual with annual dormancy/nondormancy cycle. Freshly matured seeds in some species are dormant (———) and others are conditionally dormant (- - - -). (b) Spring-germinating summer annual with annual dormancy/nondormancy cycle. (c) Facultative winter annual (———) and spring- and summer-germinating summer annual (++++) with annual conditional dormancy/nondormancy cycles. (d) Perennials with no change in dormacy state after seeds come out of dormancy. (From Baskin and Baskin, 1989.)

to enhance species survival through the use of differing habitats. Some examples illustrate the importance of this process in weed species.

A thoroughly examined case is that of the two seed types of cocklebur (*Xanthium* spp.) which are encased in the fruit and dispersed together. The upper seed, in contrast to the lower one, often fails to germinate when wetted so that at least a year separates the germination of the two types. According to Taylorson and Hendrix (1977), the rates of oxygen diffusion in the two types of seed are similar. Apparently, dormancy in *Xanthium* involves the presence of a different water-soluble germination inhibitor in each seed type to which the testa (seed coat) is impermeable. The presence of oxygen causes the degradation of these two inhibitors and subsequent rupture of the seed coat, but apparently at very different rates in the two seed types. Thus at least two batches of seed are present in each generation to assure germination in the event that the immediate environment happens to be unsuitable.

Another example of somatic polymorphism is found in common lambsquarters (*Chenopodium album*) and was studied by Williams and Harper (1965). These researchers found that an individual plant of common lambsquarters produced different types of seed that are categorized into two groups on the basis of seed color, size, coat characters, proportion of total seed production, and germination requirements (Table 4.8). Karssen (1970a,b) observed that the degree of dormancy of *C. album* seed was inversely related to size and depended on the thickness of the outer seed coat layer. This finding was supported by Williams and Harper (Table 4.8), who found large seed to have thin seed coats. Based on these and other data, Cook (1980) indicated that seed persistence in soil should be favored by a decrease in seed size and a relative increase in seed coat thickness, because a proportional decrease in seed size greatly increases the strength of the seed coat relative to the growth force of the embryo during germination. Harper (1977) indicated that the ratio of black to brown seed in *C. album* is probably environmentally controlled, as polymorphism is in other genera, since a gradient in proportion of the two seed color morphs exists across Great Britain. However, brown seed from *C. album* rarely exceed three percent of the total seed produced by a plant and they are the first seed to be produced. The brown seed of *C. album* represent a highly opportunistic strategy, whereas the black seed are more seasonal and predictable in behavior. The combination of germination responses allows the species considerable buffering against sudden selective forces, such as tillage or frosts, that might substantially disfavor a single phenotype.

Harper (1977) believes that seed and dormancy polymorphisms are so common among weed species that it is misleading to ascribe to a species any particular germination regime. This seed variability must be considered when one is attempting to stimulate maximum dormancy breaking of weed seed for subsequent seedling control, since usually only one portion of the seed germinates even under optimal conditions. Furthermore, earlier or later germinating phenotypes may be favored inadvertently by measures taken to stimulate germination and control of the most typical or abundant polymorph. In determining seed response to control measures, weed scientists should be cautious because there may be polymorphic differences within the species they are testing.

TABLE 4.8 Seed Characteristics of Individual Common Lambsquarters (*Chenopodium album*) Plants

Seed Color	Wall Thickness (μ)	Seed Size[a] (g)	Reticulation of Seed Coat	Production/ Individual (%)	Germination Requirements		
					Cold	Nitrate	Season
Black	60	1.13–1.33	+ or –	97	+	+	spring only
Brown	16	1.55–1.59	+ or –	3	–	–	none

Source: Williams and Harper (1965).

[a]1000 seed weight.

Combinations of physiological and physical dormancy. Some seed have a combination of physiological and physical dormancy, for example, an impermeable seed coat and a dormant embryo (Table 4.7). Clearly dormancy breaking in this situation is a function of the internal conditions of the embryo rather than external environmental conditions. However, the internal causes of dormancy may have been created by severe external constraints. The so-called hard seededness of many legume species that results in response to drought demonstrates this concept. The hilum of legume seed acts as a hygroscopic valve which is activated during dry conditions to allow water loss (Figure 4.18). Hyde (1954) transferred seed of white clover (*Trifolium repens*) to chambers of differing relative humidity and measured the moisture content of the seed following each transfer. He found that under humid conditions water was not able to enter the seed but that under dry conditions water vapor could escape. The embryo, therefore, dried to nearly the same water content as the driest environment to which the seed were exposed. Following this treatment, the seed could not imbibe water until the seed coat was broken. By scarifying "hard seed," Hyde found that imbibition would occur (Figure 4.18), thus causing drought-induced dormancy to be broken.

Seed with underdeveloped embryos. In some plants the embryo is underdeveloped but not dormant and completion of embryo development occurs after the seed are dispersed from the parent plant (Table 4.7). After the embryo becomes fully developed in such species, germination usually proceeds. Baskin and Baskin (1989) indicated that this type of dormancy exists in both tropical (e.g., Magnoliaceae, Degeneriaceae, Winteraceae, Lactoridaceae, Canellaceae, and Annonaceae) and temperate (e.g., Apiaceae and Ranunculaceae) plant families.

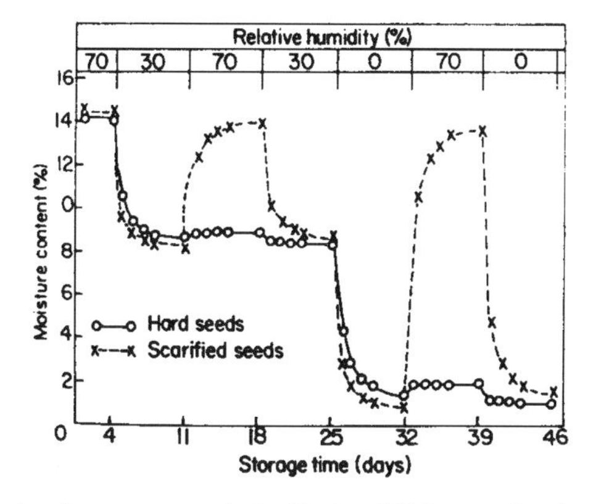

FIGURE 4.18 Changes in moisture content occurring in white clover (*Trifolium repens*) seeds transferred successively into chambers of different relative humidity. (From Hyde, 1954.)

USING SEED DORMANCY TO MANAGE WEED POPULATIONS

With adequate knowledge about the dormancy characteristics of specific weed populations, situations can be identified where very low or no seedling recruitment should occur, even if a high density of seed is present in the soil. For example, weed seed often fail to germinate under dense plant cover (Figure 4.19) or deep in the soil profile. Such locations could be used as weed seed "sinks" (places or conditions where dormant seed remained in the dormant state) by planning crop sowing, rotation, or tillage operations for that purpose. Several examples of how weed seed dormancy might be used to manage weeds are discussed below.

No tillage and night tillage. A light requirement for seed to be released from dormancy is a widespread mechanism found in many weed species (also see seedling

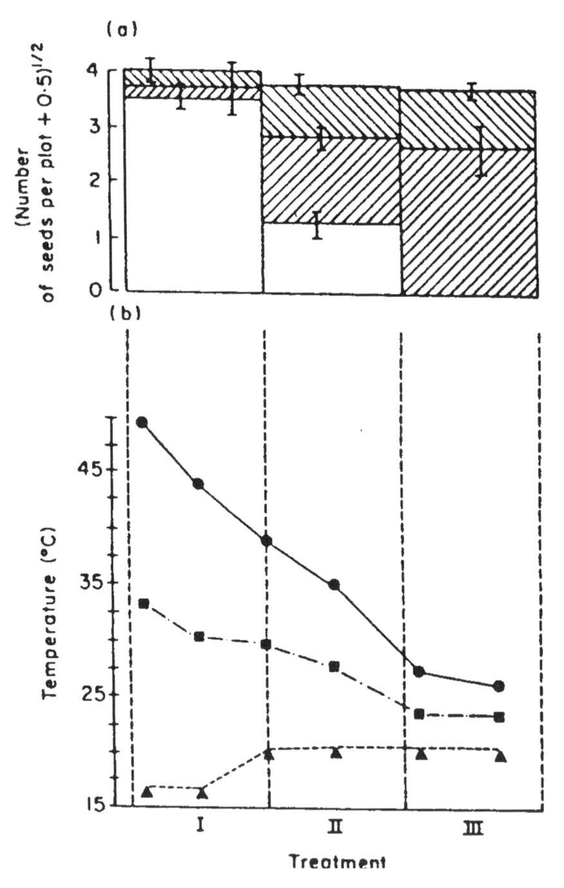

FIGURE 4.19 Influence of canopy height on soil temperature and seed germination. (a) *In situ* germination (□), germination at 20–30°C of recovered seed (▨) and viable non-germinated seeds at the end of the experiment (▧) in the three treatments of experiment 1 (listed below). (b) Thermal conditions measured in different points of three plots with different canopy height; I = bare ground; II = 30 cm; III = 110 cm (●) mean maximum, (▲) mean minimum and (■) mean temperature. Vertical bars represent least significant difference, P ≤ 0.05. (From Benech Arnold, et al., 1988.)

recruitment, this chapter). This requirement can be satisfied even after a brief exposure to light, such as that resulting from soil tillage. Because of the very-low-fluence (VLF) response mechanism (Scopel et al., 1991), light-requiring seed that are buried may detect microsecond exposures of sunlight while the soil is being disturbed by tillage. Such exposure is sufficient to trigger seed germination (Scopel et al., 1994). Thus, germination of species whose seed require light (Chancellor, 1964; Wesson and Wareing, 1969a,b; Roberts and Potter, 1980) should be impeded if a no-tillage system were implemented for crop production. The light-requiring seed of such weed species would remain dormant in the soil and eventually be lost due to predation, senescence, or decay. This tactic is, in part, a component of the no-till systems of weed control now popular in the midwestern United States. Changes in composition of the weed flora favoring species that produce seed without a light requirement for germination also have been observed following the implementation of no-tillage systems (Tuesca et al., 1995). Alternatively, it is possible that night cultivation also could result in significant reductions in weed density. Indeed, Scopel et al. (1994) confirmed earlier observations (Feltner, 1967; Woolley and Stoller, 1978; Jensen, 1991) and found that the normal practice of cultivating agricultural land during daytime can increase germination of buried seed populations by up to 500 percent over that which occurs following nighttime cultivation (Figure 4.20; see also the section on light requirement for germination later in this chapter).

Manipulation of plant canopies. Plant cover can be used to reduce weed abundance by manipulating weed seed requirements to break dormancy (Figure 4.21). Ghersa et al. (1994b) pose the following hypothetical example of this process. In Oregon, wheat is sown through the end of September and October. Within a month after crop sowing, *Lolium* spp. seedling recruitment usually occurs. However, *Lolium* seed have particular thermal and light requirements for germination, which can be disrupted by the presence of a plant canopy. Ghersa et al. believe that this response of *Lolium* seed to plant cover can be used to control its population in the winter wheat cropping system. They suggest that a fast-growing summer annual species could be sown in Oregon in late August or early September (e.g., Italian millet, sorghum, or *Panicum*) to produce a dense initial canopy. Wheat then could be sown into this canopy using zero tillage. Within a month after wheat sowing, frosts—or, in some cases, herbicide—would kill the summer annual cover. By the time the summer annual species dies, wheat should have sufficient green canopy to filter the penetrating sunlight, thus preventing germination of ryegrass. With this type of management, Ghersa et al. (1994b) believe that *Lolium* seedling recruitment could be so reduced and delayed that insignificant interference with the crop would be expected. In a preliminary experiment carried out to study the effect of short-duration plant shading on biomass production of ryegrass and wheat, a reduction of ryegrass vegetative biomass production of nearly fifty percent was obtained where winter wheat was sown under sparse canopies of *Echinochloa crus-galli* as the cover crop.

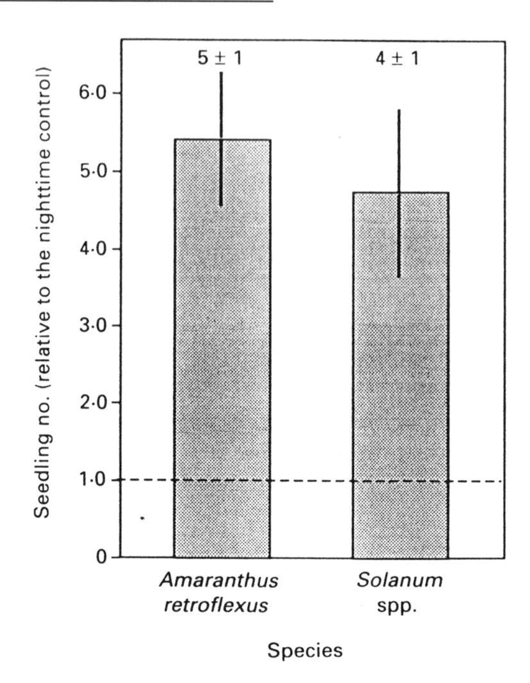

FIGURE 4.20 Effect of daytime cultivation in early summer on seedling emergence compared with the nighttime (no light) control. Absolute densities after nighttime cultivations are indicated at the top in plants m^{-2}. Error bars indicate ± 1 SE; *n* (number of blocks) = 10. The weedy flora (surveyed one month after cultivation) was dominated almost exclusively by *Amaranthus retroflexus, Solanum nigrum,* and *S. sarrachoides.* (From Scopel et al., 1994.)

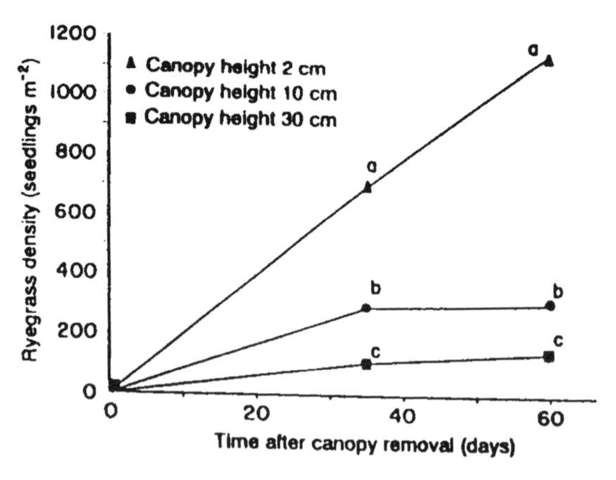

FIGURE 4.21 Time course for ryegrass seedling appearance in canopies dominated by *Cynodon dactylon* and defoliated at different heights. Different letters indicate significant differences ($P < 0.05$). (From Deregibus et al., 1994.)

Predictions of emergence. Success of control tactics is usually based on the ability of tools to suppress effectively high numbers of individual weeds. Therefore, knowledge of which proportion of a weed seed population has already germinated and under what conditions the fraction that has not germinated will germinate would be useful information to improve weed control. Unfortunately, it is practically impossible to determine which proportion of a weed population is being reached by a control method. Certainly the number of emerged seedlings can be counted, but there is generally no way to be sure of which fraction of the seed bank population they represent. The identification of specific environmental factors involved in dormancy release or induction, combined with establishment of solid functional relationships between those factors and rates of germination, would allow formation of predictive models of weed emergence in the field. Benech Arnold et al. (1990) produced such a demographic model to predict *S. halepense* seedling emergence in the field as a function of soil thermal conditions (Figure 4.22). This model considers the effect of temperature on both dormancy release and germination of non-dormant seed. Several other models also have been proposed to predict weed emergence of particular species (Alm et al., 1993; Benech Arnold and Sánchez, 1994; Wilen et al., 1996).

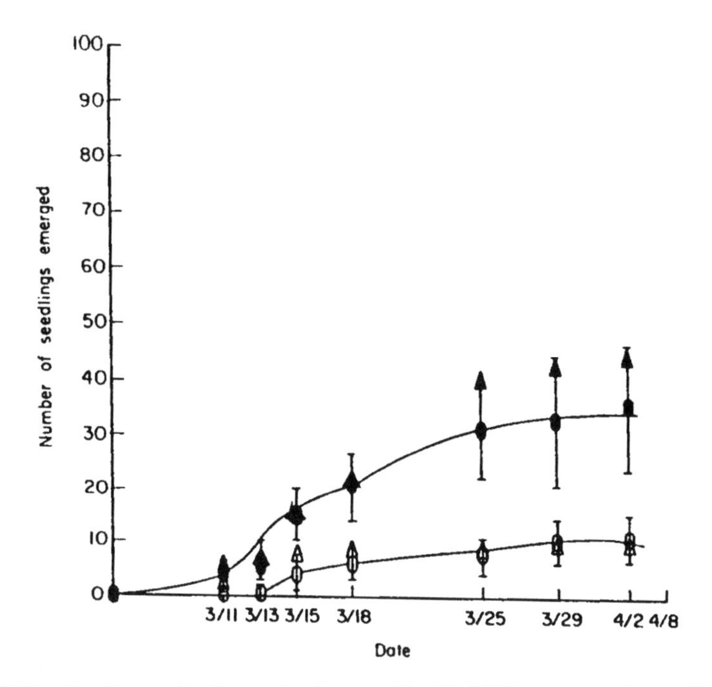

FIGURE 4.22a Cumulative number of emerged seedlings recorded in March field experiments under two different thermal regimes. Bare soil (●) and shaded soil surface (○). The observed data are compared with data obtained from simulation (▲,△). Vertical bars represent s.e. of observed data.

(continued)

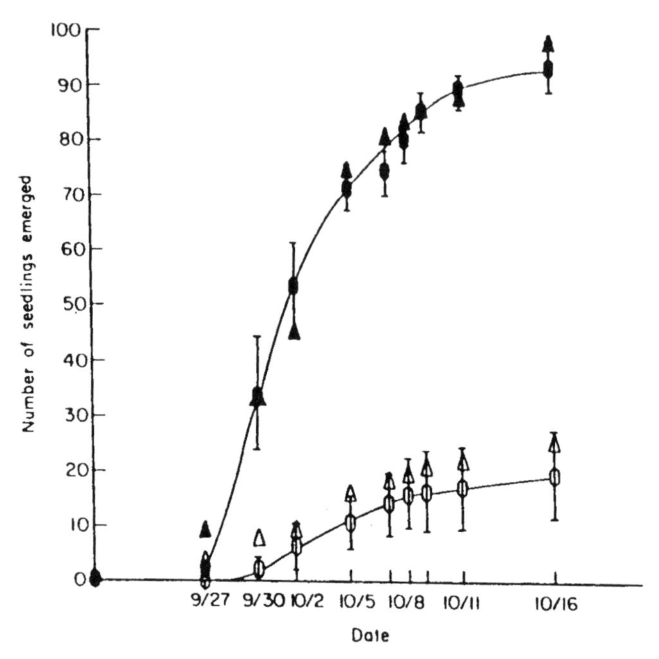

FIGURE 4.22b Cumulative number of emerged seedlings recorded in September field experiments under two different thermal regimes. Bare soil (●) and shaded soil surface (○). The observed data are compared with data from simulation (Δ). Vertical bars represent s.e. of observed data. (From Benech Arnold et al., 1990.)

SEEDLING RECRUITMENT: GERMINATION AND ESTABLISHMENT

Because of the large reproductive allocation common to most weed species, a large number of seed and/or vegetative propagules is usually produced and dispersed to the soil. Unquestionably, many of those propagules become seedlings or ramets. *Germination*, the transition from seed to seedling (or bud to ramet), is the most significant way seed are lost from the soil seed bank. It is also the most critical phase in the development of a plant. Certainly in terms of stand development, germination of both crops and weeds is of critical importance. Most pest management strategies against insects or plant pathogens are directed at the critical (most susceptible) life phase of the pest organism. This approach should also be taken with weeds. The timing and amount of weed seed germination undoubtedly influence the spectrum of species within a weed community. In the event that weed control is imposed, germination also may influence the amount of control received, as well as the composition of the weed population afterward.

Seed Germination

The germination of seed involves the initiation of rapid metabolic activity, embryo growth, radicle emergence and, finally, emergence of aerial portions of the plant. Radicle emergence is used most often as a reliable indicator that germination has begun. Before the shoots emerge from the soil, considerable underground elongation usually has begun. This growth pattern is important for weeds because seed of many weed species are adapted for shallow or surface germination. The survival of the seedling, therefore, depends on the ability of the primary root to extract moisture from increasingly lower levels in the soil profile.

Patterns of shoot emergence from the soil are varied, but two principal types are recognized. *Hypogeal* emergence, typical of Fabaceae and Poaceae, occurs when cotyledons remain below the soil, and *epigeal* emergence, such as in Asclepiadaceae and Apiaceae, occurs when cotyledons are carried above the soil surface during emergence. Both methods of emergence are common in weed species. Monocots, in contrast to dicots, emerge from the soil with the shoot apex encased in a sheath, called a *coleoptile*. The position of the cotyledons in dicots and the degree of hypocotyl, mesocotyl, or coleoptile extension in monocots can influence the survival of seedlings (Figure 4.23). These factors also are important in herbicide placement and differential selectivity among weed and crop species (Chapter 10).

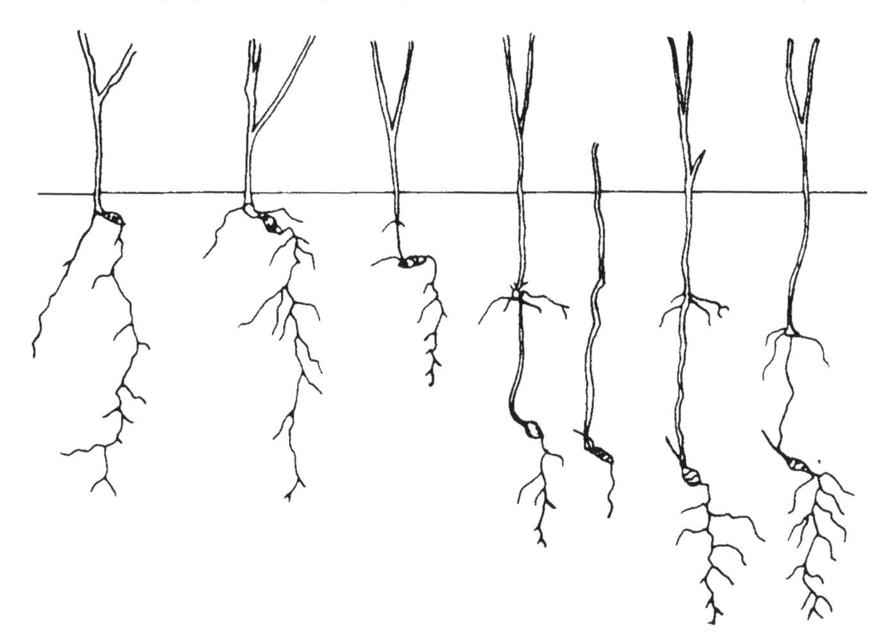

FIGURE 4.23 Underground growth of *Alopecurus myosuroides* (blackgrass) 30 days after germination. Seed was planted at depths of from near surface to 3 cm. Note the elongation of the mesocotyl to regulate the depth of planting and to provide for the crown of roots at the coleopile node nearer the surface of the soil. Seed have been blackened to show up better. (From Hanf, 1944, in King, 1966. Reproduced by permission of International Textbook Co., Glasgow.)

When a seed germinates in natural conditions, the individual plant essentially takes a chance on the soundness of the environmental conditions of a site for seedling establishment (Probert, 1992). Hence, natural selection probably favors mechanisms that decrease the probability of a seed encountering unfavorable conditions for growth after germination. No doubt these patterns of seedling survival involve dormancy mechanisms, polymorphism, and environmental constraints already discussed. Also obvious under some circumstances and with some weed species are flushes of germination. These often occur after tillage or some other disturbance such as fire. Popay and Roberts (1970a,b) attempted to determine the cause(s) for the flushes of germination of two weed species, common groundsel (*Senecio vulgaris*) and shepherdspurse (*Capsella bursa-pastoris*). They were able to characterize germination behavior in terms of seed age, light requirement, alternating temperatures, gas ratios, and soil fertility. However, these factors only partially accounted for the pattern of seedling emergence observed in the field. The physiological status of the seed and environmental conditions during the afterripening process are clearly important features of seed germination.

LIGHT REQUIREMENT FOR GERMINATION

It has been known for some time that many plant species have a light requirement that must be met before germination can occur. This criterion is especially true for seed of weed species. Saur and Struik (1964) speculated on the ecological significance of this phenomenon after noting the differential emergence of weed species from soil samples collected at night beneath various stages of oldfield succession. Since open habitats produced abundant seedlings, Saur and Struik suggested that a light-flash mechanism for dormancy breaking may assist pioneer plants in exploiting disturbed environments.

Wesson and Wareing (1969a) conducted a similar experiment by taking soil samples at night from beneath an established pasture. After discarding the top two centimeters of soil, they separated the samples by depth and attempted to germinate the seed in them under regimes of light and darkness. Little germination occurred in the dark, whereas the soil in the light produced many weed seedlings. They also dug small pits (5, 15, 30 cm deep) in the same pasture at night and covered some with opaque asbestos, covered some with glass, and left some uncovered. The numbers of seedlings that emerged from each treatment are shown in Figure 4.24. In these experiments, germination absolutely required light. An additional difference in germination response was observed between glass-covered and uncovered holes, indicating that germination could be enhanced by increasing temperature but that it was not essential for dormancy breaking. In further studies using freshly collected seed, Wesson and Wareing (1969b) observed that most of the weed species tested displayed no light requirement when seed were fresh (Figure 4.25). However, if seed from the same species were buried for 50 weeks, a light requirement was always needed for germination (Figure 4.25B and C). Some initial germination was

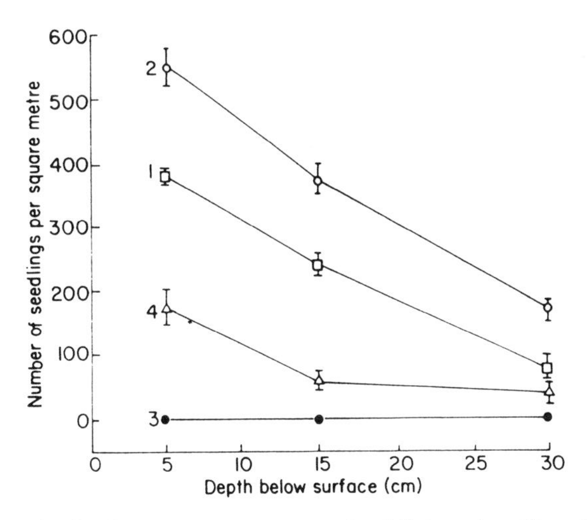

FIGURE 4.24 The number of seedlings to emerge per square meter from field plots at three different depths. (1) Plots uncovered; recorded after 5 weeks. (2) Plots covered with glass; recorded after 5 weeks. (3) Plots covered with asbestos (dark); recorded after 5 weeks. (4) Asbestos covers removed from treatment 3 and replaced with glass. Emergence recorded for further 3 weeks. (From Wesson and Wareing, 1969a.)

always evident following burial, but weed seed that did not germinate within the first few weeks after burial would not germinate until after a light requirement had been fulfilled. Thus, burial actually appears to induce a light requirement in many weed seed soon after dispersal that maintains them in a condition of dormancy for as long as they are buried.

It is interesting to speculate on the ecological reasons for the light requirement of weed seed. It is well known that most weedy species are most abundant and germinate best at very shallow soil depths. Furthermore, a major cause of mortality is deep germination, which prevents a seedling from reaching the soil surface. It seems likely, therefore, that the light requirement may act as a highly predictive indicator of disturbed areas suitable for colonization. The presence of light might indicate proximity to the soil surface, bare mineral soil, or the absence of an overstory canopy.

The light-stimulated germination of seed is known to involve the phytochrome (P) system. In the majority of cases, it is the photoconversion of P_r (absorbs wavelengths of light in the red region) to P_{fr} (absorbs wavelengths of light in the far-red region) that stimulates germination; that is, red light stimulates P_r ———> P_{fr}, whereas far-red light stimulates P_{fr} ———> P_r. Gradually in the dark P_{fr} declines to a level below that needed for germination; thus an input of red light would be required to increase the level of P_{fr} for germination to occur following a prolonged period of burial. In the case of seed, burial appears to induce a dramatic increase in light sensitivity, which is called the very-low-fluence (VLF) response mechanism (Scopel et al., 1991). The VLF response (Figure 4.20) is mediated by phytochrome and triggered by light expo-

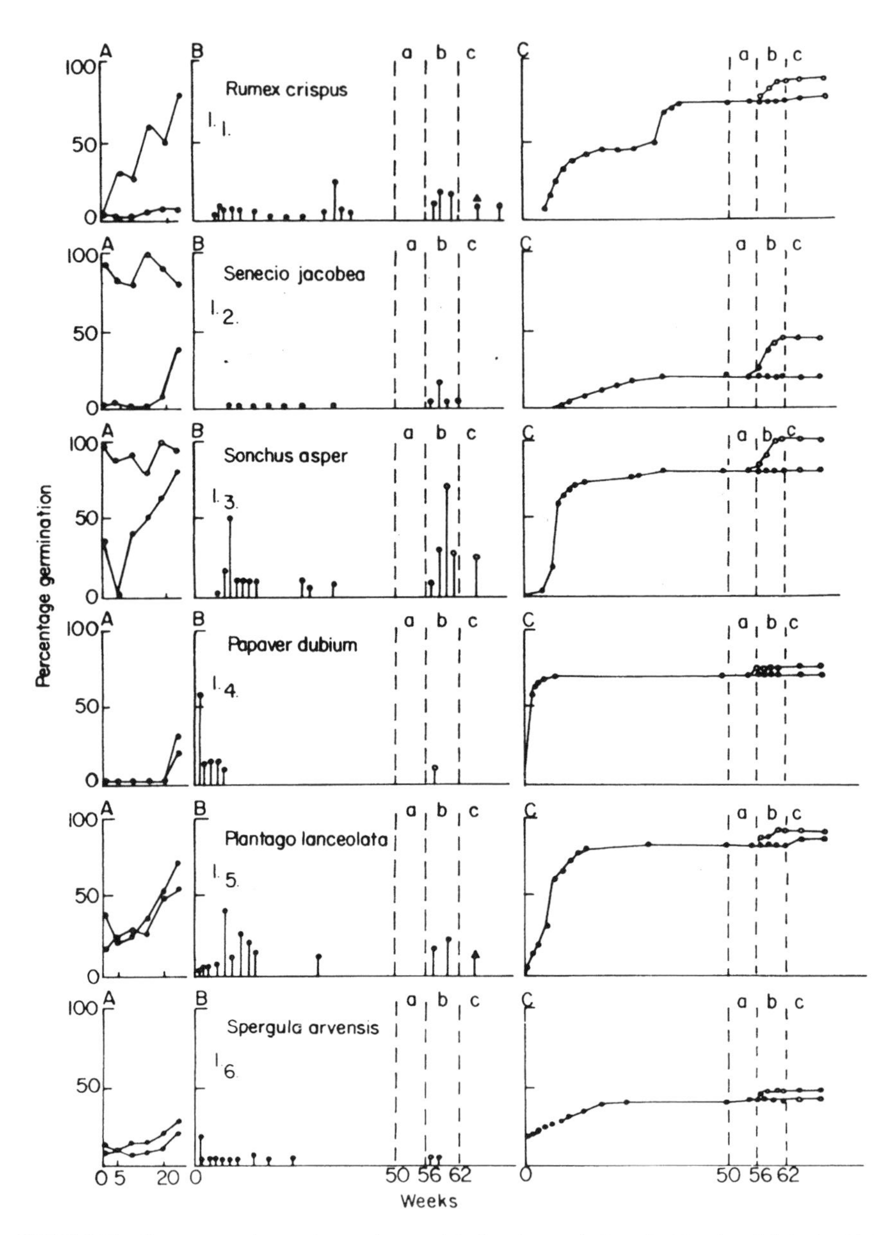

FIGURE 4.25 The germination of six weed species. A, in petri dishes from the time of collection for 25 weeks, in light (○) and dark (●). B and C, buried in soil, out of doors for 50 weeks and then disturbed under controlled conditions. (a) Germination when the soil was brought into a greenhouse but not disturbed; (b) germination when soil was disturbed in either light (○) or dark (●); and (c) germination when seeds disturbed in the dark were moved to the light. In each case, B shows weekly germination expressed as a percentage of the ungerminated seeds and C shows the same results as a cumulative percentage. (From Wesson and Wareing 1969b.)

sures that form very small amounts of P_{fr} (i.e., between 10^{-4} and 10^{-2} percent of the total phytochrome). This response allows seed to detect microsecond exposures to sunlight when the soil is being disturbed, as by tillage (Scopel et al., 1991).

In addition to providing a potential germinant with a clue about its position in or on the soil, the VLF phytochrome response may "suggest" the degree of plant community openness or the presence or absence of an overstory canopy. Leaves in a canopy transmit considerably more far-red than red light (Figure 4.26) Most of the violet, blue, green, and red light that is intercepted by leaves is absorbed or reflected, and that which is transmitted is generally far-red. Under those "enriched" far-red conditions, most seed on the soil surface would exist in the P_r condition and thus not germinate. This could account for the absence of continued weed germination in various crops after a canopy has begun to close. It also may explain why more weed seedlings often are found between rather than within established crop rows.

With many agricultural crop species, in contrast to weeds, a light requirement for germination is not apparent. For example, when planting large-seeded crop species, it is advantageous to place seed into the soil where water or other resources are more readily available than on or near the soil surface. It is likely that a light requirement for germination has been bred out of many crop species. Also possible, however, is that once harvested from the parent, crop

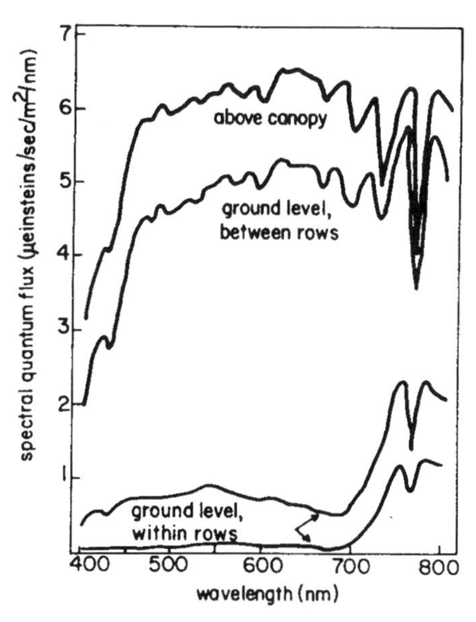

FIGURE 4.26 Influence of shading upon wavelengths of sunlight present in various regions of a sugarbeet field. Within the rows (bottom two curves) there is much less attenuation of far-red than the other wavelengths, so shaded plants contain a higher proportion of phytochrome in the P_r form than do unshaded plants. (From M.G. Holmes and H. Smith, 1975. Reprinted by permission from *Nature* 254: 512–514. Copyright 1975 Macmillan Journals, Ltd.)

seed never receives enough darkness to induce dormancy fully, or even if such seed are stored in darkness, a sufficient amount of red light is received just prior to planting to induce germination.

SAFE SITES

Regeneration from seed usually occurs in an intensely hostile environment for many plant species (Fenner, 1985). In this respect, Silvertown and Lovett Doust (1993) indicate that many studies have demonstrated seed germination to be highly responsive to fine-scale differences in the physical environment, especially at the soil surface. Harper and others (Harper, 1977; Harper et al., 1965; Sheldon, 1974) derived a concept of *safe sites* based on the observation that most seed in the seed bank do not germinate, and of those that do, few survive. Harper (1977) describes a safe site as a zone that provides the following:

- stimuli for dormancy breaking
- conditions for germination to proceed
- availability of resources for growth
- absence of hazards

Factors included in the definition of a safe site are the placement of seed in relation to the microtopography of the soil surface and the availability of water and other conditions necessary for germination. Also contributing to creating a safe site are various adaptations for acquiring resources or being buried in the soil. Thus, seed placement and germination in an appropriate safe site should enhance the survival of the resulting seedling. Unfortunately, it is difficult to determine the specific criteria for a safe site until the seedling is actually present there. However, a variety of techniques have been employed successfully to demonstrate the occurrence and variability of safe site requirements among species. These methods range from studies of artificial media to studies in field situations.

With crops, humans create safe sites to enhance germination and seedling survival. Weeds, however, must either adapt to the safe sites of crops or develop mechanisms to avoid mortality. Some of these mortality-avoiding features have been discussed already; they include the light requirement for dormancy breaking and its relationship to soil disturbance, polymorphism of seed, and hard seededness of many weed species. The abundance of *nitrophiles* (species that occur in nitrogen-rich habitats) among weeds also may indicate an adaptation to increase the probability of seedling survival on disturbed sites. There are numerous examples in which potassium nitrate stimulates germination of weedy species (Anderson, 1968). This is interpreted as a case of enzyme induction, or a signal for germination to proceed, rather than as a requirement for nitrogen fertility. Harper (1977) describes in considerable detail studies that demonstrate the occurrence of safe sites and specific seed characteristics that increase survival in these sites. As he says, the ability of a seed to find and secure a safe site can be modified somewhat (by selection) to enhance success on a species-to-species basis but, in general terms, it is a chance event. The ability of any species to occupy a site in abundance seems to be predominantly a function of the success of its ancestors in leaving well-adapted progeny (Harper, 1977).

Risk of death. In relation to weed management, it is important to consider all chances for mortality as a seedling develops from a seed. The probability of mortality is sometimes called the *critical risk of death*, and it usually determines the initial weed density experienced in a field. As just discussed, each species of weed has a certain set of requirements for germination and survival that are adapted to a particular safe site. Placement in an appropriate safe site minimizes the risk of death. Certain cultural practices, such as annual tillage, may be viewed as a significant selective force because microsites (i.e., safe sites) are continually being formed or modified. For example, a certain number of weed seedlings should survive each year after an initial disturbance clears and stirs up the soil surface. However, some species, or populations within a species, would be favored and others disfavored by the same management practices. Eventually all the safe sites generated should become filled with surviving individuals of the best-adapted genotypes. At that point, seedling saturation of the habitat has occurred, and no further recruitment of seedlings from the soil seed bank would be likely. Further germination would most likely result in death of the seedling due to unsuitability of the environment (no safe sites would be available).

Adopting a cultural practice designed to affect the safe sites of a particular component of weed germinants may have considerable utility in reducing weed densities below the saturation density. In some cases, an economic weed threshold (Chapter 5) may even be obtained. An experiment by Selman (1970) serves as an example of how safe site modification can influence weed density. Over an 11-year period, Selman compared the levels of wild oat (*Avena fatua*) seedling survival in barley that resulted from either fall or spring tillage practices. During the first seven years of his study, he observed that annual tillage in the fall prior to barley planting caused a progressive increase (1 ——> 400 plants/m^2) in wild oat density. For the next four years the same field was cultivated in the spring and then planted with barley. The wild oat density decreased to 85 plants/m^2 the first growing season after spring tillage was implemented, and eventually stabilized at about 5 plant/m^2 by the end of the study. These results reflect the value of a timely tillage as a means of reducing weed seedling density and the value of changing a cultural practice to disfavor existing adapted ecotypes. Since a flush of germination of wild oat occurs in both spring and fall, an eventual build-up of the spring-germinating wild oat ecotype might be expected, similar to that which occurred with the fall ecotype. For this reason, alternating tillage practices (and therefore barley planting dates) between fall and spring might maintain the density of both ecotypes of this weed at a reasonable economic (low) level.

For seedlings of perennial weeds, the risk of mortality should be similar to that of seedlings of annual plants. However, the risk of death for ramets differs considerably. For regrowing ramets of perennial plants the most critical time for survival is during active growth immediately following new shoot initiation. During ramet production, new shoots are not completely self-supporting; consequently, carbohydrate reserves in the rootstocks are lowest at that time. Thus, tillage, especially repeated tillage, at appropriate intervals when carbohydrate reserves are continually being depleted can have a substantial negative effect on ramet density. Agriculturists have recommended for years that it would be a good practice to fallow and to till repeatedly areas infested with perennial weeds. Similar recommendations are made by foresters to suppress unwanted coppice growth of some tree and shrub species. In the case of agriculture, the interval between cultivations varies among species, but cultivation is usually necessary every three weeks during the growing season for maximum carbohydrate depletion and ramet mortality to occur. Too great an interval between cultivations may, in fact, have a detrimental effect on weed control by stimulating ramet production and growth. Interactions between environment and disturbance can enhance ramet mortality as well. For example, considerable density reduction can result from properly timed tillage that takes advantage of both carbohydrate starvation and adverse environmental conditions. In Table 8.3 we see nearly 60 percent reduction in Johnsongrass (*Sorghum halepense*) ramet density from a single properly timed tillage in contrast to an untilled treatment. In that experiment, July tillage allowed sufficient time before fall rains for desiccation of severed rhizomes to occur.

NATURE OF WEED INFESTATIONS

In most weed species, natural dispersal of propagules produces a distribution pattern in which the density of seed is very high near the parent plant and dramatically reduced a short distance away from it (Figure 4.8), generating patchy distributions. This pattern may be completely modified under agricultural conditions, however, by tillage, herbicide use, and harvesting operations (Figure 4.27). Such spatial modification in the distribution of seed in an agricultural field reduces patchy distribution of seed and creates an overall reduction in seedling density (Ballaré et al., 1987a; Cavers and Benoit, 1989; Ghersa et al., 1993). Dispersal of seed in the soil profile by tillage also reduces seedling density, changes the rate of seed release from dormancy, and regulates the rate of seedling emergence through time. It is likely that this reduction in seedling density has a dramatic impact on the epidemic nature of weed infestations in cropped fields (Auld and Coote, 1990; Ghersa and Roush, 1993; Maxwell and Ghersa, 1992) because it removes density dependent regulation on seed production and thus allows exponential growth of the remaining, "escaped" weed populations that will then produce abundant seed, even under high efficacies of weed suppression.

In agricultural land, soil tillage and herbicides are the main hazards for weed seedlings to overcome. Under these conditions, weed mortality is frequently very high—80 percent or more—and occurs during the seedling stage both before and after seedling emergence from the soil (Sagar and Mortimer, 1976;

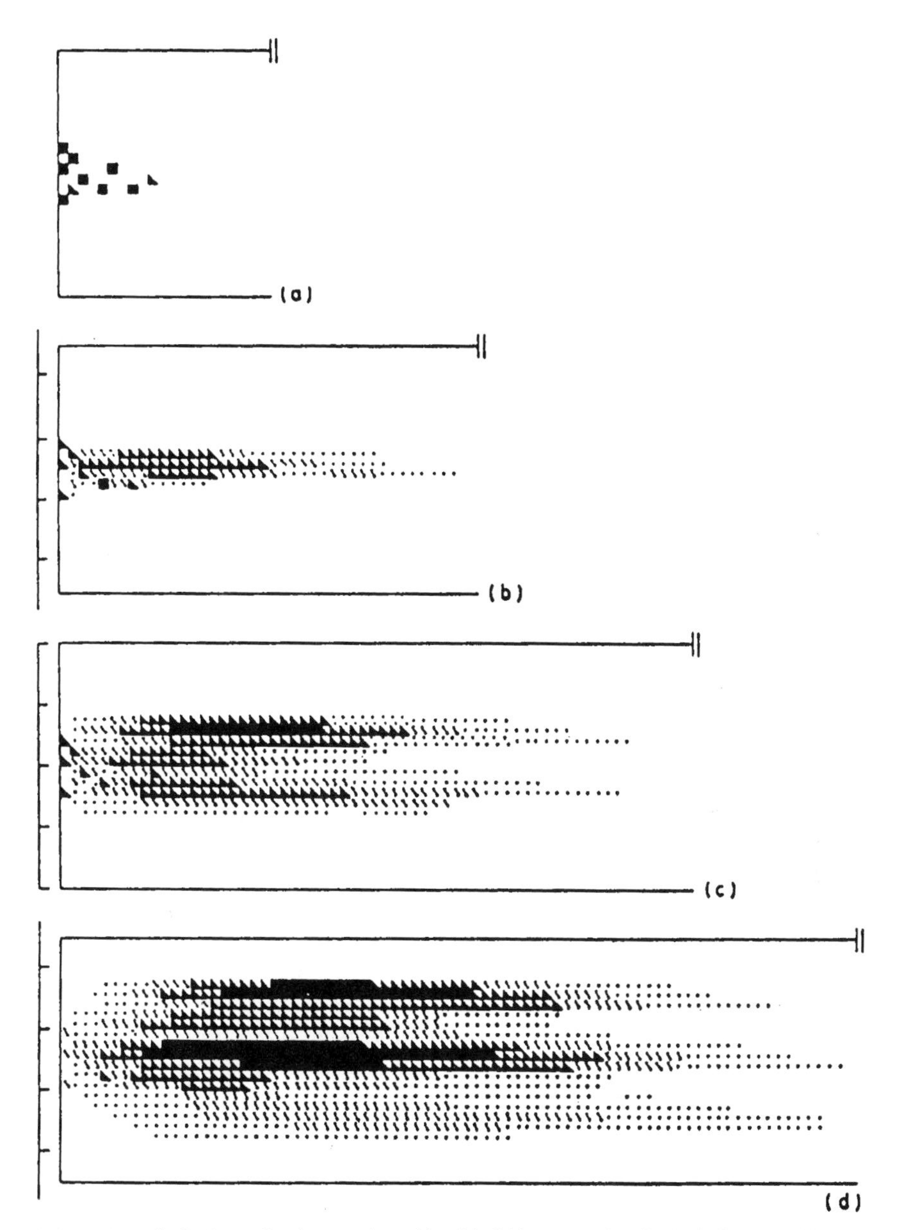

FIGURE 4.27 The distribution of seeds among the modules of the field. (a) = initial seed bank, (b–d) = the seed bank at the end of first, second, and third growing season, respectively. The beginning of the imaginary runs of the combine harvester are indicated by □ = 0, ⊡ = 1–5, ⧅ = 6–15, ◣ = 15–45, ■ = 40–130 seeds per module. (From Ghersa et al, 1993.)

Aldrich, 1984; Ballaré et al., 1987a). In spite of the drastic reductions in seedling density that occur soon after emergence, density dependent effects can still be observed in the growth and reproduction of many weed species. For example, Ballaré et al. (1987a) observed that Chinese thornapple (*Datura ferox*) plants originating from high-density patches produced less biomass and seed than those generated in low-density patches. Diseases and predation also may be modified by seedling density during early stages of weed growth (Fenner, 1985; Burdon, 1987). These factors of density dependence are generally believed to assure a constant seed rain to the soil (Chapter 5).

Being able to predict the spatial and temporal patterns of seedling weeds is a key problem facing weed scientists today. This knowledge is needed to adjust control measures to areas of fields and times of development when weeds are likely to be present, and to evaluate their risk to crop yield reduction. Benech Arnold and Sanchez (1994), Spitters (1989), and Ghersa and Holt (1995) have reviewed how such predictions might be achieved through models of germination and survival that use ecophysiological approaches. These reviews stress the importance of increasing the risk of death of the weed seedlings or delaying growth sufficiently so that weeds are competitively suppressed by crops.

Prediction and Management Using Models of Weed Reproduction, Dispersal, and Survival

The value of determining the reservoir of weed seed in the soil and the early fate of seedlings lies in being able to predict potential weed infestations. Using the demographic parameters of seed production and dispersal, seed reserves in the soil, rate of seedling recruitment, and expected mortality, it should be possible to determine the expected density of weed species likely to occur on a site. Information about the ages of seed that give rise to the majority of new seedlings also would contribute to such predictions of weed presence.

An early attempt to predict the intensity of weed infestation was conducted by Naylor (1970a, b) with blackgrass (*Alopecurus myosuroides*), a serious annual weed problem in Great Britain. The species was chosen for study because of its typically large seasonal fluctuations in density. Since a previous study had shown that 90 percent of new weed seedlings were derived from seed found 2.5 cm deep in the soil, Naylor sampled the soil for weed seed at that position. He then constructed a weed predictive index (WPI) to estimate potential densities for the individual fields in his study. The WPI could account for 84 to 98 percent of the variation in field densities of blackgrass. Thus, potential or expected weed density could be predicted with reasonable accuracy. Although density is not the only criterion that determines interspecific competition (Chapter 5) and presumably the need for weed control, the degree of weed infestation is most easily described and control measures evaluated in terms of density. Therefore, it is of considerable value to know how many of each weed species might be expected to emerge.

Thus, weed researchers and population ecologists have developed and are still developing various types of models to describe the demography of plant populations and the competitive interactions among weeds and crops (Auld, 1984; Kropff and Lotz, 1992; Maxwell and Ghersa, 1992; Sagar and Mortimer, 1976; Silvertown and Lovett Doust, 1993; Spitters, 1989). All such models can be excellent tools to understand the components of a species' life cycle, to determine the relationships among these components, and ultimately to predict changes in weed population numbers from one generation to the next as a result of environmental variation, environmental manipulation, or management tactics. As an example of such analyses, we use the classic model developed by Sagar and Mortimer (1976) to examine the population dynamics of weeds.

PREDICTIONS OF CHANGES IN WEED ABUNDANCE IN FIELDS

Sagar and Mortimer (1976) developed a working model to examine the population dynamics of higher plants, especially weeds. In the simplest form of the model, depicted in Figure 4.2, two routes are presented by which an initial population of individual plants of size A_1 in generation G_t may occur in the next generation (G_{t+1}). The population in generation G_{t+1} would be of size A_2, and would arise from genets of population A_1 (Route I) or sometimes from survivors of the earlier population (Route II). Population A_2 may be larger, smaller, or equal in size to that of population A_1. Route I represents plant establishment from seed, and Sagar and Mortimer divide it into six phases (B through G): total number of seed produced by population A_1 (B), total number of seed falling to the ground (C), total number of seed present on the soil surface (D_1), and buried (D_2) seed bank, the number of seedlings that germinate from the seed bank (E), and the number of plants that become established (F). They also indicate that invasion of seed dispersed from elsewhere (G) may occur and contributions to phases C and D may result from that source. The model (Figure 4.2) also recognizes seven interphases (*a* through *g*) which represent the probability of, or number of individuals, proceeding to the next phase. Thus, interphase *a* represents multiplication by seed, interphase *b* represents loss that occurs between seed production and arrival of seed on the soil surface, interphase *c* is the fate of seed on the ground, interphase *d* is the probability of a seed in the seed bank germinating, interphase *e* represents the seedling fate after emergence, and interphase *f* is the fate after seedling establishment. Interphase *g* is the level of seed invasion from plants that occur off-site. Route II does not normally occur in annuals, but is important for biennials and perennials that usually do not reproduce by vegetative means.

Best-case/worst-case scenarios of weed management. In order to demonstrate the value of such a model and its use for weed management, Sagar and Mortimer (1976) presented several possible schemes for population regulation in wild oat (*Avena fatua*). Beginning with a hypothetical population of 10 plants and using data from other sources, Sagar and Mortimer considered both best and worst management options for this weed species. Figure 4.28A represents a situation resulting from effective management in which severe population reduction of wild oat would occur. At each interphase, effective regulation of the weed population is achieved by (a) control of seed output by choosing a competitive crop (Chancellor and Peters, 1970), (b) harvest before seed is shed and subsequent removal of straw (Thurston, 1964), (c) maximum exposure of seed on the soil surface (Wilson, 1972), (d) sparse emergence from the soil (Thurston, 1961), and (e) high postemergence mortality (Chancellor and Peters, 1972). The combination of the interphases in Figure 4.28A was predicted by Sagar and Mortimer to cause dramatic decreases in population size (from 10 to 0.018 plants) and the contribution to the soil seed bank was only 0.06 percent of the previously existing seed reserve. In contrast, Figure 4.28B represents a series of poor management options. According to Sagar and Mortimer, the most rapid rate of population expansion arises from (a) minimal crop competition (Chancellor and Peters, 1970), (b) poor attempts at seed collection and trash destruction during harvest (Wilson, 1972), (c) incorporation of seed into the soil (Marshall and Jain, 1967), (d) maximum emergence (Marshal and Jain, 1967; Wilson, 1972) and (e and f) no subsequent mortality. This interphase combination for the wild oat life cycle

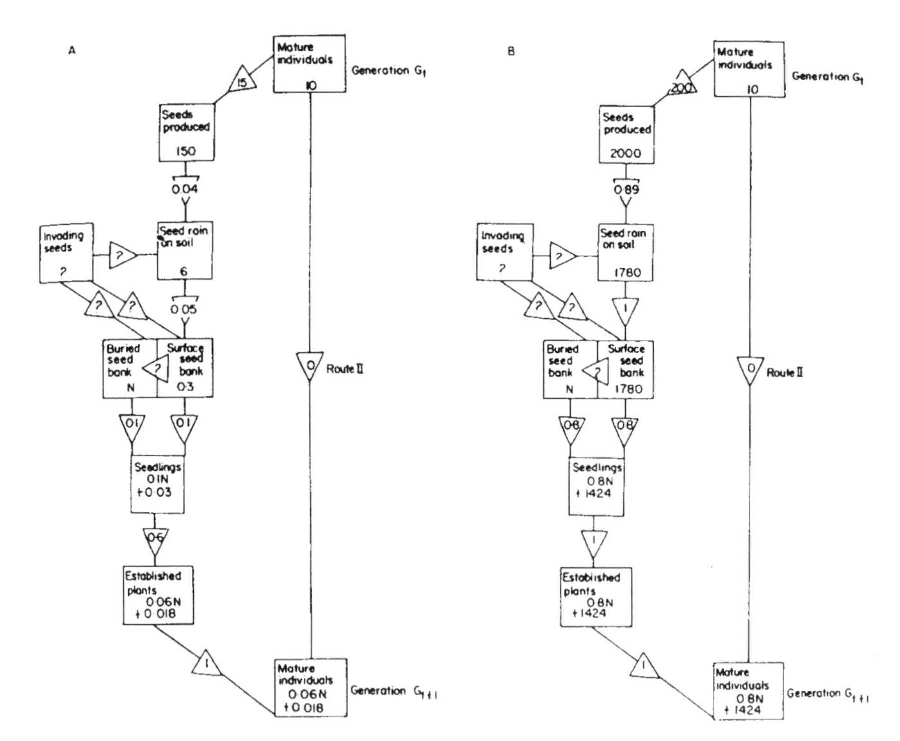

FIGURE 4.28 Life table for wild oat (*Avena fatua*) in which (A) the population increase is projected to be low and (B) the population increase is projected to be high. N represents the buried seed bank. The sources of values are acknowledged in the text. (From Sagar and Mortimer, 1976.)

(Figure 4.28B) resulted in a considerable increase in population size (+1424 plants) and the contribution to the seed bank rose by 80 percent. By comparing the two life tables for wild oat (Figures 4.28A and B), it is possible to see that the greatest control (regulation) occurs at crop harvest (b), when seed lie exposed on the soil surface (c), and by failure to emerge from the buried seed reserve (d). Clearly the best management is that which is directed at those interphases of the weed's life cycle.

SUMMARY

Weed community composition and density are assumed to be unpredictable and chaotic. Uncertainty about weed species density, composition, and changes in population and community dynamics often intensifies the perception of risk to crop systems and can lead to inappropriate management decisions. Plant population ecology, an area of plant science that deals with the numerical aspects of a particular group of individuals, can ease some of the uncertainty involving weed and crop associations. These tools can be used to describe the critical life stages of a weed species and examine how weed populations might be expected to change over time, in different environments, or from various management tactics. Natality, mortality, immigration, and emigration are the four fundamental components of plant demography. However, the complexities of weed populations are best described using diagrammatic life tables and, in some cases, difference and transition models.

Seed are important for the maintenance and growth of existing weed populations and for the initiation of new populations. Thus, seed must be both produced and dispersed for a weed species to be successful. Seed dispersal can be either over space or through time. Wind, water, animals, and humans are agents of spatial dispersal. Dormancy is the method by which seed disperse through time. The seed bank is a useful concept often applied to the dynamics of seed on or in soil. This concept views the soil as a reservoir of seed to which inputs and withdrawals are made. Most inputs (seed rain) to the seed bank are large for weed species and other colonizing plants. However, the size and longevity of seed banks vary. Although some weed seed may exist dormant in the soil for a very long time, most agricultural seed banks are of short to intermediate duration. The most common forms of withdrawal from the seed bank are germination, followed by predation, and seed senescence and decay. There is substantial evidence that if weed seed inputs can be stopped, the number of viable seed stored in soil of cultivated and abandoned fields will decline. Thus, it is important to remove late-germinating cohorts or survivors of weed control tactics from fields to avoid continually reestablishing a reserve of weed propagules in the soil.

Dormancy is the temporary failure of viable seed to germinate under environmental conditions that later allow germination to occur. Seed may be physiologically dormant, maintain physical dormancy, or have underdeveloped embryos.

: dormant do not pose an immediate threat of further weed infesta- omehow dormancy is broken. Thus, it may be best to leave dormant _tate or to create "sinks" in which to cache dormant seed through management activities. Other ways to manage weeds using dormancy are to till soil at night or to maintain green canopies that alter the light environment necessary to break dormancy.

Most weed seed require light to germinate. This is a mechanism to assure the presence of bare soil and the absence of other vegetation, since germination is the most critical stage in a plant's development. Many plants also have specific physiological or physical adaptations to assure survival in certain hospitable locations, called safe sites. Safe sites are usually created for crops, and weeds must either use those safe sites or develop mechanisms to avoid mortality. It is possible to alter management practices to disfavor weeds that have previously become adapted to the tillage, sowing, or harvesting practices of a crop system. Being able to predict the spatial and temporal patterns of weed seedling distribution is one of the most pressing problems facing weed scientists today.

Prediction of the abundance and changes in weed populations is possible using demographic models. However, such models usually require large amounts of empirical information about weed life histories in order to be constructed. Nevertheless, such models do exist for many weed-crop associations. These models demonstrate how improved knowledge of environment and biology can assist weed control decisions and either enhance direct weed suppression or eliminate the need for it.

Chapter

5

Associations of Weeds and Crops

Associations of various members of the plant kingdom are readily observed in nature. Indeed, the interactions occurring within these associations are believed to be the force that shapes the morphology and life history of individual plants and the structure and dynamics of plant communities. Often, when a plant association occurs between a crop and a weed, a loss in crop yield results. It is upon this observation about crop-weed interactions—that is, the potential negative impact of weeds on crops—that modern weed science is built. Many experiments document crop yield losses from particular weed associations with crops. As a result, the necessity for weed suppression in most crops is well founded. However, only recently have experimental methods become refined enough to distinguish among the types of weed-crop interactions. Most studies of weeds in crops deal primarily with the negative aspects of the association, especially crop loss, but there may be neutral or even positive aspects of plant-to-plant interactions, as well. By examining these neutral, negative, and positive aspects of weed-crop associations, it might be possible to make choices about weed levels and cultural practices that optimize crop production while minimizing costs.

NEIGHBORS

Soon after germination, a seedling plant must become independent of its parental resources in the seed. It must begin existence as an individual and extract from its surroundings the resources necessary for life. Plants, however, rarely exist in isolation. Instead, they usually exist with other plants in associations of the same or differing species. Therefore, it is important to consider plants as neighbors within the environments in which they exist. It is generally recognized that there are three distinct types of plant neighbors:

- parts of the same plant; leaves and branches that extend over other leaves or branches of a plant
- individuals of the same species but arising from different seed (genets)
- individuals of different taxa

Management practices used in agriculture, forestry, and urban environments are often designed to change plant associations in order to improve plant growth or appearance. *Pruning* is a cultural practice that is used to optimize the association among parts of the same plant, while *spacing* is used to optimize the interaction between like individuals. *Weeding* is most often used to influence the relationship between neighbors of different taxa.

INTERFERENCE

Burkholder (1952) categorized the possible interactions that occur among plant species growing in proximity. He used a scheme in which the interaction is symbolically (+ or –) described as an effect on two plant populations. When the two populations are close enough to respond to each other, the interaction is "on." The interaction is "off" when the two populations are apart. In most cases, an "off" interaction has no effect, but not always. Table 5.1 lists all the logically possible types of interactions between populations. As seen in Table 5.1, interactions may be positive, negative, or neutral. However, only the relationship between plants in the association is named. The actual causes of interactions may include production or consumption of resources, production of stimulants or toxins, predation, or protection, which are not described by this scheme.

The general term for interactions among species, or populations within a species, is *interference*. It is the effect that the presence of a plant has on the growth or development of its neighbors. Interference can be expressed as an alteration in growth rate or form which results from a change in the plant's environment due to the presence of another plant. Generally, no attempt is made to determine how the plant's environment has been altered; instead, only the plant's responses to it are described.

Of the ten possible interactions listed in Table 5.1, three represent the negative effects of association. These are competition, amensalism, and parasitism. These forms of interference are considered in more depth in Chapters 6 and 7. *Competition* (– –) is defined by Barbour et al. (1987) as the mutually adverse effects of organisms (plants) that utilize a resource in short supply. *Amensalism* (0–) is similar, but refers to the interaction in which only one of the plants is depressed, whereas the other remains stable. Allelopathy, the inhibition of one

TABLE 5.1 A Complete List of All Biologically Possible Types of Interactions[a]

	On		*Off*	
Name of Interaction	*A*	*B*	*A*	*B*
Neutralism	0	0	0	0
Competition	–	–	0	0
Mutualism	+	+	–	–
Unnamed	+	+	0	–
Protocooperation	+	+	0	0
Commensalism	+	0	–	0
Unnamed	+	0	0	0
Amensalism	0	–	0	0
Parasitism, predation, herbivory	+	–	–	0
Unnamed	+	–	0	0

Source: Burkholder, 1952. Cooperation and conflict among primitive organisms. *Am. Sci.* 40: 601–631. Reprinted by permission of *American Scientist*, Journal of Sigma Xi.

[a]When organisms A and B are close enough to participate in the interaction, the interaction is "on"; otherwise it is "off." Stimulation is symbolized as +, no effect as 0, and depression as –.

plant species by another through the release into the environment of selective metabolic byproducts, is considered to be a form of amensalism. *Parasitism* (+– and –0) is a special form of negative interference because one plant lives in or on another and thus derives resources directly from its host. Neutral and positive interactions among plants are also possible. These include neutralism, mutualism, protocooperation, and commensalism (Table 5.1). Usually, positive associations among two species involve a third organism or factor (Chapter 7).

Effect and Response

How can interference be positive, negative, or neutral? Vandermeer (1989) describes interference as a double transformation problem, whereby a plant and the environment in which it exists affect, or transform, each other. In other words, a plant (or any other organism) lives according to the dictates of its local environment, yet it is also an important participant in effecting change on that local environment. As stated by Harper (1977):

> A plant may influence its neighbors by changing their environment. The changes may be by addition or by subtraction and there is much controversy about which is more important. There also may be indirect effects, not acting through resources

or toxins but affecting conditions such as temperature or wind velocity, encouraging or discouraging animals and so affecting predation, trampling, etc.

This idea of effect and response is presented in diagrammatic form in Figure 5.1a in which the environment-organism transformation is shown as a crop-crop interaction involving corn and bean. One crop, for example, has an *effect* on the environment—it might remove water or certain nutrients like potassium from the soil, leaving it partially depleted, or it could enhance the environment by leaving deposits of nitrogen that it fixed from the air. Both crops must *respond* to this effect, thus setting up the dichotomy of "effect" and "response" indicated in Figure 5.1 (Goldberg and Werner, 1983).

Vandermeer notes that both the effect and response have "modifiers" (Figure 5.1b) that influence the direction and extent of the negative or positive interaction observed. In the first case, an organism (e.g., a plant) may affect the environment in a negative way with respect to other organisms (including other plants)

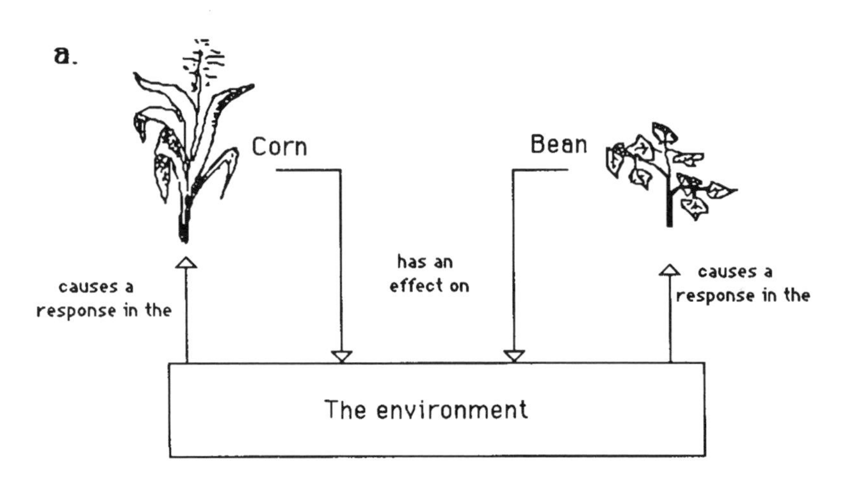

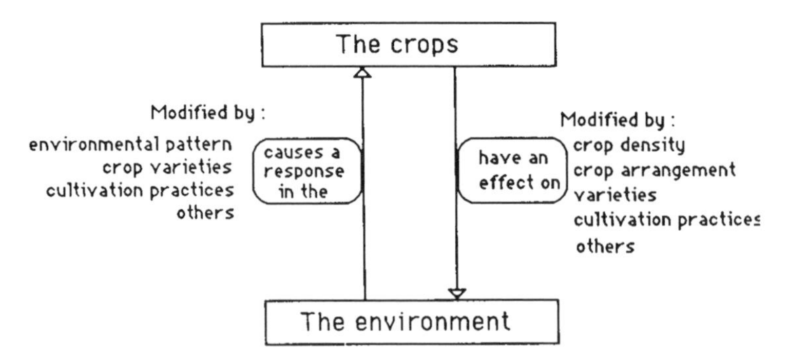

FIGURE 5.1 Diagrammatic representation of the effect-response formulation. (From Vandermeer, 1989.)

through nutrient or water extraction or the production of shade or allelochemicals. Thus, a benign environment could become more hostile to other plants and organisms. This is generally called *competition*. In another option, the environment may be affected positively by an organism, such as a plant. For example, pollinators attracted to one flowering individual may create a pollinator-filled environment for other individuals of another plant species (Vandermeer, 1989), or shrubs (e.g., sagebrush in the desert) may collect snow and thus provide more water for grasses beneath them. Vandermeer calls this effect *facilitation*, which may refer to all of the positive interactions described in Table 5.1.

Two principles emerge from this discussion that describe how plant communities develop:

- *the competitive production principle*—when one species has an effect on the environment which causes a negative response in the other species. Yet there are instances when species exist together in such a way that the negative response of one on another is less than would result from growing all species separately in pure stands.
- *the facilitative production principle*—when the environment of one species is modified in a positive way by another species such that the growth or development of the first species is enhanced.

These principles are restatements, in an agricultural (production) context, of the competitive exclusion principle and niche theory already discussed in Chapter 2.

Figure 5.2 demonstrates how the competitive production and facilitative production principles might act together in different plant communities to produce different outcomes. Each situation in Figure 5.2 is modified by plant density. In the first case (a), competition dominates, while in the second (b) facilitation dominates. In the third instance (c), both competition and facilitation dominate but at different plant densities or times.

Competition versus Other Types of Interference

There is a lack of consistency in the literature and among weed scientists in general concerning the terminology used to describe interference among plants, especially negative interference. For example, competition, according to Burkholder (1952), describes only one possible type of negative interference (Table 5.1). Implied in the term competition is the assumption that the supply of some environmental resource is insufficient for the unrestricted growth of the species in question. A further implication of the term is that the species occupy similar niches so that each is affected by the availability of the limiting resource (Figure 2.6). In this case, the implied cause of the interaction is the differential abilities of the plant species to usurp environmental resources. Unfortunately, competition has been used to describe the negative influence of one species upon another (e.g., weeds on crop yield) without consideration of resource availability or the presence of other organisms. Thus, it is possible that

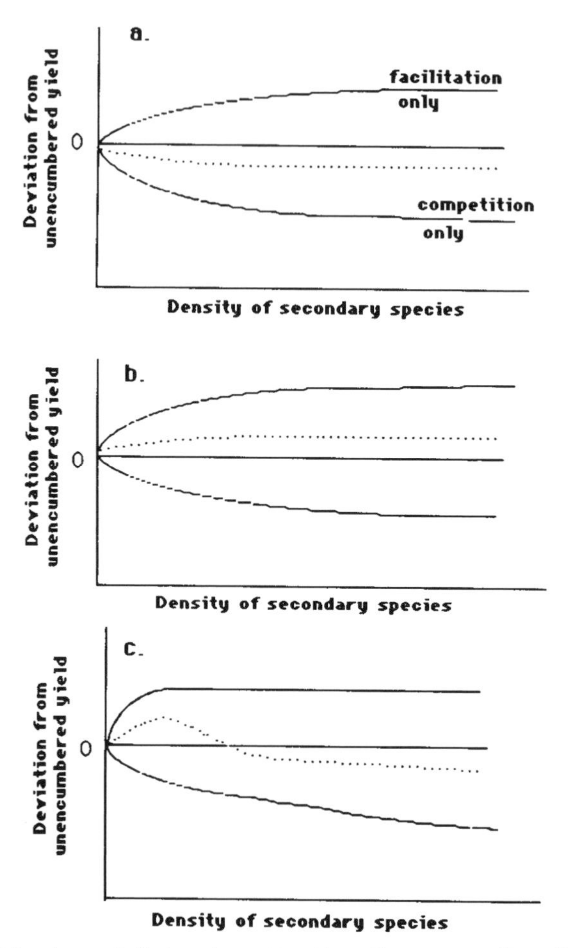

FIGURE 5.2 The balance between facilitation and competition: (a) net effect competitive, (b) net effect facilitative, (c) net effect competitive or facilitative depending on density of second species. Vertical axis represents the yield of the primary species. (From Vandermeer, 1989.)

forms of interference other than competition (for example, amensalism) could actually be measured unknowingly in some experiments. It is also possible that the negative effects of competition in some studies are "averaged" with positive effects if the system is complex enough to contain organisms other than just a single crop and weed species.

An abundance of information has accumulated that implicates plant density, proportional relationships among the species in an association, and the spatial arrangement of individual plants as important parameters contributing to competition. These parameters are most important to consider in competition studies, but they also influence the outcome of other forms of interference, as well. Because these factors of plant proximity (density, proportion and spatial arrangement) so dramatically influence the outcome of experiments designed to demonstrate the impact of plant-plant interactions, they are considered here under the general topic of interference.

Modifiers of Interference

CONCEPT OF SPACE

At some point in development after germination, the plant either exhausts its parental supplies or becomes independent of them. Further growth depends on the seedling's ability to extract the resources it needs from its local environment. The consumable *resources* are light, water, nutrients, oxygen, and carbon dioxide. These resources are in contrast to *conditions* such as temperature, which are necessary for plant growth but are not consumed. The supply of resources may be unlimited in some environments, but limitation is more common. Resource limitation can be caused by unavailability, poor supply, or proximity to neighboring plants. The presence of neighbors can aggravate an already insufficient resource or create a deficiency where there was ample resource for a single individual.

Because resource use is integrated within an individual and among plants in pure and mixed stands, some ecologists have chosen to consider the impact of resources on growth as a single conceptual unit, called *space*. Thus, space refers to the composite of all resources necessary for plant growth, as well as their interactions. The concept of space as an integrative resource allows the study of the effect of proximity between individuals without concern for the actual cause of the interaction. In this manner, the effect upon each other of individuals that share the same environment can be measured, since each must act as the biological indicator of "space" utilization by the other. However, whether to consider resource availability as a composite factor, such as space, or to consider resources independently depends on the information desired. For example, there are situations in which the identification and supplementation of a single limiting resource has corrected a growth deficiency that may have been aggravated by the presence of neighboring plants. Thus, at times it is advantageous to consider space and its implied resources as the object of negative interference, whereas at other times the influence of individual resources should be considered.

DENSITY

Density is the number of individuals per unit of area. Likely units of measure are plants per square meter, plants per hectare, or plants per pot. Density is often used to describe the number of plants in a crop, tree, or weed stand. As density increases, a certain level of individual plants is reached at which interference occurs among neighboring plants. Plants respond to density stress in two ways: through a plastic response of growth and/or an altered risk of mortality.

Effect of density on growth. Figure 5.3(a) represents the typical growth response of a plant population to increasing density. Such data are obtained by sowing various densities of a single species and harvesting the total plant biomass (yield). With the passage of time, plants that are growing at high density quickly meet the stress created by the proximity of neighbors, whereas plants at low density do so only as the neighboring plants get bigger. At harvest, it is apparent that total yield per unit of area has become independent of density, that is, the yield per unit of area is equivalent over a range of sowing densities. This phenomenon occurs because the amount of growth by individual plants decreases as the density increases (Figure 5.3b). In its initial phase or at very low densities, the yield of the population is determined by the number of individuals, but eventually the resource-supplying power of the environment becomes limiting. This ultimately determines yield, irrespective of the plant density. Such a relationship between density and productivity is repeatable and occurs for a wide range of plant species and mixtures of species. It is known as the *law of constant final yield.*

Mathematically, the law of constant final yield (Figure 5.3a) would take the form:

$$Y = N\,[(fN)] \qquad \textit{(eq. 5.1)}$$

where Y is the total yield (biomass) of the population, N is the density of the population, and (fN) is a nonlinear function of N. Figure 5.3b depicts the same relationship on an individual plant basis, that is, at high density the total yield

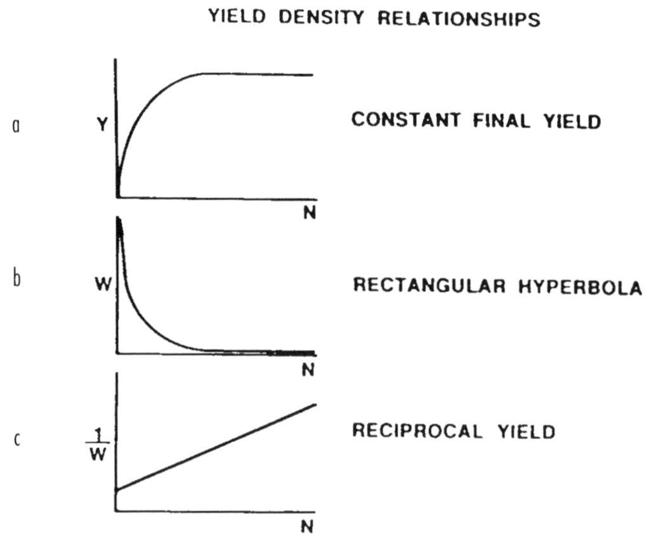

FIGURE 5.3 Diagrammatic representation of basic relationships between plant yield and plant density. Y is yield of a stand of plants (biomass area^{-1}). W is yield of an individual plant (biomass plant^{-1}). N is plant density. (From Radosevich and Roush, 1990.)

is determined by many small plants, while at low density it is determined by fewer larger ones. This relationship is normally written as:

$$w^{-\rho} = w_0 + BN \qquad \textit{(eq. 5.2)}$$

where w is the individual plant weight, w_0 is the theoretical maximum plant size in the absence of competitors, N is the density, $-\rho$ is the hyperbolic relationship between w and N, and B is a constant for all densities. Often ρ is set equal to -1 and the above equation is written as:

$$1/w = 1/[w_0 + BN] \qquad \textit{(eq. 5.3)}$$

where 1/w is the reciprocal of individual plant size (weight), $1/w_0$ is 1/theoretical maximum weight of an individual plant, and B is the slope of the line reflecting the relationship between w and N (Figure 5.3c), which may be thought of as a competition constant. This equation, depicted in Figure 5.3c, is known as the *reciprocal yield law*, which is derived from the same parameters as the law of constant final yield. This derivation will become important when we examine interference in mixed stands.

It is worth noting that altering the level of available resource does not alter the relationship of density to yield (Figure 5.3). As seen in Figure 5.4, either increasing or decreasing the amount of resource may determine the ultimate amount of biomass production but does not affect its relationship to density. Furthermore, under high initial density, ultimate, final yield may be determined by many small plants or, because of density-determined mortality, fewer larger ones. In either case, the yield per unit area is a constant feature of the environment. This environmental constancy is due to the characteristic nature of plant growth, often referred to as *plasticity*, which is the ability of plants to alter their size, mass, or number in relation to density or other environmental stress. In Table 5.2, the dry weight of individual redroot pigweed (*Amaranthus retroflexus*) plants that were grown in a greenhouse at densities ranging from 1 to 35 plants per pot is shown. The total dry weight per pot (unit of area) was relatively constant, but weight per plant decreased dramatically (6.2 to 0.2 g)

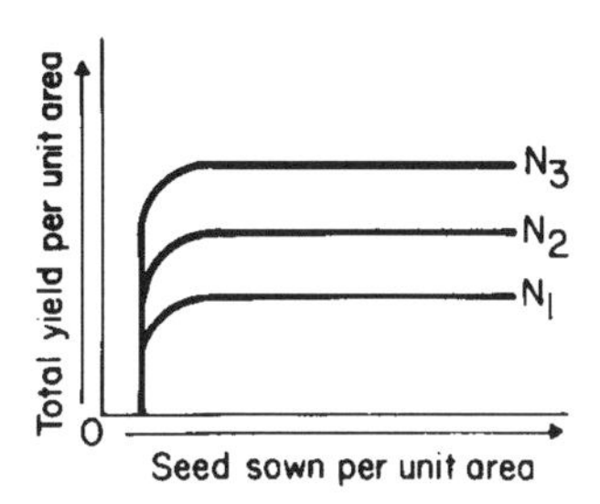

FIGURE 5.4 Influence of increasing amount of a resource on the relationship of total yield per unit area to density of seed sown. N_1, N_2, and N_3 represent increasing increments of a limiting resource—for example, nitrogen fertility.

TABLE 5.2 Dry Weight Yield of Redroot Pigweed (*Amaranthus retroflexus*) as a Function of Density[a]

Density (plants/pot)	*Dry Weight (g)*	
	Per Pot	Per Plant
1	6.2	6.20
5	6.9	1.38
15	6.2	0.41
25	6.2	0.25
35	6.8	0.19

[a]Mean dry weight of four replicates grown in a greenhouse. Davis, California.

as density increased. Consequently, few plants per unit of area generally means that they will be large plants.

Effect of density on size distribution. At any given density of a plant population, a characteristic size distribution of individuals is expected. One way to express this distribution would be to average the size or weight (total weight per number of individuals). An average value would be quite misleading, however, because normally very few plants would be found that reflect the average size. In most plant populations, a size distribution of plants arises in which most are suppressed and small and a few are large and dominant. This distribution was demonstrated by Ogden (1970) with several annual plant species (Figure 5.5) which had not yet experienced density-dependent mortality. In all cases, relatively few plants make up most of the plant biomass.

A similar phenomenon to that of annual plants is observed in stands of trees. The typical stand usually starts with a relatively large number of small trees per unit area, often thousands in natural stands and hundreds if the stand is planted. In both instances, the number of trees decreases over time and those trees that are most vigorous or best adapted to the local environment are most likely to survive. Growth in height is usually the critical factor for tree survival, although taller trees are usually largest in other dimensions of growth as well, especially crown size. As weaker (smaller) trees are crowded by their taller companions, their crowns become increasingly misshapen and restricted in size (Figure 5.6). Such trees gradually become overtopped and eventually die. Foresters recognize four standard size classes of trees (Smith 1986, Figure 5.6).

Dominant. Trees with crowns extending above the general level of crown cover and receiving full sunlight from above and partly from the side; larger than average trees in the stand with crowns well developed but possibly crowded at the sides.

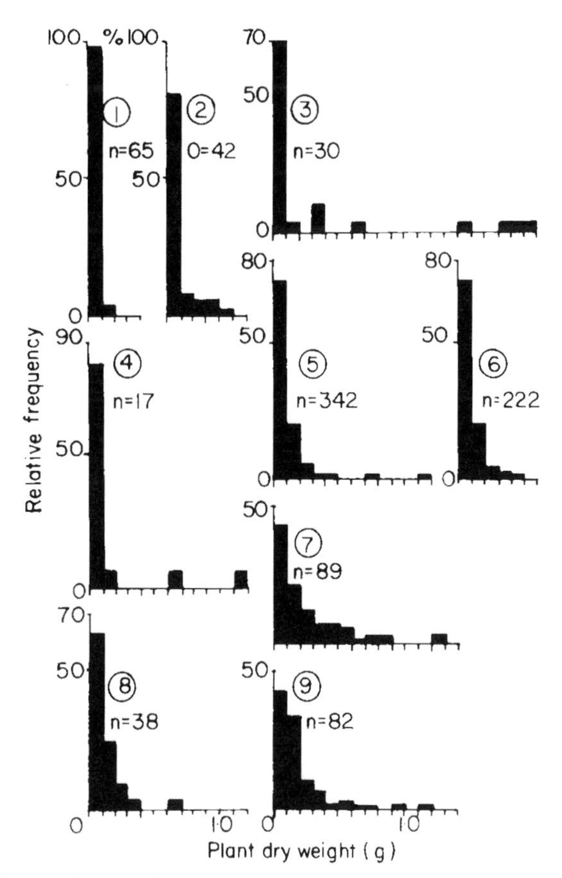

FIGURE 5.5 The frequency distribution of individual plant dry weight in some mixed annual weed populations in an arable field in North Wales. n = density of individuals per 0.5 m^2 approx. 1. Gramineae—mostly *Poa annua*; 2. *Atriplex patula*; 3. *Polygonum aviculare*; 4. all other species; 5. *Stachys arvensis*; 6. *Stellaria media*; 7. *Spergula arvensis*; 8. *Senecio vulgaris*; 9. *Polygonum persicaria* and *P. lapathifolium*. (From Ogden, 1970.)

Codominant. Trees with crowns forming the general level of the crown cover and receiving full sunlight from above but comparatively little from the sides; usually with medium-sized crowns more or less crowded on the sides.

Intermediate. Trees shorter than those in the two preceding classes but with crowns extending into the crown cover formed by dominant and codominant trees; receiving little direct light from above and none from the sides; usually with small crowns considerably crowded on the sides.

Overtopped (suppressed). Trees with crowns entirely below the general level of the crown cover, receiving no direct light from either above or the sides.

The place that an individual plant occupies within this apparent hierarchy of size classes, regardless of the plant's life cycle, is determined at a very early

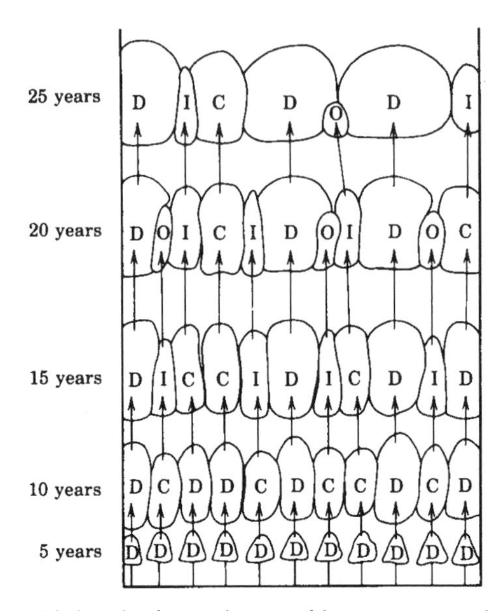

FIGURE 5.6 The relative spatial relationship of trees in the canopy of the same pure even-aged stand at successive intervals of age showing the differentiation of trees into crown classes. This figure illustrates the suppression, as a result of competition, of some trees that were initially dominants. (From Smith, 1986.)

stage of development. This concept, often called *space capture*, has been recognized as a principal component affecting interference among neighboring plants. Space capture is discussed in more depth in the following chapter about competition and limiting environmental resources.

Effect of density on mortality. Plants have an innate capacity for self-thinning as space available to them becomes more and more limited. This phenomenon was first noted by Yoda et al. (1963). It has been termed the *3/2 power law* after the mathematical relationship (slope in Figure 5.7) between plant weight (size) and density that occurs in response to thinning. It may be viewed as a lowered probability of survival as plant numbers at germination increase (Figure 5.7). In fact, growth suppression and the occurrence of weak individuals are probably less severe cases of the thinning phenomenon, since death is the most extreme response to stress. If the response to increasing density is mortality, the reciprocal yield relationship is altered. Thus, as seen in Figure 5.7, after self-thinning, the reciprocal of mean individual plant weight remains the same though density has decreased. Eventually, however, the survivors will respond to the lower density by growing more, and the reciprocal yield law will hold again.

The fact that self-thinning is also an important factor determining yield is reflected in numerous agricultural and forest management studies. Thus, agronomists often recommend certain seeding rates of annual crops and foresters

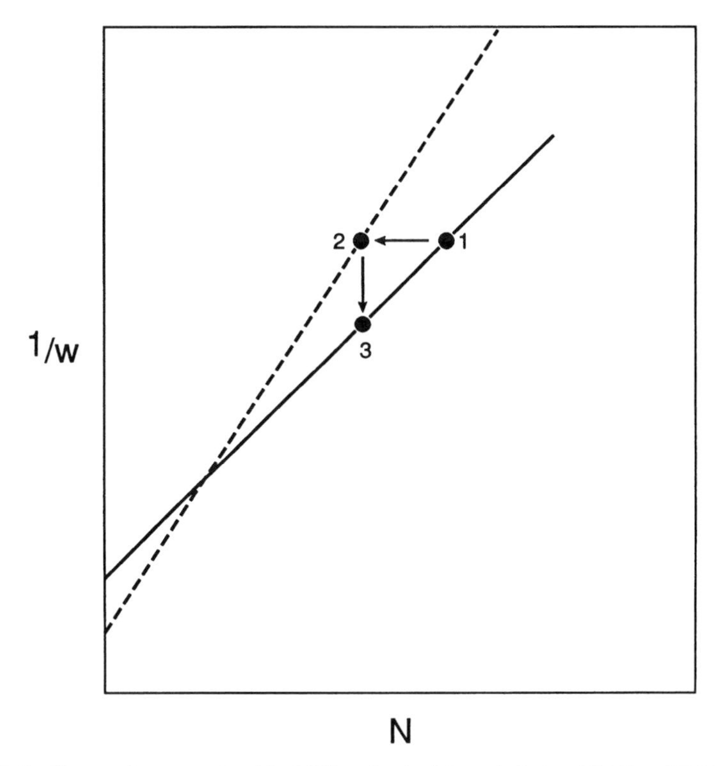

FIGURE 5.7 Diagrammatic representation of the 3/2 Power Law in relation to the Reciprocal Yield Law. W is individual plant weight (biomass plant^{-1}) and N is plant density. Solid line represents the Reciprocal Yield Law; dashed line represents plant weight after thinning.

suggest spacing distances between planted or naturally regenerated trees to avoid self-thinning or growth suppression. For example, Oliver and Powers (1978) predicted that 50 percent suppression of ponderosa pine growth will occur in 40 years if trees are planted at 3.9 m^2 as compared to 11 m^2 spacing (Figure 5.8). Furthermore, tree mortality increases significantly as trees become more confined due to close planting. Foresters have developed the relationship between stand density, density-dependent mortality, and individual tree size (Figure 5.9) for every merchantable tree species. The optimum stand density to achieve a particular size class and the amount of time between thinning or harvests can be determined from such relationships.

It has also been observed that increasing the amount of a limiting factor, such as fertility, often enhances density-dependent mortality. This happens because the dominant plants in the density-dependent hierarchy of size classes (Figures 5.5 and 5.6) continue to capture most of the resource. Consequently, larger plants become more dominant, whereas smaller plants become more suppressed or even die. A notable exception to this generalization is light. With

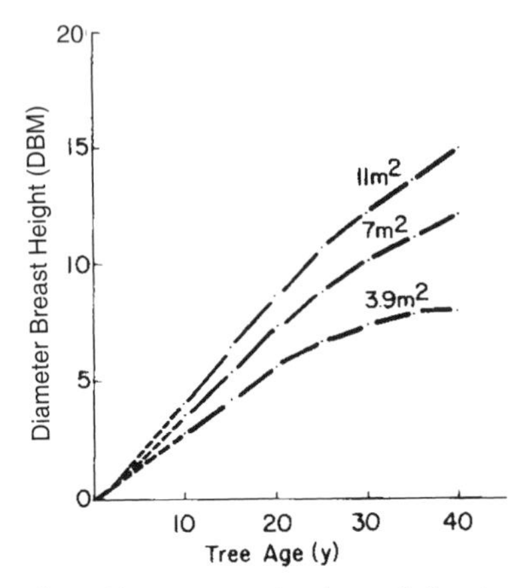

FIGURE 5.8 The effect of increasing density (decreasing spacing) on the growth of ponderosa pine trees. Tree spacing has been corrected for density-dependent mortality. (Drawn from data from Oliver and Powers, 1978.)

this environmental resource, increased levels of irradiance allow increased survival of plants in all size classes.

Effect of density on reproduction. The ultimate success of a plant species in colonizing a site must eventually be expressed in reproductive output, in order to maintain the population continually through time. The reproductive output of annual weeds is especially important since the seed represents the only link to the site between generations. Annual weed species apparently can utilize the density-related responses of plant growth (plasticity) and mortality to regulate and maintain a relatively stable reproductive output.

In a greenhouse study, Palmblad (1968) examined the density responses of eight weed species: *Bromus tectorum* (downy brome), *Capsella bursa-pastoris* (shepherdspurse), *Conyza canadensis* (horseweed), *Plantago lanceolata* (buckhorn plantain), *P. major* (broad-leaved plantain), *Senecio sylvaticus* (woodland groundsel), *S. viscosus* (sticky groundsel), and *Silene angelica.* Seed were sown by dropping them onto bare soil in 15-cm pots at densities ranging from 5 to 200 seeds per pot (55 to 11,000 seed/m^2). The following observations were made:

1. From germination to flowering, all species except Silene angelica showed density-dependent death (Table 5.3). Although the magnitude of response varied by species, the greatest mortality always occurred at high densities.
2. All plants at the lowest density survived and produced fruit (Table 5.3). However, the probability that a surviving seedling would both flower and produce fruit decreased with increasing density. In addition, *C. canadensis, P. lanceolata*, and *P. major*, which survived the high planting densities, often remained vegetative, and thus did not flower or produce fruit until the following growing period.

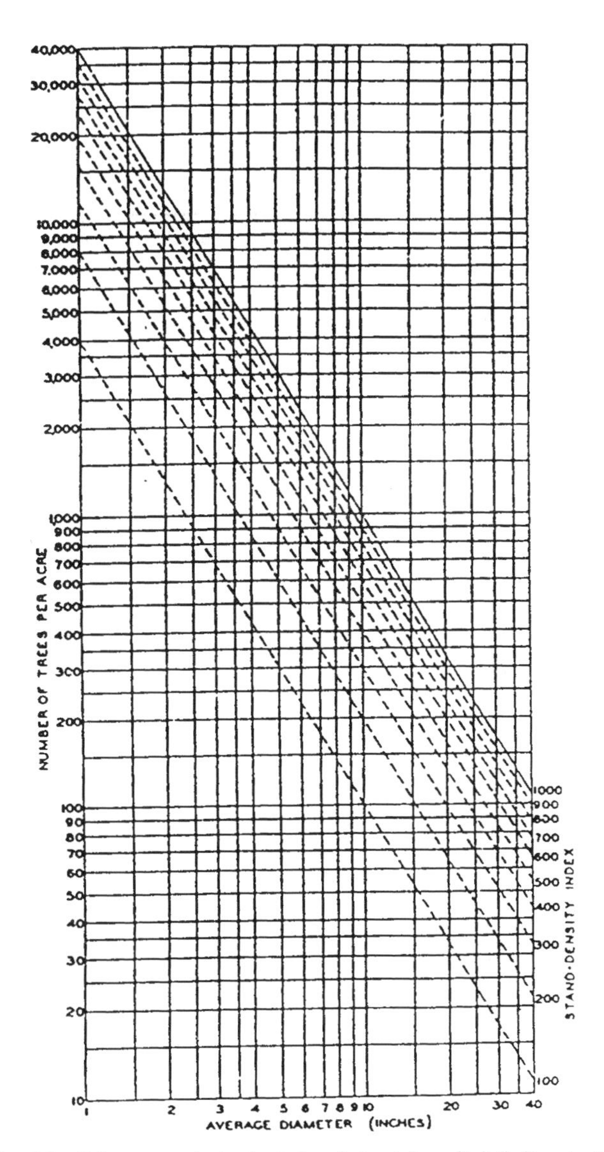

FIGURE 5.9 The relationship between tree density, density-dependent mortality, and individual tree size. Reference curve is represented by the solid line. The stand index of each of the broken-line parallel curves is the number of trees indicated by each at 10-inch average diameter. (From Reineke, 1933.)

3. The amount of seed produced per species was relatively constant and independent of density (Table 5.3). At low density, most of the plants survived to produce an abundance of seed. At high seedling densities, self-thinning generally occurred and the surviving plants were fewer but larger than if there had been no mortality. The combination of more and rela-

TABLE 5.3 The Effect of Density on Mortality and Reproduction in Selected Weed Species[a]

	Percent Mortality Before Flowering					*Percentage of Individuals That Produced Fruit*					*Number of Seed Produced (10^3 Seed)*				
	(plants/pots)					*(plants/pots)*					*(plants/pots)*				
Species	*1*	*5*	*50*	*100*	*200*	*1*	*5*	*50*	*100*	*200*	*1*	*5*	*50*	*100*	*200*
Bromus tectorum	0	—	3	5	12	100	—	81	81	75	2.4	—	2.6	2.6	2.9
Capsella bursa-pastoris	0	0	1	3	8	100	100	82	83	73	23.7	30.5	40.3	37.2	30.1
Conyza canadensis[b]	0	0	1	4	8	100	87	51	42	36	52.6	59.6	40.8	35.3	38.4
Plantago lanceolata[b]	0	0	0	0	1	100	53	8	2	1	2.2	1.4	0.4	0.03	0.06
Plantago major[b]	0	7	6	10	24	100	93	72	52	34	12.0	12.7	8.2	6.6	4.4
Senecio sylvaticus	0	0	3	7	8	100	80	83	49	67	3.8	3.6	4.3	3.7	5.1
Senecio viscosus	0	0	2	2	5	100	100	79	69	68	4.3	11.4	7.7	7.9	7.3
Silene anglica	0	0	0	0	0	100	67	28	25	28	10.3	10.6	18.6	25.4	19.7

Source: Modified from data from Palmblad (1968).

[a]Data from percent mortality and fruit production were obtained from Table 5, while numbers of seed produced were obtained from Table 6 in Palmblad (1968). Only first-year data are represented.

[b]At high densities, approximately 8, 50, and 30% of the plants in each pot for *Conyza canadensis*, *Plantago lanceolata*, and *Plantago major*, respectively, remained vegetative and did not flower or produce fruit the first year.

tively larger plants at high density plantings usually resulted in similar seed production regardless of planting density. In the case of *C. canadensis*, *P. lanceolata*, and *P. major*, subsequent reproduction the next growing season by survivors that had remained vegetative enhanced the reproductive stability of those species. Even though seed output varied dramatically from species to species, the total amount of seed produced per species was remarkably uniform across a 200-fold range of densities.

This study by Palmblad illustrates how mortality and growth plasticity due to density can work together to ensure reliable seed output from a weed community. There is also an important implication for weed management hidden within the example. Observations indicate that rarely are all of the weeds in an area removed as a result of any control measure. Therefore, the few weeds that usually escape weed control have the potential, through survival and subsequent plastic growth, to maintain a relatively constant seed rain. If the seed rain is incorporated into the soil, the potential to continually replenish the soil seed reserve, or seed bank, is also maintained. Because of constant reproductive output, therefore, the acceptable density of many weed species may be very low indeed. This demonstrates the value of "rouging" fields to remove the few exotic species or survivors following weed control. It also suggests that a realistic approach to weed management may be the determination of optimum weed densities based on desired crop productivity, combined with tactics to reduce the seed rain and dispersal of weeds, rather than maximum crop output and attempts to attain perfect weed control.

SPECIES PROPORTION

So far, we have been concerned with density as it affects a single species, although all of the relationships discussed above also apply to populations in mixed stands. In fact, mixtures of species are more common in nature than population monocultures. In agriculture, forestry, and range management, "invasion" by species other than crop species is a normal event and species diversity appears to be the rule.

When competition between at least two species is studied, proportion becomes another factor to consider. Proportion is the relative density or ratio of each plant species in a plant stand. Spitters (1983a) demonstrated the importance of species proportion to competition by expansion of the reciprocal yield law (Figure 5.3) to include a mixture of several plant species:

$$1/w_i = B_{i0} + B_{ii} N_i + B_{ji} N_j \qquad \textit{(eq. 5.4)}$$

where w_i is the weight of individual plants of species i, B_{i0} is the theoretical mean weight of individual plants of species i under competitor-free growing conditions, B_{ii} is the regression coefficient quantifying the intraspecific effect of density (N_i) of species i on the reciprocal of individual plant weight of species i, and B_{ji} is the regression coefficient quantifying the interspecific effect of density (N_j) of species j on the reciprocal of individual plant weight of species i. A similar equation can be written for species j or the equation can be expanded to include more than two species. The regression coefficients describe the effects of intraspecific (B_{ii}) and interspecific (B_{ji}) competition on individual plant weight. They also indicate that the density of each species relative to that of the other in the stand influences the yield of both species. The values of density (N_i, N_j) in these equations also reflect the need to establish the effects of all species in a mixed stand on the final yield (total outcome) of each species. This effect of proportion or relative density is often overlooked in weed-crop competition studies.

SPATIAL ARRANGEMENT

Spatial arrangement is the horizontal pattern of aggregation of plants (Figure 2.3) that reflects dispersal patterns. Fischer and Miles (1973) developed several

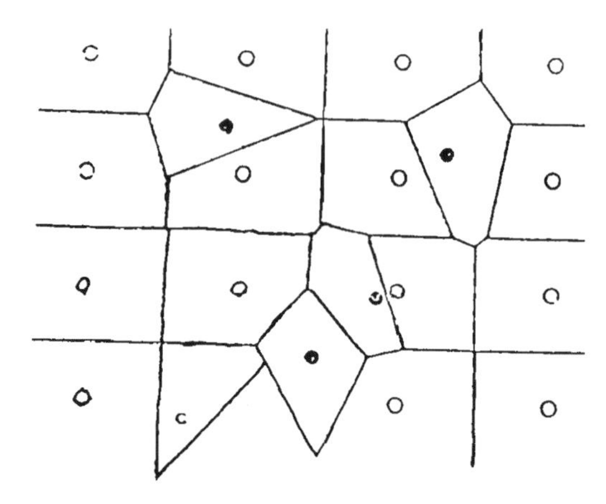

FIGURE 5.10 Plant size in relation to arrangement and "emergence" time. (From Fischer and Miles, 1973.)

theoretical stochastic models for the interference between crop plants, arranged as a grid of points, and randomly located weeds. They assumed that, in the absence of neighbors, a plant expands from emergence until it meets another plant, whereupon expansion ceases. Ultimately, each plant establishes a zone of resource exploitation (Figure 5.10) and theoretically many nonoverlapping weed and crop domains would occupy a field.

Fischer and Miles determined that plant arrangement in the field could be an important factor in determining the outcome of competition, with weeds gaining least advantage if the crop is planted in square or triangular patterns. In most competition studies, spatial arrangement among individual crop plants is held constant, usually in a square or rectangular arrangement, and is assumed to have little effect on the study's outcome. Unfortunately, there have been few experiments performed to test this assumption.

METHODS OF STUDYING COMPETITION

Several methods have been developed to study plant competition. Each method considers the factors of density, proportion of species, and spatial arrangement to varying degrees (Radosevich 1987; Cousens, 1991; Cousens and Mortimer, 1995). The methods fall into four general types of experiments: additive, substitutive, systematic, and neighborhood. In each method, total or individual plant yield, plant growth rate, or plant mortality is measured. Each method is a form of a bioassay in which the response of one species is used to describe the influence of the others in the mixture.

Additive Experiments

Additive experiments are perhaps the most common approach used to study weed-crop competition. More than two plant species can be grown together in additive experiments, but most studies are conducted with only two species, a crop and a weed. The density of one species, such as the crop, is always constant, while the density of the other is varied (Figure 5.11). The additive experimental design is relevant to many agricultural and forestry situations in which at least one species of weed infests an area already occupied by a fixed density of crop or where various weed densities occur from different weed control treatments. In this approach, crop yield usually improves markedly as weed densities diminish until weed levels are reached that do not significantly decrease crop production further (Table 5.4 and Figure 5.12).

The additive method has been criticized because it does not account adequately for the influence of total density and species proportion on the outcome of competition (Harper, 1977; Radosevich, 1987). In the additive approach, the total plant density always varies among treatments, and the proportion among species changes simultaneously with total density. Thus, two factors in the experiment vary, making it difficult to interpret the effects of either factor alone. Spatial arrangement among plants in additive experiments is assumed to be uniform, since the crop usually is planted in a grid pattern (Figure 5.11). Therefore, the influence of intraspecific crop competition is assumed to be constant. However, weed placement often is unreported or is unknown. Because of the uncertainty of proximity factors and the inability to differentiate between intra- and interspecific effects, determining the causes of interspecific interactions is difficult using this experimental approach.

Substitutive Experiments

Harper (1977) believes that many of the criticisms of additive experiments can be overcome by the substitutive approach to competition study. There are three general types of substitutive experiments: replacement series, Nelder, and diallel designs.

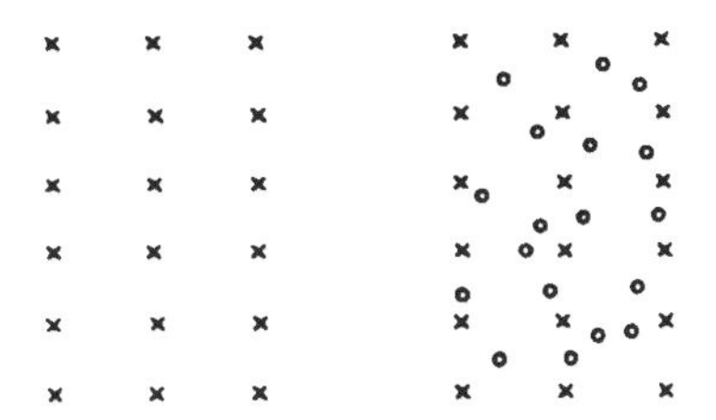

FIGURE 5.11 Additive scheme with crop plants (x) and weed plants (o). (From Spitters and Van den Bergh, 1982.)

TABLE 5.4 Selected Studies Showing the Effect of Increasing Weed Density on Crop Yield

Crop	*Weed*	*Weed Density*	*Percent Yield Reduction from Control*	*Reference*
Sugarbeet	*Kochia scoparia* (kochia)	0.04/ft of row	14	Weatherspoon and Schweizer, 1971
		0.1	26	
		0.2	44	
		0.5	67	
		1.0	79	
Soybean	*Brassica kaber* (wild mustard)	1/ft of row	30	Berglund and Nalewaja, 1971
		2	26	
		4	42	
		8	50	
		16	51	
Soybean	*Xanthium pensylvanicum* (common cocklebur)	1335/A	10	Barrentine, 1974
		2671	28	
		5261	43	
		10,522	52	
Wheat	*Avena fatua* (wild oat)	70/yd^2	22.1	Black, 1958
		160	39.1	
Wheat	*Setaria viridis* (green foxtail)	721/m^2	20	Alex, 1967
		1575	35	
Cotton	*Sida spinosa* (prickly sida)	2/ft of row	27	Ivy and Baker, 1972
		4	40	
		12	41	
Rice	*Echinochloa crus-galli* (barnyardgrass)	1/ft^2	57	Smith, 1968
		5	80	
		25	95	

TABLE 5.4 Continued

Crop	*Weed*	*Weed Density*	*Percent Yield Reduction from Control*	*Reference*
Corn	*Setaria faberii*	0.5/ft of row	4	Knake and Slife,
	(giant foxtail)	1	7	1962
		3	9	
		6	12	
		12	16	
		54	24	

Source: Zimdahl, 1980.

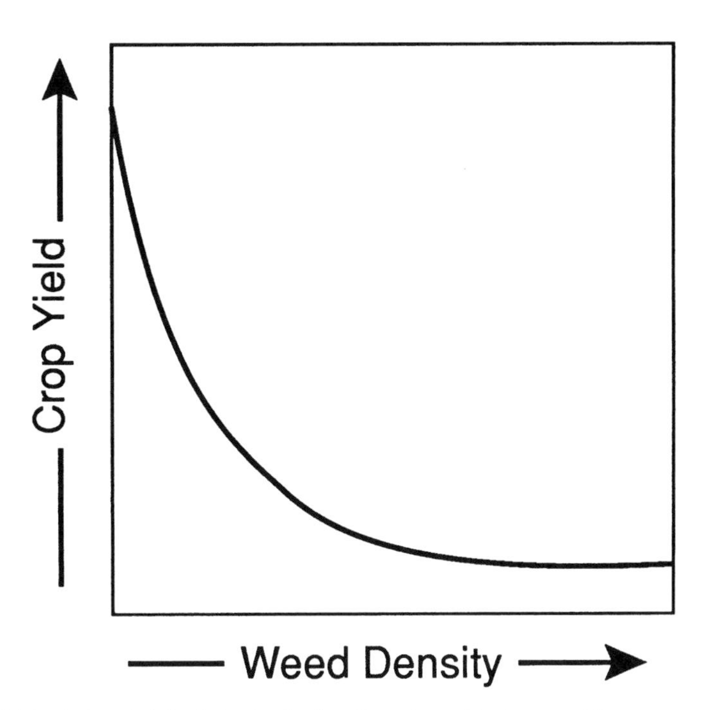

FIGURE 5.12 A schematic relationship depicting the effect of increasing weed density on crop yield.

```
x x x     x o x     o o o
x x x     o x o     o o o
x x x     x o x     o o o
x x x     o x o     o o o
x x x     x o x     o o o
x x x     o x o     o o o
```

FIGURE 5.13 Replacement scheme with crop plants (x) and weed plants (o). (From Spitters and Van den Bergh, 1982.)

REPLACEMENT SERIES

The premise of all substitutive experiments is that the yields of mixed stands can be determined by comparison to monoculture yields (Figure 5.13). Thus, a replacement series experiment includes pure stands as well as mixtures in which the proportion of the two species studied is varied. The total plant density is a constant over all treatments in such experiments (deWit, 1960). Figure 5.14 presents the four possible outcomes for the interaction of two species when grown in a replacement series experiment. In Figure 5.14, the vertical axis indi-

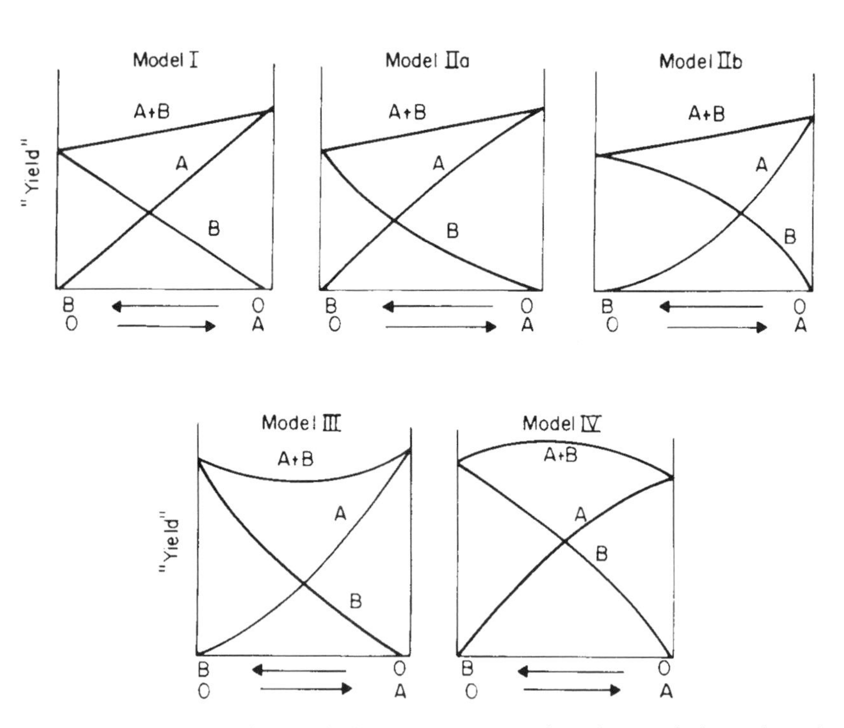

FIGURE 5.14 A variety of models for results of replacement series experiments for interference study. The vertical axis indicates some measure of plant yield and the horizontal axis represents the proportion (0 to 1.0) of the two species in the mixture. See text for explanation of the models. (Modified from Harper, 1977. Copyright 1977 by John L. Harper.)

cates some level of plant yield and the horizontal axis represents the proportion (0 to 1) of the two species in the mixture.

There are two possible interpretations for the yield versus proportion response depicted in Model I (Figure 5.14). One interpretation is that the two species are located so far apart that no interaction can occur between them. In order to detect competition, experiments using this approach must be conducted at sufficiently high densities and/or for long enough time periods to fall within the range of constant final yield (Figure 5.3). The second interpretation for Model I (Figure 5.14) is that the ability of each species to interfere with the other is equivalent, that is, each species contributes to the total yield in direct proportion to its presence in the mixtures.

In some situations, two species make similar demands upon the environment but differ in their response. In Model II a and b (Figure 5.14), one species is more aggressive than the other and contributes more than expected to the total yield, while the other contributes less than expected. This is the model for competition. In each combination, one curve is always concave while the other is always convex, indicating that the interaction between species is for a common resource(s) and that one species gains more than the other.

In Model III (Figure 5.14), neither species contributes its expected share to the total yield. The yield of the two species in any mixture is less than that achieved when either is grown in a pure stand at the same total density. This model represents mutual antagonism such that maximum productivity results from the monocultures. Mutual benefit is depicted in Model IV (Figure 5.14) since both species in the mixtures produce more than would be expected from their yields produced in pure stands. This model depicts symbiosis, but it also may indicate that each species fails to harm the other as much as expected. In such situations, each species escapes from some measure of competition with the other. Mutual benefit may occur between certain weeds and crops, but it is most important in multi- or polycropping situations.

In a substitutive experiment, the total plant density is held constant while species proportions are varied; thus, these two experimental variables are not confounded during the experiment. The spatial arrangement among individual plants in the experiment usually is nonrandom (Figure 5.13). The value of replacement series experiments is their predictiveness. There are four models to interpret neutral, negative, or positive effects between species (Figure 5.14). Predictions of shifts in species composition over time also can be made. Consider Figure 5.15 in which species A is more aggressive than species B (Model II). The dotted line indicates the predicted number of generations for one species to replace the other in a mixed stand. For example, when 2% of species A and 98% of species B is the starting proportion, the output is 9% species A. If the input toward the next generation were 9% species A, around 20% A would be the output, and so on. Consequently, six to seven generations would be required to replace species B, given the greater competitiveness of species A. The replacement of one species by another may be seen by following the stair-step arrangement of dotted lines in Figure 5.15. This type of replace-

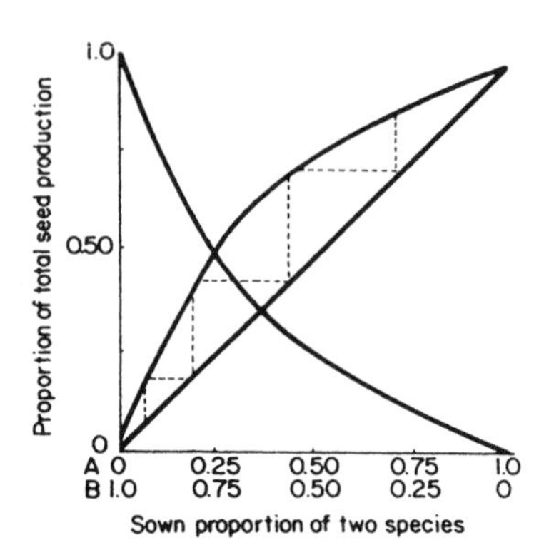

FIGURE 5.15 The use of the replacement series to predict the number of generations it will take for one species to displace another as a result of interference. The predicted number of generations for the population to change is indicated by the dashed line. The solid diagonal line indicates the yield if both species had identical competitive abilities.

ment event is important in determining dominance or species shifts under changing cultural practices.

It is possible to determine the relative effects of intra- and interspecific interference using the replacement series design, but partitioning the absolute effects cannot be accomplished readily. Jolliffe et al. (1984) developed a system to evaluate quantitatively the results of replacement series experiments. The replacement series design is also limited in that actual and expected monoculture yields, and thus the outcome of any particular experiment, will vary according to the plant density selected for study. Jolliffe et al. (1984) suggest including several monoculture densities in the experiment and calculating relative yield responses of the species in mixture to alleviate this problem. The replacement series design has also been criticized for being artificial for field implementation because a constant density rather than a variable one is used to grow most crops. The method is also cumbersome for studying mixtures of species of different life strategies or growth forms. Proportions expressed as ratios of biomass may be more appropriate than ratios based on density when life forms of the species differ markedly.

NELDER DESIGNS

Nelder experiments have been restricted predominantly to the study of competition among individuals of a single species (Nelder, 1962). These designs usually consist of a grid of plants, often planted as an arc or circle (Figure 5.16) The area per plant or the amount of space available to each plant changes in a consistent manner over the different parts of the grid. The influence of another species can be introduced into the Nelder experiment by overseeding the entire experimental area with a second species. In this case, the effect of intraspecific interference under the constant influence of a "background" species would be determined. A qualitative assessment of interspecific competition may be made

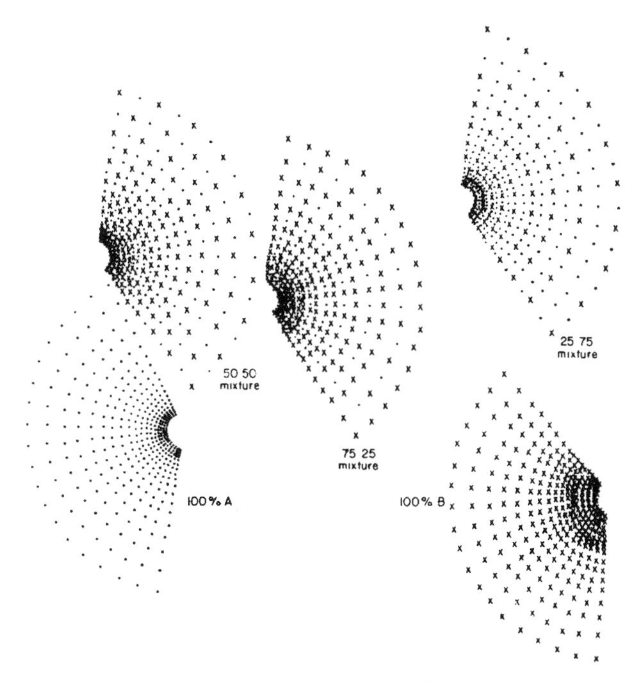

FIGURE 5.16 Examples of the distribution of various species proportions within a Nelder experiment. (From Radosevich, 1987.)

by comparing arcs with and without the presence of the background species. Interspecific effects also may be examined by alternating the placement of the species along an arc or spoke, so that differing ratios or proportions of the species result (Cole and Newton, 1987). Usually every other plant is alternated, giving a 1 : 1 ratio, or species proportion of 0.5.

The advantage of the Nelder experimental design is that an array of densities can be studied without changing the pattern of plant arrangement. In addition, only a small area is required to examine the effect of many densities, which is not the case with most square or rectangular designs. This economy of space allows considerable flexibility in dealing with possible environmental gradients in the field. However, arcs often are difficult to plant in the field, especially when spacing along or between crop rows is dictated by equipment or for other reasons. This disadvantage may not be serious, since several parallel-row arrangements for Nelder experiments have been proposed (Bleasdale, 1967). Another disadvantage is that only individual plants can be measured, causing difficulty in obtaining the "stand" effect from association with neighbors that is possible with rectangular designs. The method also does not allow for the partitioning of density and proportion effects on the interaction unless more than one species proportion is used (e.g., proportions of 1.0 vs. 0.5). Nor can the effects of intraspecific competition be separated readily from those of interspecific competition.

MECHANICAL DIALLEL EXPERIMENTS

Most plant communities are composed of several to many plant species. The diallel experiment combines individuals of each species under study into all possible pairs to examine their interactions (Harper, 1977; Radosevich, 1987). The experimental design uses only one or two individuals of each species per "treatment." In an experiment involving two species, an individual of each species is grown alone (A and B), two individuals of each species are grown together (AA and BB), and one individual of each species is grown in mixture (AB). This design allows an examination of both intraspecific (compare A to AA or B to BB) and interspecific (compare AB to either AA or BB) interactions within the framework of a substitutive experiment. The yield of the species mixtures and monocultures are somewhat analogous to the performance of genetic hybrids and inbred lines. Combinations of more than two species also may be examined. The design below uses three species.

A	AA	AB
B	BB	BC
C	CC	CA

The advantage of this approach is the simple design, which can be combined with intensive destructive data collections to determine biomass partitioning among the species under a regime of interference. The researcher, however, is restricted to working with individual pairs of plants, probably under greenhouse or growth chamber conditions. Pot or plot size may be critical in such experiments, since resources may be relatively unlimited in a system that uses only one or two individual plants. In addition, the influence of density from more than a single neighbor cannot be determined.

Systematic Experiments

Because of the joint influences of proximity factors in weed-crop competition, another approach has been developed that systematically varies both total and relative plant densities (proportion) (Spitters, 1983a,b; Firbank and Watkinson, 1985; Cousens, 1991). This approach provides a better basis for quantifying competition than either conventional additive or substitutive experiments (Roush et al., 1988a) because it provides a broader array of relative densities to examine (Cousens, 1991). Cousens indicates that two designs have been used to describe density response surfaces systematically: *addition series*—a combination of several replacement series over a range of total densities (Spitters, 1983a,b), and *additive series* or *factorial design*—a combination of additive experiments at different total densities (Rejmanek et al., 1989). The addition series encompasses a triangular portion of a matrix of density combinations (Figure 5.17a), while the additive series includes all possible combinations of several densities of each species (Figure 5.17b). Since both approaches appear to

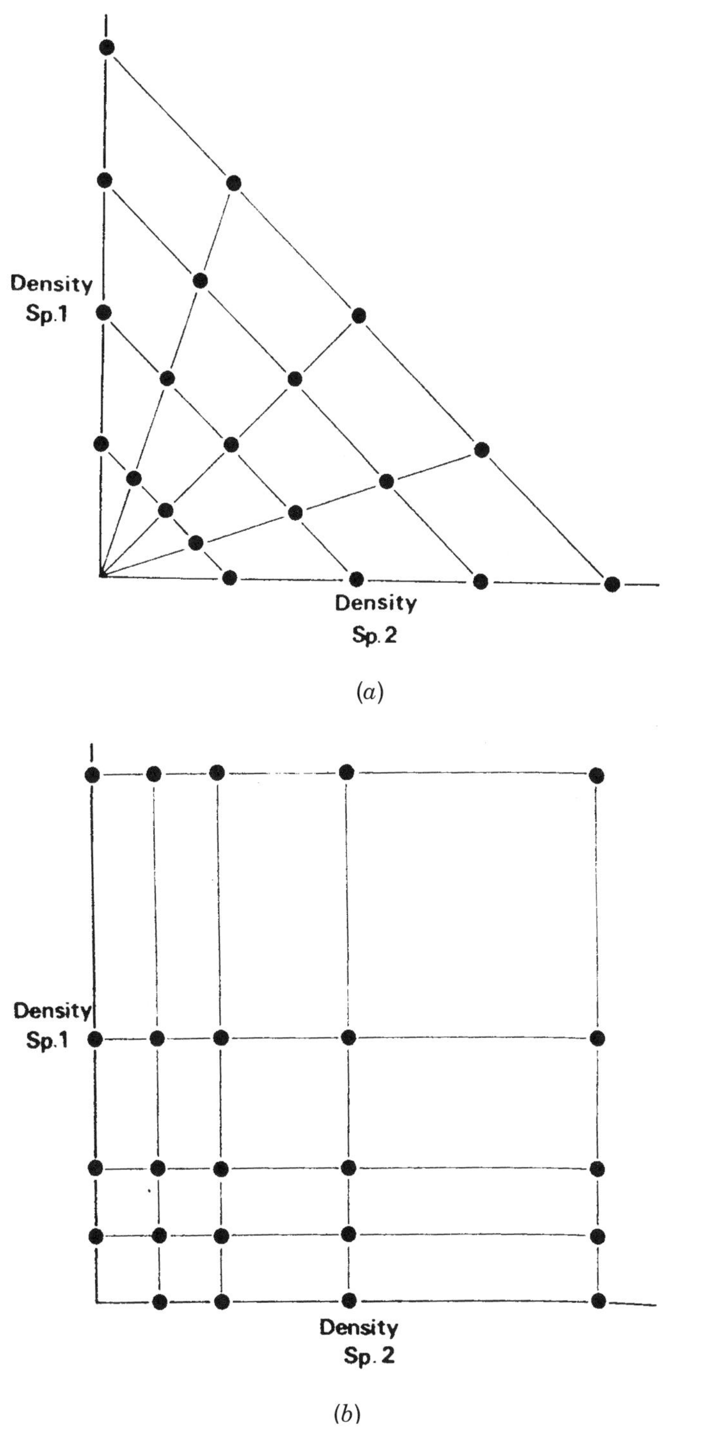

FIGURE 5.17 (a) Density combinations in an addition series design. (b) Density combinations in a factorial design. (From Cousens, 1991.)

explore a range of total and relative densities systematically, they are considered together in this section.

ADDITION SERIES AND ADDITIVE SERIES

Spitters (1983a,b) used the reciprocal yield law (Figure 5.3 and equation 5.3) as the basis for describing plant interactions. As stated earlier, multispecies reciprocal yield (equation 5.4) defines the relationship between individual plant weight and plant density for more than one species. This equation can be used to determine the yield of a species as a function of the total and relative densities of all other species in a plant mixture. By manipulation of the equation, it is possible to describe yield of a single species (e.g., i, j, or k) or to describe yield of each species as affected by interference from any other species in the plant mixture. Thus, the addition series and additive series designs, above all, represent an approach to examine competitive relationships using an array of species densities and proportions.

The experimental design for an addition series may take the form of a simple matrix of two species as in the following example:

0	1A	4A	8A	16A
1B	1/1	4/1	8/1	16/1
4B	1/4	4/4	8/4	16/4
8B	1/8	4/8	8/8	16/8
16B	1/16	4/16	8/16	16/16

where species density per plot ranges from 0 to 16 plants in monoculture and 2 to 32 plants in mixture, A and B are species A and B, respectively, and 1/1 . . . etc. are mixtures with one A and one B per plot . . . , etc. When two species are considered, the addition series is simply a group of replacement series experiments. However, more complex arrangements are necessary with more than two species. Figure 5.18 depicts the possible arrangement for an experiment involving three species. The densities of two species (x and •) increase along perpendicular gradients from zero to a high density; then a similar range of densities of another species (□) is superimposed upon species x and • (Figure 5.18). In this manner, a range of monoculture densities, total densities, and proportions can be varied systematically throughout the experiment. Miller and Werner (1987) and Roush and Radosevich (1987) have expanded this method for associations of four plant species. The regression coefficients derived from equation 5.4 or Table 5.5 are then used to determine and separate the effects of intra- and interspecific competition.

Neither the addition series nor the additive series method accounts for spatial arrangement, so arrangement of the plants either must be constant or must be assumed to have a constant effect. Hashem and Radosevich (1987, 1996) recently included various levels of crop rectangularity as another factor in their design and

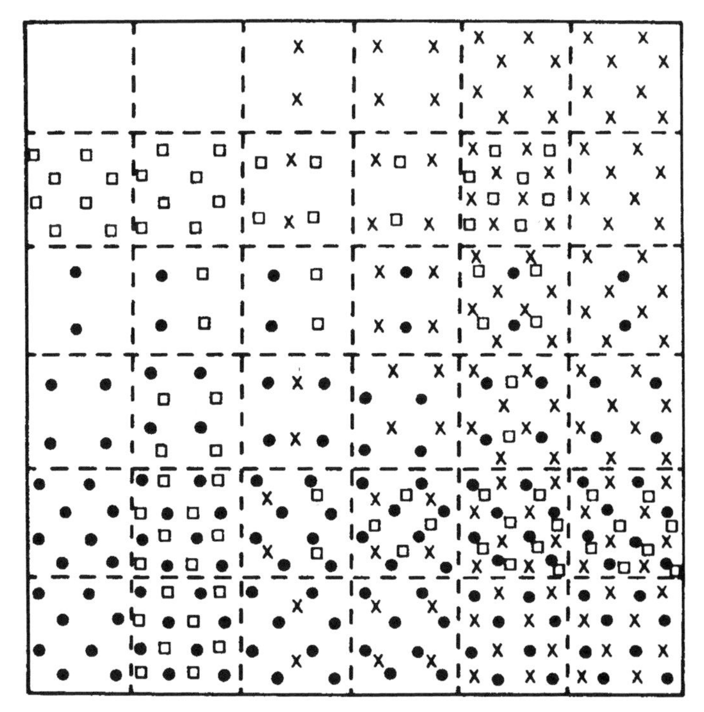

FIGURE 5.18 An example of an addition series experiment using three plant species. Densities range from 0 to 24 plants (symbols/unit of area) and proportions among species vary systematically throughout the design. In this figure, densities of x and • range from 0 to 8 plants in perpendicular directions. Densities of □ also range from 0 to 8 plants and are superimposed on x and • in a systematic manner. (From Radosevich, 1987.)

analysis of competition between wheat and *Lolium multiflorum* in order to account for the nonregular spatial arrangement usually found in the field.

Neighborhood Experiments

Most competition studies concentrate on stand yields. However, the yield of individual plants may be influenced by nearness to other individuals and by local variation in environment (microsites). When individual responses to the proximity of other plants is of primary interest, a neighborhood method may be appropriate to assess interference. In neighborhood experiments, performance of a target individual is recorded as a function of the number, biomass, cover, aggregation, or distance of its neighbors. Many generalized equations have been developed to represent the relationship of target individual (species) performance to the proximity of neighboring plants (Goldberg and Werner, 1983; Pacala and Silander, 1990; Tremmel and Bazzaz, 1993).

TABLE 5.5 Statistical Components of Competition and Their Interpretation

Statistic	*Component of Competition*	*Parameter Estimate*
Intercept	Maximum potential plant size	B_{jo}, B_{io}
Regression coefficients	Intensity of intra- and interspecific competitive effects on plant size	B_{jj}, B_{ji}, B_{ii}, B_{ij}
Interaction term	Interaction between intra- and interspecific density effects on plant size	B_{pi}, B_{pj}
Ratio of coefficients	Competitive effects of species j density relative to species i density	B_{ii}/B_{ji}, B_{ij}/B_{jj}
Model R^2	Overall importance of competition in determining plant size relative to other factors	R^2_{ii}, R^2_{ij} R^2_{jj}, R^2_{ji}
Partial R^2	Importance of competitive effect of each species density	ρR^2_{ii}, ρR^2_{ij}, ρR^2_{jj}, ρR^2_{ji}

Source: Modified from Shainsky and Radosevich, 1989b.

Goldberg and Werner describe an experimental approach in which the performance of a single target species is evaluated over a range of densities of a neighboring species. The target species either is grown alone or is surrounded by individuals of the neighboring species (Figure 5.19). The spatial arrangement of individual plants also can vary among target and neighboring species. The effect of the neighboring species on the target species is defined as the slope of the regression of performance (e.g., growth rate, survival, or reproductive output) of target individuals on the amount (e.g., density, biomass, or leaf area) of the neighboring species. The mathematical expression of this relationship is:

$$P(T) = Y + X_{TN} [A(N)] \qquad \textit{(eq. 5.5)}$$

where P(T) is the performance of the target individual and A(N) is the "amount" of neighbors. A refinement of the neighborhood method is to measure distance of neighbors from the target individuals, so the diminishing effects of more distant neighbors can be incorporated into the regression equations. In that case, the term for amount of neighbors [A(N)] can be replaced by $\Sigma_1(A_i/d_i^2)$, where A is the biomass or cover of an individual plant of the neighboring species i at a distance (d_i) from the target plant. Several possible interactions can be explored with neighborhood experiments (Figure 5.20). Because observations are based on single target individuals, however, many treatments (densities) must be examined to quantify effects of neighbors accurately using the regression approach.

A. INITIAL FIELD CONDITIONS

B. AFTER TREATMENT

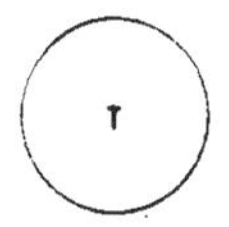

FIGURE 5.19 Example of the experimental design for evaluating competitive effects of one neighbor species (N) on a target (T). R, Q, and S represent individuals not belonging to the neighbor species selected for study. Only four steps of the neighbor density gradient (after treatment) are shown. The experiment must include a much wider range of densities to estimate accurately the slope of (X_{TN}) of the regression equation $P(T) = Y + X_{TN}[A(N)]$ as described in the text. (From Goldberg and Werner, 1983. With permission.)

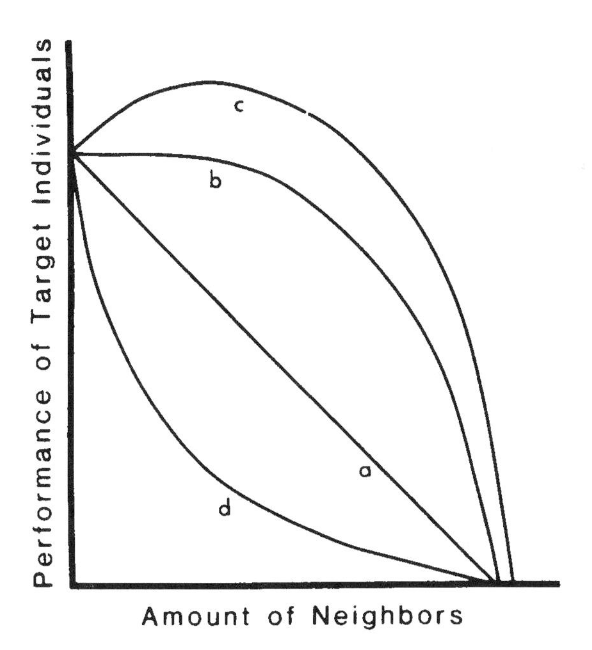

FIGURE 5.20 Examples of some possible relationships between performance of individuals of a target species and abundance of neighbor species. Curve a corresponds to the equation given in the text (linear relationship). Curve b represents a relationship in which competitive effects are minimal at low neighbor density but increasingly severe as density increases. Curve c represents a quadratic function with a peak at some intermediate neighbor density, indicating beneficial effects at low density but competitive effects at high density. Curve d represents a negative exponential function, indicating decreasing per amount competitive effects as density increases. (From Goldberg and Werner, 983. With permission.)

RESPONSES OF CROPS AND WEEDS TO COMPETITION

Competition among plants involves interactions among both biological and environmental factors. Grace and Tilman (1990) indicate that two important distinctions exist when considering the influence of competition on the outcome of plant-plant interactions. These are (1) the *intensity* of competition and (2) its *importance* (Weldon and Slausen, 1986). Intensity integrates physiological and morphological responses of individual plants when in the presence of neighbors of the same or different taxa. Importance, in contrast, describes the role of competition in relation to other processes that may influence the future abundance, density, or species composition of a plant community (Table 5.6). This distinction between intensity and importance of competition is only now being recognized by agricultural and natural resource scientists. For example, most competition experiments and models in agriculture and forestry only consider the degree of crop yield loss due to competition (intensity), without concern for its role in future weed composition or abundance (Radosevich and Roush, 1990).

Intensity of Competition

Historically, competition experiments performed in agriculture and forestry have documented levels of crop yield loss rather than the population or community implications of those interactions among crops and weeds. Empirical studies usually have been either additive or substitutive experiments. Zimdahl (1980), Cousens (1985), and Hakansson (1988) have summarized numerous experiments of these types (e.g., Table 5.4), which were conducted over an array of cropping systems and environments. Stewart et al. (1984) have provided a similar summary of experiments in young forest plantations. Crop yield response to weed density or weed cover (Auld and Tisdell, 1988; Cousens, 1985; Alstrom, 1990) is generally described by a rectangular hyperbola or similar (e.g., exponential)

TABLE 5.6 Organizational Relationships of the Intensity of Competition versus the Importance of Competition

	Intensity	*Importance*
Research focus	mechanisms	implications
Levels of organization	environment, plant population	plant populations, community
Management implications	establish yield responses and economic thresholds	forecast densities and species shifts, implement thresholds, integrate management

Source: Modified from Roush et. al., 1989b.

function (Figure 5.21). A clear law of diminishing returns exists for this relationship between crop yield and weed density (Figures 5.12 and 5.21). As weed density increases, crop yield diminishes markedly until the density of weeds is reached that does not decrease crop production further.

Weed Scientists often face a dilemma in accounting for the influence of proximity factors, especially total and relative plant densities, on the outcome of their experiments. Crops are usually grown at a constant density, determined experimentally or intuitively to maximize economic yield, while weeds create conditions in which both total and relative plant densities vary. Many of these problems are overcome by more complex experimental designs, such as replacement series, addition series, and additive series studies. The main advantage of these approaches is their ability to examine many possible interactions among the interfering species. A particularly valuable calculation that can be derived from such experiments is the *relative yield total* (RYT) because it describes how the species use resources (that is, space) in relation to each other.

$$RYT = \frac{\text{yield of species A in mixture}}{\text{yield of species A in pure stand}} + \frac{\text{yield of species B in mixture}}{\text{yield of species B in pure stand}} \qquad (eq. 5.6)$$

RYT values near 1 indicate that the same resource is being used, while values less than 1 indicate mutual antagonism (overall loss of space or resources), and values greater than 1 suggest avoidance or symbiosis (overall gain in space, see Chapter 7). Another similar calculation is *aggressivity* (A):

$$A = \frac{\text{yield of species A in mixture}}{\text{yield of species A in pure stand}} - \frac{\text{yield of species B in mixture}}{\text{yield of species B in pure stand}} \qquad (eq. 5.7)$$

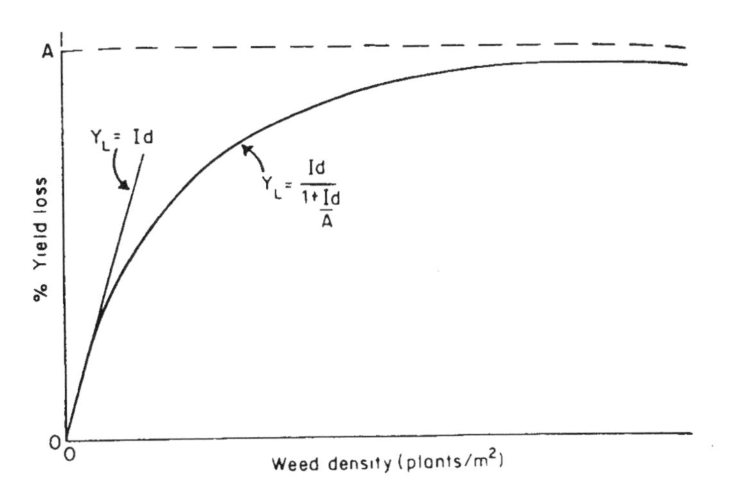

FIGURE 5.21 The rectangular hyperbolic model for relating yield loss to weed density. Y_L is percentage yield loss; A and I are the parameters that determine the shape of the curve for response of yield loss to weed density. (From Cousens, 1985.)

This calculation defines the relative success of the two species in using resources and provides a means to evaluate interference among an array of species. For example, Roush and Radosevich (1985) examined the competitiveness of four annual weed species by combining them as pairs in a replacement series experiment (Figure 5.22). In each combination, one curve is always concave while the other is always convex, indicating that the species were competing for a common resource. By calculating both RYT and A values, the following hierarchy of competitiveness was established among the four weed species: *Echinochloa crus-galli* > *Amaranthus retroflexus* > *Chenopodium album* > *Solanum nodiflorum* (Roush and Radosevich, 1985). Similarly, the replacement approach has been used successfully to assess both perennial weeds (Holt and Orcutt, 1991) and multicropping systems (Trenbath, 1976). In Figure 5.23, for example, each component of the crop mixtures is affected less by interspecific competition than by intraspecific competition, allowing for overyielding (RYT > 1) when the crops are grown in combination.

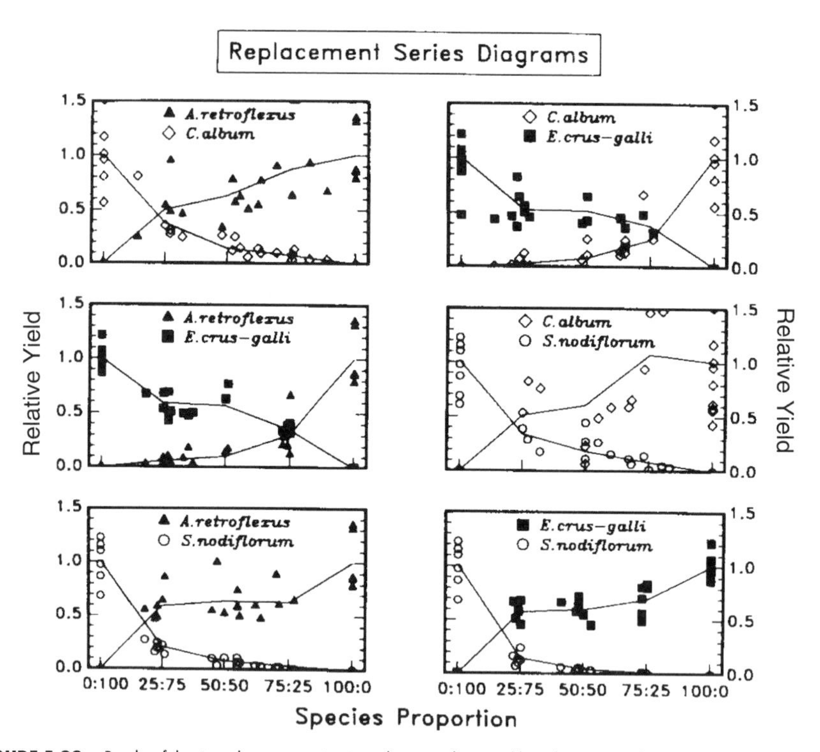

FIGURE 5.22 Results of the six replacement series: in each series relative yields at the various relative proportions are presented for both species. Relative yields represent dry weight yields of each species relative to the mean dry weight for the monoculture treatment (100%) of that species. (▲), *A. retroflexus*; (◇), *C. album*; (■), *E. crus-galli*; (○), *S. nodiflorum*. (From Roush and Radosevich, 1985. With permission.)

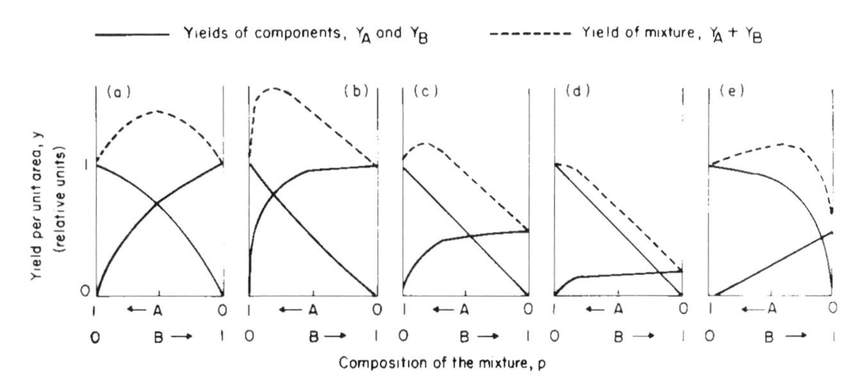

FIGURE 5.23 Effect of differences in the relative aggressiveness and sole crop yields of the components on intercrop yield. All yield curves were generated using the deWit (1960) model. In all the 1 : 1 mixtures the LER = 1.3, but overyielding by the intercrop is found at this proportion only in (a) to (c). Yield of components, Y_A and Y_B are indicated by solid lines; dashed lines represent the yield of mixture $Y_A + Y_B$. (Reproduced from Trenbath, 1976, p. 126. With permission.)

Advances in experimental designs for quantifying the intensity of competition have come primarily through the use of addition series experiments. Statistical analysis of such studies, using a multispecies form of the reciprocal yield model (equation 5.4), can describe readily the effects of intra- and interspecific competition as regression coefficients (Roush et al., 1989a). Table 5.5 summarizes these statistical components of competition. Particularly useful in assessing relative competitive ability is the ratio of competition coefficients, for example, B_{ii}/B_{ji} in equation 5.4. Generally, intraspecific competition among crop plants is more severe than the interspecific effects of weeds on crop yields. For example, Westra and associates (1991) conducted two-year competition experiments with corn and wild proso millet, using an addition series design. Analysis and coefficients indicated that the influence on corn grain yield of one corn plant was equivalent to that of 14 wild proso millet plants. Although wild proso millet is considered to be a serious weed in corn production, Westra's data suggests that corn yields are much more sensitive to corn density than to the presence of wild proso millet. Similar responses have now been found for associations of many crops and weeds, such as the wheat-ryegrass mixture summarized in Table 5.7.

Results from weed-crop experiments suggest that, at the global scale, weeds may be relatively equal in competitiveness and in the intensity of competition experienced by crops. However, the existence of competitive hierarchies at the local scale seems to be tied closely to environment and other aspects of biology such as physiology, morphology, and carbon allocation (Chapter 6). Therefore, predictions of weed-crop competition will continue to require an understanding of biological and physiological mechanisms of competition over a range of environments.

TABLE 5.7 Multispecies Reciprocal-Yield Models for Interactions Between Spring Wheat and Italian Ryegrass (*Lolium multiflorum*)[a]

Species	$1/W = B_{i0} + B_{ii} N_i + B_{ij} N_j$	R^2	B_{ii}/B_{ij}
Wheat	$1/W = 10.72 + 1.18 Nw + 0.17 N_r$	0.90	6.70
Ryegrass	$1/W = 41.64 + 3.21 N_r + 4.51 N_w$	0.43	0.75

Source: From Concannon (1987), in Radosevich and Roush, 1990.

[a]B_{i0} is the reciprocal of the theoretical maximum size of an individual. B_{ii} describes influences of intraspecific competition. B_{ij} describes influences of interspecific competition, and B_{ii}/B_{ij} predicts relative competitive ability of each species. $P < 0.01$ for B_{i0}, B_{ii}, and B_{ij} in each model.

Importance of Competition

The importance of competition in a crop-weed plant community cannot be fully understood from investigations of the intensity of competition. Competition is important in a plant community when it contributes to its organization and dynamics over time—for example, species composition, species shifts, relative fitness among populations or changes in population densities (Roush et al., 1989a). Weldon and Slausen (1986) proposed that the coefficient of determination (R^2) from regression equations relating plant responses to competition is a suitable measure of the importance of competition. They describe equations, derived from neighborhood experiments, in which the slope of the regression quantifies the intensity of competition on plant yield. The R^2 value for those equations suggests how important competition was relative to all other processes that influence plant yield, such as disease, genetics, microclimate, and herbivory (Table 5.5).

Roush (1988) applied an empirical approach to quantifying the role of competition in a community of four annual weed species. In that study, an addition series experiment was performed and the resulting plant community was measured over a two-year period. Regression models were constructed to relate changes in weed population density during the second year to competition among the weed species during the first year (Table 5.8). Results indicated that competition did have a significant influence on the population growth rates of species, especially during the first year; however, the majority of the later variation in population growth was not explained by competition (Roush, 1988). She concluded that much of the variation in population dynamics for the species studied must be attributed to seed bank and seedling emergence processes. It appears that a more thorough understanding of non-competitive as well as competitive interactions will be necessary in order to predict the dynamics of crop-weed associations.

TABLE 5.8 Coefficients of Determination for 1985 and 1986 Competition Models[a]

	R^2	
Species	***1985***	***1986***
Amaranthus retroflexus	.54	.39
Chenopodium album	.76	.22
Echinochloa crus-galli	.55	.19
Lolium multiflorum	.41	*

Source: Modified from Roush, 1988.

[a]The models were two forms of multispecies yield-density model (Spitters, 1983a).

$$1985: 1/W_i = B_{i0} + B_{ii}N_i + B_{ij}N_j \ldots + B_{in}N_n$$

$$1986: \ln W_i = B_{i0} + B_{ii}N_i + B_{ij}N_j \ldots + B_{in}N_n$$

**L. multiflorum* was excluded from the 1986 models because it was not a significant component of the1986 community.

Competition in Mixed Cropping Systems

One of the most inexpensive and widely practiced means around the world of increasing crop production is through the simultaneous culture of two or more crops on the same piece of land, which is called *intercropping* (Liebman, 1988). In Figure 5.23, Trenbath (1976) uses a replacement diagram to demonstrate how each component of a crop mixture can be affected by the other, allowing for an overyield when two crops are grown in combination. In each diagram (Figure 5.23) each crop is affected less by interspecific than by intraspecific competition, resulting in the overyield. In each case, the RYTs (equation 5.6) are greater than 1. Liebman states that such values of intercrop yields can, in fact, be quite high—for example, 1.38 for a mixture of maize with bean, 1.67 for maize with pigeonpea, 1.53 for millet with sorghum, 1.85 for barley with fava bean, 2.08 for maize with cocoyam and sweet potato, and 3.21 for maize with bean and cassava. These values represent very substantial increases in production over what would occur if the crops were grown separately as monocultures. While intercropping techniques are most often used on small farms that employ a minimum of mechanization and other technological inputs, Liebman states that they by no means are restricted to such situations. In fact, interest in the use of intercrops under the fully mechanized cultural practices of temperate-area agriculture is increasing (Horwith, 1985).

Weed suppression is often cited as one of the benefits of intercropping (Moody and Shetty, 1981; Liebman, 1986). Vandermeer (1989) states that the presumed mechanism of this phenomenon is that, through competition with the weed, one crop in the mixture provides an environment of reduced weed biomass for the other crop. Perhaps the best known example of this type of weed suppression is the use of cover crops, which are solid-grown crops grown primarily to protect and cover soil between crop rows or between periods of regular crop production (Aldrich, 1984). Liebman (1986,1988) reviewed studies of twenty-three crop and cover crop combinations and found that twenty of them provided significant weed suppression.

While these findings with cover crops are impressive, Vandermeer (1989) states that weed suppression by combinations of two crops is more equivocal. For example, Liebman (1986), through an extensive literature review, found that the suppressive effect of weeds was stronger in intercrops than in the monocultural components in eight cases, intermediate between monocultural components in another eight cases, and weaker than all monocultural components in two cases (Table 5.9).

TABLE 5.9 Strength of Weed Suppression Effects by Intercrops in which All Component Crops Are Considered "Main Crops"[a]

Intercrop Combination	*Weed-Suppression Effects*		
	Stronger than Monocultures of All Components	*Intermediate between Monocultures of Components*	*Weaker than Monocultures of All Components*
Maize–bean	Fleck et al., 1984	Soria et al., 1975	
Maize–cassava	Soria et al., 1975		Soria et al., 1975
Maize–bean–cassava	Soria et al., 1975	Soria et al., 1975	
Maize–mung bean	Bantilan et al., 1974		
Maize–sweet potato		Bantilan et al., 1974	
Maize–peanut		Bantilan et al., 1974	
Maize–sunflower	Fleck et al., 1984		
Maize–cowpea			Ayeni et al., 1984
Bean–cassava	Soria et al., 1975		
Bean–sunflower	Fleck et al., 1984		
Flax–wheat		Arny et al., 1929	
Flax–oats		Arny et al., 1929	
Sorghum–pigeonpea	Shetty and Rao, 1981	Shetty and Rao, 1981	
Pearl millet–peanut		Shetty and Rao, 1981	

Source: Liebman, 1986.

[a]References cited are found in original source.

The addition of weeds to an intercropping situation creates an interesting ecological system of three or more interconnected competitors. It seems likely that the new approaches for competition study, such as the addition series and additive series, could help unravel the complexity of interactions which no doubt occur in those systems. Vandermeer (1989) believes that such interactions implicitly involve a positive modification of the environment of one species by another, especially in the case of cover crops. As such, they represent a working example of the facilitation production principle.

THRESHOLDS OF COMPETITION

Weed management is recognized as an essential component of almost every crop production system because crop yields are affected so markedly by weed presence. In addition, factors other than yield, such as crop quality, ease of harvest, and populations of other pests or beneficial organisms, are also affected by weeds. Often the impact of weeds on one of these other factors is so significant that weed control is accomplished solely for that purpose. For example, crop quality standards in some vegetable or seed crops may be sufficiently high that very few, if any, weeds may be tolerated in those crops. Nonetheless, cost-effective weed management requires that an assessment of possible or real damage from weeds to crops be made prior to employment of weed control tactics. Thus, the concept of thresholds for weeds is central to weed management.

Coble and Mortensen (1991) indicate that the concept of thresholds has many applications in weed science, the most common being damage, period, economic, and action. *Damage* threshold describes the weed population at which negative crop impact is detected, while *period* threshold implies that there are times in a crop life cycle when weeds are more or less damaging than at other times. Such thresholds usually are expressed in biological terms, such as plant density or weed biomass per unit of area. Glass (1975) and Coble and Mortensen (1991) define *economic* threshold as the pest (weed) population density, or damage level, at which control measures should be taken to prevent economic injury to the crop from being incurred. This definition of economic threshold also implies that the cost of control should be less than the loss that would have occurred had nothing been done. The establishment of an *action* threshold (weed population level at which some action is needed to preclude crop yield loss) necessarily includes predictions of direct effects on crop yield or other forms of economic loss due to the weed's association with the crop. As we have already seen, most of the literature on crop-weed interactions attempts to quantify this negative effect of weeds on crop yields. Economic and action thresholds attempt to answer the question, "How much will a given amount of weeds reduce both crop yields and profitability?"

Density/Biomass Thresholds

The extent to which crop yields are reduced by weeds depends on many factors, such as crop species and cultivar, weed species present, location or site, and practices used that modify site conditions. Weather differences from year to year also cause annual variations in crop yield losses, affect weed competitive ability (Figure 5.24), and confound data interpretation. For these reasons, it seems nearly impossible to determine empirically the yield reductions for even the major crops in association with particular weeds in a region. In addition, experiments of weed-crop competition are rarely conducted at the entire-field scale, and results from small-plot experiments may not reflect accurately actual field-level crop responses to weeds. For example, Auld and Tisdell (1988) and Mortensen et al. (1993) demonstrate that weed populations may vary by up to 200-fold across an agricultural field. These observations indicate that weed densities can be extremely variable and that predictions of crop response based on small-plot bioassays may greatly overestimate the value of weed control at the field and farm levels.

Nevertheless, many investigations of weed-crop interactions have been conducted over the years and most demonstrate that weed plants are harmful to agricultural crops, even at low densities (Figure 5.12). The shape of the curve in Figure 5.12 implies that the negative impact of each weed plant on crop yield increases as the weed population declines, that is, even very low weed densities cause substantial losses in most crops, with greatest yields occurring where weeds are absent. The question most relevant to farmers that is raised by these data is whether it is economically reasonable to control weeds to such very low densities. Alstrom (1990) suggests that it is not, unless the additional cost of weeding is equal to the value of the weeds' marginal effect on crop yields. In other words, the additional revenue gained from crop yields achieved by weed control must equal the cost of attaining it. Obviously, the economically opti-

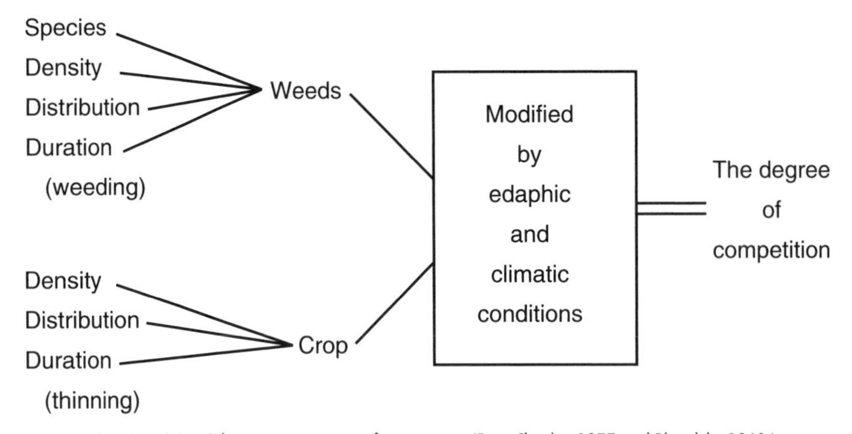

FIGURE 5.24 Schematic presentation of competition. (From Chisaka, 1977, and Bleasdale, 1960.)

mum amount of weeds must be more than zero in this case, unless the crop is infinitely valuable or weed control costs are nothing.

Although the general relationship depicted in Figure 5.12 is also true for forestry situations (Stewart et al., 1984), both tree survival and subsequent size are important yield components in young forest plantations. Wagner et al. (1989) examined the patterns of survival and stem-volume growth for planted ponderosa pine (*Pinus ponderosa*) competing with various levels of woody and herbaceous vegetation. They found that negative hyperbolic curves with opposite concavity described the relation between the abundance of "weedy" vegetation and survival and stem volume (Figure 5.25) of the pine seedlings. From

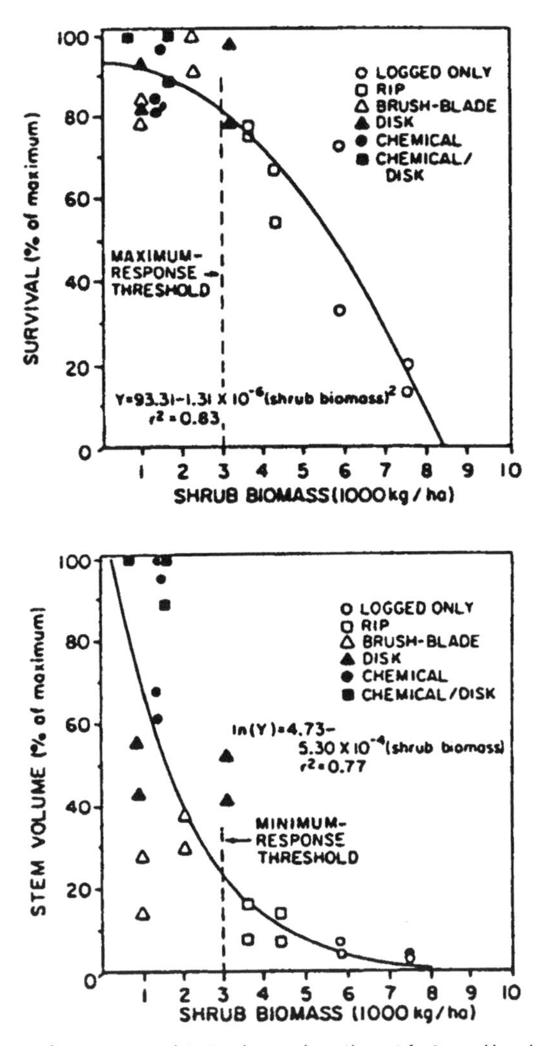

FIGURE 5.25 Percent of maximum survival (*top*) and stem volume (*bottom*) for 8-year-old ponderosa pine seedlings growing with various levels of shrub biomass in south-central Oregon after six site preparation treatments. (From Wagner et al., 1989.)

these curves, two types of competition thresholds were identified: (1) maximum-response threshold, a level of competing vegetation abundance at which additional control would not yield an increase in tree performance and (2) minimum-response threshold, a level of competing vegetation that must be reached before additional control measures will yield an appreciable increase in tree performance (Figure 5.26). The thresholds for pine stem-volume growth occurred at lower competing vegetation abundance than the threshold for tree survival, suggesting that foresters should consider tree survival and tree growth as separate silvicultural objectives when managing competing vegetation.

Critical Period Thresholds

It is the conventional wisdom of many farmers and weed scientists that early weed competition is most detrimental to crop yields and that early weed control is necessary. However, as Zimdahl (1988, 1993) points out, this generalization may not be entirely accurate, because it may have been dictated by the means of weed control rather than by biological necessity. For example, hoeing and cultivating are accomplished more easily when both crop and weeds are small. In addition, the development of preplant and preemergence herbicides also may have perpetuated the belief that early weed control was essential. There is evidence, however, that for certain crops a critical period exists during which weeds should be controlled to prevent yield losses, but that weed control for this reason at other times may not be necessary.

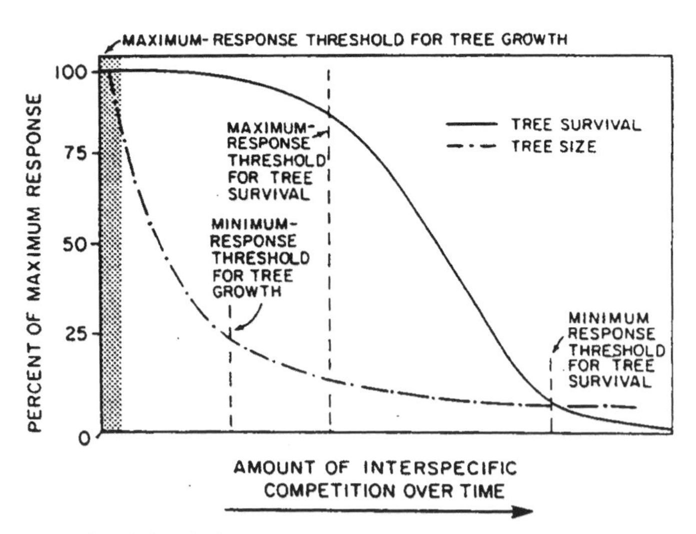

FIGURE 5.26 Hypothetical relationship between interspecific competition and tree survival and growth. The maximum- and minimum-response thresholds for tree survival and growth occur at different levels of interspecific competition. The maximum-response threshold for tree growth occurs in the shaded region under nearly vegetation-free conditions. (From Wagner et al., 1989.)

In *critical period* studies, crops are kept weed free for varying intervals of time following planting or emergence, and after this period weeds are allowed to grow for the rest of the growing season (Figure 5.27). The resulting data are compared to those of a complementary study in which weeds are allowed to grow for varying intervals of time after crop planting or emergence, with the remainder of the growing season being weed-free (Figure 5.27; Zimdahl, 1988). Although some discrepancies in published data exist, such discrepancies

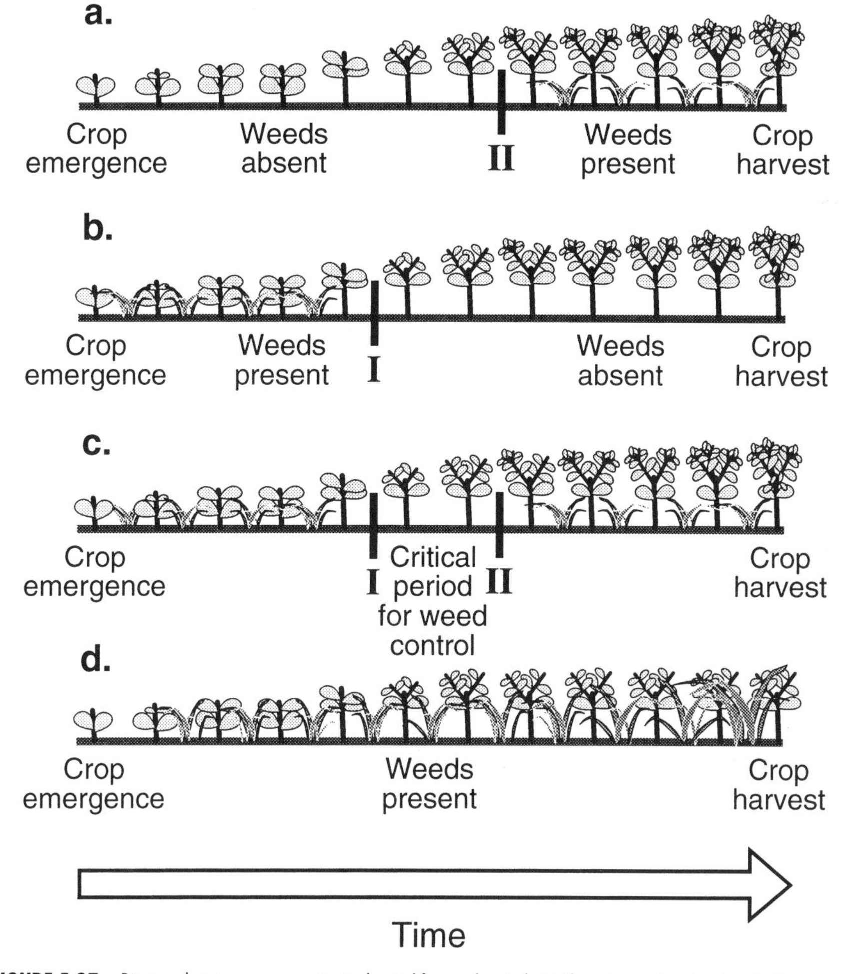

FIGURE 5.27 Diagram depicting an apparent critical period for weed control. (a) If weeds are absent up to point II, crop dominance is established and yield losses do not result, even though weeds may be present subsequently. (b) If weeds are present for a period of time following crop emergence but are absent for the remainder of the season, yield losses do not result since, presumably, early in the season weeds are too small for competition to occur. (c) The combination of results from (a) and (b) leads to the critical period between points I and II, which is a "window" of time during which weeds must be removed or suppressed to avoid crop yield loss at harvest. (d) Situation in which weeds are present throughout the growing season and crop yield loss results.

are understandable considering the wide range of locations and sources of this information. The important point here is that many crops apparently have periods when weeds may be tolerated, as well as periods when weed control is apparently mandatory to avoid crop losses (Figure 5.27). Clearly, one must consider each crop and weed situation separately to determine when weeds may be tolerated or must be controlled.

As seen in Figure 5.27a, the initial weed-free period up to point II results in crop dominance, which diminishes any subsequent competitive effects by weeds on crop yields. If weeds are not controlled, however, a period of increasing interference between crop and weed plants follows after emergence, and crop yields are reduced (Figure 5.27d). Early in their life cycles both weed and crop seedlings are small and far enough apart so that no interaction occurs, but eventually interference between the species develops. For example, canopies of both weed and crop species would be developing continuously after emergence, but they may not overlap until much later. At this point (I in Figure 5.27b) weed control for the remainder of the season prevents crop yield loss. Once the canopy of the weed is superimposed upon that of the crop, however, a loss of crop productivity most likely will result throughout the rest of the season (Figure 5.27d). Extrapolation from Figure 5.27a and b suggests a "critical period" of time (from I to II in Figure 5.27c) during which control measures are necessary in order to avoid continuing interference between the crop and weed. Weed removal any time up to the end of the critical period for control, during which crop dominance is being established, would result in no significant crop yield reduction. Further weed control after the critical period most likely is unnecessary to prevent yield loss.

Zimdahl (1980,1993) points out that the concept of a critical period for weed control has been challenged for a number of crops. In these cases, either the crops were susceptible to weed competition for most of the growing season (Hewson and Roberts, 1973b) or a single weeding at an intermediate growth stage was sufficient to avoid yield reduction (Hewson and Roberts, 1973a; Roberts et al., 1976, 1977). It also is apparent that critical periods, if they exist, can be determined only for individual weed/crop associations.

The complication of environmental changes that can occur from growing season to growing season also must not be overlooked (Figure 5.24), because environmental conditions and resources may affect the germination and growth rate of the weed and crop differentially. In other words, although the environment would be the same for both weed and crop, the relative ability of the species to respond to it may not be identical (Chapter 6). Thus, differences in either environmental conditions or the availability of resources from one year to the next could affect the length of the critical period when weeds would need to be controlled.

It seems appropriate to consider time in relation to weed control as two phases, as Dawson (1970) has done. One phase occurs after emergence when competition does not exist yet, and the other occurs later in the season when competition is more likely. In the first phase, there is little value in weed removal from the standpoint of crop yield reduction, since the plants are not big enough to interfere with the crop. However, in many situations weed control during this early time period may be most feasible for reasons of convenience and economics, to establish crop dominance early, and to reduce potential weed impacts. During the later phase, crop yields are reduced if the weeds are allowed to coexist with the crop beyond a certain point. The critical period during which competition occurs and yield reductions result may be a certain number of weeks after emergence, as suggested by much of the data on this subject (Dawson 1970; Zimdahl, 1980, 1988). However, it is more likely that this critical period relates to particular stages of crop and weed development that are relative to one another (Figure 5.27). Weed removal before that particular stage of development of crop or weed occurs, during the critical period for control, is necessary in order to avoid reductions in crop yield.

Models Using Weed Thresholds

The establishment of an economic threshold necessarily includes predictions about the direct effects of weeds on crop yields or other forms of economic loss due to weed association. Norris (1982) notes that the concept of an action threshold is basic to most insect and disease management programs but that the concept is not widely employed for weed control. He points out, rightly, that weed densities much lower than those often found in agricultural fields can result in significant crop losses. Thus, in many cropping systems the apparent economic or action threshold indicates such an overwhelming need for weed control that thresholds or acceptable levels of weeds may have been derived intuitively. On the other hand, the ready availability of relatively low-cost, effective tools, such as preplant and preemergence herbicides, may have precluded the apparent need for much greater insight.

EXAMPLES OF WEED MANAGEMENT MODELS

Weed/crop models based on threshold levels of weeds that reflect yield or other forms of economic loss are proving to be valuable additions to available weed management tools. Norris and Fick (1981) have outlined several cropping systems in which the development of weed/crop models would be useful to predict yield improvement or other economic gains from weed management. In addition, other scientists, such as Wilkerson et al. (1991), Shribbs et al. (1990), Kropff and van Laar (1993) and Knowe et al. (1995), have developed simulation models of weed and crop dynamics in an array of crop and forest plantation systems.

HERB. Wilkerson et al. (1991) have developed an interactive microcomputer program to help evaluate potential crop damage from multispecies weed complexes in soybean and determine the appropriate course of action. In this model seventy-six weed species are rated on a scale from 0 to 10 according to their competitiveness with soybean. Potential crop yield loss is estimated from these rankings, from the number of weeds of each species present in the field, and from expected weed-free yield. The recommendation whether to apply a herbicide and, if so, which one is based on herbicide cost and efficacies under different conditions and expected soybean selling price. HERB is a trade name for this model registered by North Carolina State University.

Bioeconomic crop models in corn and sugarbeet. Schweizer and his associates (Shribbs et al., 1990a,b; Lybecker et al., 1991) have developed bioeconomic weed management models for several crops, including sugarbeet and corn. These "personal computer spreadsheet models" examine the economic feasibility of preplant, postemergence and layby herbicides and late-season hand weeding decisions. The models are based on number of weed seed in the soil, field survey of weed populations, growth stages of weeds and crop, expected crop yield loss from weeds, herbicide weed control, weed control costs, and crop price.

SELOMA (Sistema Experto per la LOtta alle MAlerbe). Stigliani and Resina (1993) have developed a practical expert system for postemergence weed control in herbicide intensive crops such as cereals, sugarbeet, corn, and sorghum. The model uses a step-by-step problem-solving procedure that resembles what a weed management expert would follow. It is based on field surveys of weed density, and crop and weed growth stage and height. SELOMA suggests whether to intervene, indicates chemical and mechanical weed control treatments, and selects the best herbicide for use.

INTERCOM. The model INTERCOM was developed by Kropff and associates (Kropff and van Laar, 1993) to provide a tool to analyze complex interactions between plants that compete for the resources light, water, and nitrogen. Focus is especially on crop-weed interactions and understanding of differences in effects of various weed species on crops and variations of such effects in different environments. The model accounts for effects of temperature, radiation, rainfall, and soil hydrological characteristics on plant growth and competition situations.

Regional vegetation management model for plantation forests in western Oregon and Washington. Individual-tree and stand-level growth models have been developed by Knowe et al. (1995) to project Douglas-fir growth in the Pacific Northwest region. This model complements already existing rotation age models of Douglas-fir and guides site-specific vegetation management operations. Nearly 200 vegetation monitoring sites established in young Douglas-fir plantations have been included in the empirical database of the model. Measurements of size, distribution, and species composition of conifers, hardwoods, and other associated shrub and herbaceous species are also included. The model predicts tree and other vegetation responses to intervention treatments.

FACTORS TO CONSIDER AS WEED MANAGEMENT MODELS ARE DEVELOPED

Several factors should be considered as such models continue to be developed for weed management:

- Most threshold values in other pest control disciplines refer to the number of individuals per unit of area, per plant, or per capture device (density). Since weeds usually germinate in large numbers, difficulty in counting can be encountered. Furthermore and more important, plasticity and mortality of weeds result in continual changes in plant size and density throughout the growing season. Thus, the number of individuals may be less useful than other parameters of plant development, such as leaf area index (Chapter 6), in assessing weed impacts. Perennial weeds present a special problem since many individual ramets actually may comprise a single genetic unit.
- The adjustments in plant size that occur with changes in plant density indicate that, given enough time, weeds can become highly competitive, even at low densities. Interference occurs when root systems or canopies begin to impinge and is dependent on plant size and the spatial distribution of weeds in the field. Thus, the time and location when interference will occur is not necessarily predictable from measurements of plant density.
- The seed rain and seed bank of weeds in agricultural land and some forest and rangeland systems are large and made up of many species. Most weed seeds also exhibit some form of dormancy in the soil. These factors combine to make the seed bank a rather constant feature of agricultural systems and to make the prediction of future weed seedlings possible (see Chapter 4). Therefore, the seed bank and net reproductive output of weeds, as well as the above-ground vegetation, should be considered when assessing weed abundance.
- Most weed communities are a mixture of species. Predictions of interference within one weed species/crop combination may not be directly applicable to predictions of crop response to the entire weed complex. However, there is some evidence that a degree of equivalency exists among weed species potentially present in a cropping system (Pickett and Bazzaz, 1978; Roush and Radosevich, 1985, 1987; Coble and Mortensen, 1991). Information concerning possible equivalency or hierarchies among weed species would assist in development of predictive models.
- Most cropping systems involve some form of crop rotation. Threshold values of weeds determined for one crop in the rotation may impact a subsequent crop. For example, surviving weeds in one crop that had little impact on crop yield or quality may produce abundant seeds that germinate later and seriously affect a different crop in the rotation.
- Threshold values may have to be adjusted when other criteria besides crop loss are used to measure economic effects of weeds. These criteria include crop quality, harvesting ease, impacts on other pests or beneficial organisms, and social implications and costs associated with weed control.

It is clear that in order to develop economic models using weed thresholds, improved methods of weed community measurement, including spatial distribution of weeds and prediction of crop responses, will be necessary. In addition, the impacts of weed reproductive output and seed dormancy in relation to crop rotational patterns must be considered when developing these models. It is also apparent that we know far too little about the economics and social implications of weed control and improved crop production. These factors should be included explicitly in our management models since they are fundamental in making informed weed control decisions.

ECONOMICS OF WEED CONTROL: WHETHER TO CONTROL WEEDS

Most decisions about weed management are based on three elements: (1) weed responsiveness to tools, (2) the opportunity to improve crop productivity, and (3) profitability. However, these factors do not act independently, because each factor influences the relative importance of the other two elements in an iterative, interactive manner (Holling, 1978; Levins, 1986). Radosevich and Shula (1994), using a simulation model based on empirically derived functions from the weed science literature for each factor, mathematically integrated the three elements of a weed management decision. In this way, they examined the relationship between weed control cost and efficiency and crop productivity/profitability. This analysis demonstrates the importance of economic thresholds for weed management and questions the value of high input costs to achieve very high levels of weed control.

Weed Response to Control Tools

The search for cost-effective methods to control weeds has usually focused on physical soil disturbance by tillage, fire, grazing (Chapter 8), or herbicides (Chapters 9 and 10) as a means to prepare land for planting or to suppress vegetation during crop, tree, or forage growth. Experiments on weed control usually describe the degree of disturbance or levels and combinations of herbicide doses or times of application to kill or suppress weeds. These experiments are bioassays in which target or test plant species are subjected to various treatment intensities, such as levels, doses, timings, combinations, or forms of tools to inhibit weed growth. Injury to the target (weed) species, usually measured as reductions in occupancy, cover, height, or biomass, is then compared to untreated plants.

As a point of caution, however, it should be noted that experiments to compare tools usually assume the context of an existing crop-production system that can mask the importance of environmental and biological factors on tool performance. Therefore, adopting new technologies exclusively from such experiments may result in higher than optimum herbicide rates, number of applications, or levels of soil disturbance, because higher intensity of tool use usually compensates for a more thorough understanding of environment and biology. In contrast, lower herbicide rates, levels of disturbance, and so on are usually possible when local information about environmental variation or species responses are incorporated into such experiments.

Both logistic and sigmoidal functions are used to describe plant responses to tool intensity and herbicide dose (Figure 5.28 [top]) (Streibig, 1988; Streibig et al., 1993). Input costs for weed control also are reflected by such curves because costs generally increase proportionally to the intensity of the control measure used (Figure 5.28 [bottom]).

Crop Response to Weeds

As already discussed earlier in this chapter, the growth of crops can be affected seriously by the presence of weeds. Many experiments, which collectively examine a wide range of crops, weed species, locations, environments, and control tactics, demonstrate that the amount of crop growth is generally related to the degree of vegetation suppression. Maximum crop production always results when crops grow without weeds and lower yields occur if weeds are abundant. Cousens et al. (1984) and Alstrom (1990) indicate that results of these crop-weed experiments generally conform to the same mathematical function, shown in Figures 5.12 and 5.29. The actual equation that describes this relationship depends on the function fitted, but often a negative exponential, rectangular hyperbolic, or other monotonic function is used (R. Cousens, personal communication).

Value of Weed Control

The economics of weed control depend not only on the gain in crop yield from the absence of weeds (Figures 5.12 and 5.29) but also on the monetary value of the extra yield, and the costs of achieving the weed reduction (Alstrom, 1990). However, this subject is not straightforward and many of the considerations involved have been reviewed comprehensively by Auld et al. (1987). Auld et al. indicate that three alternatives are possible to describe how revenues are obtained from weed control. Revenues may (a) follow the law of supply and demand (that is, increase initially and then slow as more of the product is produced), (b) increase as the amount of product increases, and (c) increase initially and then decline as social or environmental costs become evident. These three scenarios are depicted in Figure 5.30a, b, and c.

The second alternative (Figure 5.30b) is generally regarded as most appropriate for agricultural systems of production, whereas alternative (a) (Figure 5.30a) is also appropriate for forestry where, for example, a premium price is attained for large or high-quality logs. Under both of these alternatives, the

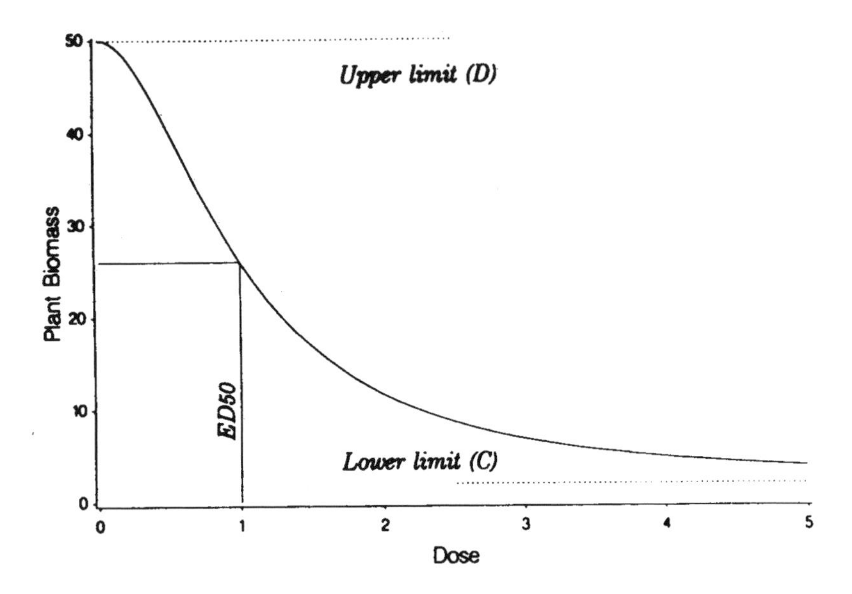

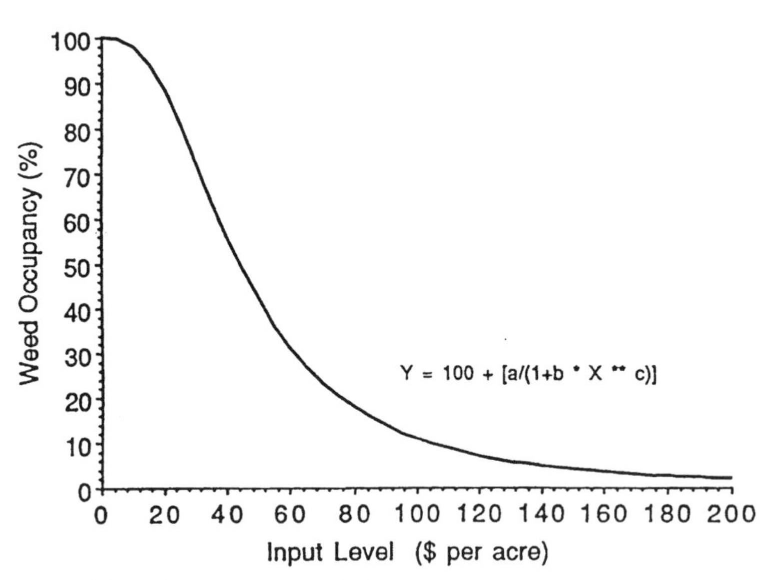

FIGURE 5.28 The logistic dose-response curve. *Top*, Plant biomass plotted against herbicide dose (untransformed; Streibig et al., 1993). *Bottom*, Weed occupancy plotted against input level (cost per acre). (From Radosevich and Shula, 1994.)

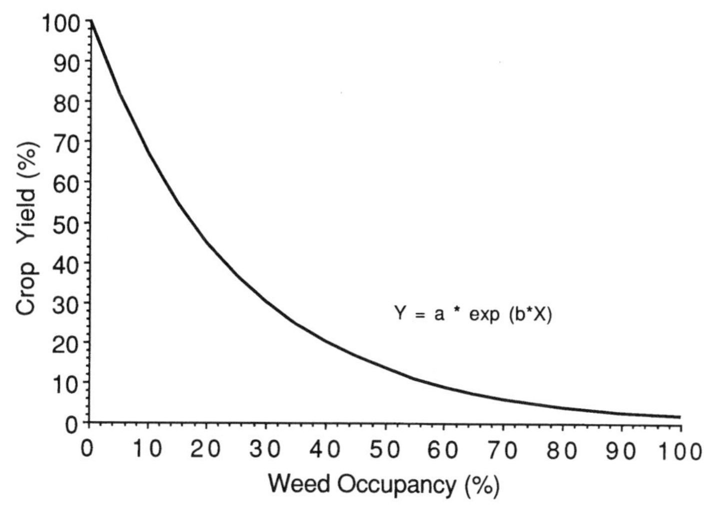

FIGURE 5.29 The relationship between crop yield and weed occupancy. (From Radosevich and Shula, 1994.)

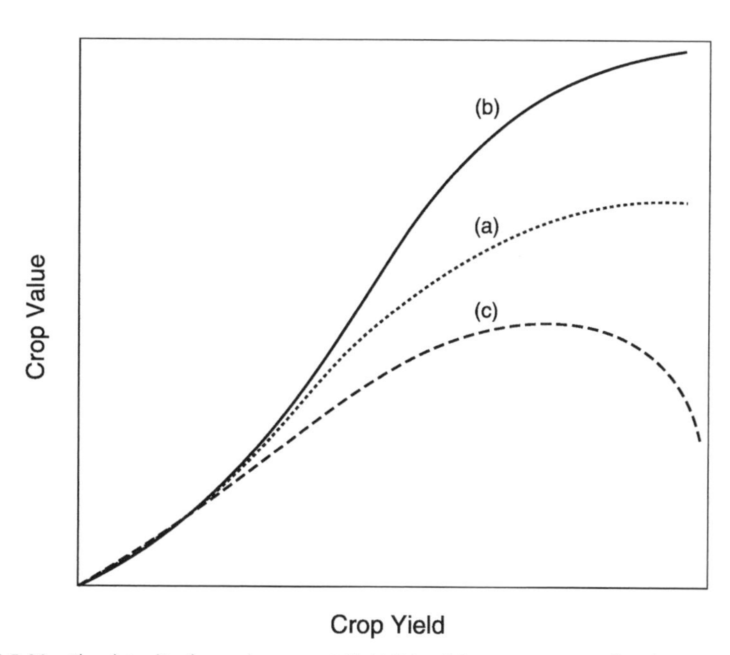

FIGURE 5.30 The relationship of crop value to crop yield. (a) Value of the crop increases initially with increasing yield, then remains constant according to supply and demand. (b) Value of the crop increases with increasing yield. (c) Value of the crop increases initially as in (a) and (b), but eventually declines as a result of "external" environmental or social costs.

value of further weed control generally rises as the price for the commodity rises because, to an extent, a greater amount of a more valuable commodity is produced. Alternative (c) (Figure 5.30c) is only now becoming evident as social and environmental costs are being expressed for increased pesticide regulation and registration, environmental impact reports, mitigation or restrictions resulting from pesticide residues, reforestation, and other restoration efforts (see Chapter 11) .

Using the economic alternative shown in Figure 5.30b, each element of a weed control decision was combined to construct the relationship shown in Figure 5.31, which demonstrates the profitability arising from various hypothetical treatment intensities or input costs. Efficiency, or inputs versus outputs, is shown as the diagonal line in Figure 5.31 and can be expressed as any common currency, for example, money, energy, or time. Presumably farmers, foresters, and land managers will not make inputs that cannot be recovered as outputs. Thus, management tactics that create greater benefits than the input costs to accomplish the task are "profitable" and probably would be continued. Strong motivation would be expected against management practices in which output benefits did not at least equal input costs (Figure 5.31).

An evaluation of gross output (revenue) in this model reveals that most intensities of weed control fall within the profitable range. However, when net benefit (revenue – cost) is plotted against input costs, many intensities of weed control result in a loss (Figure 5.31). When control measures are so low and,

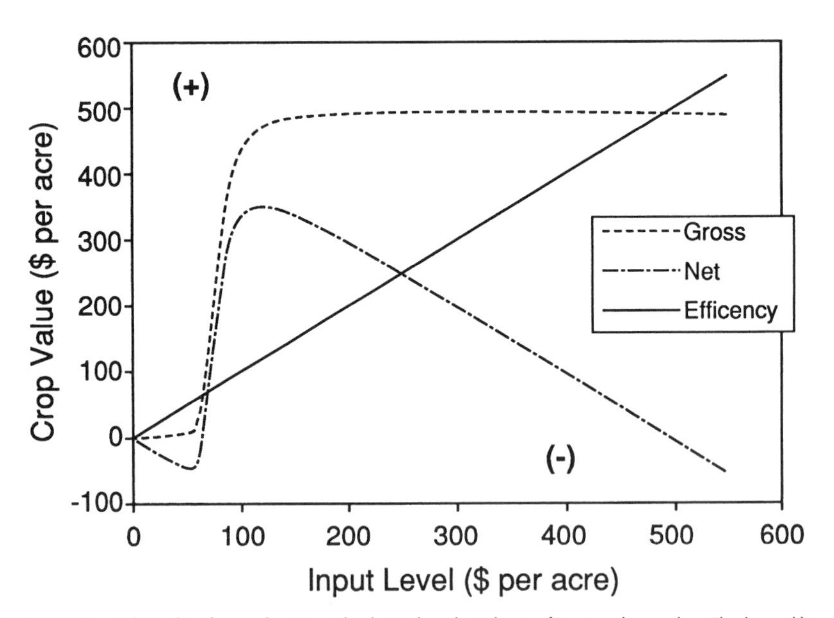

FIGURE 5.31 Relationship of crop value to input level using hypothetical costs of inputs and crop values. The diagonal line represents efficiency (i.e., costs of inputs = value of outputs); values above the diagonal line are profitable, while those below are unprofitable. (From Radosevich and Shula, 1993.)

therefore, weed abundance is so high that substantial crop losses occur, losses in net revenue obviously result; but even greater losses in net revenue can happen if input costs are too high (Figure 5.31, and Auld et al., 1987), such as the use of expensive tools or tactics for weed control or attempts to maintain extremely low weed densities. In this analysis, the threshold level of input costs is achieved at about 80 percent weed control, with greatest benefit arising when weed control is between 80 to 90 percent (compare Figures 5.29 and 5.31). These rather high levels of weed control also were achieved in this analysis at relatively low projections of input cost. Weed control levels beyond 90 percent hypothetically would result in losses in net revenue even though the practices may still be within the profitable range.

This analysis questions the value of high input costs to achieve very low weed occupancy. It is technologically feasible to create nearly vegetation-free and, therefore, highly productive agricultural fields and forest plantations. However, the economic and environmental desirability of doing so requires careful consideration. Farmers and foresters cannot afford to risk continual erosion of the financial, environmental, and biological ingredients necessary for long-term productivity of their cropping systems. Management for maximum production (Figure 5.12) may require greater investments of money, energy, and time and provide less net return than managing weeds at more optimal levels (Figure 5.31). However, optimal levels of weed management will require much greater knowledge and understanding of the environment and the biology of weed and crop species.

SUMMARY

Associations among various members of the plant kingdom are common in nature. Often when such associations occur between crop and weed species, crop yields are reduced. Most studies in weed science deal with this negative aspect of plant association, crop loss. However, other aspects of plant-plant associations also are possible. Interference is the general term used for interaction among species, or populations, that is, the effect that the presence of one plant has on another. There are many forms of interference, ranging from negative, to neutral, to positive. Competition is the most studied form of negative interference that occurs between weeds and crops.

Density is the most common plant factor that affects interference. Plants respond to density stress in two ways, plastic growth and an altered risk of mortality. Species proportion and the spatial arrangement among individuals in a plant stand are other factors influencing interference. There are many different methods that can be used to study plant competition. The various methods consider the importance of density, proportion, and arrangement to different degrees. The methods include additive, replacement series, Nelder, diallel, addition series, and additive series experiments. Most experiments in agriculture and forestry document well the levels of crop yield loss from competition with weeds. However, the importance of competition on the long-term dynamics of plant community structure and composition are less well understood.

There are several types of thresholds relevant to weed management: damage, period, economic, and action. The threshold concept is also being introduced into simulations and bioresource models to evaluate the cost effectiveness of weed control tools and tactics. Although the presence of weeds almost universally reduces crop yields, the profitability of weed control measures also must be evaluated in light of the costs to achieve acceptable levels of weed control.

Chapter

6

Physiological Aspects of Competition

At some point in the development of plants the supplying power of the environment often becomes limited. The limitation present in the environment may be aggravated by the proximity of neighboring plants. This limitation can result in the development of a competitive relationship between neighbors of identical or different species. Most studies of negative interference effectively demonstrate that such interactions occur. However, the greatest understanding of the mechanisms underlying those interactions arises when the limiting factors are identified. In this chapter we consider limiting factors as they influence the competitiveness of weeds and crops.

RESOURCES AND CONDITIONS

Factors in the environment that may influence plant growth and competition are usually divided into two categories, resources and conditions. Environmental *resources* are consumable and include light, CO_2, water, nutrients, and oxygen. In contrast to resources, environmental *conditions* such as temperature, soil pH, and soil bulk density (compaction) are not directly consumed but nonetheless affect plant growth. The general relationships of plant responses to both types of environmental factors are shown in Figure 6.1. Plant responses to environmental resources increase through a resource-limited phase to a saturation level, at which point another factor generally becomes limiting. For example, light response curves of photosynthesis generally have the shape shown in Figure 6.1a. At saturation, further addition of the resource does not increase the

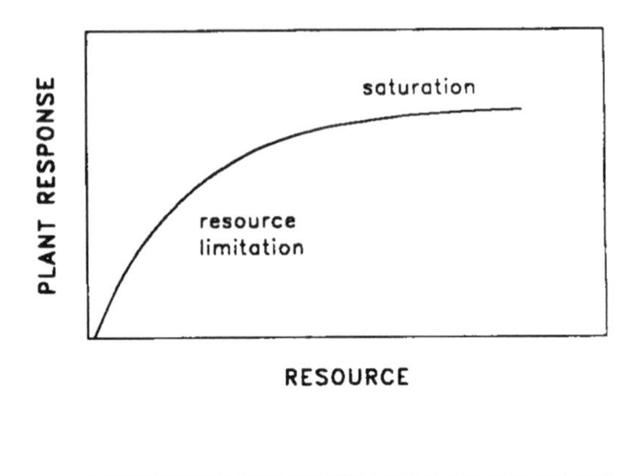

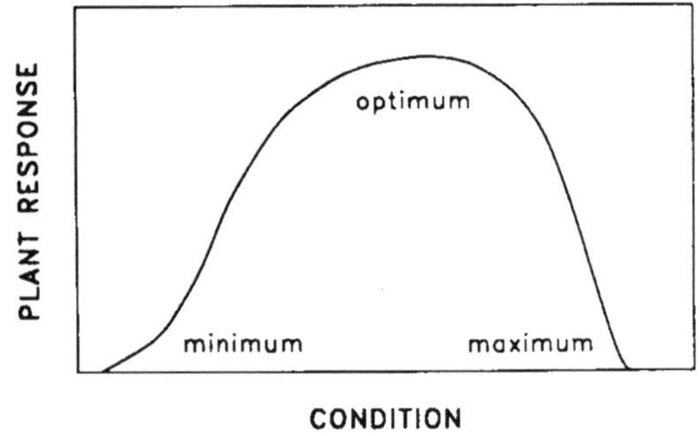

FIGURE 6.1 Idealized curves showing typical plant responses to varying levels of an environmental resource (a) and condition (b). In (a), the resource levels at which limitation and saturation occur are shown. In (b), the three cardinal points are shown: minimum, optimum, and maximum. (From Holt, 1991.)

response significantly. Higher than optimum levels of resource actually may cause the response to decline. Conditions, in contrast, usually limit plant responses either by absence or abundance until a threshold level is reached. Plant responses to environmental conditions generally have a bell shape and three cardinal points can be identified (Figure 6.1b). The minimum and maximum points occur where the process ceases, while the optimum range is where the highest rate can be maintained, if no other factors are limiting. A temperature optimum for seed germination is an example of an environmental factor that markedly influences a plant response but is not consumed. In Figure 6.1b, for example, if percent germination were plotted on the vertical axis, no germination would occur until a low threshold temperature was reached. Optimum germination would occur at an intermediate temperature and germination would cease at an upper threshold temperature. Although resource utilization by plants is strongly influenced by environmental conditions, this chapter focuses primarily on the competitive relationships between species for resources. We also discuss plant processes and functions that are affected by limitations of resources, often due to association with neighboring species.

Space Capture

As previously discussed in Chapter 5, some researchers consider resources as a single entity called space. The concept of space refers to the composite of resources necessary for growth of the plant and thus does not consider effects of single resources on plant response. While this approach is useful for evaluating the overall effects of density, proportion, and proximity of neighboring plants, it precludes analysis of mechanisms of competition in species mixtures.

A field study by Dawson (1965) demonstrated the utility of evaluating space capture, or the capture of implied resources that constitute space, as opposed to evaluating single resources. In his study Dawson grew sugarbeets and lambsquarters (*Chenopodium album*) together, and beginning at the time of sugarbeet germination, manually removed the weeds at weekly intervals until crop harvest. Treatments (plots) consisted of different durations of weeding beginning at crop germination, and different durations with weeds present before weeding was initiated, which resulted in competition treatments with progressively longer weed-free or weedy periods, respectively. Because the sugarbeet emerged before the weed and space was initially plentiful, no effect was observed on either species for the first three weeks after planting. Thereafter for approximately six weeks, each increase in the weed-free period resulted in an increase in sugarbeet yield until a maximum yield was reached. After nine weeks all the space was captured by the sugarbeets and further weeding was unnecessary. These observations suggest that critical times of interference (Figure 5.27) occur among neighboring plants based on the time of emergence and physiological factors that govern growth. Also indicated is that, if control measures are considered for weed density reduction, the control should occur while there is still enough space available for capture so that unrestricted crop growth can be realized.

A model developed from experimental work by Ross (1968), and later presented by Harper (1977), demonstrates this point more graphically. A diagram of the model in Figure 6.2 illustrates the effect of the sequence of seedling emergence on space capture. In this figure the space occupation of 10 grass seedlings randomly placed in a flat over 10 consecutive days is shown. The space each seedling preempts is assumed to be proportional to its weight. The diagram in Figure 6.2 shows that each plant stops growing when the space adjacent to it is captured by its neighbors such that the last transplants are able to grow very little. This experiment by Ross supports field observations in which the most competitive weeds appear to be the earliest to emerge. These findings indicate that the timing of emergence of a seedling population is more important than the spatial arrangement of the seedlings in determining resource utilization. While the studies by Dawson (1965) and Ross (1968) do not indicate which individual resource is in short supply in these plant systems, the space capture approach is valuable in directing further, more mechanistic research at specific times or growth stages when competition is most severe.

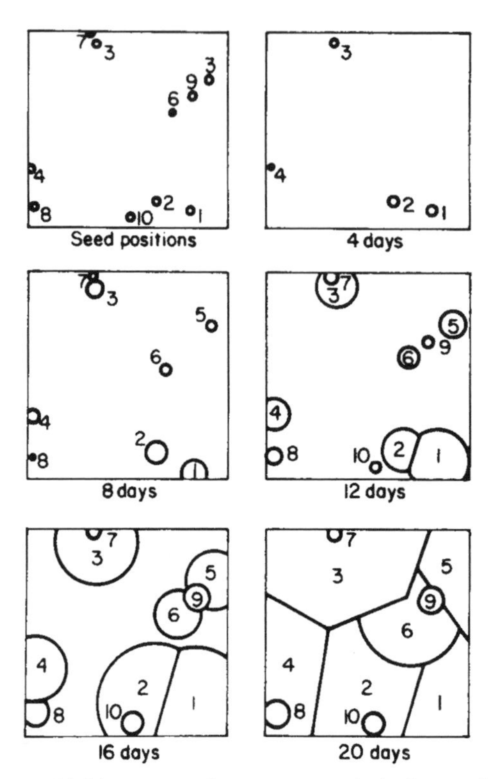

FIGURE 6.2 Diagrammatic model of the preemption of space (= resources) by developing seedlings. (From Ross, 1968.)

Mechanisms of Competition

Most experimental studies of plant competition have focused on the phenomenon and its effects in terms of plant size and yield without examining the underlying mechanisms of competition. As stated by Shainsky and Radosevich (1992), mechanisms of competition for resources must be demonstrated by:

- resource depletion associated with the presence and abundance of neighbors
- changes in morphological and physiological growth responses that are associated with changes in resources
- correlations between the presence of neighbors, resource depletion, and growth responses

Thus, mechanisms of plant competition consist of both the effect that plants have on resources and the response of plants to changed resources (Figure 6.3; Goldberg, 1990). While most definitions of competition encompass these criteria, several different theories have been developed to explain the relative importance of these components of competition and the characteristics of plants that confer

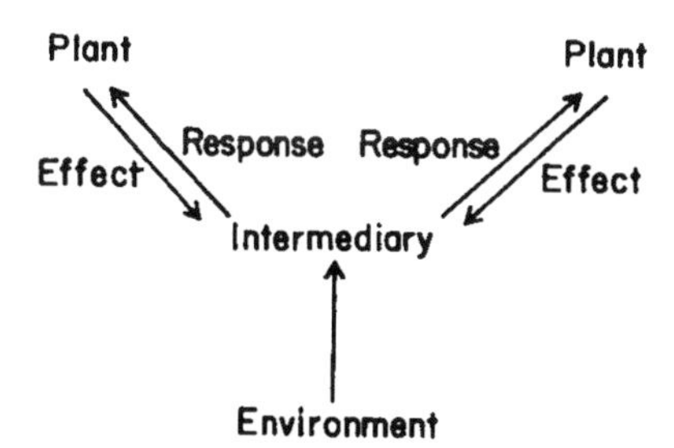

FIGURE 6.3 The effect and response components of indirect interactions between plants. The intermediary could be resources, mutualists, natural enemies, or even toxins. (From Goldberg, 1990.)

superior competitiveness. Two theories in particular that have received widespread attention are those of Grime (1979) and Tilman (1982, 1988) (Grace, 1990, 1991).

THE THEORIES OF GRIME AND TILMAN

As described in Chapter 2, Grime explains plant life histories in terms of the processes of disturbance and stress, which select for syndromes of plant characteristics. According to Grime, competition is the tendency for neighboring plants to utilize the same resources and success in competition is largely due to the capacity for resource capture (Grace, 1990; Grime, 1979). Thus, a good competitor has a high maximum relative growth rate (RGR) and can utilize resources rapidly. Tilman proposed a mechanistic, resource-based theory of plant competition (Tilman, 1982, 1988) that predicts competitive success as a function of the concentration of limiting resources (Grace, 1991). Thus, competitive success in this theory is the ability to draw resources down to a low level and to tolerate those low levels. A good competitor in this case would be the species with the lowest resource requirement. Although debate continues about the validity and relevance of these two theories (Grace, 1990, 1991), some of their differences may be explained by the time frames and associated semantics related to their definitions of competition. For example, Grime's "stress-tolerator" might be compared to Tilman's "competitor" (Grace, 1990). Furthermore, while Grime focuses on the role of particular plant traits in competitiveness, Tilman's theory deals with the dynamics of a population and thus does not focus on individuals. Nevertheless, both theories help explain how plant species compete for limiting resources and the role of plant traits in conferring competitive ability.

Plant Traits and Competition

As just described, plants can be good competitors either by rapidly depleting a resource or by being able to continue growth at depleted resource levels.

Unfortunately, most studies of crop/weed competition have focused only on the occurrence and impact of competition on crop yield. Relatively few studies have analyzed mechanisms in a quantitative way to determine specific plant traits that correlate with competitive ability. The mechanistic studies that have been published focus on plant responses to limitations in resources rather than patterns of resource depletion by plants, especially below ground. Plants affect resource availability primarily by depletion due to uptake (Table 6.1; Goldberg, 1990). In addition, for below-ground resources nonuptake mechanisms can also affect availability (Table 6.1). Both of these categories of plant effects on resources are strongly influenced by plant size. Plant responses to resources can be grouped into mechanisms for increasing resource uptake, decreasing resource loss, or increasing the efficiency of conversion of internal storage material to new growth (Table 6.1; Goldberg, 1990). The rest of this chapter presents more details on specific traits that confer a competitive advantage in resource utilization.

A few researchers have examined plant traits that are associated with competitiveness in experiments that describe both individual plant performance and

TABLE 6.1 Examples of Processes and Traits that Determine Magnitudes of Effect on Resources and Response to Resources on a Per-Unit Size Basis

Effect on Resources	*Response to Resources*
Uptake	Uptake
Physiological activity rates	Physiological activity rates
Allocation to resource-acquiring organs	Allocation to resource-acquiring organs
Architecture of resource-acquiring systems	Architecture of resource-acquiring systems
Nonuptake	Conversion efficiency
Direct addition of available forms	Loss
Association with N-fixing symbionts	Respiration rate
Leaching and throughfall	Transpiration rate
Addition in organic compounds	Tissue longevity
Litter quality and quantity	Leaching
Modification of physical environment	Translocation from senescent tissues
Temperature amelioration	
Reduce evapotranspiration	
Modification of microbial activity	
Temperature and moisture effects, root exudates, and root death	

Source: Goldberg, 1990.

biomass production in mixtures. Simultaneous growth analysis (see pages 280–285 for greater discussion) and replacement series experiments performed with pairwise mixtures of four annual weeds revealed that total plant weight, Unit Leaf Rate (ULR, an estimate of photosynthetic production on the whole plant level), and Leaf Area Ratio (LAR, a ratio of leaf area relative to dry weight) were positively correlated with competitiveness (Roush and Radosevich, 1985). Similar research with perennial weeds and cotton showed that the best predictors of competitive success in mixtures of these species were height, ULR, relative growth rate (RGR), and initial vegetative propagule weight (Holt and Orcutt, 1991). It is not surprising that, for perennial weeds growing with an annual crop, parameters of early establishment (RGR, initial propagule weight) as well as light utilization (height, ULR) were important in determining competitiveness. Gaudet and Keddy (1988) also advanced this approach in experiments that measured relative competitive ability of 44 herbaceous plant species and tested whether competitiveness was correlated with simple measurable plant traits. In their experiments, there was a strong relationship between plant traits and competitive ability (Table 6.2). Plant biomass, height, canopy diameter, canopy area, and leaf shape explained most of the variation in competitive ability of these 44 species against an indicator species, *Lythrum salicaria*. Some caution is needed in interpreting studies where traits of plants grown alone are used to predict mixture performance, since the same traits implicated in competitive-

TABLE 6.2 Correlation (r) between Traits of 44 Test Species and Biomass of the Phytomer (*Lythrum salicaria*)

Plant Traits	*Correlation with Phytomer Biomass*
Biomass, total (g)	–0.775**
Biomass, above ground (g)	–0.791**
Biomass, below ground (g)	–0.710**
Height (cm)	–0.659**
Leaf length (cm)	–0.084
Leaf width (cm)	–0.179
Leaf area (cm^2)	–0.302
Leaf number	–0.244
Leaf shape (length : width)	0.356*
Canopy diameter (cm)	–0.455*
Canopy area (cm^2)	–0.593**
Shoot to root ratio (g/g)	–0.016

Source: Gaudet and Keddy, 1988.

*$P < 0.05$; **$P < 0.001$. Correlations are simple linear correlations with phytomer biomass.

ness are often altered by the presence of neighbors (Tilman, 1990). Nevertheless, these reports provide a predictive approach for studying mechanisms of competition among plant species.

LIGHT

Sunlight is the primary energy source for all life on earth and it is made available through the process of photosynthesis. In addition to regulating photosynthesis, light regulates many other aspects of plant growth and development, such as seed dormancy and germination, phototropism, photomorphogenesis, and flowering. Likewise, light in agricultural ecosystems regulates many aspects of weed and crop growth, development, and competition. It is well known that plant responses to light are not entirely a function of the quantity of total light energy. Plants also respond to the spectral quality of light and to changing or transient light environments. In addition, the energy balance of plants is largely determined by radiation. Because photosynthesis is the major determinant of biomass production, it also plays a major role in the interactions of weeds and crops. Therefore, this section on light will focus primarily on photosynthesis and other physiological mechanisms of plant competition for light.

Light and Photosynthesis

Photosynthesis is the process whereby plants, algae, and some bacteria trap and then transform the energy of sunlight into a more useful form of energy. The elementary equation for photosynthesis is:

$$CO_2 + H_2O \xrightarrow[\text{light}]{} (CH_2O) + O_2 + H_2O \qquad \textit{(eq. 6.1)}$$

In this way plants convert light energy into chemical energy for utilization or storage. It is not surprising, therefore, that light, CO_2, and other factors that are required for photosynthesis are the objects of competition among plants. Photosynthesis usually is divided into two reaction sequences, the electron transfer reactions of photophosphorylation ("light reactions") and the carbohydrate fixation reactions of the Calvin cycle ("dark reactions"). Photophosphorylation requires light to operate and occurs in the pigment-containing chloroplast membranes (thylakoids) of green plant tissues. The Calvin cycle is a set of enzyme-catalyzed reactions that convert CO_2 into carbohydrate. While these reactions do not specifically require light to proceed, several enzymes in this pathway are activated by light so that this process does not occur in darkness. The Calvin cycle occurs in the liquid stroma of plant chloroplasts. In photophosphorylation (Figure 6.4), photons of light (represented as hv) are absorbed by chlorophyll molecules (depicted in the figure as LHCs, or light-harvesting complexes), which

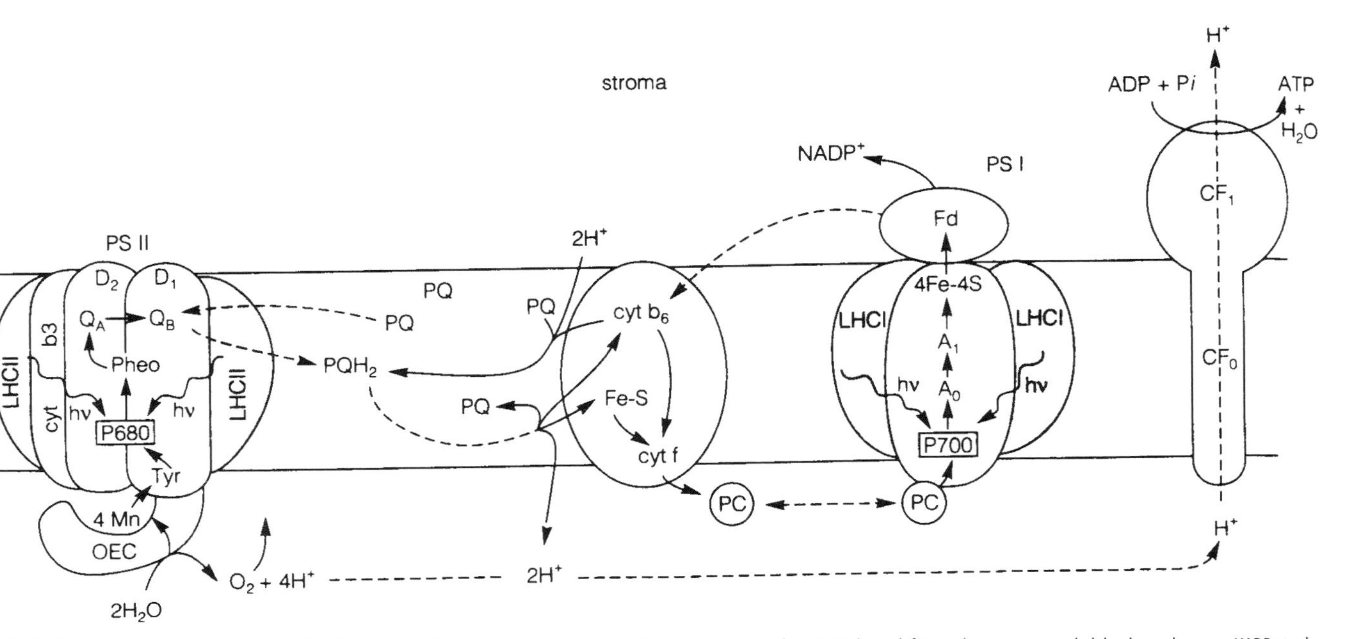

FIGURE 6.4 Cooperation of photosystem II (left), the cyt b_6-f complex, and photosystem I in carrying electrons from H_2O (lower left) in a lumen across a thylakoid membrane to $NADP^+$ in the stroma. The ATP synthase (far right, shown as CF_0 and CF_1) transports H^+ back from the lumen to stroma and converts ADP and P*i* to ATP and H_2O. For visual clarity, membrane lipids are omitted, as are most of the polypeptides in each of the four major complexes. Plastoquinone (PQ), plastocyanin (PC), and ferredoxin (Fd) are mobile and carry electrons as indicated by dashed lines. The role of cyt b_3 in PS II is still uncertain. (From Salisbury and Ross, 1992.)

drives a flow of electrons from water to NADP+. Consequently, water is oxidized to yield oxygen and the high-energy products ATP and NADPH are produced. In the Calvin cycle, CO_2 is fixed into three-carbon sugars, and subsequently into starch or sucrose, using the high-energy products of the light reactions. The basic process of photosynthesis is virtually identical in all higher plants, thus it is considered to be evolutionarily conservative. However, in certain plants that originated in hot, dry environments, including many weed species, an alternate route for synthesizing CO_2 into sugars is found. Species that initially fix CO_2 into four-carbon acids, prior to the entry of carbons into the Calvin cycle, are commonly called C_4 species (Figure 6.5). In contrast, species that initially fix CO_2 into 3-phosphoglyceric acid (3-PGA), a three-carbon acid, via the Calvin cycle are termed C_3 species. This apparently subtle change in the beginning of carbon fixation is an important physiological distinction among plant species, and it is discussed in more detail in subsequent sections of this chapter.

If photosynthesis by a single leaf is measured in response to irradiance, a typical light saturation curve can be derived (Figure 6.6a). The irradiance at which photosynthesis just balances respiration and net photosynthesis is called the *light compensation* point. This value varies among species, but in general is about 2 percent of maximum sunlight. Thereafter, with increasing irradiance, photosynthesis increases rapidly until CO_2 supply begins to limit carbon fixation. The photosynthetic response of a canopy of leaves to irradiance is usually quite different from that of a single leaf; that is, a much more linear relationship exists for a canopy (Figure 6.6b). This linearity occurs because light tends to penetrate deeply into a canopy on bright days, and therefore most of the light resource can be intercepted and used for photosynthesis. In other words, because of self-shading, higher irradiance is required to saturate an entire canopy of leaves than is required to saturate a single exposed leaf. Since most crops are planted or exist with weeds in relatively dense stands, it is important to consider photosynthesis on a canopy basis rather than an individual plant or leaf basis. Because of the linear response of canopy photosynthesis to irradiance, productivity per unit of area also is best thought of as a function of canopy rather than of individual plant performance.

LEAF AREA INDEX

A useful tool for examination of plant productivity and light availability is *leaf area index* (LAI), or the ratio of canopy leaf area to the area of land covered by that canopy. LAI is often expressed as canopy area per square meter of ground. LAI is therefore a dimensionless value indicating the amount of foliage cover in an area, which is usually arranged in several layers. LAI values are usually greater than one and vary according to the species present and their stage of development. A mathematical relationship that holds for most types of plant communities exists between foliage density and light interception. Usually irradiance decreases nearly exponentially with increasing plant

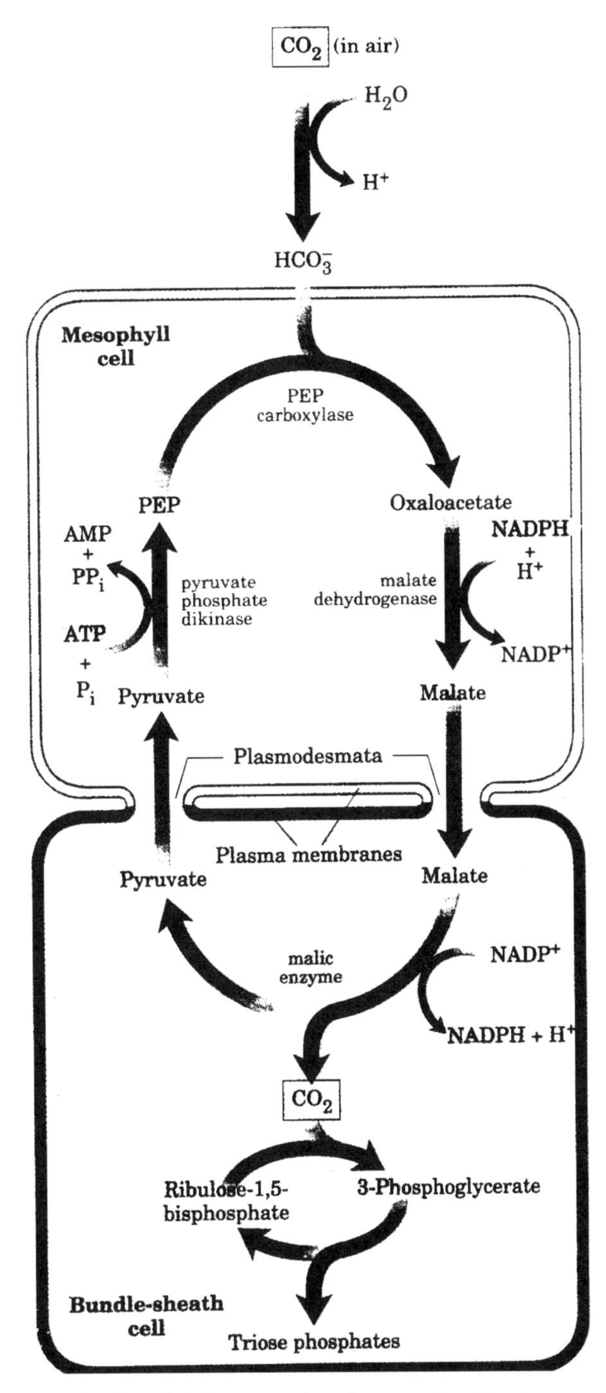

FIGURE 6.5 The Hatch-Slack pathway of CO_2 fixation, via a four-carbon intermediate. This pathway prevails in plants of tropical origin (C_4 plants). (From Lehninger et al., 1993.)

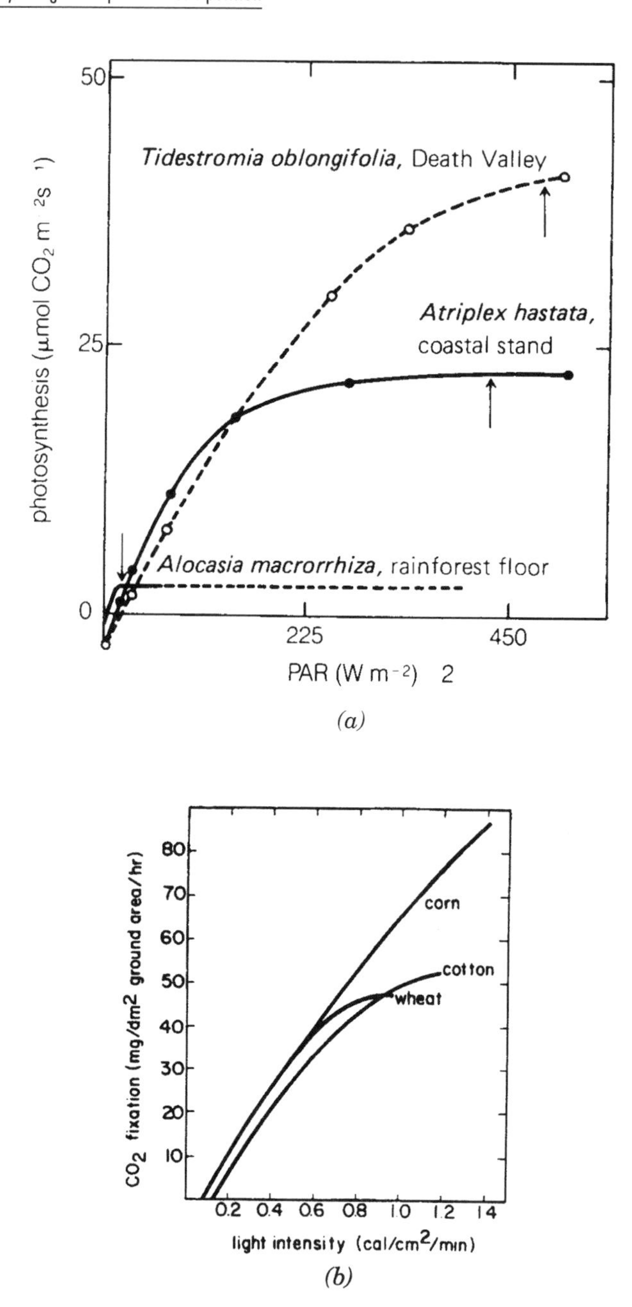

FIGURE 6.6 Influence of light on photosynthetic rates in single attached leaves (a) and plant canopies (b). In (a), leaves of three species native to different habitats are shown, and (b) depicts net photosynthesis of three plant canopies. In (a), maximum irradiances to which the plants are normally exposed (except for sunflecks that irradiate *Alocasia*) are indicated by arrows. The light compensation points are indicated on the graph where the lines cross the abscissa. (From Salisbury and Ross, 1978. Copyright 1978 by Wadsworth Publishing Company, Belmont, CA. Data for corn, wheat, and cotton were obtained from Baker and Musgrave, 1964; Pukridge, 1968; and Baker, 1965; respectively.)

cover. Thus, a plant's ability to project its canopy over that of a neighbor could impart a considerable competitive advantage with respect to light capture, if it occurred early in the life cycle or at a particularly critical developmental stage.

Effects of Light on Plant Growth and Development

As just described, for most plants maximum growth and photosynthesis occur in full light and rates decrease as irradiance is reduced. Many plants acclimate to reduced light conditions by plastic responses, including redistribution of dry matter, altered leaf anatomy, and decreased respiration rates, enzyme activities, and electron transport capacity. Numerous comparative studies have documented these now well known physiological, anatomical, and morphological characteristics of sun and shade plants. The adaptation of a leaf to light is apparently more responsive to the sum total of light energy received during the day than to peak irradiance (Chabot et al., 1979). Thus, an assessment of the light available to a plant and of the productivity attainable by that plant in that light environment must include diurnal measurements, not just instantaneous ones taken at midday.

Crops and weeds show varying degrees of shade tolerance. For example, in a comparison of soybean and weeds commonly associated with it, eastern black nightshade (*Solanum ptycanthum*), tumble pigweed (*Amaranthus albus*), and common cocklebur (*Xanthium strumarium*) were the most photosynthetically efficient under low-growth irradiance due to a combination of physiological and morphological adaptations (Regnier et al., 1988; Stoller and Myers, 1989). Many other weeds acclimate to low-growth irradiance by means of plastic responses that reduce the growth-limiting effects of shading and allow restoration of high rates of photosynthesis when the plant is subsequently exposed to high irradiance.

Weeds are often well adapted to high-light environments as well as capable of adapting to extreme variation in the light environment, particularly deep shade. Bazzaz and Carlson (1982) generated photosynthetic response curves for 14 early, mid, and late successional species grown in each of two light environments: full sunlight and 1 percent of full sunlight (Figure 6.7). Early successional species, which were all common annual weeds, had the highest photosynthetic flexibility (difference in response between sun- and shade-grown plants) to respond to different growth light conditions. The magnitude of photosynthetic flexibility decreased in plants from later successional stages. Thus, all species studied were able to change their photosynthetic output in response to growth irradiance, but the change was larger for early successional annuals (Bazzaz and Carlson, 1982). Weed plasticity in photosynthetic response to light level may result in survival and reproduction even in extremely low light, to such an extent that managing the light environment in a crop field to control weeds will be difficult.

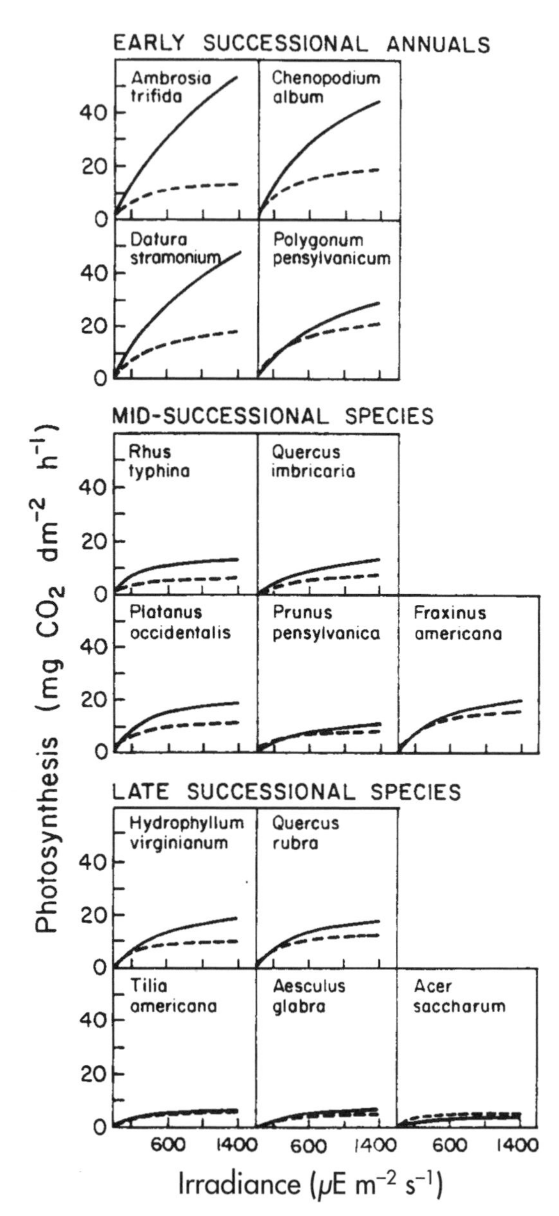

FIGURE 6.7 Photosynthetic response curves to light intensity for early successional annuals, mid-successional species, and late successional species grown in full sunlight (solid lines) and in deep shade equal to 1 percent of full sunlight (broken lines). (From Bazzaz and Carlson, 1982.)

CHANGING AND TRANSIENT LIGHT ENVIRONMENTS

Many plants, including many weeds, possess the ability to adapt quickly to changes in light levels that occur during their life cycle. Physiological and morphological adaptations that may occur in response to an increase in light during growth include increases in photosynthetic rate, leaf thickness, chlorophyll and protein contents, enzyme activities, as well as alterations in dry weight allocation, biomass, and height. When plants are transferred from high to low light, the opposite effects are generally seen. Both the capacity to adapt to changing light conditions and the type of physiological and morphological changes that occur in mature leaves vary widely among species. For weeds, such plasticity in acclimation to changing light enables plants that developed in shade to reach high rates of photosynthesis and growth if they are suddenly exposed to high light (Patterson, 1985). When transferred from high light to a shaded environment, which occurs when a canopy closes, many weeds respond with adaptations that ameliorate the growth-limiting effects of shading. Increased plant height and leaf area ratio (LAR, amount of leaf area per unit of plant biomass) are common responses to shade that may offset decreases in photosynthetic rate and biomass production that commonly occur when shade is imposed. However, seed production and vegetative reproduction by tillers, rhizomes, and tubers are generally reduced when reductions in light are extreme.

The contribution of sunflecks (in the 1 to 60 sec range) to the light environment of forest understory plants is well known (Pearcy, 1990). In contrast, little is known about the importance of sunflecks to canopy photosynthesis of crops. Most studies of crop canopy dynamics focus on the upper canopy layers and on steady-state photosynthesis. In the only report on sunflecks in crop canopies, sunflecks contributed 20 to 93 percent of the total daily irradiance in a soybean canopy at locations receiving sunflecks (Pearcy et al., 1990). The importance of this source of light to crop yield and its effects on weeds are unknown, however. If the actual contribution of sunflecks to total light in weed/crop canopies is significant, then photosynthetic production from lower leaves in canopies may be more important than previously thought. Research is needed on the nature of transient light in understories of mixed weed/crop canopies, on whether sunflecks are important in weed/crop competition, and on the potential for increasing crop photosynthesis and suppressing weeds through manipulation of canopy architecture.

Competition for Light

In his comprehensive review, C.M. Donald (1961) states that competition for light occurs in almost all cropping situations. The only exceptions to this statement are found when plants are very young or plants are sparse; in these cases there may be no apparent competition of any kind. As will be discussed later, however, even young or sparsely occurring plants may affect the light environment and consequently alter the development of their neighbors. In many cropping systems where nutrients and water are supplied, light is often the only factor that becomes limiting to plant growth. Most crop canopies are characterized by an exponential decrease in light in conjunction with an increasing depth in the canopy, and heavy shade at lower canopy levels (Figure 6.8). The extreme reduction in irradiance from full sunlight to the level under a plant canopy is readily obvious in a dense forest; however, almost complete darkness is often found under herbaceous canopies, as well. For example, irradiance in many forest environments ranges from approximately 10 to 25 percent of full sunlight, whereas irradiance of only 1 to 4 percent of full sunlight is common in many herbaceous or densely planted crop communities. As described above, there are significant reductions in the productivity of individual plants when light is limiting.

Conditions of light limitation are obviously intensified by the presence of weeds in a crop field. Many studies have quantified the effects of competition

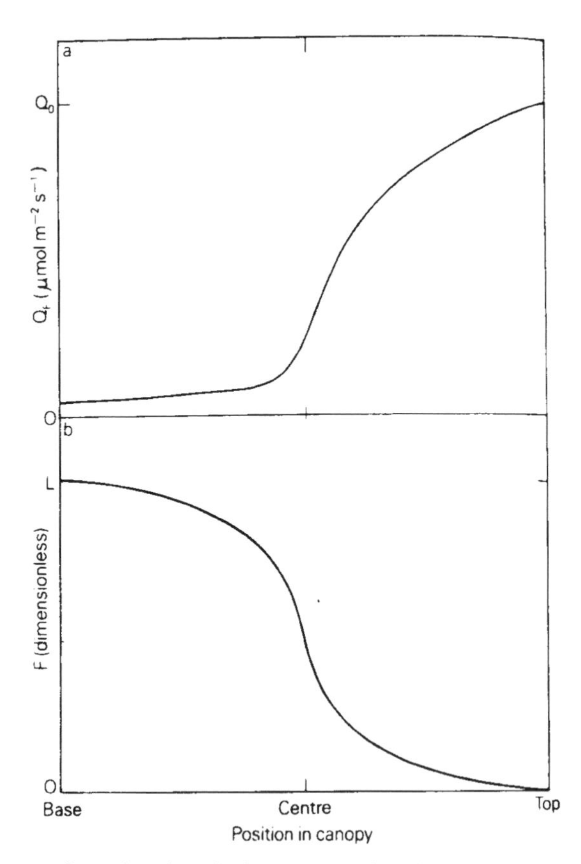

FIGURE 6.8 Changes in photosynthetic photon flux density (Q_F) (a) and cumulative leaf area index (F) (b) through a canopy with most of its leaves near the center. Q_F has its maximum value of Q_0 at the top of the canopy, and F has its maximum value of L (equal to the Leaf Area Index) at the canopy base. (From Nobel et al., 1993.)

for light between weeds and crops, yet only a few have examined the mechanisms of these effects. In one such mechanistic study, Cudney et al. (1991) showed that wild oat (*Avena fatua*) reduced light penetration and growth in mixtures with wheat by producing greater height growth than wheat. When wild oat was clipped to the height of wheat, light penetration in a mixed canopy was similar to that in monoculture wheat (Figure 6.9). Interference from wild oat planted at low densities was due to reduced leaf area of wheat at early growth stages, while high wild oat densities reduced light penetration to wheat leaves at later growth stages (Cudney et al., 1991).

The importance of weed height to competitiveness was also demonstrated in studies of competition between velvetleaf (*Abutilon theophrasti*) and soybean. In these studies, velvetleaf intercepted more light than soybean due to its greater height growth and dry weight allocation to more branches in the upper layers of the canopy (Akey et al., 1990). In similar studies, reductions in yield of processing tomato fruit were greater when tomato plants were grown in competition with eastern black nightshade (*Solanum ptycanthum*) compared to competition with black nightshade (*S. nigrum*), due to the greater height of eastern black nightshade (McGiffen et al., 1992). These results were confirmed by an analysis showing that irradiance reaching the top of the tomato canopy

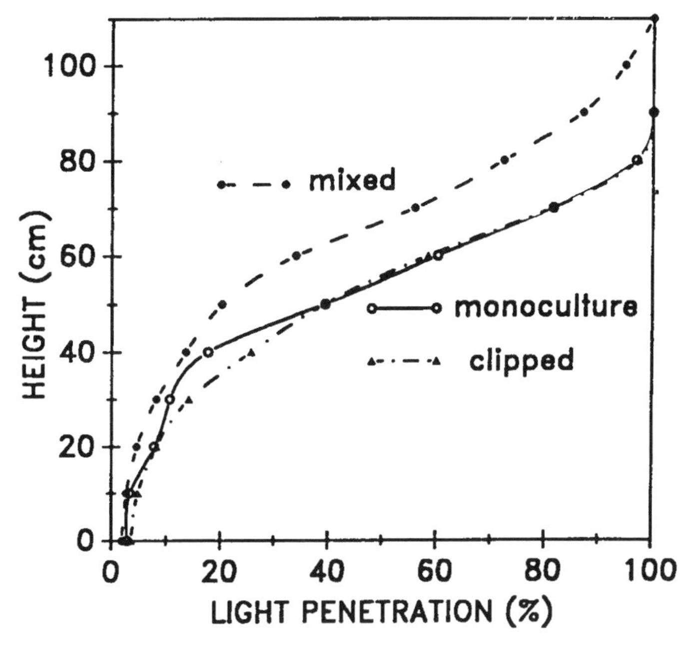

FIGURE 6.9 Light penetration into the canopy of clipped wild oat in mixed stands compared to unclipped and monoculture plots (January 1986 planting). Clipping wild oat to wheat canopy height in mixed stands resulted in increased ($P = 0.05$) light penetration when compared to unclipped mixed stands above 50 cm. Light penetration for clipped stands was not different from wheat monoculture. (From Cudney et al., 1991.)

was positively correlated with tomato yield and negatively correlated with eastern black nightshade density (McGiffen et al., 1992).

Destructive stratified clip harvesting in conjunction with measurements of light and photosynthesis in canopies has been used to evaluate mechanisms of light competition by crops and weeds. Graham et al. (1988) used this approach to develop canopy profiles of absorbed light and leaf area index by pigweed (*Amaranthus* spp.) and grain sorghum. As shown in Figure 6.10, sorghum LAI was concentrated in the 0.3 to 0.6 m canopy layer, while pigweed had the greatest LAI above 0.6 m in the canopy. Sorghum LAI and light absorption were reduced as density of pigweed in mixtures increased. Thus, by absorbing light in the upper canopy, pigweed reduced light interception by sorghum.

These and other studies show that weed and crop canopy architecture, especially plant height, location of branches, and height of maximum leaf area determine the impact of competition for light and thus have a major influence on crop yield. One approach to improving crop productivity in light-limiting environments that has received considerable attention is selection of crop genotypes with increased competitiveness with weeds, perhaps through more efficient light utilization by the crop canopy. However, such efforts will succeed only if a better understanding is reached of the physiological and morphological mechanisms of competition for light between weeds and crops (Holt, 1995).

C_3 versus C_4 Photosynthesis

The differences among plant species in terms of their ability to fix carbon have been well documented. For most weeds this process is by either the C_3 (Calvin cycle) or the C_4 (Hatch-Slack) pathway. In normal C_3 photosynthesis, O_2 competes with CO_2 for the binding site on the enzyme ribulose-1, 5-bisphosphate (RUBP) carboxylase (rubisco), the initial CO_2-fixing enzyme in the Calvin cycle. This phenomenon is called photorespiration and results in absorption of O_2 and loss of CO_2 in the light. The C_4 pathway is considered to be an adaptation to minimize this wasteful loss of CO_2 in hot, dry environments where the affinity of rubisco for O_2 is high. In C_4 photosynthesis, the first enzyme that takes up CO_2, phosphoenolpyruvate (PEP) carboxylase, has a stronger affinity for CO_2 than does rubisco, and complexes CO_2 into four-carbon acids (Figure 6.5). These acids are produced in the mesophyll cells and then move into bundle sheath cells where they are decarboxylated, yielding CO_2. This extra step in C_4 plants serves as a means of concentrating CO_2 so that high rates of photosynthesis can continue even under conditions when stomata are nearly closed or when external CO_2 concentrations are low. As a result, little or no photorespiration occurs in C_4 plants to reduce the efficiency of CO_2 fixation. In C_4 plants, it seems that recycling of CO_2 is so efficient through initial fixation to four-carbon acids that little or no CO_2 is lost.

C_4 plants fix CO_2 by the normal Calvin cycle after its initial fixation into four-carbon acids. This added C_4 pathway confers a definite advantage to

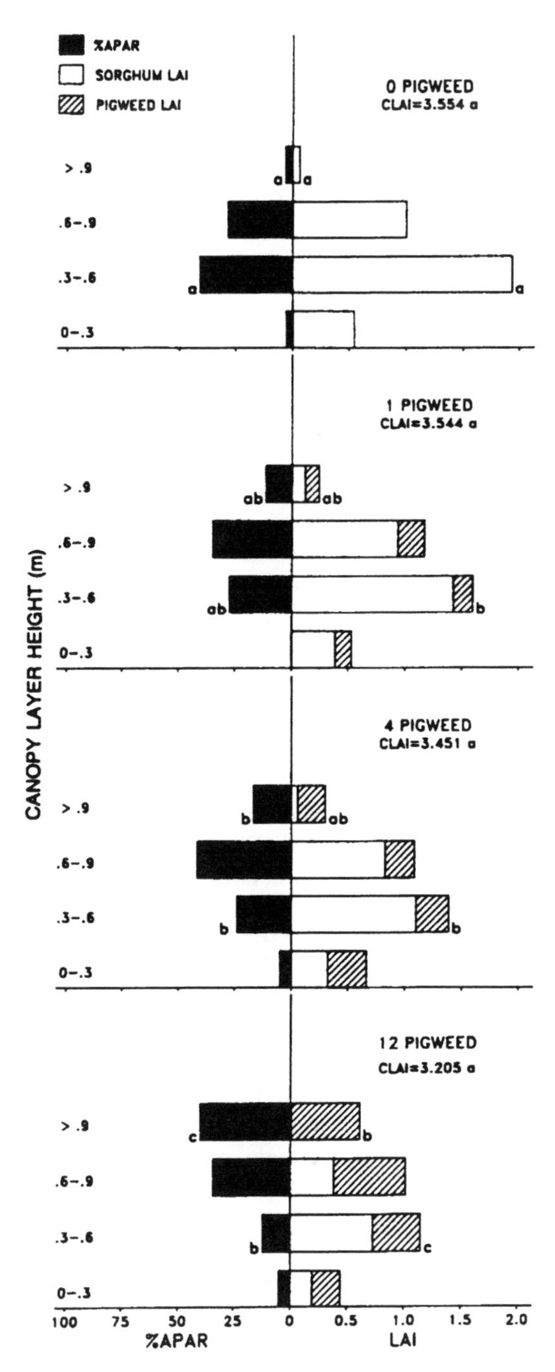

FIGURE 6.10 Light absorption and LAI as affected by height (m) within the canopy. Letters following bars indicate significant differences ($P < 0.05$) between treatments by layer for CLAI (sorghum LAI + pigweed LAI) and APAR, determined by Student-Newman-Keuls multiple range test. (From Graham et al., 1988.)

plants that possess it in terms of photosynthetic efficiency under conditions of high irradiance, high temperatures, and limited moisture availability. Under more moderate conditions, however, this advantage over plants possessing solely the C_3 pathway is much diminished due to the extra energetic costs of producing enzymes for C_4 photosynthesis. Table 6.3 lists several characteristics of the C_3 and C_4 methods for CO_2 fixation in higher plants.

C_4 PHOTOSYNTHESIS AND COMPETITIVENESS

In 1969, Black et al. evaluated the method of photosynthetic CO_2 fixation as a physiological basis on which to explain plant competition. They divided weed and crop species into two categories, efficient and nonefficient, based on CO_2 compensation point, photorespiration, presence or absence of a C_4 cycle, and enhancement of CO_2 fixation in the absence of O_2. From their examination of a wide range of plant species (Table 6.4), Black et al. concluded that the most competitive weeds and highly productive crops fell into the efficient category and possessed the C_4 pathway. Crops typically grown in more temperate zones and cool season (winter annual) weeds were in the nonefficient C_3 category. Although this scheme is an oversimplification to account for the competitive nature of weeds and crops in all environments, the physiological differences in CO_2 fixation between weed and crop species may be of considerable importance in hot, dry agricultural areas. For example, most of the troublesome summer annual weeds in the Central Valley of California are C_4 species, whereas few C_4 plants occur there in the winter or in the more moderate Salinas Valley to the west.

Several types of crop/weed interactions are possible based on the carbon fixation pathway. Depending on the climate to which these species combinations are exposed, slight to severe competition could be expected.

- Cool season crop (C_3) versus C_4 weed—for example, sugarbeet versus barnyardgrass (*Echinochloa crus-galli*) or redroot pigweed (*Amaranthus retroflexus*).
- Cool season crop (C_3) versus cool season weed (C_3)—for example, wheat versus wild oat (*Avena fatua*) or mustard (*Brassica* spp.).
- C_4 crop versus C_4 weed—for example, corn versus barnyardgrass (*E. crus-galli*)
- C_4 crop versus C_3 weed—for example, corn versus common lambsquarters (*Chenopodium album*).

In a now classic study of the importance of C_4 photosynthesis in species interactions, Pearcy et al. (1981) evaluated photosynthetic performance and competition of two weed species, common lambsquarters (*Chenopodium album*, C_3) and redroot pigweed (*Amaranthus retroflexus*, C_4), grown under different temperature regimes. For both species, photosynthetic rates at measurement temperatures of 25°C were similar for plants in all growth temperatures. However, at higher measurement temperatures *A. retroflexus* had higher rates of photo-

TABLE 6.3 Comparison of Photosynthetic Pathways

Trait	*C_3 Heliophyte (adapted to a high light environment)*	*C_4*
Taxonomic diversity	Very wide: algae to higher plants	Perhaps some algae, no lower vascular plants or conifers, wide among flowering plants
Typical habitat	No pattern	Open, warm saline (some exceptions)
Leaf anatomy	Palisade + spongy parenchyma	No mesophyll differentiation; large bundle sheath; Kranz-type
Light saturation point	6000 ft-c	8000–10,000$^+$ ft-c
Optimum temperature	20–30°C (lower in tundra)	30–45°C (as for light above, can be lower for C_4 species in different habitats)
Maximum photosynthetic rate ($mg/dm^2 \cdot hr$)	30	60
Maximum photosynthetic rate ($mg/g \cdot hr$)	55	100
Maximum growth rate ($g/dm^2 \cdot day$)	1	4
Water use efficiency (g CO_2/g H_2O)	300	600
Photorespiration	High	Low
Na required	No	Yes
Fixation path and enzyme	CO_2 + 5-C —> 2 3-C PGA; ribulose-1,5-bisphosphate carboxylase	CO_2 + 3-C —> 4-C acids; PEP carboxylase
Stomatal behavior	Open in day, closed at night	Open in day, closed at night
Space-time relations	Entire Calvin cycle in any mesophyll cell	Initial fixation in mesophyll, then transfer of acid to bundle sheath for Calvin cycle
CO_2 compensation point	55 ppm	5 ppm

Source: Black, 1973. Adapted from the *Annual Review of Plant Physiology*, vol. 24. Copyright 1973 by Annual Reviews, Inc.

TABLE 6.4 Plant Species Divided into Efficient and Nonefficient Groups Based on Their CO_2 Compensation Concentration, Photorespiration, Presence or Absence of C_4 Cycle for CO_2 Fixation, and Enhancement of CO_2 Uptake in the Absence of Oxygen

Plant Species	*CO_2 Compensation Concentration (ppm)*	*Photorespiration (mg $CO_2/dm^2 \cdot hr$)*	*C_4 Cycle*	*CO_2 Uptake Enhancement (%)*
Efficient Dicotyledons				
Amaranthus albus L. (tumble pigweed)	3	—	—	—
Amaranthus palmeri S. Wats. (Palmer amaranth)	—	—	yes	0
Amaranthus retroflexus L. (redroot pigweed)	—	0	—	—
Amaranthus edulis Speg. (grain amaranth)	—	0	—	—
Gomphrena globosa L. (globe amaranth)	2	—	—	—
Atriplex spongiosa F.v.M. (pop saltbush)	—	—	yes	—
Atriplex rosea L. (red orach)	<5	—	—	—
Atriplex semibaccata R. Br. (Australian saltbush)	—	—	yes	—
Portulaca oleracea L. (common purslane)	2	—	—	—
Portulaca grandiflora Hook.	2	—	—	—
Kochia scoparia (L.) Roth (kochia)	3	—	—	—
Salsola kali L. (Russian thistle)	0	—	—	—
Nonefficient Dicotyledons				
Atriplex patula var. *hastata* L. Gray (halberdleaf orach)	35	—	—	—
Phaseolus vulgaris L. (common bean)	—	—	no	45
Glycine max Merrill (soybean)	>37	—	no	—
Beta vulgaris L. (sugarbeet)	42	2.4	no	—

TABLE 6.4 Continued

Plant Species	*CO_2 Compensation Concentration (ppm)*	*Photorespiration (mg $CO_2/dm^2 \cdot hr$)*	*C_4 Cycle*	*CO_2 Uptake Enhancement (%)*
Nonefficient Dicotyledons				
Spinacea oleracea L. (spinach)	40	—	—	—
Nicotiana tabacum L. (tobacco)	60	—	no	56
Gossypium hirsutum L. (cotton)	—	2.2	—	38
Helianthus annuus L. (sunflower)	40	3.1	—	45
Chenopodium album L. (lambsquarters)	35	—	—	—
Lactuca saliva L. (lettuce)	>37	—	no	—
Efficient Monocotyledons				
Zea mays L. (corn)	0	0	yes	0 to 6
Cynodon dactylon (L.) Pers. (bermudagrass)	5	0	—	—
Paspalum notatum Flugge (bahiagrass)	—	—	—	2
Paspalum distichum L. (knotgrass)	<5	—	—	—
Echinochloa crus-galli (L.) Beauv. (barnyardgrass)	<5	—	—	—
Echinochloa stagnina Beauv.	—	—	—	−3
Digitaria sanguinalis (L.) Scop. (large crabgrass)	<5	—	—	—
Setaria italica (L.) Beauv. (foxtail millet)	—	—	—	0
Setaria glauca (L.) Beauv. (yellow foxtail)	<5	—	—	—
Chloris gayana Kunth. (rhodesgrass)	<5	—	yes	3
Sorghum vulgare Pers. (sorghum)	—	0	yes	4
Sorghum halepense (L.) Pers. (Johnsongrass)	—	—	yes	—

TABLE 6.4 Continued

Plant Species	*CO_2 Compensation Concentration (ppm)*	*Photorespiration (mg CO_2/dm^2 · hr)*	*C_4 Cycle*	*CO_2 Uptake Enhancement (%)*
Efficient Monocotyledons				
Panicum capillare L. (witchgrass)	<5	—	yes	—
Panicum miliaceum L. (proso millet)	<5	—	yes	—
Panicum bulbosum H.B.K. (bulbous panicgrass)	<5	—	yes	—
Andropogon gayanus Kunth.	—	—	—	−2
Eragrostis chloromelas Steud. (Boer lovegrass)	—	—	—	0
Eragrostis pilosa (L.) Beauv. (India lovegrass)	<5	—	—	—
Eragrostis brownei (Kunth.) Nees.	—	—	yes	—
Eleusine coracana (L.) Gaertn. (African millet)	—	—	—	8
Saccharum officinarum L. (sugarcane)	7	—	yes	—
Cyperus rotundus L. (purple nutsedge)	—	—	yes	—
Cyperus eragrostis Lam. (tall umbrellaplant)	<5	—	—	—
Nonefficient Monocotyledons				
Cyperus alternifolius var. *gracilis* (L.) Hort. (slender umbrella flatsedge)	—	—	no	—
Dactylis glomerata L. (orchardgrass)	>37	—	—	46
Lolium multiflorum Lam. (Italian ryegrass)	—	—	—	50
Triticum aestivum L. (wheat)	>37	—	no	23, 50
Phalaris arundinacea L. (reed canarygrass)	>37	—	—	45
Avena sativa L. (oat)	40	—	no	—
Agrostis alba L. (redtop)	>37	—	—	—

TABLE 6.4 Continued

Plant Species	CO_2 Compensation Concentration (ppm)	Photorespiration ($mg\ CO_2/dm^2 \cdot hr$)	C_4 Cycle	CO_2 Uptake Enhancement (%)
Nonefficient Monocotyledons				
Hordeum vulgare L. emend. Lam. (barley)	>37	—	—	—
Poa pratensis L. (Kentucky bluegrass)	—	—	—	37
Oryza sativa L. (rice)	>37	—	—	45
Panicum commutatum Schult. (variable panicgrass)	>37	—	no	—
Panicum lindheimeri Nash (Lindheimer's panicgrass)	>37	—	no	—

Source: Black et al., 1969.

Note: A dash indicates that data are not available.

synthesis than *C. album*, whereas at lower measurement temperatures *C. album* had the photosynthetic advantage (Figure 6.11). Furthermore, while the photosynthetic temperature response curves were similar in all three growth regimes for the C_3 species (*C. album*), the C_4 species (*A. retroflexus*) had much higher rates of photosynthesis when grown under elevated temperatures. Under differing growth temperature regimes, the competitive abilities of the two species in mixture closely paralleled their photosynthetic relationships. In each case, the species with the photosynthetic advantage also had a higher initial growth rate prior to canopy closure, which resulted in eventual overtopping and shading of the other species. The importance of factors influencing plant size at the initiation of any competitive interaction has already been stressed. The photosynthetic pathway seemingly acts just as any other factor does by increasing plant size, and therefore competitive advantage, at the time of interaction.

PLANT RESPONSES TO CO_2

Direct competition for CO_2 is theoretically possible among weed and crop species growing in proximity. To examine this possibility, Oliver and Schreiber (1974) compared net carbon exchange of redroot pigweed (*Amaranthus retroflexus*), prickly sida (*Sida spinosa*), and birdsfoot trefoil (*Lotus corniculatus*) grown together and separately. On a total leaf area basis, CO_2 utilization changed as the canopy of the mixture developed. However, direct competition for CO_2 did not occur even though the plants studied had different photosyn-

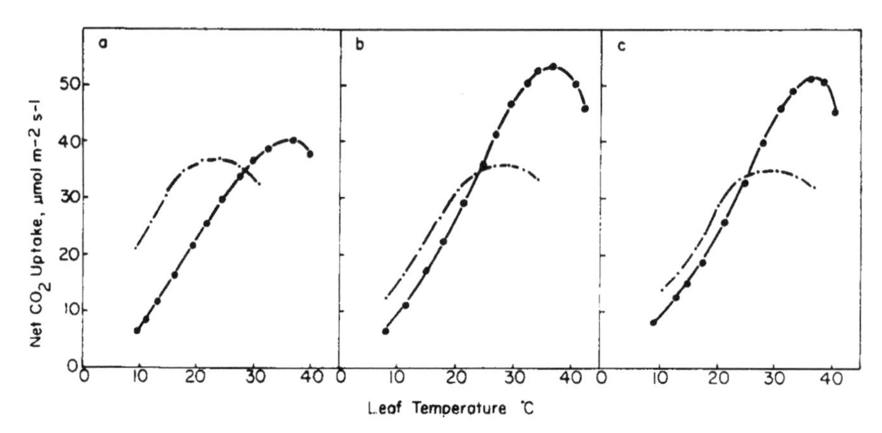

FIGURE 6.11 Temperature dependence of photosynthesis for *Chenopodium album* (•) and *Amaranthus retroflexus* (●) measured at a quantum flux density of 1.8 mmol/cm^2 · sec, 300–330 µbar CO_2 pressure and a VPD of 5–10 µbars. The growth conditions were (a) 17/14°C, (b) 25/18°C, and (c) 34/28°C. (From Pearcy et al., 1981.)

thetic rates. Under field conditions, zones of CO_2 depletion may occur within a canopy. However, competition for CO_2 probably would not be as critical as competition for other resources because CO_2 concentrations in the atmosphere are always greater than the CO_2 compensation point of plants (see Table 6.3).

Weeds and global climate change. Of more importance than CO_2 limitation are possible long-term effects on plants of potential increased CO_2 concentration in the global atmosphere as a result of global warming. Human activities are rapidly changing the global environment by altering the atmosphere's chemical composition and restructuring the surface of the earth (Lippold, 1995). By the middle of the next century, the mean global temperature is expected to be higher than it has been for perhaps millions of years, due primarily to increased concentrations of CO_2 and other "greenhouse" gasses. Changes in global rainfall patterns and storm intensity are also likely results of the warming process (Breymeyer and Melillo, 1991). Global climate change could result in broad scale redistribution of vegetation across the planet (King and Neilson, 1992). However, there continues to be serious discussion about possible distribution of even major vegetation types, depending upon which models are used to simulate climate change scenarios. In general, boreal forests are expected to decrease under all climate scenarios, while temperate and tropical forests and grasslands/shrublands are expected to increase in areal extent. Depending upon the magnitude of the warming, up to 60 percent of the earth's vegetated surface changes from one biome to another in these simulations (King and Neilson, 1992). Whether CO_2 concentrations will continue to rise through the next century or stabilize as new carbon sources and "sinks" are generated through worldwide shifts in vegetation pattern remains uncertain. Uncertain also is the potential for fire to increase under warming scenarios and its role in vegetation replacement in either the presence or absence of elevated carbon dioxide levels.

As discussed in Chapter 3, weeds continue to evolve to human disturbances. They also have proven to be highly adaptable organisms in the environments they inhabit and sometimes to be symptomatic of more serious environmental problems. Thus, it is interesting to speculate on the generalized impacts and responses that weeds might have under various scenarios of global climate change.

Weeds, warming and CO_2 elevation. If CO_2 levels and temperatures continue to rise, weed species with the C_3 mechanism of CO_2 fixation should be favored, since photorespiration would be reduced by the higher atmospheric CO_2/O_2 ratio (Patterson and Flint, 1980). Other beneficial effects of CO_2 enrichment—for example, on water use efficiency (WUE) and plant water status—would enhance growth of both C_4 and C_3 plants (Patterson, 1995). Experimental evidence suggests, however, that effects of enhanced CO_2 are in turn affected by water status of the plants (Patterson, 1986). These data demonstrate the dangers of studying responses to resources in isolation and without regard for their interactive effects on plants. Unfortunately, such studies are often difficult to conduct due to complexities both in experimental design and in the equipment that would be required.

In contrast to the above scenario, the possibility also exists for a feedback mechanism in which initial CO_2 enrichment results in higher biomass of C_3 species, creating a carbon "sink" and stabilizing of the atmospheric CO_2 concentration. If atmospheric CO_2 stabilized, physiological advantage could be imparted to C_4 species as the climate warms. Fire also might exacerbate the current rise in atmospheric CO_2 by creating a short term but significant elevation in CO_2 while disfavoring non-sprouting species, especially shrubs and trees. For example, Billings (1990) suggests that the widespread distribution of downy brome (*Bromus tectorum*) throughout much of the Great Basin of North America might now be aided by atmospheric CO_2 enrichment and, therefore, superior competitive ability over native C_4 grasses. He also observes that the increased incidence of range fires following the unintentional introduction of downy brome has resulted in the nearly total replacement of the native sagebrush (*Artemisia tridentata*)/perennial grassland by this introduced grass in many areas of the Great Basin.

Light Quality

An important aspect of light that has implications for weed/crop ecosystems is the role of spectral quality of light in plant interactions. Leaves absorb light in the blue and red regions and reflect or transmit in the green and far-red regions of the spectrum (Figure 6.12). As a result, the spectral photon distribution of radiation within a canopy is severely depleted in the 400 to 700 nm range and

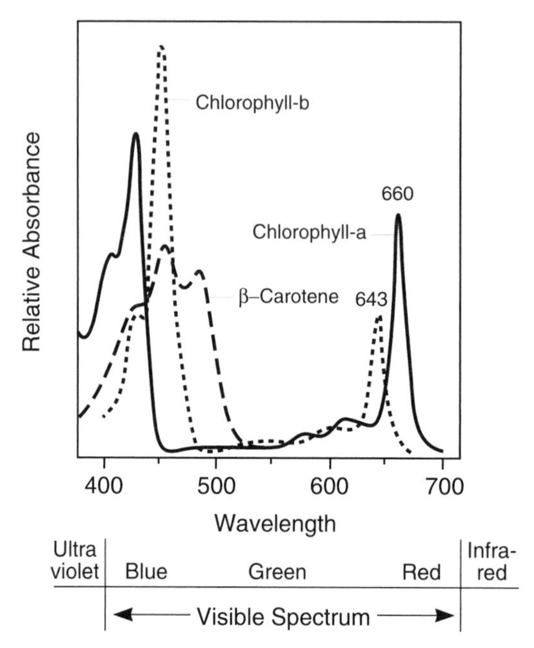

FIGURE 6.12 Absorption spectra of chlorophyll-a, -b, and carotene. (Modified from Halliwell, 1984; Hipkins, 1984; and Avers, 1985.)

enhanced in wavelengths over 700 nm (far-red light; Figure 6.13a). Thus, with increasing depth in a plant canopy, light is enriched in far-red wavelengths and has a lower red : far red (R : FR) ratio. This effect of canopies on light quality is seen for both vertically and horizontally propagated radiation (Figure 6.13b). With increasing distance from a canopy, even in nonshaded conditions, the horizontal R : FR ratio rises only gradually and is accompanied by a rise in the ratio of far red : red forms of phytochrome (Pfr : Pr; Holmes and Smith, 1975; Smith et al., 1990). Thus, the effects of a canopy on light quality can be measured even at some distance from the canopy.

PLANT RESPONSES TO QUALITY OF LIGHT

Light quality regulates many plant processes, especially germination, morphogenesis, and flowering (Ballaré et al., 1992; Casal and Smith, 1989). Morphological responses to shade are also well known, and include increased stem elongation, decreased leaf/stem dry weight ratio, and decreased tillering. These responses have been shown to be triggered by phytochrome in response to low R : FR ratios under canopies (Smith, 1982). Research by Ballaré, Casal and coworkers showed that these responses to altered R : FR ratios occur before the onset of mutual shading among neighbors. Based on this work, it was proposed that far-red radiation reflected by nearby leaves is a means of early detection of neighbors that signals oncoming competition during canopy development (Ballaré et al., 1990, 1992).

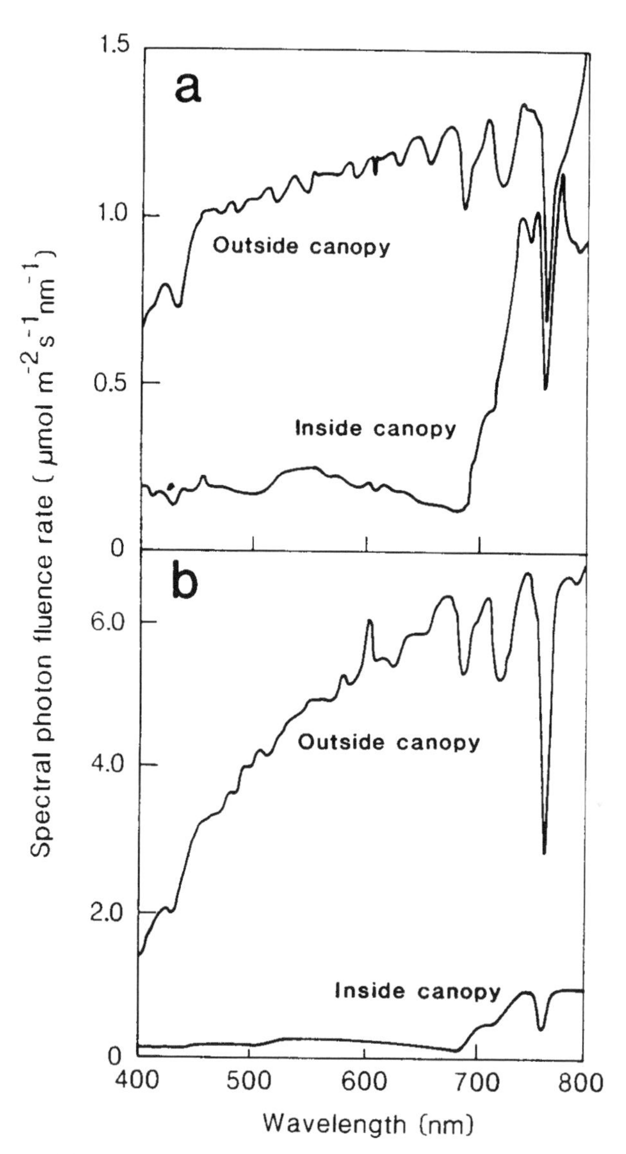

FIGURE 6.13 Spectral photon distribution of vertically and horizontally propagated radiation within and outside an artificial mustard canopy: (a) vertically-propagated radiation; (b) horizontally propagated radiation. (From Smith et al., 1990.)

An elegant demonstration of plant responses to non-shading neighbors was made by Novoplansky et al. (1990). Seedlings were placed in the vicinity of various objects, including some that altered the spectral quality of light. *Portulaca oleracea* seedlings that confronted a green shading object grew away from the object in a nonrandom manner (Figure 6.14). By doing this, seedlings avoided growth in the direction with higher far-red light. This work showed that *P.*

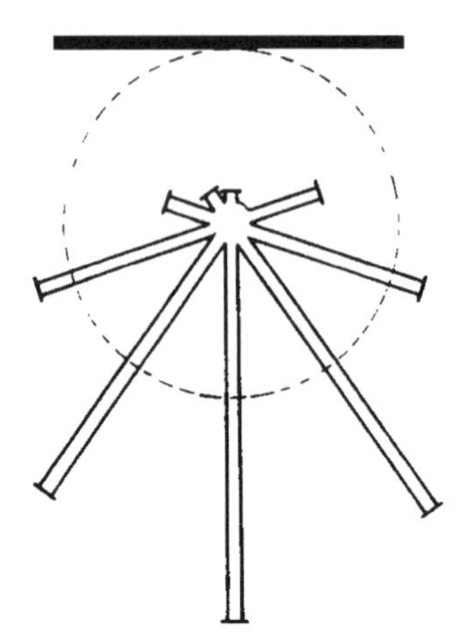

FIGURE 6.14 Mature *Portulaca* plants confronting a large shading object made of green polyethylene. The scheme shows the relative frequency of branches that reached the same distance as the shading object by the third week of the experiment. The broken circle represents the expected frequency of branches if their distribution were random. The analysis was based on a total of 50 plants and 872 branches. There were about ten times more branches growing away from the half circumference of the green sleeve as there were on the opposite side ($P < 0.0001$). (From Novoplansky et al., 1990.)

oleracea perceived the spectral composition and direction of light as clues for the probability of the direction of future shade, and grew away from it.

Plant responses to decreased R : FR ratios differ from their responses to decreased irradiance. To separate these responses, seedlings of *Impatiens capensis* were grown in environments differing in light quantity and quality, and morphological responses of the seedlings were compared (Schmitt and Wulff, 1993). Under decreased light quantity, seedling dry weight and leaf area decreased relative to growth in full sun, which is a typical response that occurs during competition for light. Under leaf shade, however, in which both light quantity and quality were altered, seedling dry weight and leaf area did not change. Instead, shoot length and internode length increased relative to growth in full sun, a typical response to a reduced R : FR ratio. Similar results were found for *Datura ferox* and *Sinapis alba* (Ballaré et al., 1991), *Eichhornia crassipes* (Methy et al., 1990), *Veronica* spp. (Dale and Causton, 1992), and *Trifolium repens* (Solangaarachchi and Harper, 1987).

The sensing by plants of local changes in R:FR ratios influences subsequent growth patterns, assimilate allocation, and therefore future light interception. Because of these effects, light quality must play an important role in competitive interactions of plants. For example, the detection of neighbors prior to mutual shading suggests that the traditional view of the critical period concept needs reevaluation (see Chapter 5). If early R : FR responses are detrimental to crop seedlings, then there may be no period of time during which weed presence is acceptable. New findings on responses to light quality may also have implications for breeding crops for specific photomorphogenic traits. However, it is not yet understood how responses to a reduced R : FR ratio affect overall growth and productivity of plants in canopies. As discussed in Holt (1995), further research is needed to determine the energy costs to the plant of early R : FR responses, whether weeds and crops differ in response to altered R : FR, whether there are additional effects of reduced R : FR such as increased flower and fruit abscission, and most important, whether early light quality changes have long-term effects on crop yield.

WATER

Water is considered to be one of the most variable resources necessary for plant growth. This is especially true if rainfall is the only source of the water supply. If irrigation is used to supplement the water available to plants, some of the variation in supply is reduced. Competition for water usually is considered to be the most severe in rangeland, pasture, and forestry situations, and under the conditions of dryland agriculture. There are three factors that govern water availability for plant growth: (1) the amount of seasonal water supply, (2) plant morphology and root development, and (3) physiology, in particular the water use efficiency of the species. Since the amount of the water supply in the absence of irrigation is rarely able to be controlled, competition for water is likely to occur among species in agricultural environments where water is limited.

Plant Water Relations

The plant provides a pathway for water movement between the soil and the atmosphere. This continuous path of water begins in the soil and moves through the roots, stems, and leaves, and then into the atmosphere. As seen in Figure 6.15, numerous biotic and abiotic factors can influence this process. The movement of water occurs in response to differences in *water potential.* Water potential (ψ) is a thermodynamic measure of the ability to do work, that is, a measure of the free energy of water in the biosphere compared to the free energy of pure water. Pure water has a ψ of 0 MPa (megapascals, a unit that equals 10 bars); the ψ of water in the biosphere is negative. Water movement occurs down a gradient of ψ, from higher (less negative) to lower (more negative). Therefore, water movement occurs from the soil through the plant to the atmosphere when the following gradient is established:

$$\psi \text{ soil} > \psi \text{ root} > \psi \text{ stem} > \psi \text{ leaf} > \psi \text{ air}$$

Common values for water potential could be –0.1 MPa in soil, –1.0 MPa in stems, –1.2 to –1.5 MPa in leaves, and –100 MPa in the atmosphere (Slatyer, 1967). The stomata exert the major control over water flux from leaves to the

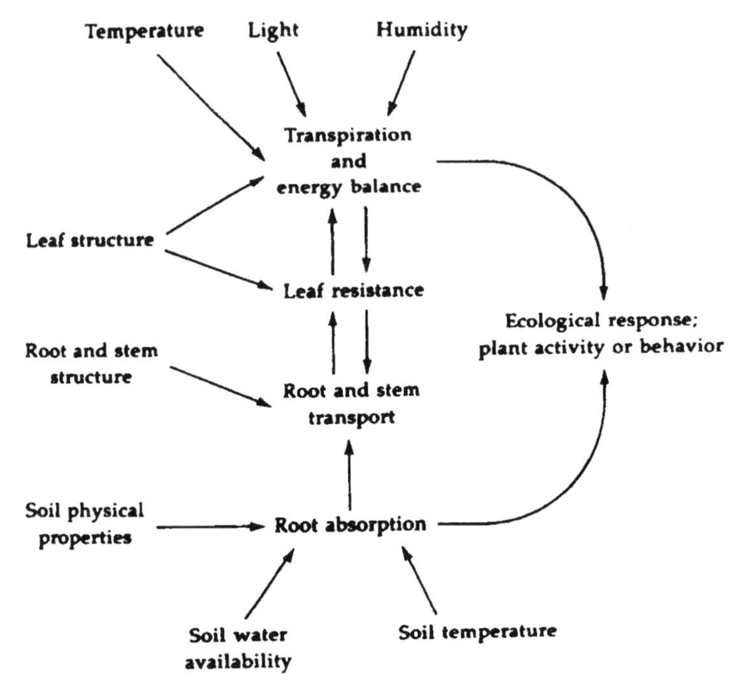

FIGURE 6.15 Diagram showing the interactions of the physical and biotic factors that are the most important in determining the ecological response of plants to water. (From Barbour et al., 1987.)

air in the process of transpiration. In addition to providing turgor, a fluid matrix, and a necessary component of many physiological processes, water is important in maintaining the energy balance of the leaf. The evaporation of water from leaves via transpiration serves to lower leaf temperature. Because stomata regulate this process as well as the uptake of CO_2 fixed in photosynthesis, a dynamic balance must be reached in plants between minimizing water loss and maximizing photosynthetic efficiency.

The primary methods used for measuring ψ of plant tissues of higher plants under field conditions are the psychrometric and pressure chamber techniques (Koide et al., 1991). Thermocouple psychrometers are used when excised tissue samples can be transported to a laboratory for analysis. The pressure chamber technique, in contrast, is particularly well suited for field measurements of plant ψ (Scholander et al., 1964). In this technique, a stem or twig is removed from a plant, sealed in the chamber, and pressure is applied until xylem sap is forced from the cut surface of the sample. The pressure exerted is equivalent to the xylem potential (ψ xylem) of the stem or twig at the time of its removal from the plant. By sampling plant xylem potential periodically throughout the day and throughout a growing season, one can determine the plant's response to the environment in terms of water stress. Usually, predawn values of xylem potential are highest and correspond to maximum available soil moisture.

Midday values are lower than those taken predawn and can represent conditions of highest temperature and least available soil moisture or maximum water stress. Furthermore, xylem potential at any time of day tends to decrease as the growing season progresses and soil moisture becomes less plentiful.

Effects of Water Stress on Plant Growth and Development

When the evaporative demand on a plant exceeds the moisture supply in the soil environment, the plant experiences some degree of water stress. Since CO_2 uptake in photosynthesis is inevitably accompanied by water loss, all plants experience some degree of water stress throughout the day. One immediate effect of water stress is stomatal closure due to reduced turgor in guard cells, which results in a restriction, or increased resistance, in the movement of water, CO_2, and O_2 through the leaves. The degree to which water loss is regulated at the expense of photosynthetic capacity is determined by the stomatal resistance to gas flow that develops as ψ of the plant decreases. This relationship between stomatal resistance, transpiration, and CO_2 assimilation is called *stomatal control.* Water stress can also have a detrimental effect on photosynthesis by causing decreases in electron transport and photophosphorylation and by interfering with the synthesis of chlorophyll and carboxylating enzymes (Hsiao, 1973). In addition, because of the importance of positive turgor to leaf expansion, an early symptom of water stress is often a reduction in the rate of leaf expansion. A list of plant processes that are affected by water stress is shown in Figure 6.16.

Given the potential severity of the effects of water stress on plant survival and growth, it is not surprising that many species possess one or more mechanisms for withstanding times and situations of water shortage. In general, these mechanisms fall into one of three categories:

- adaptations leading to avoidance of water stress
- adaptations leading to the conservation and efficient use of water
- adaptations for tolerance to water stress, usually at the cellular level, which protect cells and tissues from injury during severe desiccation (Fitter and Hay, 1987)

Avoidance of stress is often determined by the morphology and distribution of the root system and by leaf characteristics that favor retention of water. Conservation and efficient use of water are optimized in certain plants through physiological adaptations such as high water use efficiency associated with the C_4 pathway of photosynthesis, as discussed below. Other mechanisms for water conservation include alterations of canopy and leaf characteristics, such as the ratio of root length to leaf area and duration of leaf retention (e.g., deciduous growth habit), tight stomatal control, and xerophytic adaptations such as leaf pubescence and water storage in fleshy structures. Tolerance of des-

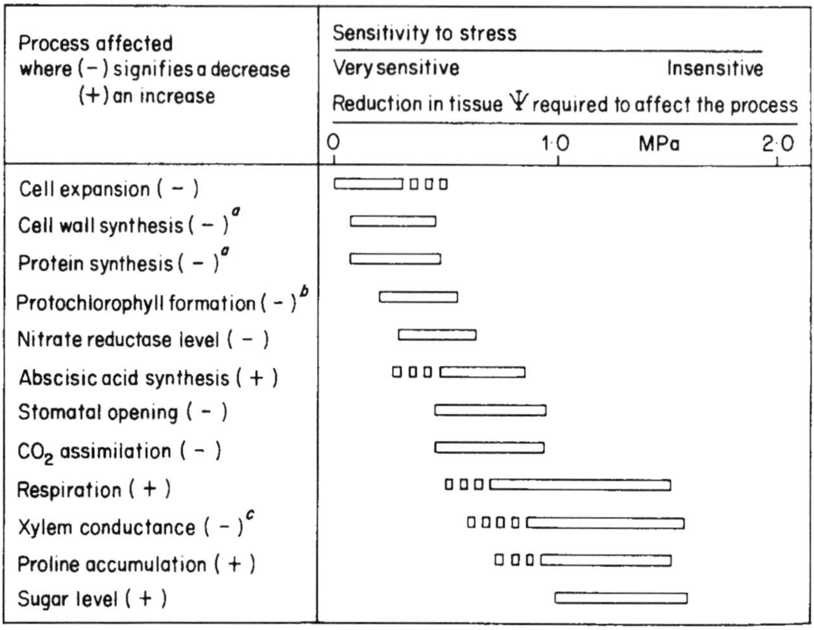

Process affected where (−) signifies a decrease (+) an increase

Sensitivity to stress
Very sensitive — Insensitive
Reduction in tissue Ψ required to affect the process
0 — 1·0 — MPa — 2·0

Cell expansion (−)
Cell wall synthesis (−)[a]
Protein synthesis (−)[a]
Protochlorophyll formation (−)[b]
Nitrate reductase level (−)
Abscisic acid synthesis (+)
Stomatal opening (−)
CO_2 assimilation (−)
Respiration (+)
Xylem conductance (−)[c]
Proline accumulation (+)
Sugar level (+)

[a]Rapidly-growing tissue; [b]etiolated leaves; [c]should depend on xylem dimensions

FIGURE 6.16 The influence of water stress on the physiology of mesophytic plants. The continuous horizontal bars indicate the range of stress levels within which a process is first affected, whereas the broken bars refer to effects which have not yet been firmly established. The reductions in tissue ψ used are in relation to the ψ of well-watered plants under mild evaporative demand. (From Hsiao, 1973.)

iccation is usually achieved by osmotic adjustment, which occurs in some tissues either by the synthesis or absorption of solutes or by dehydration, both of which effectively decrease plant water potential.

Root Function and Distribution

The root systems of plants have received considerably less attention from scientists than have the aboveground parts. This is because roots are relatively inconspicuous, often do not have direct economic value, and are not easily available for study. As a result, our knowledge of roots is largely inadequate. It is generally agreed, however, that root systems serve two distinct functions for the plant. The first is purely mechanical and consists of providing support and anchorage to the growing medium. The second function is physiological; it is only through the root system that minerals and water from the soil solution are supplied to the plant and food reserves previously manufactured are stored. If either function is disrupted, a general decline in plant vigor is usually noted, whereas when roots are healthy a surprising amount of injury to the aboveground portions of the plant can be sustained.

T.K. Pavlychenko, during much of the third and fourth decades of this century, compiled a series of quantitative studies on the rooting patterns of weeds and crops grown in both competitive and noncompetitive associations. In dryland agriculture typical of the Canadian plains, Pavlychenko found that the most severe competition between weeds and crops centered around the supply of available soil moisture. His comments concerning root interactions and their influence on shoot development are especially incisive.

> Competition begins as soon as the root system of one plant invades a feeding area of another, and usually takes place long before tops are developed sufficiently to exert serious competition for light. Therefore, in dry climates roots actually decide the success or failure in competition between species otherwise equally adapted to a region. The top growth is then developed in proportion to the extent of the root system. (T.K. Pavlychenko, 1940)

By using the tedious soil-block washing method, Pavlychenko was able to study quantitatively the root distributions of individual weeds and crops and mixtures of the two. Some of his findings are truly impressive. Consider Figure 6.17, in which a single wild oat plant excavated after 80 days of growth had developed a total root system measuring more than 50 miles! Such underground allocation of carbohydrate and apparent ability to extract resources from the soil environment must have important implications for the competitive relationships among species.

Results from field studies of Pavlychenko on the root distributions and subsequent yields of barley and wheat, grown both individually and in competition with wild oat (*Avena fatua*) or wild mustard (*Sinapis arvensis*), are shown in Table 6.5 and Figures 6.18 to 6.20. In these experiments the cereal crops were grown in 15-cm rows and the weeds were planted between the cereal rows. Both weeds and crops emerged together. Wheat and barley also were grown without weeds for comparative purposes. At harvest, the root systems were separated under water from those of associated plants by the soil-block washing technique, individually analyzed, dyed, spread to their natural conditions according to field charts, and mounted together.

These data vividly demonstrate several points about roots that are significant in competitive interactions. Early after emergence of both crop species, their root systems in competitive and in weed-free situations were nearly equal in size and much larger than those of the weeds growing with them (Table 6.5). However, later in the season the sizes of the crop roots were smaller when grown with a weed, in comparison to the weed-free condition. Crop yields were also reduced in proportion to the decreased size of the root systems, owing to the presence of weeds (Table 6.5). Since planted crops are expected to have a more uniform pattern of germination than weeds, and since these two crop

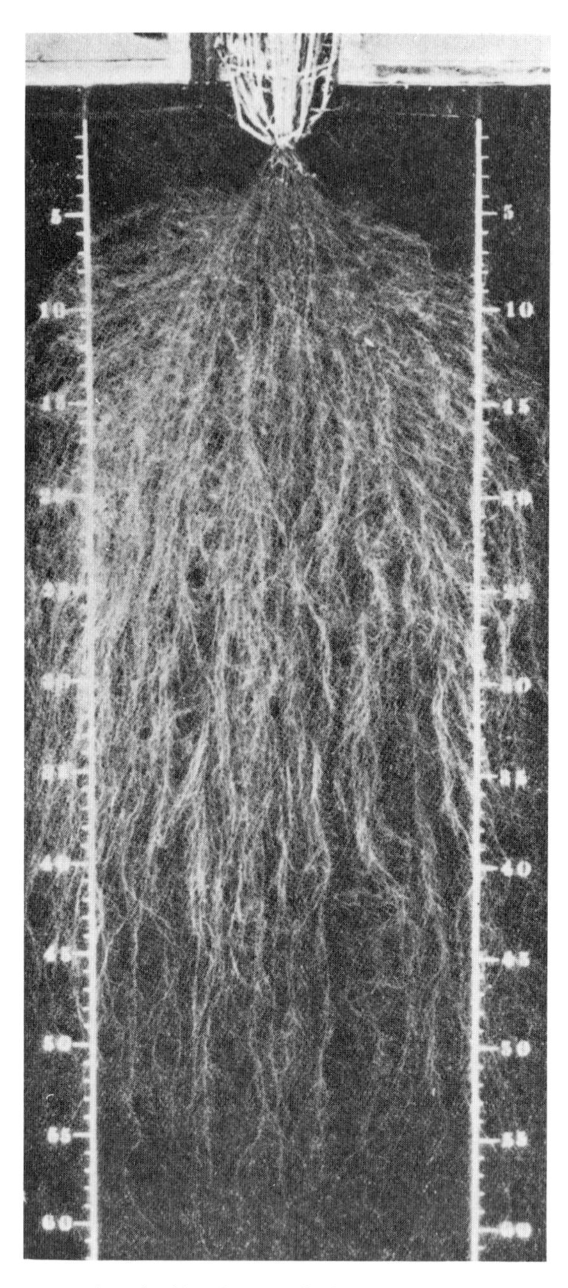

FIGURE 6.17 The root system of a single wild oat plant, grown free from competition and excavated 80 days after emergence. Total length of roots was 54 miles. (From Pavlychenko, 1937b.)

TABLE 6.5 A Comparison of the Yields and the Lengths of the Root Systems per Plant of Barley and Wheat Grown Free of Weeds and in Competition with Wild Oat (*Avena fatua*) and Common Mustard (*Sinapis arvensis*)

				Length of Roots							
				Intraspecific Competition				Interspecific Competition			
	Time from Emergence (days)	Cereal (in.)	Cereal (%)	Cereal (in.)	Cereal (%)	Wild Oat (in.)	Wild Oat (%)	Cereal (in.)	Cereal (%)	Common Mustard (in.)	Common Mustard (%)
Barley	5	112	100	98	87	16	14	99	88	22	20
	22	5849	100	4200	72	752	13	3814	65	2850	49
	40	9865	100	7232	73	1974	20	6446	65	4623	47
Yield at maturity	—		100		85		—		74		—
Wheat	5	68	100	56	82	23	33	57	84	28	41
	22	3534	100	2820	80	1725	49	2711	77	4118	116
	40	5342	100	4169	78	2046	38	3449	65	6053	117
Yield at maturity	—		100		72		—		69		—

Source: Pavlychenko, 1937a, 1940.

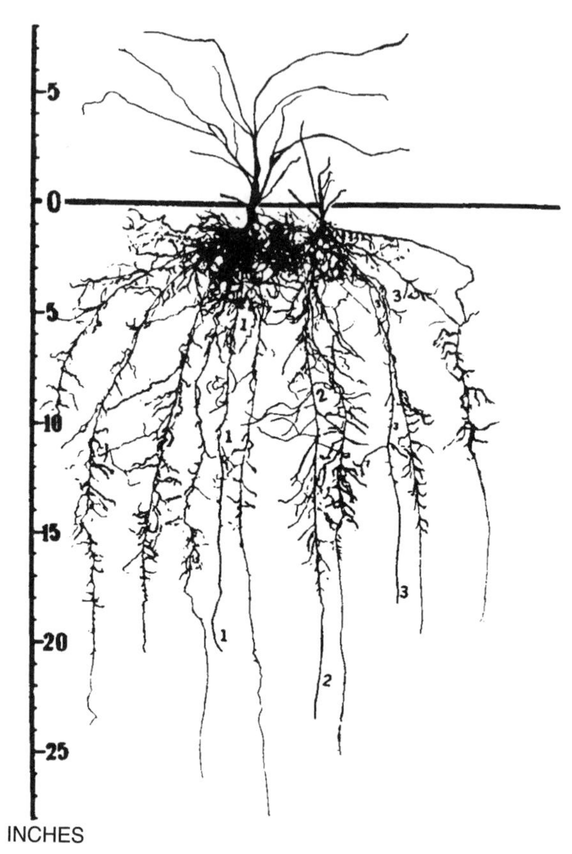

FIGURE 6.18 Root competition between Hannchen barley (left) and wild oats (marked 1,2,3), 22 days after emergence. (From Pavlychenko, 1937a.)

species were able to develop an extensive root system at the early stages of seedling development, some implications for management are suggested.

Successful weed control might be achieved in some crops by establishing the plants in the absence of weeds to allow development of vigorous, early-developing root systems (compare weed sizes and crop roots in Figures 6.18 and 6.19). The advantage of these crops over competing weeds that emerge later should be evident throughout the growing season. In the event that the crop and weeds emerge together, the superior competitor will be determined by each species' ability to grow and usurp resources (especially water; Figures 6.18 to 6.20). In this case, competition may aggravate the limitation of a resource in short supply, which perhaps would not be limiting to a single species growing alone. Thus, increased mortality or decreased growth of the plant with poorest early root development would be the likely result.

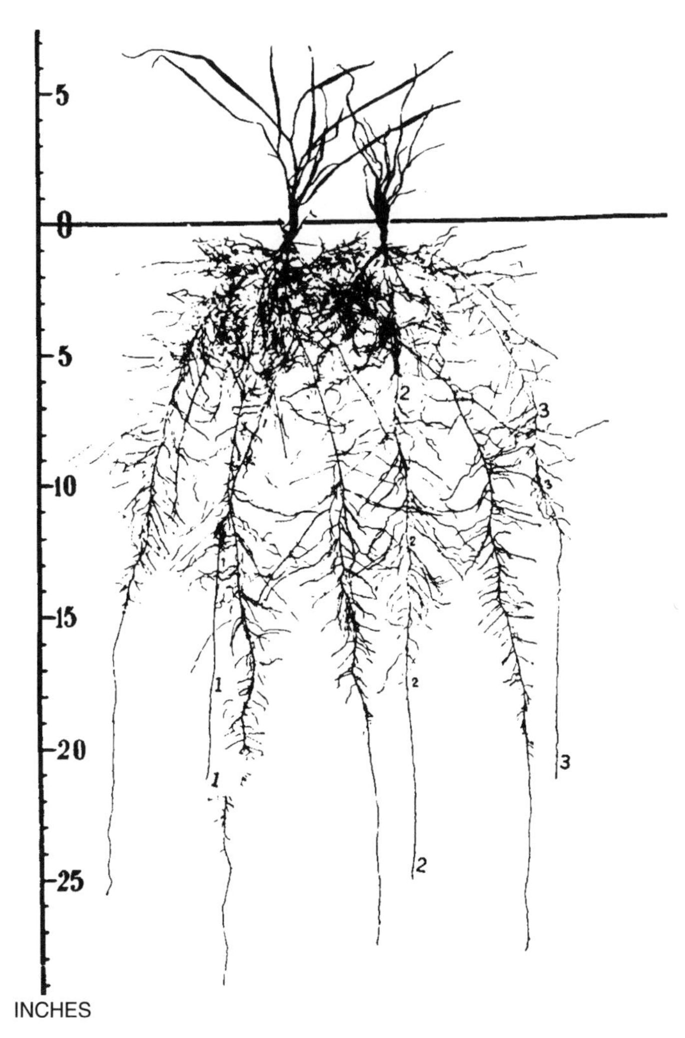

FIGURE 6.19 Root competition between Marquis wheat (left) and wild oats (marked 1,2,3), 22 days after emergence. (From Pavlychenko, 1937a.)

SOIL MOISTURE EXTRACTION BY ROOTS

Davis and his associates (1965) studied patterns of soil moisture use to determine their potential role in weed/crop competition in a cropping system that depended on summer rainfall for moisture. They determined the root moisture extraction profiles of eight weed species and sorghum grown in monoculture field plots. Prior to establishment of plots, the experimental area was flood irrigated to assure that the soil profile was at field capacity to a depth of 6 ft. Then

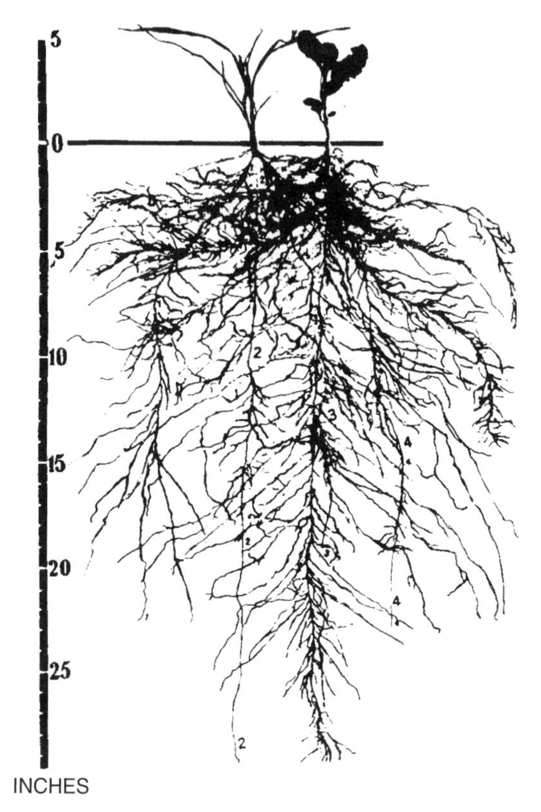

FIGURE 6.20 Root competition between Marquis wheat (left, roots marked 1,2,3,4) and wild mustard, 22 days after emergence. (From Pavlychenko, 1937a.)

two-week-old seedlings of each species were transplanted into single rows, with 6 in, between plants, in the middle of 400 ft^2 plots. Throughout the growing season, gravimetric soil samples were taken at 2-ft intervals from the middle of each plant row out to 8 ft from each row and to a depth of 6 ft. Using this procedure, Davis et al. were able to construct the water use profile for each species (Figure 6.21).

The moisture extraction profiles varied among species, with greatest extraction occurring directly under each plant. By calculating the difference between the quantity of water remaining in the root profile and that in an adjacent area outside the root zone, Davis et al. determined the total amount of water used by each species during their experiment. They found that the amount of moisture used per plant was correlated with the root moisture extraction profiles they constructed (r = 0.79), thus demonstrating the differential ability of species to extract soil moisture from various locations in the soil. These results also suggest that competition for soil moisture probably will be most intense within the crop row and that weed control should be directed mainly toward that zone.

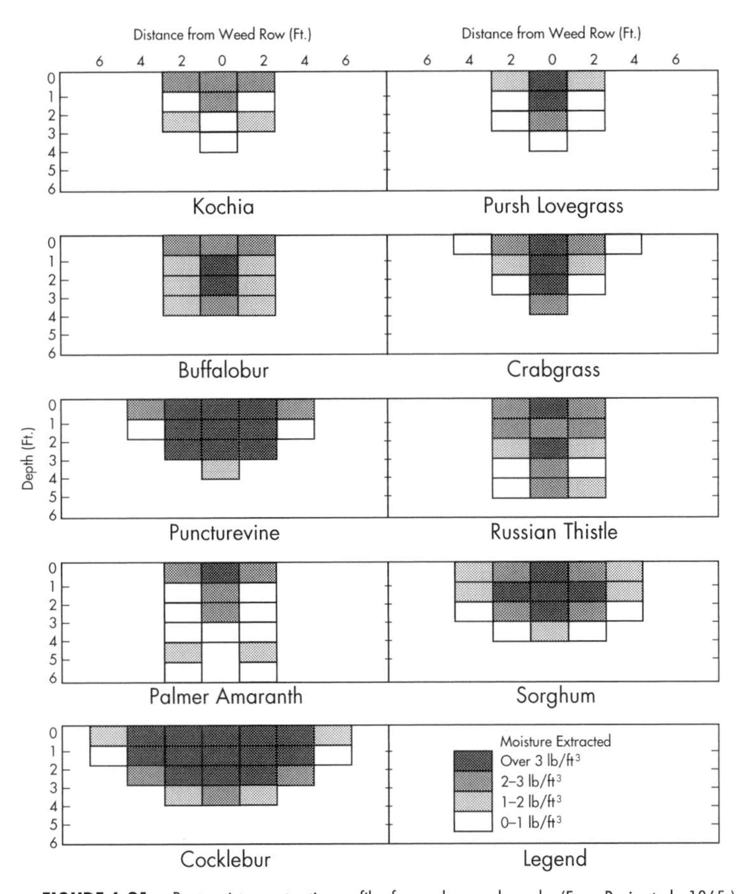

FIGURE 6.21 Root moisture extraction profiles for sorghum and weeds. (From Davis et al., 1965.)

Although reports have been published more recently on soil moisture extraction patterns of crops and weeds, none have provided an analysis as detailed as those of Pavlychenko (1937a, b; 1940) and Davis et al. (1965). As mentioned before, difficulty in working with roots in field situations is the likely reason that so little is known about the belowground functioning of plants, relative to aboveground processes. Clearly this is an important area of plant adaptation and interaction that still needs considerable attention.

Water Use Efficiency

It was noted by Black et al. (1969), based on early work by Briggs and Shantz (1913a, b), that certain plant species are able to use less water per unit of dry matter produced than others (Table 6.6). Those plants with low water requirements, or high *water use efficiencies* (WUE = g CO_2 fixed/g water used) are expected to

TABLE 6.6 Water Requirements for Weed and Crop Species at Akron, Colorado[a]

Efficient Species	*Water Requirement*[b]	*Nonefficient Species*	*Water Requirement*[b]
Monocotyledons			
Panicum miliaceum L. (proso millet)	267	*Hordeum vulgare* L. emend. Lam. (barley)	518
Sorghum spp.	304	*Triticum durum* Desf. (durum wheat)	542
Zea mays L. (corn)	349	*Triticum aestivum* Desf. (wheat)	557
Setaria italica (L.) Beauv. (foxtail millet)	285	*Avena sativa* L. (oats)	583
Sorghum sudanense (Piper) Stapf (Sudangrass)	305	*Secale cereale* L. (rye)	634
Bouteloua gracilis (H.B.K.) Lag. (blue gramagrass)	338	*Agropyron desertorum* Fisch. (crested wheatgrass)	678
		Bromus inermis Leyss. (smooth bromegrass)	977
		Oryza sativa L. (rice)	682
Dicotyledons			
Amaranthus graecizans L. (prostrate pigweed)	260	*Beta vulgaris* L. (sugarbeet)	377
Amaranthus retroflexus L. (redroot pigweed)	305	*Chenopodium album* L. (lambsquarters)	658
Salsola kali L. (Russian thistle)	314	*Polygonum aviculare* L. (prostrate knotweed)	678
Portulaca oleracea L. (common purslane)	281	*Gossypium hirsutum* L. (cotton)	568
		Solanum tuberosum L. (potato)	575
		Xanthium pensylvanicum Wallr. (cocklebur)	415
		Solanum triflorum Nutt. (cutleaf nightshade)	487
		Solanum rostratum Dunal. (buffalobur)	536
		Helianthus annuus L. (sunflower)	623

TABLE 6.6 Continued

Efficient Species	*Water Requirement*[b]	*Nonefficient Species*	*Water Requirement*[b]
Dicotyledons			
		Artemisia frigida Willd. (fringed sagebrush)	654
		Verbena bracteata Lag. & Rodr. (prostrate vervain)	702
		Brassica oleracea capitata L. (cabbage)	518
		Brassica rapa L. (birdrape)	614
		Brassica napus L. (turnip)	714
		Ambrosia artemisiifolia L. (common ragweed)	912
		Citrullus vulgaris Schrad. (watermelon)	577
		Cucumis sativus L. (cucumber)	686
		Vigna sinensis Endl. (cowpea)	569
		Trifolium incarnatum L. (crimson clover)	636
		Phaseolus vulgaris L. (common bean)	700
		Medicago sativa L. (alfalfa)	844

Source: Black et al., 1969.

[a]Data of Shantz and Piemeisel, 1927.

[b]Grams of water required to produce 1 g of dry matter.

be more productive during times of limited water availability than those with high water requirements. It is obvious from comparing Tables 6.4 and 6.6 that C_4 plants have a much lower water requirement than C_3 plants per gram of CO_2 fixed. Patterson and Flint (1983) tested this observation by evaluating water use efficiencies of soybean and seven weeds under controlled-environment conditions. WUE in their studies was measured for whole plants as milligrams of dry matter produced per gram of water lost, as well as for single leaves using the ratio of photosynthesis to transpiration (mg CO_2 fixed per g water lost). They found that WUE on both a whole plant and a single leaf basis was two- to threefold greater for the C_4 species (*Amaranthus hybridus*) than for all the other species studied. It was interesting that among the seven C_3 species studied, the crop (soybean) had the highest rates of water loss per plant, due not to high rates of transpiration but rather to its greater amount of leaf area per plant (Patterson and Flint, 1983).

It might be expected that plants with high water use efficiency would be more competitive than those with a low water use efficiency. To date this expectation has not been substantiated by experimental evidence. Instead, Pearcy et al. (1981) observed that the difference in water use efficiency between common lambsquarters (*Chenopodium album*) and redroot pigweed (*Amaranthus retroflexus*) affected the competitive relationship of those species very little. When the two species were grown together, it was obvious that the C_3 species (*C. album*), which possesses a low WUE, controlled the availability of the water resource by virtue of poor stomatal control. However, *A. retroflexus* (C_4) was not eliminated under the enforced drought conditions caused by the other species, apparently because of its higher WUE. These results provide a good example of the distinction that must be made between the ability to deplete a resource rapidly and the ability to continue growth at depleted resource levels. In this example, both mechanisms are utilized to survive competition for water.

Competition for Water

Competition for water between weeds and crops reduces soil moisture availability, which can cause water stress and ultimately reductions in both weed growth and crop yield. Adaptations for avoidance or tolerance of water stress or for conservation of moisture in tissues are found widely in weeds and crops and often determine the degree of stress experienced. Numerous studies have examined the effects of limiting soil moisture on weeds and crops both in monocultures and in mixtures (see review by Patterson, 1995). Far fewer are the studies that examine possible mechanisms of competition for water. Most of these studies have been conducted in controlled-environment conditions rather than in field environments; nevertheless, they have provided insights into possible mechanisms of competition for water.

Geddes et al. (1979) and Scott and Geddes (1979) studied the water status and competitive relationship of soybean and common cocklebur (*Xanthium pensylvanicum*) in a field experiment. Both species were grown in monocultures and in mixtures in the field. Geddes et al. (1979) observed that a 52 percent reduction in crop seed yield resulted from the association with cocklebur. Soil moisture extraction measured throughout the growing season for soybean, common cocklebur, and the interspecific mixture of the two species was 550, 620, and 650 cm of water, respectively, at a 137-cm depth in the soil. These and other data reported by Geddes et al. (1979) indicated that competition for water was occurring between soybean and cocklebur during the study.

Figure 6.22 depicts the seasonal midday water status of soybean and common cocklebur (Scott and Geddes, 1979). For both species, xylem potential

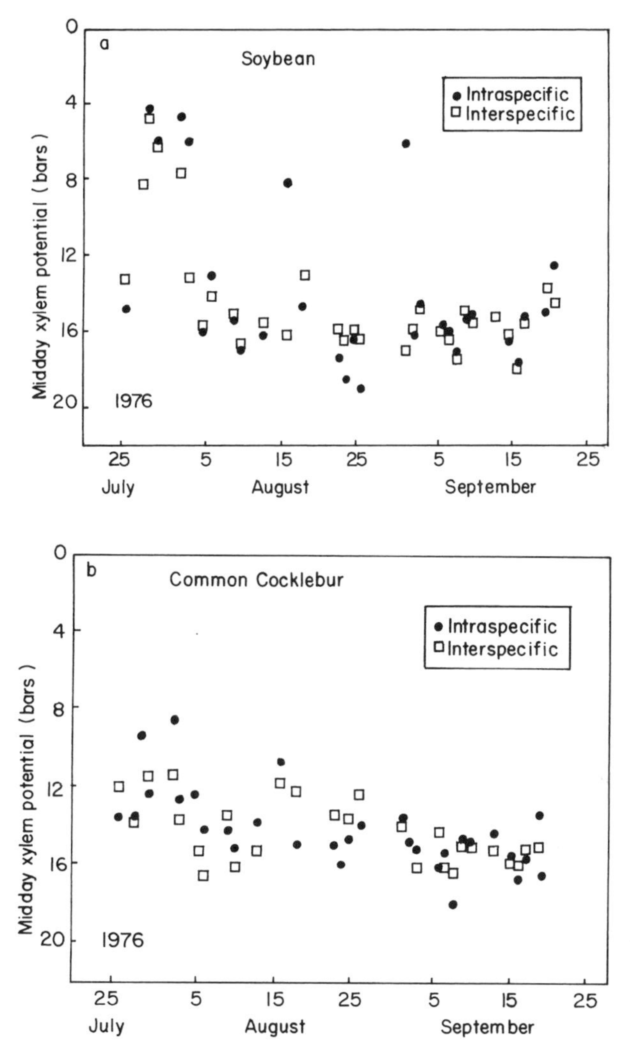

FIGURE 6.22 Seasonal midday xylem potentials of soybean (a) and common cocklebur (b) when grown under intra- and interspecific competition. (From Scott and Geddes, 1979.)

decreased as the season progressed but little difference in ψ_{xylem} was observed whether the species were grown alone or together. However, soybean ψ_{xylem} was higher in late July than that of cocklebur, whereas ψ_{xylem} of cocklebur did not decrease abruptly in early August as did that of soybean (Figure 6.22). This marked decrease in ψ_{xylem} of soybean in August coincided with the initiation of reproductive growth. Since competition for water in this study was severe (as shown by differences in water use) and yet few differences in ψ_{xylem} were observed between intraspecific and interspecific treatments, different patterns of water use between the two species were indicated (Geddes et al., 1979).

When mean midday ψ_{xylem} was examined (Table 6.7), soybean was always under less stress (higher ψ_{xylem}) than cocklebur during vegetative growth (July), thus suggesting better stomatal control by soybean, perhaps to maintain a critical level of ψ_{plant}. Although similar ψ_{xylem} was experienced by both species during the reproductive phase (August), these relatively low values of ψ_{xylem} are particularly stressful for soybean since podfill is a critical stage of development that requires adequate moisture (Scott and Geddes, 1979). When stomatal resistance to water vapor loss from leaves was examined (Table 6.8), soybean was found to moderate water loss more effectively (have higher R_S) than cocklebur, especially under interspecific competition. Greater resistance to water movement out of leaves of soybean compared to cocklebur can account for the similar midday ψ_{xylem} experienced by the crop and weed (Figure 6.22). These results also suggest that soybean is less able to tolerate water loss than cocklebur and attempts to "conserve" soil moisture.

The strategy exemplified by cocklebur, in which poor stomatal control results in relatively high water use, is particularly good for a competitor if its neighbors are water conservers. In this case, the cocklebur used (wasted) a disproportionately larger amount of the resource (water), making it unavailable for use by its neighbor (soybean), and therefore limited the growth of the crop. Cocklebur does not sacrifice any photosynthetic production by closing stomata to accommodate water loss, as soybean must. This feature of competition is especially critical for soybean during reproduction, since it is at that stage of development that photosynthate is needed for seed production.

Another mechanism for survival in mixtures of species is partitioning of resources between competitors in such a way that stress is minimized or avoided.

TABLE 6.7 Mean Midday ψ_{xylem} of Soybean and Cocklebur during Vegetative and Reproductive Growth of Soybean[a]

Growth Stage	*Competition*	*Species*	*Mean Midday ψ_{xylem} (bars)*
Vegetative	Intraspecific	Soybean	−7.46a
		Cocklebur	−11.34b
	Interspecific	Soybean	−8.92a
		Cocklebur	−11.82b
Reproductive	Intraspecific	Soybean	−15.24ab
		Cocklebur	−14.65a
	Interspecific	Soybean	−15.72b
		Cocklebur	−14.53a

Source: Scott and Geddes, 1979.

[a]Means at the same growth stage followed by the same letter do not differ at 5% level using Duncan's multiple range test.

TABLE 6.8 Values of Stomatal Resistance (R_s) of Soybean and Common Cocklebur

		R_s (s/cm)			
		Intraspecific		*Interspecific*	
Date	*Time*	*Soybean*	*Cocklebur*	*Soybean*	*Cocklebur*
July 14	0900	1.77	0.99	1.83	1.33
	1130	0.82	0.78	0.87	1.10
	1345	0.71	1.02	1.20	1.12
August 5	0900	1.05	1.49	1.79	1.45
	1100	0.82	0.47	—	—
	1330	1.27	0.91	1.93	1.12
	1600	1.42	1.40	2.57	0.97
August 17	0940	1.23	1.04	1.41	0.98
	1110	0.93	0.81	1.27	0.93

Source: Scott and Geddes, 1979.

Gordon and Rice (1992) examined rooting morphology and partitioning of water between two annual grassland species. *Erodium botrys* and *Bromus diandrus* were grown in monocultures and 50 : 50 mixtures at three densities in one-meter-tall containers outdoors. For *E. botrys*, greater competitive effects on reproductive output were caused by intraspecific competition (monoculture plots) than interspecific competition, while, for *B. diandrus*, competition had little effect on reproduction. When belowground growth was evaluated, differences in root morphology and depth were found for these two species (Figure 6.23). *B. diandrus* roots grew primarily in the upper 10 cm of soil, while *E. botrys* roots were distributed both in the surface and deep soil. In mixtures, roots of the two species had a pattern intermediate between those of the monocultures (Gordon and Rice, 1992). These underground growth patterns resulted in partitioning of soil resources so that soil moisture potential was affected by total plant density but not by species composition, as shown in Figure 6.24. Partitioning of soil moisture in this system thus minimized the effects of interspecific competition relative to intraspecific competition and allowed coexistence of the two species.

In view of the differences among plants in patterns of water use and strategies for withstanding water stress, weed management through irrigation practices may be possible. Patterson (1995) summarized numerous studies of weed/crop competition for water and concluded that, in some situations, decreased water availability favors crop growth by reducing the impact of the competing weed (Table 6.9). However, it is likely that one reason for the lower competitive effect of weeds on crops in water-limited situations is the lower

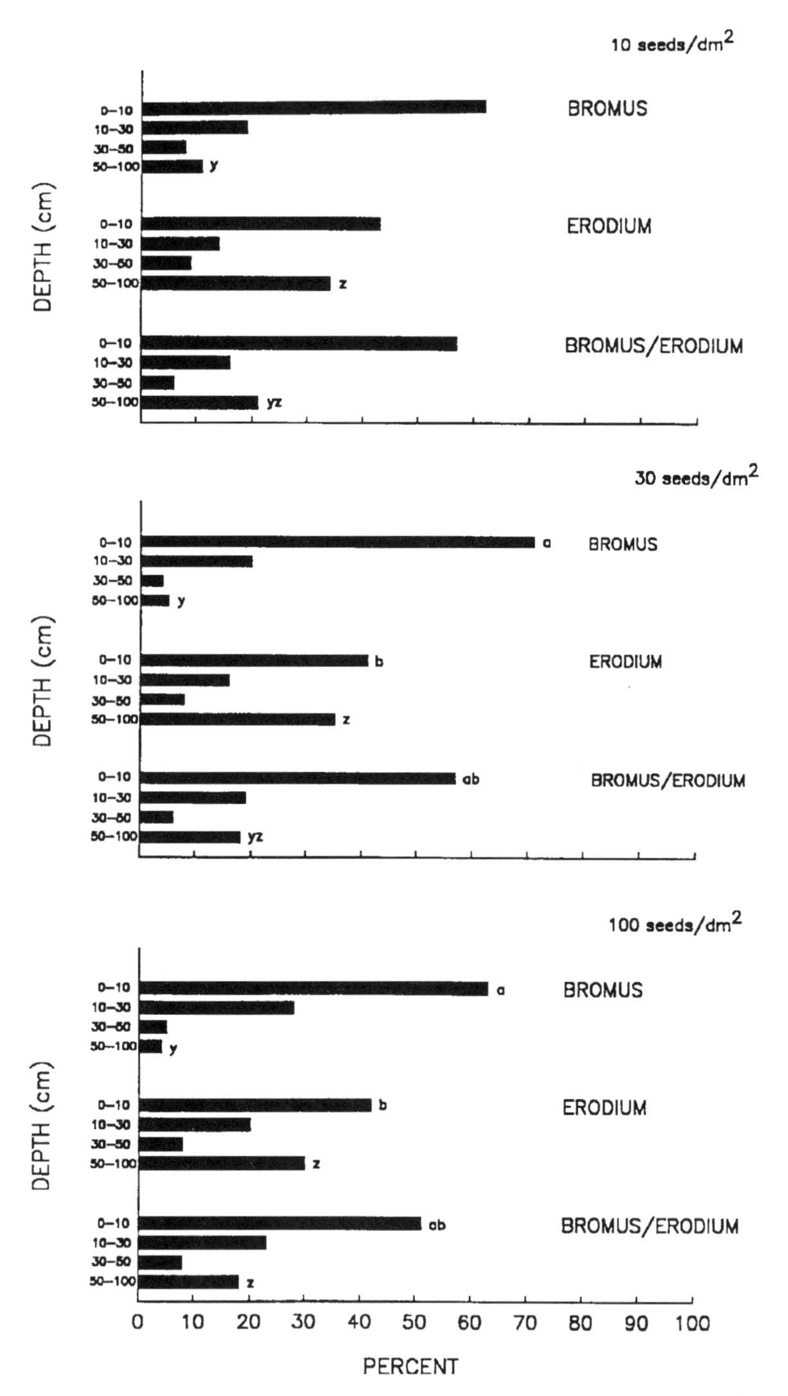

FIGURE 6.23 Percent root distribution over depth by stand composition and density. Different coefficients indicate significant differences in allocation across stands within a depth class ($P < 0.05$). (From Gordon and Rice, 1992.)

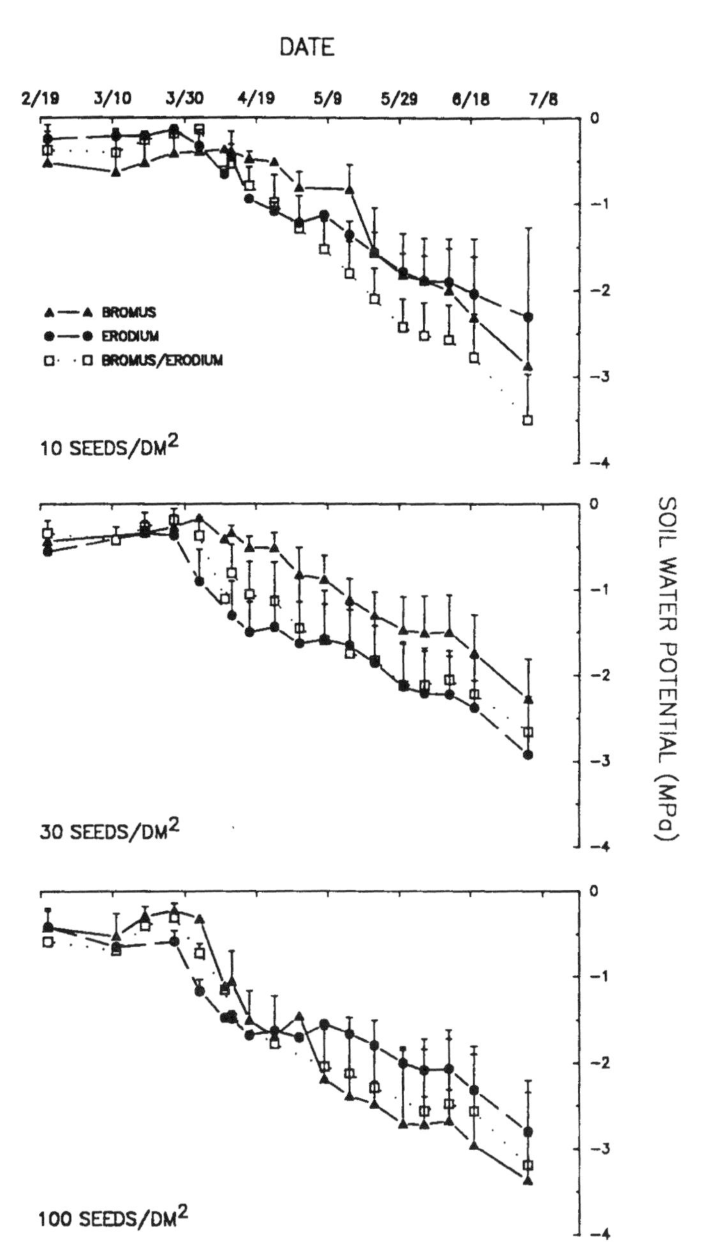

FIGURE 6.24 Soil water potential (40-cm depth) over time by stand composition and stand density. (From Gordon and Rice, 1992.)

TABLE 6.9 Effects of Decreased Water Availability on Weed Impact on Crop Growth or Yield

Crop	*Weed*	*Effect*[a]	*Reference*[b]
Soybean	Cocklebur	−	50
Soybean (tilled)	Sicklepod	+	7
Soybean (no-till)	Sicklepod	0	7
Soybean	Jimsonweed	−	34
Soybean	Jimsonweed	+	92
Soybean	Jimsonweed	0	57
Soybean	Common ragweed	−	17
Soybean	Redroot pigweed	+	58
Soybean	Pennsylvania smartweed	+	94
Soybean	Velvetleaf	−	33
Soybean	Multi-species	+	40
Soybean	Giant foxtail	−	32
Soybean	Giant foxtail	−	37
Soybean	Giant foxtail	+	94
Soybean	Giant foxtail	−	36
Soybean	Fall panicum	−	36
Soybean	Robust foxtail	+	58
Cotton	Devil's claw	−	81
Cotton	Silverleaf nightshade	+	30
Cotton	Jimsonweed	−	57
Winter wheat	Jointed goatgrass	0	4

Source: Patterson, 1995.

[a]0 = no effect, + = increased impact, − = decreased impact.

[b]References are cited in original source.

potential maximum yield of the crop, which would also be impacted by water limitations. Thus, only if water stress reduced weed growth more than crop growth would the manipulation of soil moisture for weed management be a reasonable strategy.

NUTRIENTS

Plants are thought to absorb mineral elements more or less indiscriminately from the soil, their rooting medium. A deficiency of any particular chemical element often makes it impossible for the plant to complete its life cycle. However, the mere presence of an element in a plant's tissue does not necessarily indicate that it is essential for growth or development. The elements and inorganic compounds necessary for plant growth are normally categorized according to the relative quantities required for adequate nutrition. Those required in large amounts, *macronutrients*, include carbon, oxygen, hydrogen, nitrogen, potassium, calcium, phosphorous, magnesium, and sulfur. These elements are also the basic materials from which protoplasm, membranes, and cell walls are constructed and maintained. In addition, there are other elements, *micronutrients*, that are required in lesser amounts but are essential for plant growth. Table 6.10 summarizes the functions of inorganic nutrients in plants and typical con-

TABLE 6.10 A Summary of the Functions of Inorganic Nutrients in Plants

Element	*Principal Form in which Element is Absorbed*	*Usual Concentration in Healthy Plants (% or ppm of Dry Weight)*	*Important Functions*
Macronutrients			
Carbon	CO_2	~44%	Component of organic compounds
Oxygen	H_2O or O_2	~44%	Component of organic compounds
Hydrogen	H_2O	~6%	Component of organic compounds
Nitrogen	NO_3^- or NH_4^+	1–4%	Component of amino acids, proteins, nucleotides, nucleic acids, chlorophylls, and coenzymes
Potassium	K^+	0.5–6%	Involved in osmosis and ionic balance and in opening and closing of stomata; activator of many enzymes
Calcium	Ca^{2+}	0.2–3.5%	Component of cell walls; enzyme cofactor; involved in cellular membrane permeability; component of calmodulin, a regulator of membrane and enzyme activities

(*continued*)

TABLE 6.10 Continued

Element	*Principal Form in which Element is Absorbed*	*Usual Concentration in Healthy Plants (% or ppm of Dry Weight)*	*Important Functions*
Macronutrients			
Phosphorus	$H_2PO_4^-$ or HPO_4^{2-}	0.1–0.8%	Component of energy-carrying phosphate compounds (ATP and ADP), nucleic acids, several essential coenzymes, phospholipids
Magnesium	Mg^{2+}	0.1–0.8%	Part of the chlorophyll molecule; activator of many enzymes
Sulfur	SO_4^{2-}	0.05–1%	Component of some amino acids and proteins and of coenzyme A
Micronutrients			
Iron	Fe^{2+} or Fe^{3+}	25–300 ppm	Required for chlorophyll synthesis; component of cytochromes and nitrogenase
Chlorine	Cl^-	100–10,000 ppm	Involved in osmosis and ionic balance; probably essential in photosynthetic reactions that produce oxygen
Copper	Cu^{2+}	4–30 ppm	Activator or component of some enzymes
Manganese	Mn^{2+}	15–800 ppm	Activator of some enzymes; required for integrity of chloroplast membrane and for oxygen release in photosynthesis

TABLE 6.10 Continued

Element	*Principal Form in which Element is Absorbed*	*Usual Concentration in Healthy Plants (% or ppm of Dry Weight)*	*Important Functions*
Micronutrients			
Zinc	Zn^{2+}	15–100 ppm	Activator or component of many enzymes
Molybdenum	MoO_4^{2-}	0.1–5.0 ppm	Required for nitrogen fixation and nitrate reduction
Boron	$B(OH)_3$ or $B(OH)_4^-$	5–75 ppm	Influences Ca^{2+} utilization, nucleic acid synthesis, and membrane integrity
Elements Essential to Some Plants or Organisms			
Sodium	Na^+	Trace	Involved in osmotic and ionic balance; probably not essential for many plants; required by some desert and salt-marsh species and may be required by all plants that utilize C_4 pathway of photosynthesis
Cobalt	Co^{2+}	Trace	Required by nitrogen-fixing microorganisms

Source: Raven et. al., 1992.

centrations found in healthy plants. It should be pointed out, however, that species can differ widely in their absolute ion concentrations and relative proportions of different ions.

Note that with the exception of carbon, oxygen, and hydrogen, which comprise the bulk of structural material and carbohydrates in plants, even those elements considered to be major resources for growth are actually present in whole plants in surprisingly small amounts. However, any element can be limiting, and normal plant growth occurs only when that deficiency has been corrected. Given the critical importance of mineral nutrients for plant growth and development and the high densities of crops and weeds found in most agricultural fields, it is not surprising that competition for nutrients occurs. In terms of competition, there is also a strong interaction between nutrient availability and other physiological or morphological functions. For example, adequate nutrition for root development cannot be discounted as a factor in competition for water. Furthermore, increases in soil fertility have been shown to increase aboveground plant growth and thus intensify competition for light. Realizing this complexity in the interactions of different resources, this section of the chapter considers the competitive relationships among weeds and crops for only a few of the major nutrients. Competition for minor nutrients also may be severe at times, but the process should be similar regardless of the level of requirement for the nutrient.

Effects of Soil Fertility on Plant Growth

Uptake and assimilation of nutrients by plants is the primary means of incorporation of minerals into the biosphere. Other organisms, such as mycorrhizal fungi, are often associated with plant roots and are involved in nutrient acquisition (see also Chapter 7). Following root absorption, nutrients are translocated to other parts of the plant. Under conditions where uptake of nutrients exceeds the rate of replenishment in the soil solution, a zone of nutrient depletion forms adjacent to plant roots which decreases as distance to the root surface increases (Figure 6.25). As a result, continued uptake of nutrients by roots depends not only on their intrinsic nutrient uptake capacity but also on continuous root growth into previously untapped areas of soil (Taiz and Zeiger, 1991).

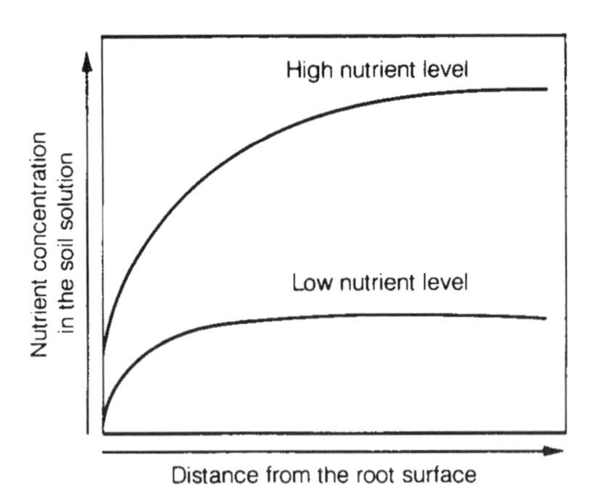

FIGURE 6.25 Formation of a nutrient depletion zone in the region of the soil adjacent to the plant root. A nutrient depletion zone is formed when the rate of nutrient uptake by the cells of the root exceeds the rate of replacement of the nutrient by diffusion in the soil solution. This depletion causes a localized decrease in the nutrient concentration in the area adjacent to the root surface. (From Taiz and Zeiger, 1991.)

The response of plants to nutrient concentrations follows the general trend described in Figure 6.1a. A more specific curve for the response of plant growth or yield to tissue nutrient concentration is depicted in Figure 6.26. As seen in this figure, low tissue concentrations of a nutrient below some critical concentration result in reduced growth, while high concentrations may result in toxicity. In the zone of adequate nutrient concentration, further increases in the mineral do not result in increased growth. Since the nutrient status of plant leaves is closely correlated with the soil nutrient supply, leaf nutrient analysis such as that shown in Figure 6.26 is often used to establish critical nutrient concentrations for specific crops. Plant tissue analysis also has become common as an indicator of nutrient deficiency rather than other, more traditional, morphological deficiency symptoms. Such tissue analysis has been the foundation of the development of crop fertilizer programs in many agricultural systems.

In recent years, studies on mineral nutrition have focused primarily on ion transport physiology. However, the kinetics of nutrient uptake and transport are important primarily as short-term responses to local fluctuations in soil ion concentrations and do not indicate overall strategies for nutrient acquisition by plants. Long-term adaptation to limitations in soil nutrients is achieved by alterations in demand for and use of nutrients and by changes in root morphology and distribution (Fitter and Hay, 1987). In general, nutrient-poor environments are characterized by high levels of resource allocation to roots and increased specific root length (SRL, length of root per unit root weight). Most agricultural situations, however, are treated with fertilizers to promote crop growth. Thus, intense competition for nutrients rather than long-term depletion should characterize these habitats. Furthermore, weed competition with crops probably

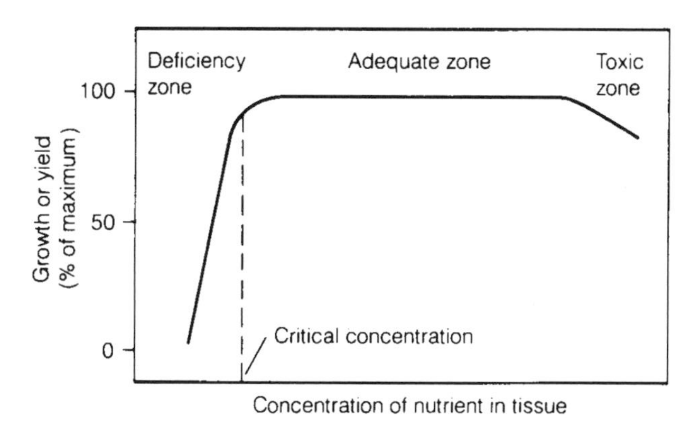

FIGURE 6.26 Relationship between yield (or growth) and the nutrient content of the plant tissue. The yield parameter may be expressed in terms of shoot dry weight or height. The deficient, adequate, and toxic zones are indicated on the graph. In order to obtain data of this type, plants are grown under conditions in which the concentration of one essential nutrient is varied while all others are in adequate supply. The effect of varying the concentration of this nutrient during plant growth is reflected in the growth or yield. The critical concentration for that nutrient is obtained by determining the concentration below which a reduction in yield or growth occurs. (From Taiz and Zeiger, 1991.)

increases as the resource base of a soil is deteriorated (Ghersa and Martinez-Ghersa, 1992a; see Chapter 11 for more discussion of this topic.).

The importance of inorganic fertilizers, in particular nitrogen, potassium, and phosphorous, in maximizing crop production is well known. As pointed out in reviews by DiTomaso (1995) and Patterson (1995), however, applications of fertilizers also benefit weeds, often more than they benefit crops. Competition for nitrogen is generally greater than for other macronutrients, so that additions of N fertilizer often both increase crop yield and intensify competition with weeds. One mechanism for this advantage of fertilization to weeds relative to crops appears to be accumulation of mineral nutrients by many weed species in excess of the critical concentration needed for growth. Thus, in designing crop fertilization strategies, it is important to be aware of the differences between crops and weeds in nutrient requirements and uptake, and of the effects of nutrients on the competitive interactions between crops and weeds.

Competition for Nutrients

J.W. Ince (1915) conducted one of the earliest studies on the impacts of weeds on soil fertility. He determined the ash, nitrogen, and phosphorus content of several weed and crop specimens. The ash content in the weed species averaged about 12 percent, which was only slightly higher than that of the four crops in his study. Percent nitrogen and phosphorus also were similar between crops and weeds. Based on these and other experiments in which certain crops and weeds were grown in association and their nutrient levels were determined, Ince concluded that weeds make a severe drain on soil fertility and deprive crops of considerable mineral nutrition.

Vengris and his associates (1955) also addressed the subject of nutrient utilization by conducting a thorough analysis of tissue nutrient contents in weeds and crops. Their objective was to assess the importance of weed removal via tillage on nutrient availability to crops. In these experiments Vengris et al. grew corn, redroot pigweed (*Amaranthus retroflexus*), common lambsquarters (*Chenopodium album*), crabgrass (*Digitaria sanguinalis*), and barnyardgrass (*Echinochloa crus-galli*) as pure stands in rows in the field and as mixtures of corn and the weeds combined. The mixture plots (corn plus weeds) were cultivated to remove weeds not planted as part of the experiment. Plant samples were taken for nutrient analysis one month after plant emergence and again at harvest (Vengris et al., 1955).

Table 6.11 compares nutrient content of corn grown with and without weeds, and the weed species when grown either with corn or in pure stands. The nitrogen and potassium contents of corn were suppressed from association with weeds, whereas the overall nutrient status of the individual weeds usually was not influenced by the presence of corn. Magnesium behaved differently from the other nutrients in that it increased in corn when weeds were present (Table 6.11). Vengris et al. concluded that weeds competed strongly for nutri-

TABLE 6.11 Nitrogen, Potassium, Calcium, and Magnesium Content of Corn and Weeds (Air-dry Basis)

	N%		K%		Ca%		Mg%	
Plant	*Young*	*Early Maturity*	*Young*	*Early Maturity*	*Young*	*Early Maturity*	*Young*	*Early Maturity*
Corn, alone	3.78	1.44	4.02	0.90	0.40	0.17	0.43	0.27
Corn with weeds	3.45[b]	1.38	3.08[c]	0.70[b]	0.42	0.20	0.57[b]	0.36[b]
Pigweed, alone	4.36	2.45	4.33	1.80	1.94	0.76	1.90	1.01
Pigweed with corn	4.06	2.52	4.05	2.42[b]	1.87	0.62	1.78	0.93
Lambsquarters, alone[a]	5.30	2.66	3.92	2.00	1.55	0.70	1.68	0.77
Lambsquarters with corn[a]	5.12	2.64	4.01	2.50[b]	1.43	0.75	1.59	0.75
Crabgrass, alone	4.26	2.11	3.96	1.98	0.50	0.34	0.86	0.92
Crabgrass with corn	4.02	2.09	4.17	1.80	0.42	0.38	0.83	0.96
Barnyardgrass, alone	3.82	1.56	2.96	0.98	0.77	0.82	1.15	1.07
Barnyardgrass with corn[a]	—	1.51	2.88	0.87	0.64	0.63	1.04	1.24

Source: Vengris et al., 1955. *Agronomy Journal* 47: 213–216, by permission of the American Society of Agronomy.

Note: Plots received 200 lb/acre of N, 200 lb/acre of P_2O, 200 lb/acre of K_2O. Data are averages of 1952 and 1953.

[a]One year's data only.

[b]Significant at 5 percent level in comparison with grown alone.

[c]Significant at 1 percent level in comparison with grown alone.

ents even though high rates of nitrogen, phosphorus, and potassium were used in their study. That different plant species could absorb and utilize nutrients differentially is not surprising. More important, however, is that these results suggest that it may not be possible to overcome nutrient deficiencies, and thus to compensate for nutrient competition by weeds, simply by increasing the level of soil fertility.

Studies by Gray et al. (1953) support the observations of Vengris et al. (1955). In greenhouse experiments, Gray and his associates found that bentgrass (*Agrostis palustris*) maintained a higher level of potassium uptake relative to that of Ladino clover when the two species were grown separately or together. The differential potassium uptake between the two species was closely correlated to their respective root cation exchange capacities. Gray et al. concluded that for the soil type used in their study, and using practical rates of potassium fertilization (265 kg/ha), it would be impossible to maintain an adequate potassium supply for Ladino clover when it was associated with bentgrass. Other findings by Vengris et al. (1955) suggest that in certain weed/crop associations, the weeds may consume nutrients luxuriant compared to the crop (Figure 6.27).

A thorough analysis of nutrient accumulation by weeds in comparison to tomato and bean crops was conducted by Qasem (1992). A variety of common weeds occurring in fields with either tomato or bean on a vegetable farm were collected and analyzed for N, P, K, Ca, and Mg content. The concentrations of these mineral nutrients in shoots of several weeds relative to associated bean or tomato plants are shown in Table 6.12. Overall, the weeds examined were better nutrient accumulators than either crop. Percentages of N, P, K, and Mg were higher in weed shoots than in the crops, while Ca concentration was generally lower. While not conducted under controlled conditions, this study provides strong evidence for the luxuriant accumulation of mineral elements by many weed species (Qasem, 1992).

The importance of nitrogen fertility to competition between Italian ryegrass (*Lolium multiflorum*) and winter wheat was the object of study by Appleby et al. (1976). In this additive experiment the influence of ryegrass density and three nitrogen fertility regimes on short- (Nugaines) and tall- (Druchamp) statured varieties of wheat was compared. Over the course of the study, Appleby et al. observed that increasing ryegrass density always decreased wheat yield (Table 6.13). High levels of nitrogen in combination with high ryegrass density decreased wheat yields the most. These data suggested that the ryegrass was bet-

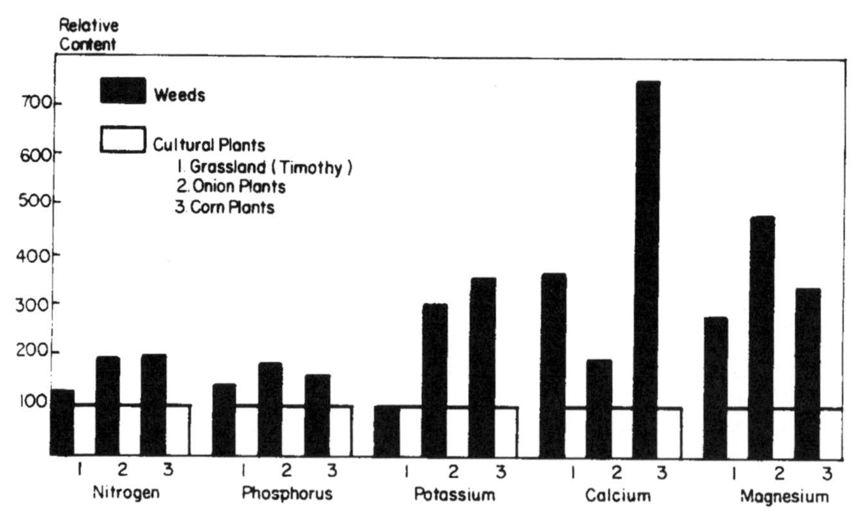

FIGURE 6.27 Mineral content of weeds and crops. Weeds are represented as having a greater mineral content than their associated cultivated plants. The cultivated plants have been given the value 100. Average data of 1950–1951. Samples collected from typical farms in the Connecticut Valley, Massachusetts. (Data from Vengris et al., 1955, and redrawn from King, 1966. By permission of the International Textbook Company, Glasgow.)

TABLE 6.12 Comparison of Relative Mineral Element Composition of Shoots of Weeds Associated with Bean and Tomato Plants (Crop Plants = 100)

	Bean					Tomato				
Plant	*N*	*P*	*K*	*Ca*	*Mg*	*N*	*P*	*K*	*Ca*	*Mg*
Bean or Tomato	100	100	100	100	100	100	100	100	100	100
Chenopodium murale	119	111	154	49	198	109	121	137	39	119
Cichorium pumilum	141	168	160	77	123	108	118	119	55	44
Malva sylvestris	172	171	139	117	114	103	113	105	65	36
Sonchus oleraceus	156	147	174	57	100	126	147	122	43	44
Portulaca oleracea	160	116	177	100	193	127	155	143	42	113
Sisymbrium irio	149	153	154	79	109	124	153	124	68	51
Rumex obtusifolius	146	168	185	42	164	108	58	79	12	34

Source: Qasem, 1992.

ter able to respond to increased nitrogen fertility than was either variety of wheat. In the absence of effective ryegrass suppression, therefore, increased fertility was of questionable value. Liebl and Worsham (1987) found similar results with both nitrogen and potassium fertilization in mixtures of the same two species.

Numerous other more recent studies have confirmed these early findings on the competitive interactions of crops and weeds for soil nutrients. For example, in mixtures with wheat, wild oat (*Avena fatua*) benefited more than wheat from additions of nitrogen fertilizer, apparently due to higher nitrogen use efficiency. Nitrogen fertilization in this mixture increased the competitive impact of wild oat and decreased wheat yield (Carlson and Hill, 1985). The form in which nitrogen fertilizer is supplied may also affect weeds and crops differentially. In competitive mixtures of pigweed (*Amaranthus retroflexus*) and corn, pigweed accumulated 2.5 times as much nitrogen as corn when high levels of nitrogen fertilizer were supplied (Teyker et al., 1991). However, in contrast to corn, pigweed preferentially absorbed NO_3^- and was unable to use nitrogen supplied as NH_4^+. In this particular crop/weed mixture, therefore, enhancing the proportion of nitrogen supplied as NH_4^+ relative to NO_3^- would restrict growth of NH_4^+-sensitive pigweed (Teyker et al., 1991).

Although all nutrients and all plant species are not expected to act identically, some generalizations can be made. Plant species growing in proximity, either weeds or crops, can interact competitively for nutrients. However, the value of increased nutrient availability will be determined by each species' ability to respond to and utilize the added resource. It is clear that weeds consume nutrient resources at least as readily and, in many cases, more readily than crops. The removal of weeds from a cropping system should therefore improve the nutrient availability to the crop and increase productivity within the limits

TABLE 6.13 Grain Yields of Two Wheat Cultivars at Three Nitrogen Levels and Four Ryegrass Densities

			1971			1972		
Cultivar	*N Level (kg/ha)*	*Ryegrass Planting Level*	*Ryegrass Density (plants/m²)*	*Wheat Grain Yield (kg/ha)*	*Grain Yield Reduction (%)*	*Ryegrass Density (plants/m²)*	*Wheat Grain Yield (kg/ha)*	*Grain Yield Reduction (%)*
Nugaines	56	0	1	3070	0	1	2260	0
		Low	12	2680	12.7	11	2143	5.2
		Med	62	2070	32.6	29	1880	16.8
		High	114	1920	37.5	73	1580	30.1
	112	0	0	3570	0	0	2640	0
		Low	13	3380	5.3	13	2180	17.4
		Med	45	2300	35.6	43	1800	28.8
		High	118	2000	44.0	87	1500	44.3
	168	0	1	3450	0	0	2640	0
		Low	14	3260	5.5	10	2260	14.4
		Med	52	2183	36.7	37	1880	28.8
		High	84	1720	50.1	76	1470	44.3
Druchamp	56	0	1	3680	0	1	2710	0
		Low	7	3340	9.2	10	2600	4.1
		Med	38	3410	7.3	36	2300	15.1
		High	95	2950	19.8	66	2300	15.1
	112	0	0	3990	0	1	3500	0
		Low	12	4220	(+5.8)	10	3200	8.6
		Med	41	3570	10.5	40	3050	12.9
		High	85	3180	20.3	75	2520	28.0
	168	0	1	4150	0	0	3650	0
		Low	9	3790	8.7	7	3280	10.1
		Med	38	3150	24.1	31	3127	14.3
		High	69	2840	31.6	73	2490	31.8
LSD, 0.05				643			446	
$s_{\bar{x}}$				230			159	

Source: Appleby et al., 1976. By permission of the American Society of Agronomy.

imposed by water or other resources. Furthermore, increased fertility does not appear to be a substitute for weed density reduction. This may be due to luxury consumption of nutrients by some weeds or it may be due to an effect of increased fertility on some process of plant growth, such as root or canopy development, with concomitant increased resource utilization.

Interactions of Nutrients with Other Resources and Plant Density

The interacting effects of limitations in more than one resource on weed and crop growth and competition are typically studied in controlled environments due to the complexity of the treatments to be examined. Similarly, the relationship between soil fertility and the eventual limitation of another resource is especially hard to determine in field competitive situations common to most crop/weed complexes. In field situations, plant density also has a significant effect on resource limitation. As seen in Figure 6.28, increasing the nutrient status of a crop at any given plant density can increase productivity to a significant

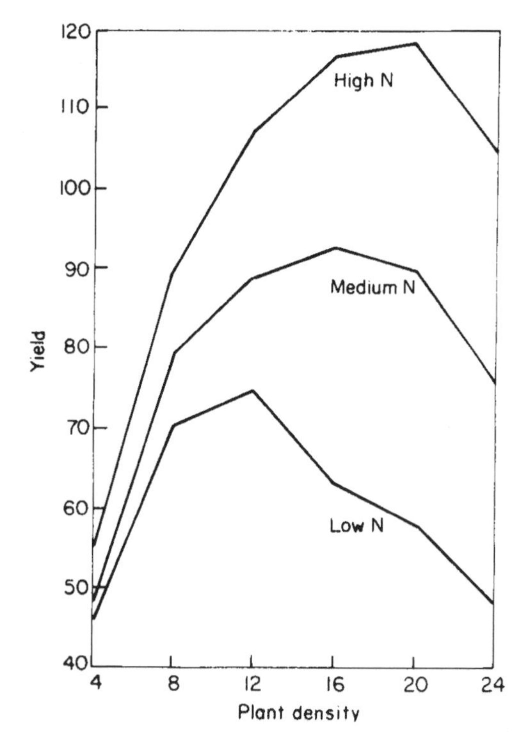

FIGURE 6.28 The relationship between density, nitrogen supply, and yield in maize (*Zea mays*). Densities are in 10^3 plants/acre and yields in bushel/acre. (Reproduced from Lang et al., 1956. By permission of the American Society of Agronomy.)

degree, but maximum productivity is obtained only when high fertility is combined with optimal density.

Often the major effect of supplying nutrients to a plant is to decrease the time before, and thus increase the likelihood that, other factors will become limiting. For example, added nutrients can allow for greater root development and therefore more thorough water extraction by a competitor. Similarly, faster aboveground growth and canopy closure can result from supplying additional nutrients, and this can be followed by increased competition for light and selective exclusion of the poorer competitor.

Harper (1977) describes a hypothetical situation involving nutrient availability that demonstrates the complexity of resource interactions. If a crop were planted in nutrient-deficient soil, isolated plants of that crop would be stunted, root systems would not interlock, and growth might be so poor that canopies would not interfere. Upon addition of nutrients to this hypothetical population, growth could be increased to such an extent that individual plants begin to shade each other. As Harper (1977) states, "At no time would the plants have to reduce the availability of nutrients to each other, but an apparent density dependent response to fertility would be seen." This relationship between plant density, soil fertility, and the limitation of other resources may explain the large yield increases in winter wheat that Appleby et al. (1976) observed when high nitrogen fertility was combined with reductions in plant density due to weed control (Table 6.13). A similar response to that of Appleby et al. was observed by LeStrange and Hill (1983) on rice yields when the combined effects of various barnyardgrass densities and nitrogen fertility levels were examined.

A better understanding of the interacting effects of resources on plant growth and competition is an area of Weed Science that needs considerably more attention (see also Chapter 7). Much more information is needed about the basic mechanisms of plant responses to single and interacting resources and how these mechanisms affect and, in turn, are altered by competition. Such information could be used to develop strategies for fertility management to achieve maximum crop growth and weed control. In some situations, however, resource interactions can be so complex that they are impossible to unravel with the conventional experimental methodology (yield studies) commonly used. Since resource limitation is intimately associated with plant growth, a better understanding of the mechanisms of competition for resources may lie in the detailed study and analysis of plant growth parameters characteristic of both weeds and crops.

PLANT GROWTH

The ability of the individual plant to obtain light, water, and nutrients for growth often determines the success of that individual in its environment. Successful individuals grow rapidly or large, develop through the various stages of their life cycle, and eventually are replaced in the environment by their progeny. The life cycle of unsuccessful individuals often is arrested before its completion, and death results. Therefore, plant growth, as well as the developmental phases that accompany growth, are fundamental to the understanding of plant function, that is, the manner of interaction within the environment. Thus it is not surprising that plant growth has been the object of extensive study by botanists and is the subject of extensive reviews.

Weier et al. (1982) define *plant growth* as the gradual increase in size by natural development. This phenomenon encompasses almost all the disciplines within the botanical sciences and is clearly beyond the scope of this book. However, numerous botanical texts are available that cover the subject of plant growth from the elementary point of view to the complex. In this text we do not intend to review the general aspects of plant growth and development that are common to all plants, nor do we intend to present a detailed analysis of plant anatomy, morphology, or physiology, since excellent texts are available on those subjects. Rather, in relation to crops and weeds, we will consider growth as an integrated feature of the whole plant that is living in its natural environment.

Growth as a Unifying Concept

Growth is a universal phenomenon of all plant life. All plants, as they proceed through their life cycles, are capable of changing their size and form if exposed to suitable conditions. The processes that result in physiological and morphological changes as a plant ages are strongly interconnected, and the term "growth" is used for any or all of them. Growth in the above context means the irreversible increase in plant size, which is often accompanied by changes in form and occasionally by changes in individual numbers. The amount of growth (however measured) resulting from the influences of various biotic and abiotic factors is what determines competitive interactions among plants. A better understanding of resource limitation and its consequences could result from the analysis of growth parameters of individual plants involved in a competitive relationship.

Because plants are effectively anchored to the soil by roots, they are required to spend their entire lives at a single location. This sessile nature of plant occupancy dictates that plant growth will be influenced markedly by the environment of that location. However, as a plant grows throughout its life cycle, it is not likely to be exposed continually to the same environment. Furthermore, as a plant grows, new tissues are laid down progressively so that while some organs are being produced, others are developing, and still others have matured. Thus, each organ of the plant is the result of past developmental events that occurred during earlier growing conditions. Because the organs have experienced and responded to different past environments, they also respond to the present environment in different ways. To interpret growth meaningfully as a unifying feature of the whole plant, its relationship to both past and present environments must be considered.

Another important feature of plant growth is also evident from the above discussion. New growth that arises within the constraints of current environmental conditions is necessarily a function of the growth that has already occurred. Thus, the growth of a plant at any time is relative to the size it has already attained. This generalization is especially important when considering the size and abundance of organs responsible for new biomass production, such as leaves, in relation to the rest of the plant (Evans, 1972). That is, the greater the size of the productive organs of a plant, the greater the new growth that will result.

The relationship of plant growth to existing size has been compared, in a financial analogy, to an initial monetary investment and, thereafter, to the compounding of interest. This analogy appears to be an accurate description of plant growth and has been extended by many authors to include other transactions. For example, Evans (1972) distinguishes between "productive investments" (leaves) in contrast to "nonproductive investments" (the rest of the plant):

> As we are accustomed to think in monetary terms it is clear, without an algebraic argument, that with a fixed distribution between productive and non-productive investment the relative rate of growth of the capital as a whole will be higher, if the rate of interest on the productive investment is higher; and that at a fixed rate of interest, the rate of growth of the capital as a whole will be higher, the larger the proportion invested in the productive securities. Converted back from our analogy into plant terms, this means that the rate of growth of a plant relative to its size at any time will be higher if either (a) the rate of increase of dry weight per unit leaf area is higher; or (b) the ratio of leaf area to total dry weight is higher; or both.

Thus it is possible to visualize, as many have done, plant growth as a result of an initial investment into biomass with the efficiency of production equivalent to the rate of interest.

Plant Growth Analysis

Beginning with the early works of Blackman (1919) and Kidd and West (1919), a set of techniques has been developed to integrate the effects of environment, development, and size on plant growth. These techniques, collectively called *mathematical growth analysis*, recognize total dry matter production and leaf area expansion as important processes in determining vegetative growth. The techniques require frequent destructive harvests of plant material at intervals throughout a plant's life cycle. The basic information collected at each harvest includes dry matter production of roots, leaves, stems, and reproductive organs, and leaf area. From these basic data, it is possible to calculate rates of dry matter production per unit leaf area or net assimilation rates, relative growth rates, relative leaf expansion rates, and partition coefficients for plant biomass and leaf area. Thus, the components of plant growth can be separated and compared under a range of environmental conditions and resource limitations. In Table

6.14, Patterson (1982) defines the parameters commonly used for plant growth analysis. These parameters can be calculated in a variety of different ways, which will be discussed later in this chapter. Formulas for calculation of growth analysis parameters using several approaches are summarized in Table 6.15. A more complete description of the derived quantities used in plant growth analysis and their mathematical definitions is given in Evans (1972), Hunt (1978) and Chiariello et al. (1991).

Although the techniques of plant growth analysis are often performed on individual species under uniform, controlled growing conditions, the comparative analysis of several species grown under similar environmental regimes is particularly useful in studies of weed and crop interactions. Such experiments provide valuable information for understanding the basis of differential success when plants are grown in mixed stands. Additionally, while more difficult to accomplish, comparisons of growth parameters of crops and weeds growing in mixtures may lead to a better understanding of the mechanisms underlying competitive success of weeds.

RELATIVE GROWTH RATE

Hunt (1978, 1982) points out that when plant biomass accumulation is examined over time, a distinction between absolute and relative growth rate must be made. Hunt considers a hypothetical experiment in which two plants have

TABLE 6.14 Growth Analysis Definitions and Formulas

$T_2 - T_1$ = length of harvest interval (days) = ΔT

W_1, W_2 = dry weight at beginning and end of harvest interval

A_1, A_2 = leaf area at beginning and end of harvest interval

$\Delta W = W_2 - W_1$; $\Delta A = A_2 - A_1$

$R_w = \bar{R}$ = Relative growth rate (g/g · day) = $(\ln W_2 - \ln W_1)/\Delta T$

R_a = Relative leaf area growth rate (dm^2/dm^2 · day) = $(\ln A_2 - \ln A_1)/\Delta T$

NAR = Net assimilation rate (g/dm^2 · day) = $(\Delta W/\Delta A) \times (\ln A_2 - \ln A_1)/\Delta T$

LAI = Leaf area index = dimensionless ratio of leaf area to land area

CGR = Crop growth rate (g/dm^2 land surface · day) = NAR × LAI

LWR = leaf weight ratio (g leaf wt/g total wt)

LAD = leaf area duration (dm^2 days) = $[\Delta A/(\ln A_2 - \ln A_1)] \times (\Delta T)$

BMD = Biomass duration (g days) = $[\Delta W/(\ln W_2 - \ln W_1)] \times (\Delta T)$

SLW = Specific leaf weight (g/dm^2)

SLA = Specific leaf area (dm^2/g), SLA = 1/SLW

LAR = leaf area ratio (dm^2/g total wt), LAR = LWR × SLA

Source: Patterson, 1982. By permission of Academic Press, New York.

TABLE 6.15 Summary of Traditional Growth Analysis Parameters, Units, Instantaneous Values, and Methods of Calculation

Parameter	*Units*	*Instantaneous*	*Classical (interval)**	*Functional*	*Integral†*
Relative growth rate, Specific growth rate (RGR, *R*, SGR)	d^{-1} w^{-1}	$\frac{1}{W}\frac{dw}{dt}$	$\frac{\ln W_2 - \ln W_1}{t_2 - t_1}$	$\frac{1}{g(t)}\frac{df(t)}{dt}$ where $W = f(t)$	$\frac{W_2 - W_1}{BMD}$
Net assimilation rate, Unit leaf rate (NAR, ULR, *E*)	$gm^{-2}d^{-1}$ $gm^{-2}w^{-1}$	$\frac{1}{A}\frac{dW}{dt}$	$\frac{1}{t_2 - t_1}\int_{t_1}^{t_2}\frac{1}{A}\frac{dW}{dt}dt$	$\frac{1}{g(t)}\frac{df(t)}{dt}$ where $A = g(t)$	$\frac{W_2 - W_1}{LAD}$
Leaf area ratio (LAR, *F*)	m^2g^{-1}	$\frac{A}{W}$	$\frac{1}{t_2 - t_1}\int_{t_1}^{t_2}\frac{A}{W}dt$	$\frac{g(t)}{f(t)}$	
Specific leaf area (SLA)	m^2g^{-1}	$\frac{A}{W_L}$	$\frac{1}{t_2 - t_1}\int_{t_1}^{t_2}\frac{A}{W_L}dt$	$\frac{g(t)}{h(t)}$ where $W_L = h(t)$	
Leaf weight ratio (LWR)		$\frac{W_L}{W}$	$\frac{1}{t_2 - t_1}\int_{t_1}^{t_2}\frac{W_L}{W}dt$	$\frac{h(t)}{f(t)}$	

Source: Chiariello et al., 1991.

W = dry biomass (g); t = time (d or w); A = assimilatory area (m^2); W_L = dry leaf biomass (g); BMD = biomass duration (g d); LAD = leaf area duration (m^2d).

*Explicit general formulas; see Table 15.2 in the original source for specific formulas for calculating net assimilation rate.

†Approximate formulas; formulas for BMD and LAD are given in Section 15.2.1.c of the original source.

grown for a week under the same environmental conditions. One plant weighed 1.0 g at the start of the experiment, whereas the other weighed 10 g. At the end of the week both plants had gained 1.0 g in biomass. Since both plants grew an equal amount over the same period of time, their absolute growth rate is the same, that is, 1.0 g/week. However, over an identical time period one of the plants doubled in biomass, whereas the other only increased by a tenth of its original weight. Clearly, the lighter plant was more efficient in biomass accumulation when initial size was also considered. This relative increase in plant material per unit of time (g/g · week) is called the *relative growth rate* (R, R_w, or RGR). Obviously, a much more informative comparison of the two plants can be made by considering relative growth rate as well as the absolute amount of biomass accumulation. RGR is considered to be one of the most ecologically significant plant growth indices.

The relative growth rate was first considered by Blackman (1919), who proposed an "efficiency index" for dry weight production. RGR is defined by the differential equation:

$$\mathrm{RGR} = \frac{1}{W} \bullet \frac{dW}{dt} = \frac{d(\ln W)}{dt} \qquad \textit{(eq. 6.2)}$$

In this equation dW is the change in dry weight over a given time interval represented by dt and W is the total individual plant dry weight. RGR represents the instantaneous value of relative growth rate. A more involved concept in theory but more useful in practice is the mean relative growth rate ($\bar{R}$) (Hunt, 1978; Evans, 1972). According to Evans (1972), $\bar{R}$ is calculated using the following equation:

$$\bar{R} = \frac{\log_e W_2 - \log_e W_1}{t_2 - t_1} \qquad \textit{(eq. 6.3)}$$

In this equation $\bar{R}$ values are derived from dry weight determinations over successive harvests. Thus, $\bar{R}$ represents a mean value for the time interval $t_2 - t_1$. W_2 and W_1 are the total dry weights at the beginning (W_1) and end (W_2) of the harvest interval (Evans, 1972).

It is well known that R is highly responsive to environmental conditions and to initial plant size. Changes in R may also occur due to *ontogenetic drift*, that is, the changes in development that occur in a plant as it grows older. However, by using short time intervals between harvests, the impact of ontogenetic drift can be minimized. Thus comparisons of R can be made among species when grown under identical environmental conditions if harvests are frequent enough.

OTHER GROWTH PARAMETERS

Evans (1972), Hunt (1978), and Patterson (1982) indicate that R is the product of two components, the net assimilation rate (NAR, also called unit leaf rate, ULR, E) and leaf area ratio (LAR). Thus, R may be expressed as:

$$\frac{1}{W} \bullet \frac{dW}{dt} = \frac{1}{L_A} \bullet \frac{dW}{dt} \times \frac{L_A}{W} \qquad \textit{(eq. 6.4)}$$

$$R = \text{NAR} \times \text{LAR} \qquad \textit{(eq. 6.5)}$$

where L_A is leaf area of the plant. LAR, the amount of leaf area per unit of total plant biomass, is a measure of the relative leafiness of the plant. NAR is the net gain in weight per unit of leaf area. Using these two components, R can be expressed in both morphological and physiological terms. LAR is a morphological index of plant form, whereas NAR is a physiological index closely connected with the photosynthetic activity of the leaves (Evans, 1972). The splitting of R into the components of NAR and LAR is advantageous because it relates dry weight increase to those organs most concerned with carbon assimilation, the leaves. Further splitting of the above growth parameters into other components also is possible. For example, Patterson (1982) and Hunt (1978) note that the LAR is the product of the specific leaf area (SLA), or area per unit leaf dry weight, and the leaf weight ratio (LWR), or leaf dry weight per unit total plant dry weight. Thus LAR = SLA × LWR. The relationships among growth parameters are particularly useful in evaluating the responses of plants to environmental changes that occur during growth. These and other plant growth parameters are explained more fully in Chiariello et al. (1991), Evans (1972), and Hunt (1978, 1982).

APPROACHES FOR CALCULATING GROWTH PARAMETERS

The parameters typically used in growth analysis may be calculated in one of several ways. The most common approaches for calculation are *classical*, *integral*, and *functional* growth analysis (Chiariello et al., 1991). These three methods use the same set of parameters to describe plant growth but differ in how the parameters are estimated. Table 6.15 presents a summary of traditional growth analysis parameters, equations for instantaneous values, and formulas based on the three calculation methods listed above.

In the *classical*, or interval, approach growth parameters are estimated either as instantaneous values or as mean values, calculated by integrating instantaneous values and dividing by the time between harvests (Evans, 1972). As mentioned above, mean R, or $\bar{R}$, can be calculated directly from dry weight determinations over successive harvests. However, problems arise when mean values for NAR and LAR are desired, because both parameters depend on the relationship between biomass (W) and leaf area (A) as they change over time. As described more fully by Chiariello et al. (1991), the equations for calculating NAR vary depending on whether the relationship between W and A is linear, quadratic, exponential, or has some other form. Thus, in classical growth analysis, these terms should be calculated as instantaneous values if the relationship between W and A is not known.

Chiariello et al. (1991) describe the *integral* approach to growth analysis as one that incorporates terms for leaf and biomass duration during the growing period. As shown in Table 6.15, calculation of R and NAR using biomass or leaf area duration terms are only approximations. This approach is valuable when canopy or seasonal performance is of interest.

The *functional* approach to growth analysis has become increasingly popular with the increasing availability of robust curve-fitting computer software. In this approach, growth parameters are calculated from functions fitted to the changes in W and A or their logarithms over time (Hunt, 1982). Since fitted equations can reveal the relationship of W and A, some of the assumptions required in the classical approach are avoided. Functions such as exponential, logistic, or polynomial equations of different orders can be used to describe plant growth (Hunt and Parsons, 1974). Care must be taken, however, to insure that the functions chosen have biological relevance. A discussion of the strengths and weaknesses of these three growth analysis approaches is found in Chiariello et al. (1991).

Relative Growth Rate of Weeds

Hunt (1978) indicates that comparisons of interspecific differences in relative growth rate were conducted first by Blackman (1919), and that the literature contains values of R_w ($\bar{R}$) for several hundred species. Unfortunately, few of the species reported in the literature were grown under strictly comparable environmental conditions, so only a limited number of interspecific comparisons of R can be made. Grime and Hunt (1975) provide one of the largest bodies of comparative data on relative growth rates that is available. In their study, Grime and Hunt calculated $\bar{R}$ and R_{max} (the maximum potential relative growth rate), for 132 species. Each species was grown under favorable conditions in controlled-environment chambers for up to five weeks. Their values for $\bar{R}$ and R_{max} for a number of weed species are shown in Table 6.16.

The weed species presented in Table 6.16 were most often associated with arable land (A), grazed meadows and pastures (P), or other disturbed but productive habitats (M). The $\bar{R}$ values for the weed species in Table 6.16 are usually significantly greater than 1.0 g/g · week. Some of the R_{max} values [lambsquarters (*Chenopodium album*), field bindweed (*Convolvulus arvensis*), velvetgrass (*Holcus lanatus*), annual bluegrass (*Poa annua*), and chickweed (*Stellaria media*)] exceed 2.0 g/g · week. These values are in marked contrast to the values of other species in their study that predominated in either stable (undisturbed) or unproductive habitats. Species from unproductive or undisturbed sites usually had $\bar{R}$

TABLE 6.16 Maximum Potential Relative Growth Rate (R_{max})[a], Mean Relative Growth Rate ($\bar{R}$)[a], and Plant Species Associated with Arable Land (A), Meadows and Pastures (P), or Manure Heaps (M)

Species	*Common Name*	*Site Association*[b]	R_{max} *(g/g · week)*	$\bar{R}$ *(g/g · week)*
Agropyron repens	Quackgrass	AMP	1.21	1.21
Agrostis stolonifera	Creeping bentgrass	APM	1.48	1.48
Agrostis tenuis	Colonial bentgrass	P	1.36	1.36
Cerastium holosteoides	—	P	1.46	1.46
Chenopodium album	Common lambsquarters	AM	2.12	1.25
Convolvulus arvensis	Field bindweed	A	2.44	1.36
Cynosurus cristatus	Crested dogtailgrass	P	1.54	1.54
Dactylis glomerata	Orchardgrass	PM	1.31	1.31
Festuca rubra	Red fescue	P	1.18	1.18
Holcus lanatus	Velvetgrass	PM	2.01	1.56
Matricaria matricarioides	Pineappleweed	A	1.17	1.17
Plantago lanceolata	Buckhorn plantain	P	1.70	1.40
Poa annua	Annual bluegrass	AM	2.70	1.74
Poa trivialis	Roughstalk bluegrass	PM	1.40	1.40
Polygonum aviculare	Prostrate knotweed	A	1.43	1.43
Polygonum convolvulus	Wild buckwheat	A	1.92	1.35
Ranunculus repens	Creeping buttercup	P	1.39	0.93
Senecio vulgaris	Common groundsel	M	1.63	0.84
Stellaria media	Chickweed	AM	2.43	2.09
Taraxacum officinale	Dandelion	P	1.19	1.19
Trifolium repens	White clover	P	1.26	1.26

Source: Grime and Hunt, 1975.

[a]R_{max} is the highest value of R obtained for each species during the periods of observation. $\bar{R}$ was calculated by Grime and Hunt from Fisher's (1920) formula ($\log_e W_5 - \log_e W_2)T$, where W_5 and W_2 are whole plant dry weights at 5 and 2 weeks, respectively, and T is the time interval, 3 weeks.

[b]Site associations for plant species were determined from Table 1 of Grime and Hunt (1975).

and R_{max} values near 0.5 g/g · week. It appears that rapid growth—that is, a large value for relative growth rate—is an important characteristic of weeds common to arable land and other productive sites. Since the amount of dry matter produced by an individual plant must be proportional to the overall resources used during growth, relative growth rate has often been used as an indicator of potential competitive ability among crop and weed species.

Poorter and Remkes (1990) provided a similar table of values for *R*, NAR, LAR, and several other growth parameters for 24 wild species, including many common weeds, all of which have C_3 photosynthesis. The data were obtained from an experiment in which the 24 species were grown in a growth chamber under optimal nutrient supply and a growth analysis was conducted. As shown in Table 6.17, values of *R* for these species range from 113 to 365 mg g^{-1} day^{-1}, which are similar to the values shown in Table 6.16 when corrections for units are made. To better understand the causes of interspecific differences in *R*, other growth parameters were analyzed to determine their relationship to values of *R*. This analysis revealed a very high correlation between LAR and *R* (Figure 6.29). Apparently, for the species and conditions in this experiment, the more a plant invested in leaf area, the faster it could grow and produce new biomass. Based on an analysis of the natural habitat of the 24 species, Poorter and Remkes postulated that their inherent variability in *R* was due to selection for high LAR in nutrient-rich environments and low LAR in nutrient-poor environments.

A review of sixty publications reporting values for several growth parameters supported the importance of LAR in determining potential *R* (Poorter, 1990). For herbaceous weeds and crops, there was a strong correlation of *R* with LAR (Table 6.18). Correlations were lower for trees, shade plants, and C_4 species, apparently because of the more important role of NAR in conditions where light often limits photosynthesis and in plants possessing the C_4 pathway of photosynthesis. Analyses such as those of Grime and Hunt (1975), Poorter (1990), and Poorter and Remkes (1990) provide valuable insight into the overall growth characteristics of many weed and crop species. They also demonstrate the value of plant growth analysis in explaining ecological relationships of these species to their environment.

Predictions of Weed/Crop Competition with Growth Analysis

Plant growth analysis is a powerful quantitative tool for studying growth and competition of weeds and crops in the field environment. This approach was used by Roush and Radosevich (1985) to describe competitive interactions among four weed species. Holt and Orcutt (1991) also used plant growth analysis to investigate the relationship of growth characteristics to competitiveness in mixtures of cotton and three perennial weeds [Johnsongrass (*Sorghum halepense*), purple nutsedge (*Cyperus rotundus*), and yellow nutsedge (*C. esculentus*)]. In the latter study, plants were grown individually and in pairwise mixtures in pots buried in the ground outdoors to allow harvest of belowground growth. Weekly

TABLE 6.17 Values of RGR (mg g^{-1} day^{-1}), NAR (g m^{-2} day^{-1}), LAR (m^2 kg^{-1}), SLA (m^2 kg^{-1}), LWR (g g^{-1}), SWR (stem weight ratio, g g^{-1}) and RWR (root weight ratio, g g^{-1}) for 24 Species[a]

Species	*RGR*	*NAR*	*LAR*	*SLA*	*LWR*	*SWR*	*RWR*
Brachypodium pinnatum	174	9.0	19.8	40.7	0.49	0.24	0.27
Briza media	157	9.1	17.5	35.1	0.51	0.19	0.30
Corynephorus canescens	113	8.3	14.2	33.1	0.43	0.27	0.31
Cynosurus cristatus	176	11.7	14.7	32.0	0.46	0.16	0.38
Dactylis glomerata	229	10.3	22.4	50.2	0.45	0.22	0.32
Deschampsia flexuosa	135	10.1	13.2	27.6	0.48	0.18	0.34
Festuca ovina	132	10.4	12.9	25.3	0.51	0.20	0.29
Holcus lanatus	268	14.1	19.4	43.7	0.43	0.22	0.34
Lolium perenne	214	11.5	19.5	38.8	0.50	0.20	0.31
Phleum pratense	227	8.9	27.2	47.8	0.54	0.17	0.30
Poa annua	272	11.8	23.9	46.7	0.50	0.21	0.29
Anthriscus sylvestris	239	11.5	21.2	40.2	0.53	0.19	0.28
Galinsoga parviflora	365	10.5	35.9	55.5	0.64	0.13	0.23
Geum urbanum	224	8.6	27.1	41.5	0.65	0.13	0.21
Hypericum perforatum	205	7.4	27.3	49.5	0.55	0.11	0.34
Lysimachia vulgaris	223	8.4	27.7	42.2	0.66	0.18	0.16
Origanum vulgare	203	8.0	25.2	40.6	0.62	0.11	0.27
Pimpinella saxifraga	171	10.2	16.5	31.2	0.54	0.13	0.33
Plantago major	240	11.8	21.1	32.8	0.64	0.12	0.24
Rumex crispus	327	10.7	32.3	49.3	0.63	0.12	0.26
Scrophularia nodosa	302	11.1	28.6	44.3	0.64	0.09	0.26
Taraxacum officinale	260	9.9	27.3	44.1	0.60	0.07	0.32
Trifolium repens	206	12.1	17.1	40.4	0.44	0.28	0.28
Urtica dioica	317	10.1	33.0	51.3	0.64	0.14	0.22

Source: Poorter and Remkes, 1990.

[a]All figures are mean values for the time period that the species had a total plant dry weight between 30 and 100 mg.

measurements were made on individual plants for growth analysis, while mixtures were harvested after ten weeks of growth and biomass production was measured.

Growth analysis of plants grown individually showed that the four species differed in measured and calculated growth parameters. As a group, the weeds had higher biomass production, RGR, and ULR (NAR) than cotton, indicating greater resource use and production efficiency (Figures 6.30 and 6.31). In con-

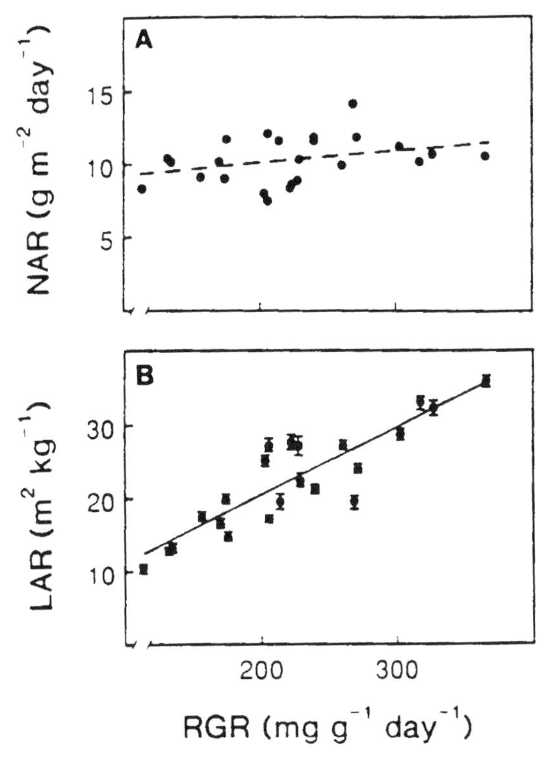

FIGURE 6.29 (A) Mean NAR and (B) mean LAR for 24 species plotted against mean RGR during the time that plants had a dry weight between 30–100 mg. Error bars indicate the mean SE at the 6 harvest days (n = 8). The continuous straight line indicates a significant linear regression ($P < 0.05$) of this parameter with RGR, the broken line a nonsignificant relation. (From Poorter and Remkes, 1990.)

trast, cotton had higher LAR and CDI (canopy density index) than the weeds as a group, indicating that it attained greater leafiness and canopy closure than the weeds during the experiment (Figure 6.32). Stepwise multiple regression analysis was performed on the data using biomass production in competition (calculated as aggressivity) as the dependent variable and mean values for growth parameters as independent variables. The analysis identified four growth parameters out of 12 that were calculated that best described competitiveness. Those four were height, NAR, RGR, and initial propagule weight (PWT). Holt and Orcutt (1991) suggested from these results that parameters of light utilization (height, NAR) and early establishment (RGR, PWT) were best predictors of competitive success in these mixtures. Higher NAR in the weeds could be accounted for by the C_4 photosynthetic pathway, while high RGR and PWT were due to the early advantage conferred by vegetative reproduction. In this system, greater LAR of cotton could not compensate for lower NAR. Thus, these quantitative results offer a credible ecological explanation for the competitive interactions occurring in mixtures of these species.

TABLE 6.18 Inherent Differences in RGR (in mg g^{-1} day^{-1}), and Their Relation with the Components NAR, LAR, SLA, and LWR[a]

Reference[b]	*Investigated Species*	*Range in RGR*	*PI*						
			NAR	*LAR*	*SLA*	*LWR*	*PS*	*SR*	*RR*
A. Trees vs. Herbs									
Coombe, 1960	1 tree + 1 crop species	53–93	++	−	+	—			
Whitmore Wooi-Khoon, 1983	1 tree + 1 crop species	109–129	++	−					
Jarvis & Jarvis, 1964	2 tree + 1 crop species	117–239	+	+					
Poorter, unpublished	1 tree + 1 grass species	152–203	++	o	+	−	++		
Coombe & Hadfield, 1962	2 tree + 1 crop species	20–118	++	+	o	+			
Oberbauer & Donnelly, 1986	6 tree + 1 crop species	19–126	+	+	+	o			
Okali, 1971	3 tree + 1 crop species	40–150	+	+					
Mean PI			10	0	3	−2			
B. Between Trees									
Loach, 1970	4 tree species	10–26	+	+					
Mooney et al., 1978	5 *Eucalyptus* species	146–178	+	+	++	o	o		
Pollard & Wareing, 1968	6 tree species	14–30	+	+					
Delucia et al., 1989	3 tree species	11–24	o	+			o		
Kwesiga & Grace, 1986	2 tree species	77–98	− −	++	++	++			
Huxley, 1967	2 *Coffea* species	19–20		na					
Karlsson & Nordell, 1987	2 tree species	3–4		na					
Popma & Bongers, 1988	3 tree species	11–15		na					
Tolley & Strain, 1984	2 tree species	46–55		na					
Mean PI			−2	11	17	4			

TABLE 6.18 Continued

Reference	Investigated Species	Range in RGR	*PI*						
			NAR	*LAR*	*SLA*	*LWR*	*PS*	*SR*	*RR*
C. C_3 vs. C_4 Species									
Bazzaz et al., 1989	1 C_3 + 1 C_4 wild species	95–127	++	–	o	o			
De Jong, 1978	2 C_3 + 1 C_4 species	21–53	++	o	o	o			
Jones et al., 1970	2 C_3 + 1 C_4 species	110–220	++	o					
Rajan et al., 1973	3 C_3 + 1 C_4 crop species	131–210	++	o					
Saxena & Ramakrishnan, 1983	2 C_3 + 2 C_4 weeds	8–36	++	o					
Warren Wilson, 1966a	2 C_3 + 1 C_4 species	188–332	+	o	–	+			
Roush & Radosevich, 1985	2 C_3 + 2 C_4 weed species	232–262	+	+					
Potter & Jones, 1977	9 crop + weed species (6 C_3, 3 C_4)	202–482	o	++	+	o			
Sage & Pearcy, 1987	1 C_3 + 1 C_4 weed species	340–410	o	++					
Sionit et al., 1982	3 C_3 + 1 C_4 crop species	76–159	o	++					
Hofstra & Stienstra, 1977	1 C_3 + 1 C_4 grass	86–91		na					
Mean PI			7	2	0	2			
D. Sun vs. Shade Species									
Pons, 1977	1 sun + 1 shade species	166–188	++	—	—	+		++	++
Corré, 1983c	1 sun + 1 shade species	245–339	+	+	+	o			
Corré, 1983b	2 sun + 2 shade species	180–232	+	+	++	o			
Corré, 1983a	2 sun + 2 shade species	165–327	o	+	+	–			
Corré, 1983a	1 sun + 2 shade species	254–414	+	+	+	o			
Mean PI			8	2	–2	0			

(continued)

TABLE 6.18 Continued

Reference	*Investigated Species*	*Range in RGR*	*PI*						
			NAR	*LAR*	*SLA*	*LWR*	*PS*	*SR*	*RR*
E. Between Herbs									
Eagles, 1969	2 ecotypes *Dactylis glomerata*	165–224	++	—					
Thorne, 1960	3 crop species	129–180	++	—					
Blackman & Black, 1951	3 *Trifolium* species	87–98	++	–					
Thorne et al., 1967	2 crop species	69–139	++	–					
Arnold, 1974	2 wild species	99–111	+	+					
Blackman & Wilson, 1951	3 crop species	77–165	+	+					
Boot & Mensink, 1990	5 grass species	36–65	+	+	++	–			
Day et al., 1986	mutant + parent *Glycine max*	63–90	+	+					
Paul et al., 1984	5 cultivars *Lycopersicon esculentum*	180–220	+	+					
Roetman & Sterk, 1986	13 microspecies *Taraxacum officinale*	117–211	+	+	+	o			
Woodward, 1979	2 grass species	86–181	+	+					
Woodward, 1983	2 *Poa* species	220–280	+	+	++	–			
Higgs & James, 1969	4 grass species	55–149	+	++					
Brewster & Barnes, 1981	8 cultivars *Allium cepa*	115–147	o	++	+	+			
Bruggink & Heuvelink, 1987	3 crop species	310–410	o	++					
Poorter & Remkes, 1990	24 wild species	113–365	o	++	+	+	o	–	–
Kriedemann & Wong, 1984	2 crop species	102–177	o	++					
Warren Wilson, 1966b	2 wild + 1 crop species	187–309	o	++					
Blackman & Wilson, 1951	2 crop species	42–92	–	++					
Delucia et al., 1989	1 shrub + 1 herb	84–154	–	++					
Dijkstra & Lambers, 1986	2 genotypes *Plantago major*	212–280	–	++	++	o	–	—	

TABLE 6.18 Continued

Reference	Investigated Species	Range in RGR	PI						
			NAR	LAR	SLA	LWR	PS	SR	RR
Gottlieb, 1978	2 *Stephanomeria* species	120–140	–	++	++	+	o		
Poorter, this article	8 wild species	136–268	–	++	++	+	o		
Shibles & MacDonald, 1962	2 strains *Lotus corniculatus*	188–207	–	++					
Woodward, 1975	2 *Sedum* species	170–203	—	++					
De Kroon & Knops, 1989	2 wild species	19–25		na					
Graves & Taylor, 1986	2 *Geum* species	39–44		na					
Holt, 1988	2 biotypes *Senecio vulgaris*	142–147		na					
Konings et al., 1989	3 *Carex* species	76–80		na					
Musgrave & Strain, 1988	2 cultivars *Triticum aestivum*	85–91		na					
Myerscough & Whitehead, 1967	4 wild species	120–132		na					
Pavlik, 1983	2 dune grasses	55–56		na					
Pegtel, 1976	2 ecotypes *Sonchus arvensis*	70–72		na					
Russell & Grace, 1978	2 grass species	124–127		na					
Watson & Baptiste, 1938	2 crop species	25–27		na					
Wilhelm & Nelson, 1978	4 genotypes *Festuca arundinacea*	20–28		na					
Wilson, 1982	2 populations *Lolium perenne*	132–141		na					
Mean PI			2	8	9	1			

[a]Also relations with photosynthesis (PS), shoot respiration (SR), and root respiration (RR) are given, all measured on whole shoots and roots and expressed per unit leaf area. For each reference a linear regression was performed with the RGR of the investigated (sub)species as independent variable and the different components of growth as dependent variable. Thereafter, the percentage increase (PI) in the fitted parameters was calculated, given a 10% increase in overall mean RGR. The following classification was made: —, PI < −8; –, −8 < PI < −2; o, −2 < PI < 2; +, 2 < PI < 8; ++, PI > 8; na, difference between lowest and highest RGR values less than 10% or 10 mg g^{-1} day^{-1}, no analysis carried out. In those cases where only NAR or LAR was given, the other component was calculated with the equation RGR = NAR · LAR. If plants were grown under a number of environmental conditions, the analysis carried out at the hightest light intensity, highest nutrient supply, the lowest windspeed, the lowest NaCl concentration, and ambient CO_2 concentration close to 350 $\mu l L^{-1}$ or mean temperature closest to 23°C was chosen. Only those analyses are given where root weight was determined as well and species were grown under identical conditions (Poorter, 1990).

[b]References are cited in original source.

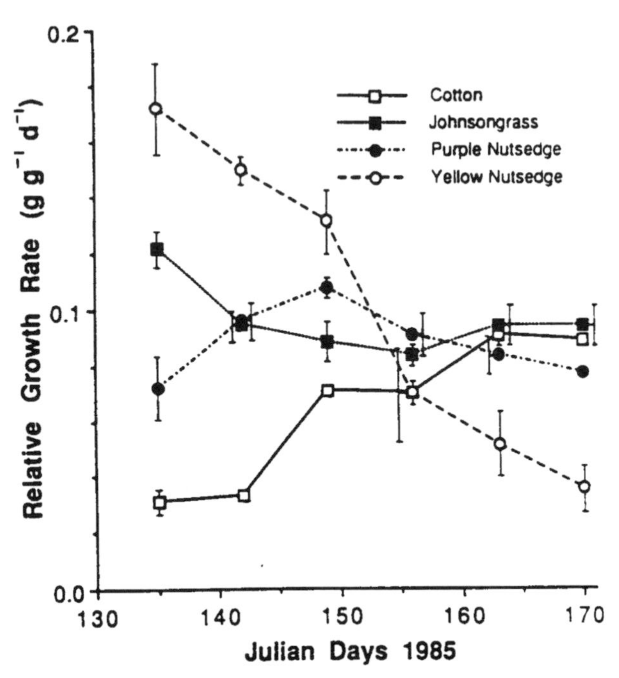

FIGURE 6.30 Relative growth rate of cotton, Johnsongrass, purple nutsedge, and yellow nutsedge from Julian day 135 (5 weeks) to day 171 (10 weeks) after planting in 1985. Error bars show one standard error above and one below the mean. (From Holt and Orcutt, 1991.)

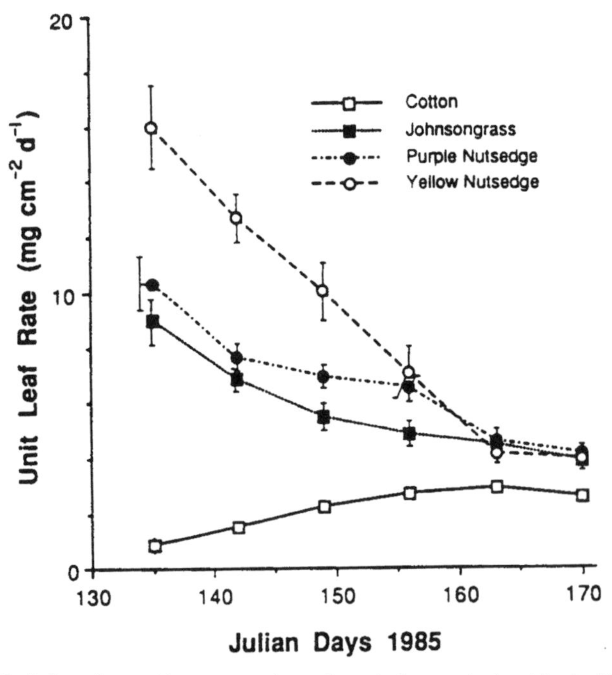

FIGURE 6.31 Unit leaf rate of cotton, Johnsongrass, purple nutsedge, and yellow nutsedge from Julian day 135 (5 weeks) to day 171 (10 weeks) after planting in 1985. Error bars show one standard error above and one below the mean. (From Holt and Orcutt, 1991.)

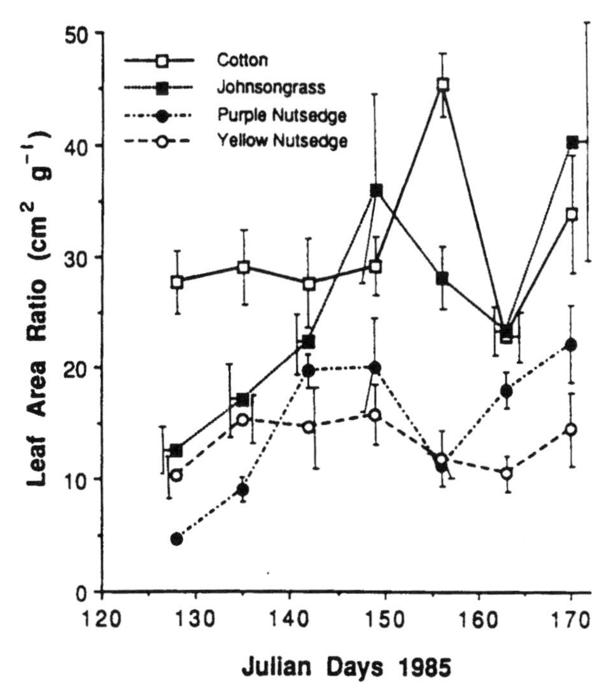

FIGURE 6.32 Leaf area ratio of cotton, Johnsongrass, purple nutsedge, and yellow nutsedge from Julian day 135 (5 weeks) to day 171 (10 weeks) after planting in 1985. Error bars show one standard error above and one below the mean. (From Holt and Orcutt, 1991.)

When plant species are grown together, they often respond differently to environmental factors than when they are grown separately. Thus, in assessing the relative competitiveness of species that exist in mixed stands, it is more realistic to study their growth responses in mixture rather than as isolated individuals. Williams (1963) demonstrated this point by growing three clover species in the field as either pure stands or as various species combinations in mixture. The three clovers—subterranean clover (*Trifolium subterraneum*), rose clover (*T. hirtum*), and crimson clover (*T. incarnatum*)—were grown separately (2500 plants/m^2) or in the various possible combinations of species as 1 : 1 mixtures. Water and nutrients were added periodically during the experiment so that light was the major resource competed for.

Williams observed that initially rose and crimson clovers had larger cotyledons than subterranean clover. However, as subterranean clover attained full canopy development, it became the dominant species when grown with either of the other two clovers. The relative growth rates ($\bar{R}$) of the three species grown in pure stands were similar throughout the growing season (Table 6.19). However, in mixed stands, $\bar{R}$ of subterranean clover always increased substantially at the expense of the other species, which declined in $\bar{R}$. A similar response also was observed with net assimilation rate (NAR). Thus, relative species dom-

TABLE 6.19 Relative Growth Rate ($\bar{R}$) and Net Assimilation Rate (NAR) on a Leaf-Area and a Leaf-Weight Basis in Simple and Mixed Communities of Clover

	Relative Growth Rate (mg/g · day)			*Net Assimilation Rate*					
				Leaf-Area Basis (mg/dm² · day)			*Leaf-Weight Basis (mg/g · day)*		
	0–58 Days	*58–79 Days*	*79–99 Days*	*0–58 Days*	*58–79 Days*	*79–99 Days*	*0–58 Days*	*58–79 Days*	*79–99 Days*
Subterranean clover alone	43	48	28	40	30	19	91	98	71
Rose clover alone	44	39	27	49	32	20	92	72	63
Crimson clover alone	44	46	29	39	28	18	88	90	66
Subterranean (with rose)	41	53	35	40	36	25	88	111	91
Rose (with subterranean)	42	44	10	51	41	8	88	87	23
Crimson (with rose)	45	48	39	42	32	27	90	94	92
Rose (with crimson)	43	40	12	49	35	9	88	79	29
Crimson (with subterranean)	45	47	24	42	30	15	91	91	56
Subterranean (with crimson)	42	52	34	44	35	25	89	108	88

Source: Williams, 1963.

inance (that is, competitive success) resulted from changing patterns of $\bar{R}$ and NAR as the species developed through the vegetative phase. That both $\bar{R}$ and NAR were similar among the species when they were grown alone indicates a common physiological response to a common resource.

Comparative studies between weed and crop species, utilizing either controlled environments or field conditions and mathematical growth analysis procedures, are a valuable tool in developing weed management strategies. Patterson (1982) indicates that weed phenology models developed through growth analysis, and the identification of limiting resources, may aid in determining critical stages in weed life cycles, in scheduling timely application of control practices, and in synchronizing biological control methods. Likewise, weed productivity models, which may ultimately interface with crop productivity models, should aid in the prediction of the competitiveness of specific weeds with specific crops and in the establishment of economic threshold levels for weed infestations. Once important factors for competition and their impacts on plant growth are known, it becomes possible to manipulate the system to facilitate crop productivity.

DESCRIPTIVE MODELS OF WEED/CROP INTERFERENCE

Given the seasonal and annual fluctuations in environmental conditions and long-term changes in cultural practices in agricultural systems, shifts in the species composition of any weed/crop community are likely. It would therefore be of great value to be able to predict the occurrence and competitive success of weed species under the constraints of a productive but constantly changing environment. Since agricultural plant communities typically maintain a full complement of weed seed in the seed bank (see Chapter 4), the competitive edge for either crop or weed species is based on the probability of an initial interspecific encounter and the subsequent ability of the individual to grow. In this respect, demographic relationships such as those demonstrated by Sagar and Mortimer (Chapter 4) should be useful in assessing the potential density of individual weeds and the likelihood that those individuals would interact with the crop or with other weeds.

In order to predict the competitive relationships in weed/crop communities, three key factors must be considered. These factors are spatial arrangement, timing of germination, and growth rate of the plants. A plant that is separated from, establishes before, or grows faster than its neighbors should have the competitive advantage when interacting with other plants. If all the resources necessary for plant growth are integrated into the concept of "space," then space capture can be viewed as a key mechanism in regulating competitive interactions between and within species. In the absence of other forms of negative interference, plants that maximize space capture relative to their neighbors should be most successful (competitive). Thus, models that describe and predict competitive relationships and outcomes of weeds and crops based on underlying physiological processes are useful tools both to direct further research and to suggest possible management strategies.

Spitters and Aerts (1983) introduced the first simulation model for weed/crop competition based on plant physiological responses to resource limitation. Subsequent work by Spitters, Kropff, and colleagues (Kropff, 1988; Kropff et al., 1984; Spitters, 1989) developed this approach more fully. As reviewed by Kropff and Lotz (1992), these models describe competition in terms of the distribution of growth-determining or -limiting resources (light, water, nutrients) and the efficiency with which each species utilizes them. Figure 6.33 shows the simplified structure of such a model for growth of a crop in monoculture. Light is the driving force for crop and weed growth in this scheme. Data on leaf area index (LAI), vertical distribution of leaf area, and light extinction properties for each species are used to calculate a light profile within the canopy. Data on the photosynthetic light response of each species can then be used with light profile

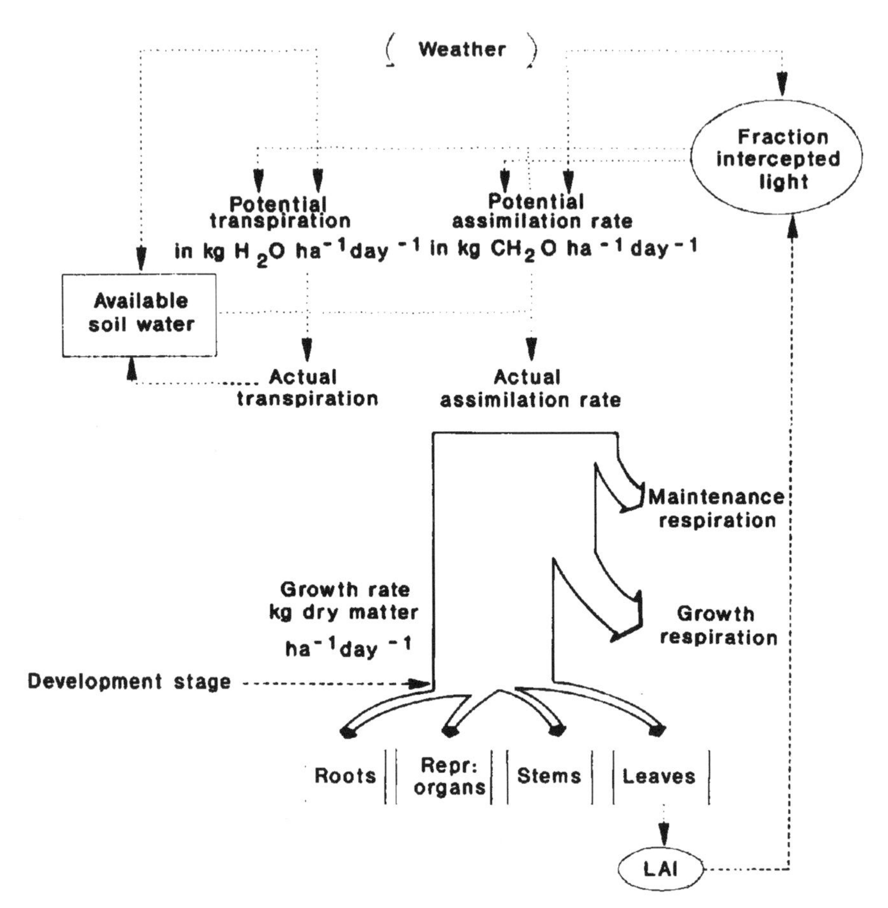

FIGURE 6.33 A simplified scheme of the daily calculation procedure in an ecophysiological model for crop growth in a monoculture. (From Kropff and Lotz, 1992.)

information to develop a profile of photosynthesis for each species in a mixed canopy. The procedure for making these calculations in order to simulate competition for light in a mixed weed/crop canopy is shown in Figure 6.34. Subsequent calculations include integration of assimilation over a day to calculate net daily growth rate (Kropff and Lotz, 1992). A set of papers describing progress in developing simulation models of competition between weeds and crops has been published by Kropff and colleagues (Kropff et al., 1992, 1995; Kropff and Spitters, 1991, 1992).

A systems approach such as that promoted by Kropff and colleagues is one way in which basic knowledge of weed and crop biology and ecology can be used to test ideas about weed management strategies. Simulation models are not necessarily constrained by current agricultural practices and equipment limitations, thus ways to manipulate the agricultural environment to favor

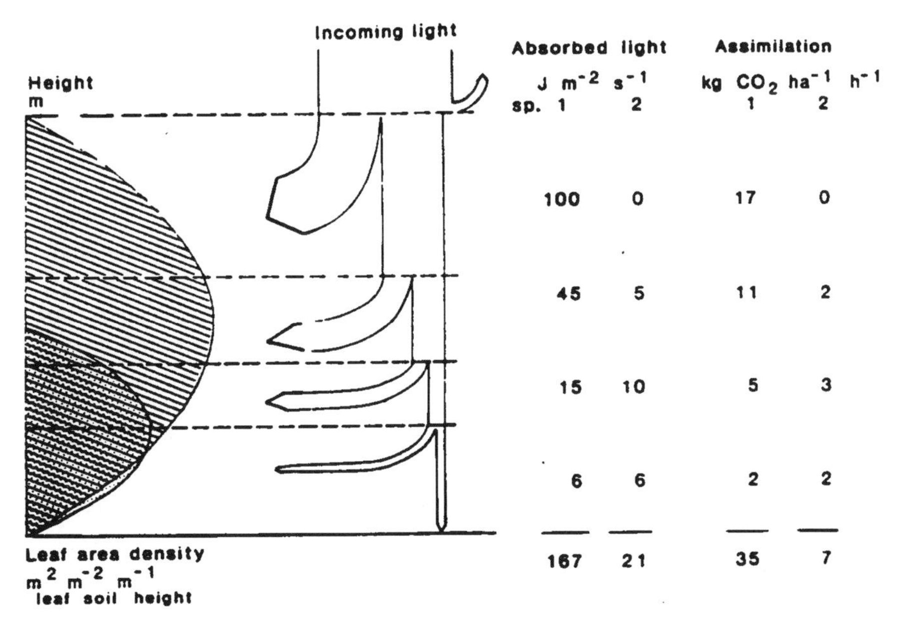

FIGURE 6.34 Principles of the simulation procedure for competition for light used in the model for crop/weed interactions. The shaded area is the LAI of spp 1 and the dashed area is the LAI for spp 2. From the amount of absorbed radiation per species and per leaf layer, the rate of CO_2 assimilation is calculated. Integration over the leaf layers results in canopy assimilation for the competing species. (From Kropff and Lotz, 1992.)

crops and disfavor weeds can easily be tested. Models of this sort also inevitably point out areas where knowledge is lacking and therefore where more research is needed.

SUMMARY

Environmental resources are consumed during plant growth while environmental conditions influence growth but are not consumed by the plant. The outcome of competition may be influenced by either resources or conditions, but it is for environmental resources that competition between neighboring plants occurs. Mechanisms of plant competition consist of both the effect that plants have on resources and the response of plants to changed resources. Considering resources as a single entity, space, is one approach to studying competition that allows evaluation of the overall effects of density, but precludes analysis of mechanisms of competition.

Photosynthesis is the only physiological process by which plants appreciably increase their biomass. The resources directly involved in photosynthesis are light, CO_2, and water. In addition to being a substrate for photosynthesis, water also is involved in leaf cooling and is a component of most metabolic processes in the plant. Nutrients likewise are an important resource needed for plant growth, as many are components of enzymes that act as catalysts for numerous physiological processes. Maximum photosynthesis and growth of most plants are dependent on the light environment. Many weeds have the ability to adapt to reduced light conditions and changes in light levels during growth through alterations in their physiology and morphology.

Competition for light differs from competition for other resources because a photon of light, which becomes available instantly, must be used immediately by a leaf or it is lost as a photosynthetic energy source for that leaf. This characteristic of light is in contrast to other resources that can be collected, transported, and stored for later use by the plant. Because of the nature of light as a resource, an important factor in competition for light among species is the physical position of leaves of one plant in relation to the leaves of a neighboring plant. Analysis of leaf area index (LAI) is a means to study the mutual effects of weed and crop species on canopy development, and to assess possible competitive advantages among species in the field. Recent research has shown that plants respond to alterations in light quality due to the presence of other plants even before mutual shading occurs. Whether these responses affect subsequent growth and productivity of plants in canopies remains to be tested.

Weed and crop species fix carbon dioxide (CO_2) into sugars by either of two major photosynthetic pathways, the C_3 or the C_4 pathway. Many summer annual weeds and crops possess the C_4 pathway of CO_2 fixation, whereas the C_3 pathway is more prevalent among winter annual weeds and cool-season crops. Under conditions of high irradiance and high temperatures, C_4 species seemingly are competitively superior to C_3 species. However, under moderate conditions the photosynthetic advantages of the C_4 pathway are diminished, and C_3 species are often superior. Due to different efficiencies in CO_2 fixation between C_3 and C_4 species, C_3 species are expected to benefit more than C_4 species from potential atmospheric CO_2 enrichment due to global warming.

Water is a major component of plant life. The amount of water available for plant use is determined by the seasonal supply of rainfall or irrigation, root structure and distribution in the soil, and the water use efficiency of individual plant species. Usually competition among species for available soil moisture is most significant in arid regions, where the water supply cannot be supplemented by irrigation. The excellent work conducted by T.K. Pavlychenko has provided considerable insight into the rooting patterns of both weeds and crops grown under competitive and noncompetitive regimes. His studies indicate that under dryland agricultural conditions, differential root development among weeds and crops ultimately determines the outcome of competition.

Plants can possess many mechanisms for withstanding water stress. These include adaptations leading to avoidance of stress, conservation and efficient use of water, and tolerance to water stress, usually by osmotic adjustment. These adaptations are generally based on the morphology and distribution of the root system, leaf characteristics, physiological mechanisms for maintaining high water use efficiency (WUE), and stomatal control. Research on root distribution in soil and patterns of soil moisture extraction is a difficult, but essential, area that needs greater attention by weed scientists.

The fact that stomata are involved both in gas exchange for photosynthesis and as a control of transpiration indicates a close relationship between plant water status and productivity. When competition for water occurs between species, a water use pattern involving relatively poor stomatal control, high transpiration rates, and high leaf productivity is a desirable strategy for a competitor. By utilizing a disproportionately large amount of water via transpiration, a competitor can limit the availability of that resource to its neighbor. The increased water use efficiency of plants with the C_4 pathway of carbon fixation does not necessarily impart competitive superiority to C_4 plants over C_3 plants.

Nutrients are also a major source of competition among neighboring plants. Often when weeds are grown in association with a crop, the nutrient status of the crop declines when compared to that of the crop growing alone. In some crop and weed combinations, the weeds may even consume nutrients luxuriantly, relative to the crop. However, increased fertility does not substitute for weed density reduction in a cropping system, especially if the limiting resource is unknown. Whether the effect of nutrient availability and plant density on crop yields is due to luxuriant consumption of nutrients by weeds or to interactions with other features of plant growth is an intriguing area for additional research.

The technique of mathematical growth analysis provides a way to examine growth processes in considerable quantitative detail. Important growth parameters that can influence competition are relative growth rate (R), net assimilation rate (NAR), and leaf area ratio (LAR). The amount of dry matter produced by an individual plant is an indicator of its overall utilization of the resources available for plant growth. Dry matter production may be expressed as mean relative growth rate ($\bar{R}$), which is the product of average net NAR and LAR. Three primary approaches for calculating these growth parameters are classical, integral, and functional growth analysis. The maximum potential relative growth rate (R_{max}) of individual plants has been used as an indicator of overall potential competitive ability. Many weeds of agricultural land are annuals with high values of R_{max}.

By determining the growth characteristics of weed species when exposed to varying levels of environmental resources or conditions, the potential range and degree of competitiveness can be determined. However, when species of either weeds or crops are grown together, the behavior of each species in the mixture is often different from when the species are grown separately. Thus, while insights about competitive mechanisms can be gained from growth analysis of individual plants, it is also important to study specific growth responses of plants grown in mixture. Simulation models of weed/crop competition based on plant physiological responses to resource limitation are a particularly useful way in which information about weed and crop ecology can be used to test potential weed management strategies.

Chapter

7

Other Types of Interference

As seen in Chapter 5, several other types of interference are possible that range from negative to positive interactions. The negative interactions, other than competition, include amensalism and parasitism, while positive interactions include commensalism, protocooperation, and mutualism (Table 5.1). Most interactions among plants occur through an "intermediary" (Figures 5.1 and 6.3), such as resources, pollinators, dispersers, herbivores, or microbial symbionts (Goldberg, 1990). Since all such interactions are, therefore, indirect, two distinct processes are possible. One or both plants in the association can have both an *effect* on the abundance of the intermediary and a *response* to the changes in the abundance of the same intermediary (Chapter 6). The type of interference that occurs depends on the identity of the intermediary and whether effects and responses are positive or negative (Table 7.1). Although the types of interference in Table 7.1 do not conform directly to the interactions described by Burkholder (1952) (Table 5.1), the concept of effect and response is valuable in that it suggests a mechanism through which such interactions might occur.

OTHER FORMS OF NEGATIVE INTERFERENCE

Allelopathy

Sometimes the depressive effect of a plant upon its neighbors is so striking that competition for a common resource does not seem adequate to explain the observation. In this case a dramatic decrease in plant biomass or mortality is

TABLE 7.1 Types of Indirect Interactions among Plants[a]

Types of Interaction	*Intermediary*	*Effect*	*Response*	*Net*
Exploitation competition	Resources	−	+	−
Apparent competition	Natural enemies	+	−	−
Allelopathy	Toxins	+	−	−
Positive facilitation	Resources	+	+	+
Negative facilitation	Resources	−	−	+
Apparent facilitation	Natural enemies	−	−	+

Source: Goldberg, 1990.

[a]In this classification, resources of plants include mutualists such as pollinators or dispersers as well as abiotic resources such as light, water, mineral nutrients, and CO_2. + and − in the Effect, Response, and Net columns indicate the effect of plants on abundance of the intermediary, the response of some "target" plant to abundance of the intermediary, and the net effect of plants on the "target" plant, respectively.

usually evident for one species but not for the other. Such a condition is termed *amensalism.* One explanation for such observations is that some plants release into the immediate environment of other plants toxic substances (allelochemicals) that harm or kill them. This phenomenon has been called *allelopathy*, and it is distinct from other forms of negative plant interference in that the detrimental effect is exerted through release of a chemical by a donor plant. The term allelopathy was coined by Molisch (1937) to describe chemical interactions among plants, including stimulatory as well as inhibitory responses. Many cases of allelopathy also involve the presence of microbes in the plant association.

In terms of plant responses, the existence of amensalism, or more specifically allelopathy, has become reasonably well-documented over the last several decades (Putnam and Tang, 1986; Rizvi and Rizvi, 1992). However, Putnam and Tang point out that few field investigations have definitively separated allelopathy from other forms of negative or positive interference because of the complexity of the problem. Putnam and Duke (1978) suggest that several reasons account for the difficulty in differentiating allelopathy from other forms of interference, especially competition. These difficulties include

- a general lack of nomenclature to describe adequately plant responses that occur in this manner
- a lack of techniques to separate allelopathic interactions from competition
- a failure to prove the existence of direct versus indirect influence via other organisms or microenvironmental modification

Nonetheless, a considerable body of information has accumulated that implicates allelopathy as an important form of plant interference (Table 7.2).

TABLE 7.2 Common Agroecosystem Weeds with Alleged Allelopathic Potential

Scientific Name	*Common Name*	*Reference*[a]
Abutilon theophrasti	Velvetleaf	Gressel and Holm (1964)
Agropyron repens	Quackgrass	Kommedahl et al. (1959)
Agrostemma githago	Corn cockle	Gajić and Nikočević (1973)
Allium vineale	Wild garlic	Osvald (1950)
Amaranthus dubius	Amaranth	Altieri and Doll (1978)
Amaranthus retroflexus	Redroot pigweed	Gressel and Holm (1964)
Amaranthus spinosus	Spiny amaranth	VanderVeen (1935)
Ambrosia artemisiifolia	Common ragweed	Jackson and Willemsen (1976)
Ambrosia cumanensis	—	Anaya and DelAmo (1978)
Ambrosia psilostachya	Western ragweed	Neill and Rice (1971)
Ambrosia trifida	Giant ragweed	Letourneau et al. (1956)
Antennaria microphylla	Pussytoes	Selleck (1972)
Artemisia absinthium	Absinth wormwood	Bode (1940)
Artemisia vulgaris	Mugwort	Mann and Barnes (1945)
Asclepias syriaca	Common milkweed	Rasmussen and Einhellig (1975)
Avena fatua	Wild oat	Tinnin and Muller (1971)
Berteroa incana	Hoary alyssum	Bhowmik and Doll (1979)
Bidens pilosa	Beggarticks	Stevens and Tang (1985)
Boerhovia diffusa	Spiderling	Sen (1976)
Brassica nigra	Black mustard	Muller (1969)
Bromus japonicus	Japanese brome	Rice (1964)
Bromus tectorum	Downy brome	Rice (1964)
Calluna vulgaris	—	Salas and Vieitez (1972)
Camelina alyssum	Flax weed	Grummer and Beyer (1960)
Camelina sativa	Largeseed falseflax	Grummer and Beyer (1960)
Celosia argentea	Celosia	Pandya (1975)
Cenchrus biflorus	Sandbur	Sen (1976)
Cenchrus pauciflorus	Field sandbur	Rice (1964)
Centaurea diffusa	Diffuse knapweed	Fletcher and Renney (1963)
Centaurea maculosa	Spotted knapweed	Fletcher and Renney (1963)
Centaurea repens	Russian knapweed	Fletcher and Renney (1963)
Chenopodium album	Common lambsquarters	Caussanel and Kunesch (1979)
Cirsium arvense	Canada thistle	Stachon and Zimdahl (1980)
Cirsium discolor	Tall thistle	Letourneau et al. (1956)
Citrullus colocynthis	—	Bhandari and Sen (1971)
Citrullus lavatus	—	Bhandari and Sen (1972)

TABLE 7.2 Continued

Scientific Name	*Common Name*	*Reference*[a]
Cucumis callosus	—	Sen (1976)
Cynodon dactylon	Bermudagrass	VanderVeen (1935)
Cyperus esculentus	Yellow nutsedge	Tames et al. (1973)
Cyperus rotundus	Purple nutsedge	Friedman and Horowitz (1971)
Daboecia polifolia	—	Salas and Vieitez (1972)
Digera arvenis	—	Sarma (1974)
Digitaria sanguinalis	Large crabgrass	Parenti and Rice (1969)
Echinochloa crus-galli	Barnyardgrass	Gressel and Holm (1964)
Eleusine indica	Goosegrass	Altieri and Doll (1978)
Erica scoparia	Heath	Ballester et al. (1977)
Euphorbia corollata	Flowering spurge	Rice (1964)
Euphorbia esula	Leafy spurge	Letourneau and Heggeness (1957)
Euphorbia supina	Prostrate spurge	Brown (1968)
Galium mollugo	Smooth bedstraw	Kohmuenzer (1965)
Helianthus annuus	Sunflower	Rice (1974)
Helianthus mollis	—	Anderson et al. (1978)
Hemarthria altissima	Bigalta limpograss	Tang and Young (1982)
Holcus mollis	Velvetgrass	Mann and Barnes (1947)
Imperata cylindrica	Alang-alang	Eussen (1978)
Indigofera cordifolia	Wild indigo	Sen (1976)
Iva xanthifolia	Marshelder	Letourneau et al. (1956)
Kochia scoparia	Kochia	Wali and Iverson (1978)
Lactuca scariola	Prickly lettuce	Rice (1964)
Lepidium virginicum	Virginia pepperweed	Bieber and Hoveland (1968)
Leptochloa filiformis	Red sprangletop	Altieri and Doll (1978)
Lolium multiflorum	Italian ryegrass	Naqvi and Muller (1975)
Lychnis alba	White cockle	Bhowmik and Doll (1979)
Matricaria inodora	Mayweed	Mann and Barnes (1945)
Nepeta cataria	Catnip	Letourneau et al. (1956)
Oenothera biennis	Evening primrose	Bieber and Hoveland (1968)
Panicum dichotomiflorum	Fall panicum	Bhowmik and Doll (1979)
Parthenium hysterophorus	Parthenium ragweed	Sarma et al. (1976)
Plantago purshii	Wooly plantain	Rice (1964)
Poa pratensis	Bluegrass	Alderman and Middleton (1925)
Polygonum aviculare	Prostrate knotweed	Al Saadawi and Rice (1982)

(*continued*)

TABLE 7.2 Continued

Scientific Name	*Common Name*	*Reference*[a]
Polygonum orientale	Princessfeather	Datta and Chatterjee (1978)
Polygonum pensylvanicum	Pennsylvania smartweed	Letourneau et al. (1956)
Polygonum persicaria	Ladysthumb	Martin and Rademacher (1960)
Portulaca oleracea	Common purslane	Letourneau et al. (1956)
Rumex crispus	Dock	Einhellig and Rasmussen (1973)
Saccharum spontaneum	Wild cane	Amritphale and Mall (1978)
Salsola kali	Russian thistle	Lodhi (1979)
Salvadora oleoides	—	Mohnat and Soni (1976)
Schinus molle	California peppertree	Anaya and Gomez-Pompa (1971)
Setaria faberi	Giant foxtail	Schreiber and Williams (1967)
Setaria glauca	Yellow foxtail	Gressel and Holm (1964)
Setaria viridis	Green foxtail	Rice (1964)
Solanum surattense	—	Sharma and Sen (1971)
Solidago sp.	Goldenrod	Letourneau et al. (1956)
Sorghum halepense	Johnsongrass	Abdul-Wahab and Rice (1967)
Stellaria media	Common chickweed	Mann and Barnes (1950)
Tagetes patula	Wild marigold	Altieri and Doll (1978)
Trichodesma amplexicaule	—	Sen (1976)
Xanthium pensylvanicum	Common cocklebur	Rice (1964)

Source: Putnam and Weston, 1986.

[a]All references are cited in the original source.

It is obvious that such a trait exhibited among weed species could have significant adverse impacts on crop yield. Since crops also may produce allelochemical substances, the opportunity also exists to exploit allelopathy for weed control.

RESPONSES OF PLANTS TO ALLELOCHEMICALS

Chemicals with allelopathic potential can be present in virtually every kind of plant tissue, including leaves, flowers, fruits, roots, rhizomes, and seeds. However, whether the substances can be released into the environment of neighboring plants in sufficient quantities to suppress growth is still a question that must be asked when investigating many alleged cases of allelopathy (Putnam and Weston, 1986). Whether these chemicals once released also persist long enough to suppress a succeeding generation of plants also remains an unanswered question in a number of cases. However, allelopathy can sometimes provide obvious and startling responses in affected plants. For example, herbaceous plants growing near black walnut (*Juglans nigra*) may either fail to germinate or suddenly die as a result of juglone, a chemical produced in the leaves and roots of the tree. Dramatic reductions in crop growth have also been attributed to quackgrass (*Agropyron repens*) residues (Table 7.3) However, Putnam and Weston (1986) indicate that the most common expressions of allelopathy may be much more subtle than the above examples and, as such, have longer-term effects.

Toxins from residues. A primary effect of allelopathy on crop production seems to result from an association with plant litter in or on the soil (Table 7.4) Numerous organic chemicals are present in plant material and, when crop or other plant residues are left on the soil surface after harvest or plowed under, chemicals can be released by rainfall or microbial decomposition. Patrick (1971), Rice (1984), and Barnes et al. (1986) all cite situations in which apparent phytotoxic substances are associated with the decomposition of plant residues (e.g., Table 7.4). In the case of plant residues, however, subsequent mortality or growth suppression do not have to be related directly to the release of a toxic organic substance from plant material. Rather, modification in microenvironment (for example, localized alteration in soil pH or other conditions as a result of litter decomposition) could account for the phytotoxic response. Also possible is the release of a phytotoxic microbial product that accumulates as residues are degraded.

TABLE 7.3 Reduced Crop Growth in Response to Quackgrass Residues Either in No-tillage or Conventional Tillage Plot (Plowed and Disk-harrowed)

	Plant Weight (Percent of Control)[a]	
Crop	*No-Tillage*	*Conventional Tillage*
Alfalfa	3.7	54.8
Cabbage	20.6	73.3
Carrot	8.9	65.2
Cucumber	35.3	83.1
Oats	50.0	76.2
Pea	40.5	79.9
Sorghum	18.7	87.3
Sweet corn	40.4	88.8

Source: Putnam and Weston, 1986.

[a]The control was an adjacent conventionally tilled plot.

TABLE 7.4 Influence of Water-soluble Substances Extracted from Different Plant Residues on Germination and Growth of Wheat Seeds

		Inhibition (%)					
		Germination		Root Growth		Shoot Growth	
Crop Residues		AC[a]	No AC	AC	No AC	AC	No AC
Sweet clover stems	C[b]	−3[d]	−8	58	7	24	21
	H[c]	−1	−8	51	12	10	10
Wheat straw	C	7	5	36	7	14	21
	H	−5	−5	18	36	7	28
Soybean hay	C	3	−3	80	30	66	45
	H	−1	−5	51	39	48	45
Oat straw	C	3	10	87	64	83	76
	H	−1	−3	84	45	79	62
Brome grass	C	1	3	71	55	48	59
	H	−1	27	71	62	52	78
Cornstalks	C	−7	89	75	87	62	93
	H	5	38	47	75	62	83
Sorghum stalks	C	9	100	87	100	86	100
	H	3	72	84	82	83	93
Sweet clover hay	C	64	3	95	82	90	83
	H	26	100	95	100	90	100
Mean	C	9.6	24.9	73.6	54.0	59.1	62.3
	H	3.1	27.0	62.6	56.4	53.9	62.4

Source: Guenzi and McCalla, 1962. By permission of the Soil Science Society of America.

[a]AC, autoclaving for 1 hr at 20 lb steam pressure.

[b]Cold-water soluble substances (extracted at 25°C).

[c]Hot-water soluble substances (extracted at 100°C).

[d]Negative sign indicates stimulation.

Toxins from leachates and exudates of plants. Another source of allelochemicals is the production and release of toxins (secondary products) by growing plants that ultimately inhibit development of adjacent plants. However, this process does not appear to have as great an effect on yield as litter leaching or decay. Several notable exceptions exist, however. For example, juglone can be leached from living black walnut foliage, as discussed earlier; sesquiterpene lactones found in foliage leachates of plants in the Asteraceae, Apiaceae, and Magnoliaceae can inhibit germination of crop and weed species (Fischer,

1986); and caffeine from coffee tree foliage can inhibit the abundance of weeds and young coffee plants in the understory of coffee plantations (Waller et al., 1986). Nonetheless, specific evidence of foliage leachates or root exudation of growth inhibitory secondary compounds is limited.

Effects of allelochemicals on seed. Numerous seed bioassays have been accomplished that demonstrate the positive and negative effects of allelochemicals on germination (see Leather and Einhellig, 1986 for references). For example, the germination of *Striga asiatica*, an important parasitic weed species, is enhanced by the presence of strigol, a substance produced by the roots of a susceptible host plant. Two functions of allelochemicals that may be present in seed are the prevention of seed decay and the inhibition of germination. Both of these processes, decay and germination, can account for substantial losses of seed from the soil seed bank (Chapter 4). Putnam and Tang (1986) indicate that research both to induce weed seed germination chemically or to inhibit it using allelochemicals would be fruitful areas of study. In addition, allelochemical enhancement of weed seed decay would be a worthwhile effort, although little research is underway in this area.

METHODS TO STUDY ALLELOPATHY

A wide array of techniques has been proposed for the study of allelochemicals. For the most part these include specific methods for toxin isolation, followed by bioassays to test for phytological activity. The methods proposed for the isolation of toxins range from organic solvent extraction to cold-water infusion. Various bioassay procedures have been used to test plants or seed treated with or exposed to the putative allelochemical. Such methodology is effective in establishing the existence of naturally occurring toxic substances, but care must be taken in interpretation of results because the toxin also must be "released" into the environment and be present at phytotoxic concentrations in order to be considered allelopathic. Most straightforward are studies that use litter or crop residues directly. In these studies, either living or dead plant material is mixed in or placed on the soil for some period of time, after which the soil is bioassayed for allelochemical activity.

Other assay techniques, in addition to toxin isolation and direct assay of litter, have been developed to detect the occurrence of inhibitory root exudates. These include using various media for growing plants, and examining the media of "donor" plants for phytotoxicity to "recipient" plants.

An appealing technique is the stair-step system for the study of allelopathy. In this system, donor and recipient plants are grown separately in sand solution with the pots alternated in stair-step fashion (Figure 7.1). The soil solution is circulated from donor to recipient and back again a number of times. An innovation to this technique was developed by Tang and Young (1982), in which an exchange column is inserted between donor and recipient plants. Thus, any substance exuded by roots of the donor can be isolated and bioassayed for phytotoxicity. In these experiments, care must always be taken to assure that

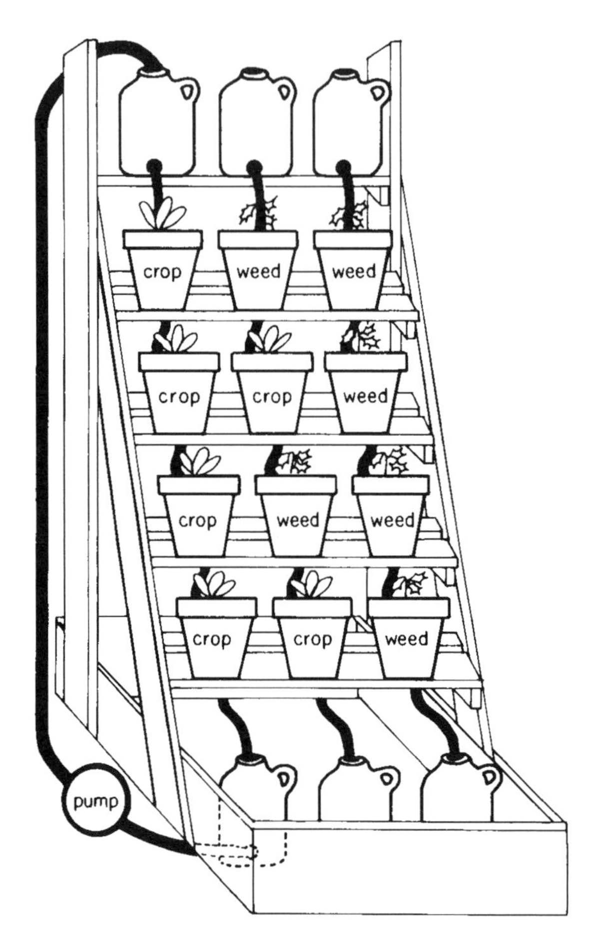

FIGURE 7.1 A schematic representation of the stair-step technique for the study of allelopathy.

water, nutrients, or other conditions necessary for growth are never limiting. For example, Putnam and Duke (1978) point out that variability in the amount of light at each step of the staircase often is not considered in such experiments.

Although most studies to date on allelopathy have been concerned with the event itself and less with the causal agent, significant progress has been made in both isolation and identification of specific allelochemicals (Putnam and Tang, 1986; Rizvi and Rizvi, 1992). Future research in this area must prove specifically that a toxic substance is produced by donor plants and that it accumulates or persists long enough at sufficient concentrations in the environment to inhibit development of other plants. With increasing sophistication in instruments for chemical detection, this goal appears near to becoming realized. It is equally important that potential limiting factors or conditions other than allelopathy are identified in situations involving negative interactions among plants, even in the event that phytotoxic substances are found.

MICROBIALLY PRODUCED PHYTOTOXINS

It is a well known biological phenomenon that various kinds of microorganisms produce substances detrimental to other organisms. These chemicals may act as repellents, suppressants, inductants, or attractants. Duke (1986) notes that many microbial phytotoxins exist in nature (Table 7.5) and that there is

TABLE 7.5 Examples of Several Non Host-specific Phytotoxins Produced by Microorganisms

Microbial Toxin	*Source*	*Effect*	*Reference*[a]
Acetylaranotin	*Aspergillis terreus*	Growth inhibition	Kamata et al. (1983)
Alteichin	*Alternaria eichorniae*	Necrosis	Robeson et al. (1984)
Altersolanol A	*Alternaria porri*	Growth inhibition	Suemitsu et al. (1984)
Botrydienol	*Botryotinia squamosa*	Growth inhibition	Kimata et al. (1985)
Brefeldin A	*Alternaria carthami*	Chlorosis	Tietjen et al. (1983)
Chaetoglobosin K	*Diplodia macrospora*	Growth inhibition	Cutler et al. (1980)
Cladosporin	*Cladisporium cladosporioides* *Aspergillus* spp. *Eurotium* spp.	Growth inhibition	Springer et al. (1981)
Colletotrichin	*Colletotrichum nicotianae*	Necrosis and growth inhibition	Gohbara et al. (1978)
Coronatine	*Pseudomonas syringae*	Chlorosis	Mitchell (1981)
Cyclopenin	*Penicillium cyclopium*	Growth inhibition and necrosis	Cutler et al. (1984)
Cytochalasins	*Phomopsis* spp.	Growth inhibition	Cole et al. (1981 a,b); Cox et al. (1983); Wells et al. (1976)
Dehydrocurvularin	*Alternaria macrospora*	Necrosis	Robeson and Strobel (1985)
Desmethoxyviridiol	*Nodulisporium hinnuleum*	Necrosis monocots	Cole et al. (1975)
Dihydropergillin	*Aspergillus ustus*	Growth inhibition	Cutler et al. (1981)
Gougerotin	*Streptomyces* sp.	Growth inhibition	Murao and Hayashi (1983)
Hydroxyterphenyllin	*Aspergillus candidus*	Growth inhibition	Cutler et al. (1978)
Mevinolin	*Aspergillus terreus*	Growth inhibition	Bach and Lichtenthaler (1983)
Moniliformin	*Fusarium moniliforme*	Growth inhibition, necrosis, and chlorosis	Cole et al. (1973)

(continued)

TABLE 7.5 Continued

Microbial Toxin	*Source*	*Effect*	*Reference*[a]
Monocerin	*Exserohilum turcicum*	Chlorosis and necrosis	Robeson and Strobel (1982)
Naramycin B	*Streptomyces griseus*	Growth inhibition	Berg et al. (1982)
Neosolaniol monoacetate	*Fusarium tricinctum*	Growth inhibition	Lansden et al. (1978)
Oosporein	*Chaetomium trilaterale*	Growth inhibition	Cole et al. (1974)
Ophiolbolin	*Helminthosporium oryzae* and *Cochliobolus miyabeanus*	Disrupts membrane function	Cocucci et al. (1983)
Orlandin	*Aspergillis niger*	Growth inhibition	Cutler et al. (1979)
Pergillin	*Aspergillus ustus*	Growth inhibition	Cutler et al. (1980)
Phomenone	*Phoma destructiva*	Wilt and growth inhibition	Capusso et al. (1984)
PR-toxin	*Penicillium roqueforti*	Growth inhibition	Capusso et al. (1984)
Prehelminthosporol	*Dreschlera sorokiana*	Growth inhibition, necrosis, and chlorosis	Cutler et al. (1982)
Radicinin	*Alternaria helianthi*	Necrosis	Tal et al. (1985)
Stemphylotoxin I	*Stemphylium botryosum*	Necrosis	Barash et al. (1982)
Toyocamycin	*Streptomyces toyocaensis*	Chlorosis	Yamada et al. (1972)
Viridiol	*Gliocladium virens*	Necrosis	Howell and Stipanovic (1984)
Zinniol	*Alternaria carthami*	Necrosis	Robeson and Strobel (1984); Tietjen et al. (1983)

Source: Duke, 1986.

[a]All references cited in the original source.

considerable interest in using either these organisms or the chemicals they produce to suppress weed growth, that is, as naturally-produced herbicides. Examples of the use of microbes for weed control purposes include:

- a soil-borne fungus *Phytophthora palmivora*, for control of strangler vine (*Morrenia odorata*)
- the aerial fungus *Colletotrichum gloeosporoides* sp. *aeschynomene* for control of northern jointvetch (*Aeschynomene virginica*) in rice and soybean

- the wilt fungus *Cephalosporium diospyri* to inhibit Virginia buttonweed (*Diodia virginiania*)
- the fungus *Chondrostereum purpureum*, which is applied to cut surfaces of *Prunus serotina* stumps in reforestation areas

Duke (1986) also notes that in many cases microbially produced substances offer novel chemistries, high efficacy, new and more desirable selectivity, and favorable environmental properties as compared to herbicides. In spite of problems in economical production and in some cases absorption by treated plants, several microbially derived chemicals are now being registered and sold. Usually these products are highly specific, as suggested by the list above, for the suppression of a single plant species or group of species.

Parasitism

A parasite is a plant or animal living in, on, or with another living organism (its host) at whose expense it obtains food, shelter, or support. Parasites can be obligate, surviving only in association with the living host, or nonobligate, living either saprophytically or on a living host. In addition, some parasitic flowering plants are hemiparasites. These plants have chlorophyll but depend on the host plant for water and mineral nutrition.

Most parasitic flowering plants occur in about ten families, but only four families contain the most troublesome parasitic weeds. These are Convolvulaceae (*Cuscuta*, dodders), Loranthaceae (*Arceuthobium, Phoradendron, Viscum*; mistletoes), Orobanchaceae (*Orobanche*, broomrapes), and Scrophulariaceae (*Striga*, witchweeds). The species in these four families represent the most important plant diseases in agricultural crops (Figure 7.2) and forest trees (Figure 7.3) that are caused by parasitic plants. Each genus is represented in North America, except *Viscum*, although *Striga* is found only in limited areas of North and South Carolina (Figure 8.33). Parasitic weeds also are of major importance in tropical agriculture (Table 7.6). In general, parasitic weeds are grouped into root parasites, such as *Striga* and *Orobanche*, and stem parasites such as *Cuscuta, Loranthus*, and *Arceuthobium*. The characteristics and economic importance of the various species within each genus have been described in detail by other authors (King, 1966; Kuijt, 1969; Musselman, 1982, 1987). For this reason we have chosen to concentrate on those features of parasitic plants that make them unique among the weed taxa.

ADAPTATIONS OF PARASITIC WEEDS FOR DISPERSAL AND GERMINATION

In order to survive, seedlings of parasitic plants must quickly find a suitable host plant. There are three methods through which parasitic plants increase the probability of successful contact with their host. In a number of parasitic species (for example, dodders), the seed are relatively large. Thus, the seed has sufficient food reserve to allow the radicle to grow extensively while it is

FIGURE 7.2 Dodder (*Cuscuta* spp.) attached to alfalfa. (Photograph courtesy of A.P. Appleby, Oregon State University.)

FIGURE 7.3 Dwarf mistletoe (*Arceuthobium americanum*) on lodgepole pine. (Photograph courtesy of W. Theis, U.S. Forest Service, Corvallis, Oregon.)

TABLE 7.6 Common Parasitic Weeds in the Tropics

Family	*Weed*	*Crops Affected*
Convolvulaceae	*Cuscuta campestris, C. chinensis, C. reflexa*	Broad beans, tomatoes
Lauraceae	*Cassytha filiformis*	Citrus, *Ocimum* spp.; many woody weed spp.
Loranthaceae	*Dendrophthoe falcata, Viscum capitellatum, Tapinanthus bangwensis, Arceuthobium* spp., *Loranthus* spp.	Citrus, cocoa, cola, mango, rubber, tea, and teak
Orobanchaceae	*Orobanche crenata, O. aegyptica, O. ramosa*	Sunflower, tomato, tobacco, beans, watermelon
Santalaceae	*Thesium* spp.	Sugarcane, onion
Scrophulariaceae	*Alectra* spp., *Buchnera hispida, Striga* spp.	Cowpea, groundnuts, maize, millet, sugarcane. Sorghum is a potential host. Slight infestation in rice.

Source: Akobundu, 1987.

seeking a host plant. Another mechanism for location of an appropriate host relies on birds for dispersal. Kuijt (1969) indicates that a precise mode of dispersal has evolved in such cases that relies on birds to deposit the seed of the parasite, for example, *Phoradendron* on host branches. A similar mechanism of dispersal that ensures the same end is demonstrated by the propulsion of seed from the fruit of dwarf mistletoes (*Arceuthobium*) into the branches of the same or a nearby tree. The third adaptation for host location requires a biochemical exudate which is produced by the root of the host plant in order to initiate germination of the parasite seed. This requirement for germination is most pronounced in *Orobanche* seed. A further adaptation for host location following germination is demonstrated by *Orobanche* as well as by *Striga*. This is the chemotropic growth of the radicle of these parasitic species toward the root of their host plants. Although this feature of germination is probably highly evolved and acts to enhance seedling survival, it also may be used to obtain some control of these weed species. For example, species of *Striga* can be induced to germinate by plant species that are not preferred hosts, that is, "trap" crops. Since the trap crop cannot support the growth of the parasite, it acts to reduce the abundance of witchweed seed in the soil through seedling mortality.

PHYSIOLOGICAL CHARACTERISTICS OF PARASITIC WEEDS

Although many parasitic species of weeds contain at least some chlorophyll, others do not. Some species that have chlorophyll—for example, *Cuscuta* spp. and *Arceuthobium* spp.—apparently photosynthesize to only a limited degree, whereas others fix carbon nearly as well as other nonparasitic members of their families. Experiments that utilize radioactively labeled elements and substances clearly demonstrate the passage of organic material, minerals, and water from the host to the parasite. Such experiments have been conducted on a wide range of species in many genera, most notably on *Striga, Phoradendron, Arceuthobium, Cuscuta*, and *Orthocarpus*. However, the degree of dependence on the host plant often varies with age and species of the parasitic weed. For example, *Striga* attaches to the roots of a host plant soon after germination but does not emerge from the soil for several weeks. During this time, *Striga* is totally dependent upon the host plant, but once they emerge *Striga* plants produce chlorophyll and begin to generate their own assimilates. Water and mineral nutrients still must be obtained from the host plant, however. *Orobanche*, on the other hand, is a root parasite that lacks chlorophyll and depends upon its host plant for its total sustenance.

The major organ of parasitic weeds for attachment and penetration of the host tissue is the *haustorium*. Haustoria appear to vary in structure according to species but all have a similar function, which is attachment and subsequent transport of materials from host to parasite. It should be noted, however, that some bi-directional movement of some materials, though rarely photosynthate, can occur between parasite and host. Figure 7.4 depicts the penetration of the haustorium of *Cuscuta campestris* (field dodder) into a species of *Impatiens*. Since the hypha of the haustoria contact both xylem and phloem of the host plant, transport of water and minerals, as well as carbon compounds, is known to occur. However, the exact mechanisms of these transfers is poorly understood. Additional studies (see Musselman, 1987) suggest that transfer of plant hormones between host and parasite also occurs. Thus flowering of many hosts and parasites is in synchrony.

Kuijt (1969) has provided an excellent taxonomic and morphological review of parasitic plants. In addition King (1966) and Musselman (1982, 1987) have summarized many of the biological characteristics of parasitic angiosperms that are of greatest economic importance. It is apparent from these reviews that parasitic weeds often have a significant detrimental effect on their hosts, since direct removal of resources from them occurs. However, in many cases a fine line seems to exist between parasitism and hemiparasitism. It is also apparent that parasitic weeds have evolved closely with their host species (also see Chapter 3). For example, the haustoria of many parasitic weeds apparently do not form if the plants are grown axenically but are rapidly induced in the presence of host roots or host root exudates (Lynn, 1985). Perhaps the study of this close evolutionary association can lead to better management of these species. Detailed studies concerning the responses of parasitic weeds to chemical signals and the mechanisms of production and transport of these chemicals also may prove useful in this respect.

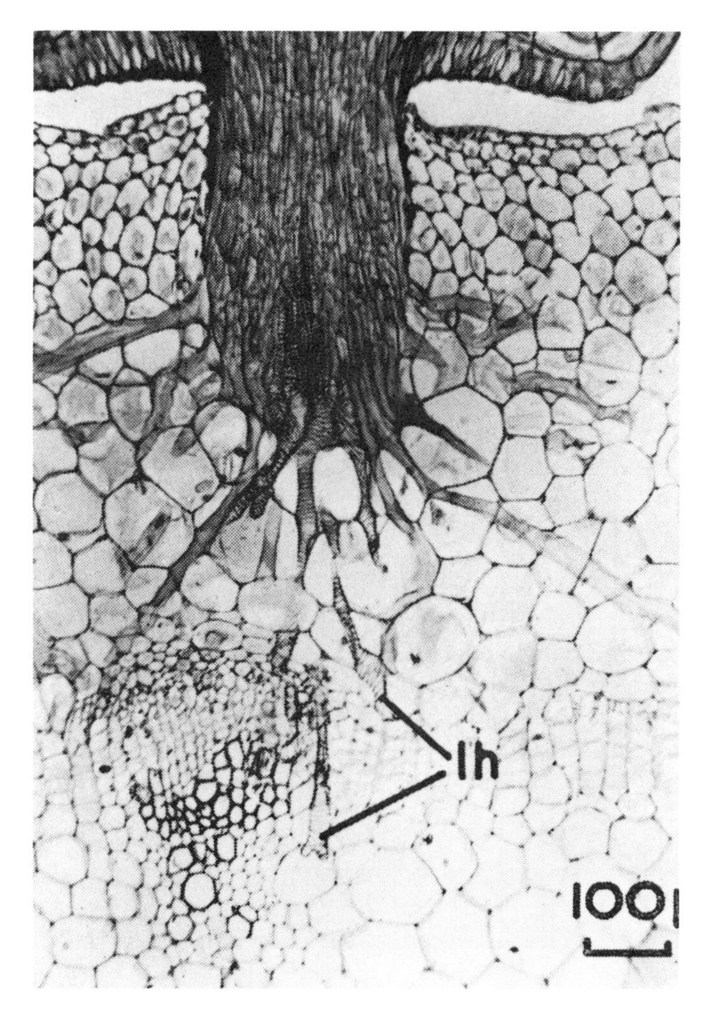

FIGURE 7.4 The haustorium of field dodder (*Cuscuta campestris*) in the stem of *Impatiens* sp. (From MacLeod, 1962. By permission of the Botanical Society of Edinburgh.)

FORMS OF POSITIVE INTERFERENCE

Commensalism

Commensalism (Table 5.1) may be thought of as a one-way relationship between two organisms. In this type of association, only one organism is stimulated by the presence of the other and inhibited by its absence, whereas the other, or host, is unaffected. Common examples of commensalism are those in which the host organism serves as a surface for attachment and support or a means of shelter for the other organism, without itself being affected (Whittaker,

1975). The organism that benefits gains physical anchorage or protection from the environment.

Commensalism between plants is often encountered in the form of epiphytes, plants that grow on other plants. The epiphyte uses the host for mechanical support rather than as a source of nutrients or water, which are supplied by humid air or rainwater. Many lower plants among the algae, lichens, and mosses commonly are found growing on the bark of trees. Examples of epiphytes among higher plants include herbaceous perennial plants such as some ferns, bromeliads, orchids, and cacti. These may be found growing attached to trees for physical support.

The nurse plant syndrome is another type of commensal relationship between plants. In these cases, one plant generally is found growing in the shade or shelter of the host, or nurse plant. The host, usually a shrub or perhaps an adult form of a seedling plant being benefited, provides shade and protection from high temperatures, soil drying and frost, and sometimes from herbivores (Barbour et al., 1987). This pattern of growth is exhibited by the saguaro cactus as well as several species of desert annual plants. In all these examples, the species that are positively associated with a host plant show no host specificity, nor is the host affected by the presence of the associated plant. In the case of desert annuals, in addition to providing protection, the shrub hosts act as a barrier to trap organic debris and therefore provide a suitable substrate for growth.

Protocooperation

The above examples are rather specialized in nature, and no proven examples are found in the literature of commensalism occurring among weeds and crops. It is much more common that two plants that interact affect each other reciprocally. Such is the case with *protocooperation*, in which both organisms are stimulated by the association but unaffected by its absence. An example of protocooperation that occurs among species in many different habitats is natural root grafts. Generally occurring between trees, these grafts allow a mutual exchange of photosynthate and other materials to occur (Barbour et al., 1987). In many instances, fungal hyphae (mycorrhizae), in addition to dramatically enhancing soil nutrient uptake, may link two or more plants together, thus facilitating the protocooperation.

Another type of plant interaction frequently associated with negative effects is the exudation of materials from the roots of one plant and subsequent absorption by the roots of another. This type of transfer, involving the one- or two-way movement of organic and inorganic metabolites, may be beneficial in some cases. For example, Neill and Rice (1971) observed that soil from the rhizosphere of western ragweed (*Ambrosia psilostachya*) markedly stimulated the growth of several species of plants growing in the same field (Table 7.7). Similarly, Gajić and her associates (Gajić et al., 1976) demonstrated that wheat grain yields were increased appreciably when grown in mixed stands with corn-

TABLE 7.7 Stimulation of Plant Growth by Rhizosphere Soil from *Ambrosia psilostachya* Collected in the Field in July[a]

Test Species	*Mean Dry Weight (mg) with Standard Error*	
	Control	*Test*
Amaranthus retroflexus	42 ± 8.0	95 ± 10.0[b]
Andropogon ternarius	25 ± 1.4	33 ± 2.1[b]
Bromus japonicus	22 ± 1.1	46 ± 3.3[b]
Digitaria sanguinalis	56 ± 6.2	117 ± 7.7[b]
Leptoloma cognatum	20 ± 1.9	36 ± 1.8[b]
Rudbeckia hirta	16 ± 0.8	25 ± 1.6[b]
Tridens flavus	27 ± 1.6	43 ± 3.2[b]

Source: Neill and Rice, 1971.

[a]The control soil was collected in the same field at least 1 m away from the *A. psilostachya* plants.

[b]Difference from control significant at 0.05 level or better.

cockle (*Agrostemma githago*) as compared with pure stands of wheat. Sterile seedlings of corncockle stimulated the growth of sterile seedlings of wheat when grown on agar, thus indicating the presence of a root exudate. Subsequent isolation of the exudate, agrostemmin, and application to a wheat field (1.2 g per ha) increased grain yields of both nitrogen-fertilized and unfertilized stands.

Leaching of metabolites from aboveground plant parts by rain, dew, and mist has also been shown to be a source of movement of beneficial substances between plants. Calcium, phosphorus, magnesium, and other inorganic nutrients, as well as carbohydrates, amino acids, and organic acids can be transferred in this manner (Table 7.8). The plants for which beneficial leaching of nutrients has been demonstrated include many vegetable crops, grains, grasses, cotton, and tobacco. Residues from water and methanol extracts of over ninety weed and crop species were found to have either stimulatory (six species) or allelopathic (18 species) effects on the test plant, purple top turnip, at concentrations ranging from 3 to 300 ppm (Nicollier et al., 1985). Although little research is available that explores the mechanism of such beneficial associations in weed/crop or weed/weed interactions, it is important to realize the potential role of root exudation and foliage leaching in maintaining soil fertility and perhaps influencing plant succession.

INTERCROPPING

Probably the most well-known form of protocooperation among plants is that which occurs in polycropping or intercropping situations. The practice of

TABLE 7.8 Metabolites Leached from Plant Foliage

	Organic		
Inorganic	*Carbohydrates*	*Amino Acids*	*Organic Acids*
Boron	Fructose	Alanine	Aconitic
Calcium	Galactans	Arginine	Adipic
Chlorine	Glucose	Asparagine	Ascorbic
Copper	Lactose	Aspartic Acid	Citric
Iron	Pectic substances	β-Alanine	Fumaric
Magnesium	Raffinose	Cysteine	Glutaric
Nitrogen	Sucrose	γ-Aminobutyric acid	Glycolic
Phosphorus	Sugar alcohols	Glutamic acid	Lactic
Potassium		Glutamine	Maleic
Silica compounds		Glycine	Malic
Sodium		Histidine	Malonic
Strontium		Hydroxyproline	Pyruvic
Sulfur		Isoleucine	Succinic
Zinc		Leucine	Tartaric
		Lysine	Acidic glycosides
		Methionine	
		Phenylalanine	
		Proline	
		Serine	
		Threonine	
		Tryptophan	
		Tyrosine	
		Valine	

Source: Tukey, 1966.

simultaneously planting two crops is in widespread use throughout the world, particularly in the tropics (Francis et al., 1976; Vandermeer, 1989). A review of the available literature on intercropping research reveals conflicting evidence about the advantages of this practice (see Figure 5.2). Certainly such factors as plant density, spatial arrangement, stage of development, time of planting, and soil fertility must be considered in planning and evaluating results of intercropping experiments, as discussed in Chapter 5 for competition studies. Depending on the particular crops and their environment, much evidence has accumulated, however, which confirms that growth and yield are greater when certain crops are grown simultaneously than when they are grown alone as monocultures (Altieri and Liebman, 1987; Vandermeer, 1989; Daniel et al., 1979). This situation is depicted in Figure 5.23 and discussed more fully in Chapter 5. In the mixtures shown in Figure 5.23*a, b*, and *e*, each component is affected less by interspecific competition than it is by intraspecific competition when grown alone; thus overyielding by the intercrops occurs.

The success of cropping mixtures usually is measured in terms of the "relative yield total" (RYT) or "land equivalent ratio" (LER). The relative yield of any crop is calculated as the ratio:

$$RY = \frac{\text{Yield in Mixture}}{\text{Yield in Pure Culture}} \qquad (eq.\ 7.1)$$

Summing the relative yield values for the two crops in a mixture gives the RYT for that mixture (equation 5.6). A term more often used by agronomists is the LER, land equivalent ratio, which is essentially the same value as RYT; however, it is defined to be the total land required by the sole crops to yield the same amount as the intercrop mixture (Yadav, 1982). Intercrop mixtures where the RYT or LER = 1 are the result of the crops making similar demands on environmental resources. For these crops there is no apparent yield advantage for planting mixtures. If the intercrop components exploit the environment in different ways, then competition is avoided or minimized and RYT or LER > 1. Joliffe (personal communication) points out reservations, however, about the appropriateness of using RYT or LER as the only basis for examining the intercrop advantage. He suggests that such an approach often leads to underestimation of the yield advantage from intercropping and suggests alternate methods of calculation.

Mechanisms for the intercrop advantage. In intercropping, many explanations for the advantage of mixtures over sole plantings have been proposed (Trenbath, 1976; Lamberts, 1980). Competition may be avoided if the crops are separated in space through canopy or root stratification, or separated in time through variation in maturity time or length of growing season. Differing growth requirements or nutrient uptake abilities also may reduce competition between two components of a mixture and allow more efficient utilization of resources. Physical support (e.g., in maize/climbing bean intercrops) and physical protection from wind and frost have been demonstrated to raise yields of mixtures over those of the sole crops combined. Some cropping mixtures benefit from weed or insect suppression, water conservation, and erosion control as a result of more complete and efficient use of the environment by the crops (Gliessman and Altieri, 1982; Altieri and Liebman, 1987).

Weeds in intercrop systems. Weed control is often cited as one of the advantages of intercropping (Moody and Shetty, 1981; Liebman, 1986, 1988). The introduction of even a single weed species to an intercropping mixture adds substantial complexity to the analysis of its performance, however. Vandermeer (1989) suggests that the most obvious method of analysis is to compare the direct and indirect effects of the three mutually competitive species on each other. Figure 7.5a depicts the simple joint interaction between the crops and weed species under the conditions of monoculture production, while Figure 7.5b illustrates the parallel situation of intercropping (Vandermeer, 1989). However, these direct effects of one plant on another are not the only ones of concern since the indirect effects shown in Figure 7.5c and d also are possible and would be added to the direct effects of C_1, C_2, and W on each other. In other words, the yield of crop C_2 could be enhanced by the lower competitive ability of crop C_1 that occurs as a result of the weeds' presence (Figure 7.5c) or the competitive ability of both C_1 and C_2 relative to each other could be lowered by the presence of W (Figure 7.5d), which would further enhance the yield of both crops.

The most obvious method of analysis of weedy intercrop situations would be a comparison of yield response of the "main" crop to the intensity (density or amount) of the "secondary" crop in a weedy and weed-free environment

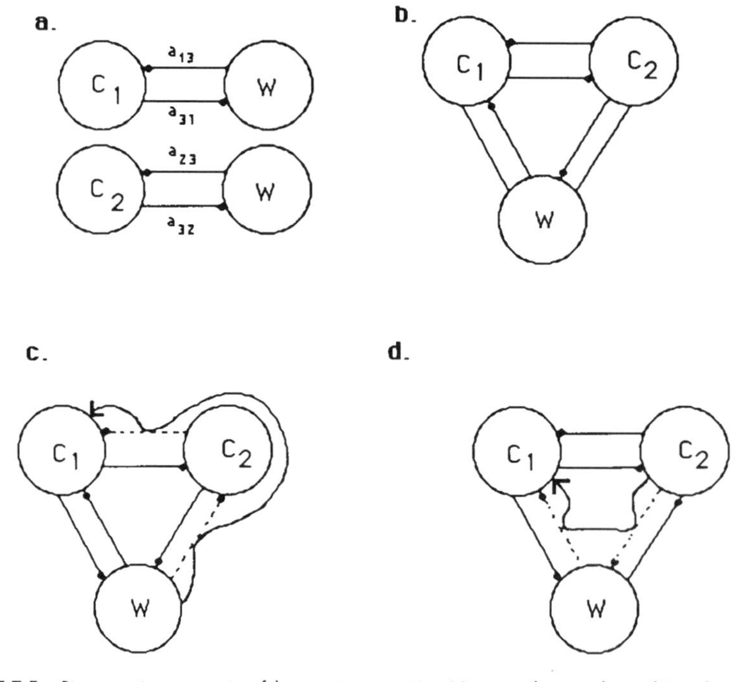

FIGURE 7.5 Diagrammatic representation of three species competition: (a) two weedy monocultures, (b) weedy intercrop, (c) illustration of indirect positive effect of weed on crop number 1, (d) illustration of indirect positive effect of the second crop species on the first. (From Vandermeer, 1989.)

(Vandermeer, 1989). This type of analysis also is appropriate to study the effects of intercropping when other organisms such as insects or diseases are involved in the interaction, that is, tests with and without the pest or disease present. However, Vandermeer (1989) cautions that often the indirect effects of weed control by intercrops are both subtle and complex. Thus, the advantage conferred to the intercrops may not be evident in the typical weedy versus non-weedy LER computations.

Mutualism

In contrast to the other types of positive interactions already discussed, *mutualism* is an obligatory type of relationship. The benefits gained by each partner link them into mutual, physiological interdependence. In the event that one partner is absent, they both suffer, or in some cases cannot even exist as free-living organisms. Mutualism should be distinguished from protocooperation such as is found in beneficial intercropping situations. The yield advantage in intercropping often occurs because the plants fail to suffer in the presence of each other rather than because of any benefits they afford each other. *Symbiosis* is another term often used for positive interactions. Symbiosis may be defined as the permanent, intimate association of two dissimilar organisms (Whittaker, 1975). Because it is generally used to refer to mutually beneficial relationships, symbiosis is often equated with mutualism.

PLANT-FUNGAL INTERACTIONS

One of the most well-known mutualistic relationships in nature occurs in the form of lichens. A lichen is a symbiotic association between an alga and a fungus. The hyphae of the fungus surround the algal cells to afford protection from the environment and a favorable environment for growth. Photosynthates and often nitrogen compounds are supplied by the alga to the fungus.

Many fungi also are found in symbiotic or mutualistic associations with roots of higher plants in the form of mycorrhizae. The fungal hyphae in these associations may form a mantle over the root surface and penetrate inter- and/or intracellular regions of the root cortex (Allen, 1990; Gerdemann, 1968; Daniel et al. 1979). Many benefits to the host plant by mycorrhizae have been suggested. Some of these include serving as root hairs and thus increasing the root absorptive surface; increasing supplies of nitrogen, phosphorus, other nutrients, and water to the root; and increasing the decomposition of organic matter in the vicinity of the plant roots, which is then available for uptake. The plant is thought to supply carbohydrates and other metabolites for the benefit of the fungus. Mycorrhizae occur in many diverse habitats and on a variety of plants including grasses, shrubs, and trees. Through this association the host plant often is able to grow in nutrient-poor soil that otherwise would be unsuitable for normal growth and development.

NITROGEN FIXATION

Another type of mutualistic relationship involving higher plants is symbiotic nitrogen fixation. This process involves the conversion of nitrogen gas to organic ammonia, which only certain prokaryotic organisms can carry out. Species of the nitrogen-fixing bluegreen algae *Nostoc* and *Anabena* are found in symbiotic association with certain liverworts, cycads, and *Azolla*, a genus of small aquatic ferns. Many legumes are found associated with the nitrogen-fixing bacterium *Rhizobium*. Certain shrub and tree species such as *Purshia tridentata* and *Alnus rubra* are associated with as yet unidentified nitrogen-fixing microorganisms. Generally, the nitrogen-fixing symbiont is found localized in morphologically specialized structures on the host plant, for example, in root nodules of legumes. Frequently the symbiont is morphologically and physiologically distinct from the free-living species of the same genus.

The *Azolla-Anabena* association is among the most agronomically important of the mutualistic relationships so far described. *Anabena* supplies *Azolla* with its total nitrogen requirement and the fern supplies carbohydrates and metabolites to the alga. This particular association is being studied intensively around the world for its potential as a fertilizer source in the culture of rice. For this purpose, *Azolla* may be grown concurrently with rice or may be used as a green manure after death of the plant occurs. It has been estimated that three-fourths of all the nitrogen required by rice in California can be met by cultivating *Azolla* in rice fields.

The symbiotic nitrogen-fixing association between legumes and *Rhizobium* also has received considerable attention for many years, from the early book by Wilson (1940) to reviews by Phillips (1980), Dilworth and Glenn (1991), and Stacey et al. (1992). By 1940 cropping mixtures involving legumes and nonlegume combinations of plants were in widespread use by many cultures. Such mixtures as clovers with grasses, peas with oats, cowpeas or soybeans with maize, and winter vetch with rye or wheat were encountered frequently. Many benefits were and still are attributed to these mixtures, mostly due to the symbiotic nitrogen-fixing association of the legume and bacterium. These benefits include excretion or exudation of nitrogen by the legume for use by the nonlegume, stimulation of soil microorganisms, and the return of nitrogen to the soil by sloughing off of roots and nodules. These practices are still in use today, either with legume/nonlegume cropping mixtures or legume/nonlegume rotations. According to the diagrams of plant-plant interaction depicted in Figures 5.1 and 6.3, the nitrogen-fixing associations discussed above affect (improve) the environment of the nonsymbiotic plant species (e.g., rice or other nonlegume) resulting in a dramatic improvement in yield (response) of the entire mixture.

ADVANTAGES OF MUTUALISM

The long-term, evolutionary advantages of mutualism in the examples discussed above are long-lived absorbing organs and tight nutrient cycling. Thus, both partners can tolerate low levels of available nutrients due to increased effi-

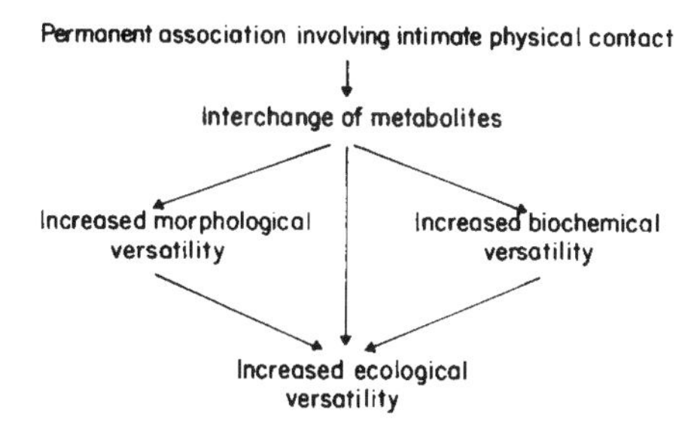

FIGURE 7.6 Characteristics of mutualistic symbioses between autotrophs and heterotrophs. (From Lewis, 1974, in Carlile and Skehel (eds.), *Evolution in the Microbial World.* By permission of the Cambridge University Press, New York.)

ciency of extracting essential minerals. Consequently, increased ecological amplitude is gained by the partners in association, as shown in Figure 7.6. Under conditions of nutrient stress, mutualistic symbioses are favored by natural selection (Lewis, 1974). In an evolutionary sense, positive interactions are very beneficial to the survival of the interacting organisms and tend to be favored over negative interactions (Barbour, 1987; Odum, 1971). In order from least to most interdependent, positive interactions are considered to have developed in an evolutionary sequence:

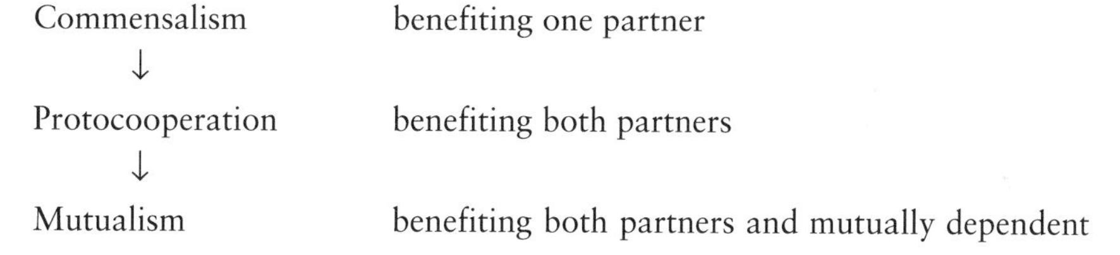

Commensalism	benefiting one partner
↓	
Protocooperation	benefiting both partners
↓	
Mutualism	benefiting both partners and mutually dependent

It is certain that many plants that we consider to be weeds have evolved into these types of positive associations with other weeds, crops, and organisms. An appropriate direction for weed management and control programs, therefore, is to begin to understand and exploit the positive effects of weeds on crops as well as the negative effects of crops on weeds.

FACILITATION

The process of *facilitation* is presented here as being complementary to competition, in order to account for the many cases in agriculture and nonagricultural systems in which one plant species provides some sort of benefit for another plant species. In some instances, two plant species (e.g., weeds or crops) use different components of an ecosystem or use the same component in different ways. Within this general mechanism, both or all partners in the association might benefit simply from reduced competition. In other cases, the plant species may alter the environment of one another positively, called facilitation or the *facilitative production principle*, according to Vandermeer (1989). For example, a hypothetical situation might exist in which two plant species compete intensely. However, they are only able to do so because of the protection provided by one species from a critical herbivore, while the other species provides nitrogen for use by both species. In this situation, the beneficial (facilitative) effects could be strong, but the competitive response sufficiently intense to offset them, leading to an RYT or LER of less than 1. Vandermeer (1989) points out that standard experimental procedures would not suggest the operation of any other process than that of competition. Thus, it is important to develop a framework that recognizes the importance of facilitative components of plant-plant interactions, as well as competitive ones (Tables 5.1 and 7.1). For example, our previous discussion about weeds in intercrop situations suggests that facilitation could play a role in intercrop performance. Facilitation also could be important in monoculture crops that are invaded by more than a single species of weed.

Within the scope of this book, discussion focuses on plant interactions, especially crop/weed and weed/weed interactions. It is certain, however, that the interactions listed in Tables 5.1 and 7.1 are not restricted to plants. In fact, as just described, many of the most biologically interesting and economically relevant examples of interactions between plants and other organisms occur as protocooperation and mutualism. In addition to the examples already given, these include the well recognized plant-pollinator and plant-seed-disperser syndromes. Other instances of plant–other organism interactions do not follow strictly the definitions in Table 5.1. For example, weed management that benefits crops often does so by affecting insect populations, either positively or negatively. A thorough review of this subject by Andow (1988) suggests ways in which weed populations can benefit crops by means of their effects on insect populations.

Weed Interactions with Other Organisms

The presence of certain weeds in and around an agricultural field often reduces specific pest populations. For example, Andow (1983, 1988) observed that 105 species of insect herbivores were more abundant in monocultures than in weedy or intercropped systems. Weeds also may serve as hosts for beneficial insects by providing pollen, nectar, or shelter or by bridging life cycle gaps. These beneficial insects are often predators or parasites of crop pests (Andow, 1988). Certain weeds also may be preferred hosts, or decoys, for crop diseases and insects, thereby reducing crop damage from them. In some cases, the decoy plants may even stimulate germination of soil-borne pathogens, which subsequently die due to unavailability of a suitable host (Charudattan and DeLoach, 1988).

In his review of integrated pest management, Norris (1982) identifies the basic principle underlying the weed-insect/pathogen interactions discussed above, by suggesting that habitat diversification leads to increased stability of insect populations, including pests, their predators, and other beneficial organisms. This argument was supported by Swift and Anderson (1993), who also suggested that both ecosystem stability and productivity should increase as a function of enhanced plant biodiversity (Figure 7.7). In many cases, weed removal has been shown to increase the incidence of pest attacks on crops, presumably due to a reduction in the diversity of plant resources available for the pest (Norris, 1982; William, 1981; Andow, 1988). However, all the examples cited in the above reviews refer to specific weed-crop combinations, and it is quite likely that at least an equal number of examples exist in which the presence of weeds actually increased crop pest populations. There are enough instances where weeds have been shown to benefit crops, however, to warrant a reevaluation of present weed control practices and to reconsider our definition of a "weed."

SEED PREDATION

As demonstrated in Chapter 4, seed predation is one of the most important processes that regulate the occurrence of seed in seed banks. For example, demographic models in which soil seed losses only include death by seed aging and germination usually give outputs where the seed population grows exponentially or reaches unrealistically high values. In spite of this fact, there are very few studies,

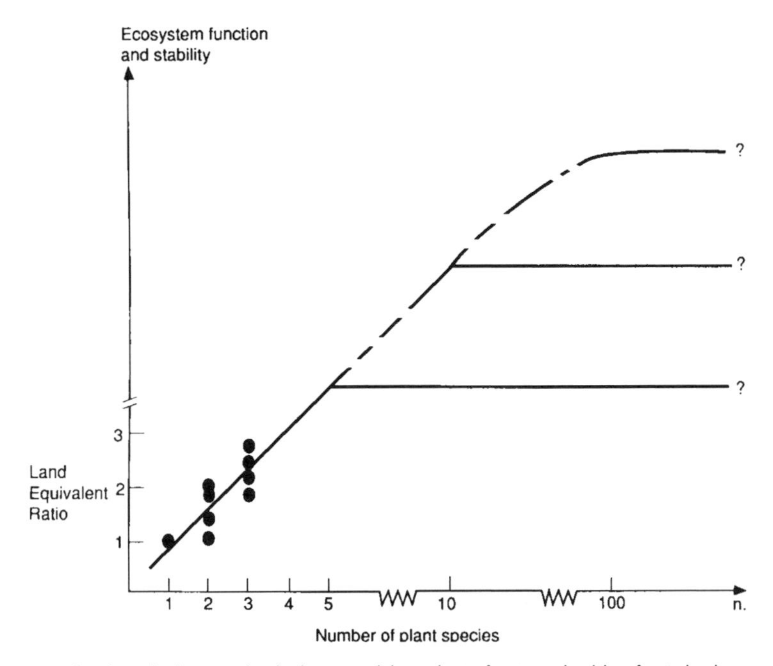

FIGURE 7.7 The relationship between plant biodiversity and the productive function and stability of agricultural systems. Data points for land equivalent ratio (LER) are for Central American multiple cropping systems. The lines are hypothetical, postulating possible relationships between plant species number and the efficiency and stability of ecosystem function. It should be emphasized that yield gain (LER) is unlikely to increase beyond three species, however. Improved ecosystem function will be invested in other aspects such as greater nutrient retention. (From Moreno and Hart, 1979.)

if any, that quantify the interaction between seed predation of the weed seed rain, seed bank dynamics, and the subsequent distribution and composition of weed populations on agricultural, forest plantation, or rangeland sites.

Sagar and Mortimer (1976) briefly explored the potential for seed predation to influence subsequent weed populations of wild oat under "worst case"/"best case" management scenarios (Figure 4.28). Seed predation by granivores (usually rodents, insects, or birds) also has been the focus of plant population studies in deserts and following masting (massive nut production) events of certain forest or woodland tree species. However, much of the information for agricultural and other managed lands remains anecdotal. For example, as discussed in Chapter 4 in the Willamette Valley of Oregon in the United States, farmers often grow alfalfa for several years following row crops to "clean up" their fields from weeds. Ghersa (unpublished) studied this phenomenon and found that an alfalfa crop provided sufficient habitat cover for a small seed-eating rodent, *Peromyscus* sp., to inhabit those fields. The rodent was so efficient in seed predation that it could eliminate nearly 99.8 percent of the weed seed rain in the alfalfa crop, thus substantially reducing the weed seed bank in the soil over a few growing seasons. Apparently, insufficient cover is present during the culture of most row crops in the Willamette Valley for the mouse to persist, allowing weed populations to build up in those crops.

HERBIVORY

According to Harper (1977), herbivory by animals can decrease growth and fecundity, stimulate compensatory regrowth or, in severe cases, cause mortality of plants. Crawley (1987) and Louda (1989) believe that herbivory influences the competitive interactions among plants by reducing the ability of grazed individuals to acquire resources or by eliminating the individuals as competitors altogether. These effects of herbivory are so important in natural systems that they often mask or change significantly the outcome of competition that would have occurred had the plants not been grazed, a phenomenon Louda calls *apparent competition* (Table 7.1). The impacts of herbivory on plant species richness (composition) probably do not act directly through the animals eating plant populations to local extinction, but through modifications in the plants' competitive abilities following the feeding activities. If this hypothesis is true, then rangeland managers, farmers, and foresters may be able to modify the grazing activities of domestic animals, wildlife species such as deer and elk, or even "pest" organisms such as insects, slugs, and snails to limit the regrowth potential of some weed species.

Habitat Management

Creating technologies for habitat management is now one of the most challenging aspects of food, fiber, and wood production because it is through development of such technologies that the ecological stability of these production systems can be attained. It is also through an understanding of habitats that the synergies of biological systems and environment can be transformed into management practices. An important advance in this challenge began two decades ago with the Integrated Pest Management (IPM) movement for pest control. As Levins (1986) pointed out then, "even though development of IPM has not been a smooth unfolding of knowledge and technique [about the role of biology and environment in pest management] but rather an uneven, erratic course, it provided the first steps to soften the hostile confrontation of human production activities with all living nature, except the crop." While searching for a method to decrease the use of nonselective methods of pest suppression (Corbet, 1981), the IPM movement began to develop an awareness of the positive and negative feedbacks present among different organisms in the agroecosystem food web (Chapter 2). In short, the approach allowed for the eventual development of more ecologically rational technologies than was possible using earlier models of pest control (Table 2.2). However, the technologies often have proven to be site specific and fine grained rather than general and broad-brush. Therefore, ways now must be found to incorporate the information that farmers and local land managers know about their surroundings into the more general concepts and approaches that are derived from research.

In any landscape where disturbance occurs, such as from logging, cropping, or grazing, habitat pattern is determined mainly by spatial and temporal arrangements of human activities. Agricultural practices such as planting pattern, crop phenology, and crop rotation, and different land uses all give particular opportunities to enhance or inhibit population growth of weeds and other organisms. The populations, likewise, respond to the habitat heterogeneity according to their dispersal range and the probability of finding food, resources, or safe sites. Thus, the habitat grain combined with the biological characteristics of any particular population establish complex interactions and patterns of behavior that are often difficult to determine empirically but can be observed over time on a landscape.

For example, many weed populations have evolved sophisticated dormancy and dispersal mechanisms to avoid unfavorable periods (patches) in a crop rotation sequence (Chapter 4). Advances might be made by taking advantage of this tendency for modern weed management to select for a stable, specialized weed flora. New weed management tactics could take advantage of this predictability through small but precise modifications in agricultural routines that would produce dramatic reductions in weed density, composition, and crop yield loss. Still another strategy might be to alternate weed management approaches. In this case, weed management might begin with several years of a usual practice to achieve predictability through high selection pressure. The manager might then switch practices to place at a disadvantage the traits selected by the conventional method. Management then would alternate between periods of selection, followed by exploitation of the selected traits.

Levins and Vandermeer (1994) suggest that empirical ecological studies combined with modeling should provide the basic tools necessary for improved understanding of habitats and landscapes, with the basic goal being to enhance the stability and predictability of the production systems that depend upon them. Whether an increase in species diversity will result from adding more crop species, such as through intercrops, or by simply managing weed abundance and composition is not really the important issue here. What is important, however, is the necessity to forge a better understanding of the links between population biology and plant community ecology by redesigning and reexamining experiments to answer the question, "What do species [even weeds] actually do in ecosystems?" (Lawton, 1994).

SUMMARY

Several forms of interference other than competition are possible among plants. These include amensalism and parasitism, which are negative plant interactions, and commensalism, protocooperation, and mutualism, which are all positive relationships among plant species. The most common form of amensalism is allelopathy, the detrimental effect exerted on a plant from the release of a chemical by another plant. The primary effect of allelopathy on crop production results from an association with plant residues, although other sources of allelochemicals can be secondary products leached or exuded from foliage or roots. Seed also can be an important source of allelochemicals. Recently, there has been considerable effort to isolate and produce specific allelochemicals or microorganisms that produce them for use as naturally produced herbicides. Parasitic weeds are plants that live on or in another plant at whose expense they obtain food, shelter, or support. Parasitism is an obligate relationship for the parasite because, without the host plant, it will die. Parasitic flowering plants can be important weeds in cropland and forests of both temperate and tropical areas. These weeds usually have specific, highly evolved adaptations that assure finding and attaching to the host plant.

Commensalism is a non-injurious relationship between two plants in which one is benefited without any harm occurring to the other. Commonly, the host plant serves as a source of support, anchorage, or protection for the other plant, without itself being affected. The nurse plant syndrome probably is the best known form of commensalism. During protocooperation the two plants in the association are stimulated by the interaction but unaffected by its absence. Root grafts, mycorrhizal associations, and chemical exudates are all ways in which protocooperation can occur. Intercropping and polycropping are much-utilized forms of protocooperation that occur among plant species. In intercrop situations, two crops are grown together, usually simultaneously, resulting in more total crop production than if the two crops had been grown separately. Weeds in an intercrop situation add significant complexity to the association and to its experimental analysis. Mutualism is an obligatory relationship between two organisms in which both partners benefit from the association. However, both suffer if one of the partners is absent. Mutualistic relationships among plant species are rare, but similar relationships between plants and microorganisms are quite common. Lichens and symbiotic nitrogen fixation are well-known forms of such symbioses. Often nitrogen-fixing and a nonnitrogen-fixing plant species are mixed to create a positive, highly productive plant association.

Facilitation is a positive ecological process that is complementary to competition. It describes the situation in which at least one plant species in a mixture provides some form of benefit to another plant species. Facilitation requires a positive alteration of the environment of at least one species in the mixture by another. However, many facilitation effects require the presence of organisms other than just plants. For example, there is considerable evidence that weeds (associated plants) in many crop and noncrop situations reduce the prevalence of insect pests and diseases. Both seed predation and herbivory of weeds also could be facilitative processes. Habitat management is the development of new technologies that strive to maintain stability in production systems by understanding the synergies of biology and environment. The IPM approach was perhaps the first effort toward habitat management of pest organisms. Empirical ecological studies that are combined with models and grower/land manager inputs provide the tools necessary to understand habitats and landscapes upon which production systems depend.

PART

III

Technology of Weed Science

Crop production usually requires that the competitive impact of weeds be minimized in order to optimize yields. This is accomplished using various kinds of weed control tools which, obviously, must not injure the crop. Weed control is a component or tactic of the more general strategy of vegetation management, which includes fostering beneficial vegetation as well as suppressing undesirable plants. Successful vegetation management depends on knowledge of plant identification, life history, biology, and associations with other organisms, and the selection of the proper weed control method or tool.

In the preceding chapters, characterestics that make weeds successful organisms in agricultural, forest, and range habitats were examined. The following chapters (8 through 10) describe various weed-control tools and discuss general principles of weed management. Implicit is the premise that better weed management—that is, greater effectiveness of tools and tactics for weed prevention and suppression—results from increased knowledge of weed biology and of the interrelationships of crops and weeds and the environments they both share. While many of the principles discussed apply generally to all cropping systems, it should be noted that much of the specific information in the following chapters may be relevant for the United States only.

Chapter

8

Methods and Tools of Weed Management

Weed management options should be considered before corrective actions for a weed problem are begun, because weeds often are controlled or contained best by practices that occur prior to crop planting. Once weeds become established in a field they rarely can be eradicated from it. For example, Norris (1992) calculated that weed reductions must exceed 99.9 percent to reduce seed inputs to the soil enough to maintain a stable seed bank that does not increase. Those estimates were empirically substantiated by Ballaré et al. (1987a,b). Furthermore, those authors found that weed control efficiencies up to 95 percent—a common level used by weed scientists to rate herbicide performance as excellent—could not prevent weed populations from increasing year by year. Significant improvements in current weed control procedures will be brought about only by a thorough understanding of weed biology in the crop environment, coupled with a management system that utilizes all suitable techniques to reduce weed populations and maintain them below levels that cause increases in the seed bank.

PREVENTION, ERADICATION, AND CONTROL

Prevention involves procedures that inhibit or delay weed establishment in areas that are not already inhabited by them. These practices restrict the introduction, propagation, and spread of weeds on a local or regional level. Preventive measures include cultural practices, such as seed cleaning, that disfavor weed dissemination into fields, the use of quarantines, and weed laws. These methods of weed management are discussed later in this chapter.

Eradication is the total elimination of a weed species from a field, area, or region. It requires the complete suppression or removal of seed and vegetative parts of a weed species in a defined area. Although several regional eradication projects have been attempted, this goal is rarely, if ever, achieved without monumental effort. Eradication usually is attempted only in small areas or areas with high-value crops because of the difficulty and high costs associated with eradication practices. *Control* practices reduce or suppress weeds in a defined area but do not necessarily result in the elimination of any particular weed species. Weed control, therefore, is a matter of degree and depends upon the goals of the people involved, effectiveness of the weed control tool or tactic used, and the abundance and tenacity of the weed species present. There are four general methods of weed control: physical, cultural, biological, and chemical.

Influence of Weed Control on Crop and Weed Associations

Weed control practices influence plant communities (crops and weeds) in two major ways—by reduced plant density and by altered species composition in the area being treated.

REDUCTIONS IN WEED DENSITY

The effects of weed control on total plant density are usually obvious, because most tools reduce effectively the abundance of weeds in a crop. Most agriculturists rely heavily on mechanical forms of soil disturbance by plows, disks, harrows, and so on and on herbicides for weed control both prior to and after planting. These tools reduce plant (weed) density very well. This reduction in total plant density generally results in favorable crop yield responses from weed control (Table 5.4).

ALTERATIONS IN SPECIES COMPOSITION

Weed control practices also influence markedly the plant species composition of a field. However, this impact often is not recognized as an important consequence of weed control. Consider the rolling cultivator in Figure 8.1. This implement reduces weeds between the crop rows effectively, but leaves both crop and weeds within the rows untouched. Thus, the cultivator affects both total plant density and the proportion and composition of species remaining in the field. Few studies exist that explore the effects of weed control, or other cultural practices, on subsequent weed stands or crop production. One study is that of Haas and Streibig (1982) that was discussed in Chapter 1 (Table 1.6). They observed that some weed species increased while others decreased, depending on method of cultivation and use of practices such as combine harvesting, which increased soil compaction, lime, nitrogen, and herbicides. Compositional changes in plant species are often dramatic during forest regeneration. In Table 8.1, the total den-

FIGURE 8.1 A rolling cultivator being used to control weeds in a sugarbeet crop. (Photograph courtesy of R.F. Norris, University of California, Davis.)

sity of the plants was reduced by weed control but the composition of the stand was most affected. As a result of the vegetation manipulation, the composition of the forest stand was changed from a mixture of red alder and Douglas-fir to almost pure Douglas-fir, which lasted for decades.

Changes in composition also may result within a weed species because of weed control practices. For example, Price et al. (1984) collected samples of wild oat (*Avena fatua*) along a transect encompassing rangeland, ditch banks, and agricultural fields. Through enzyme analysis, they determined that cultural practices for cereal production apparently allowed special wild oat genotypes to evolve, probably in response to the frequent tillage necessary to grow that crop. Herbicides (chemical control, see Chapters 9 and 10) effectively reduce weed density and also have a marked and rapid effect on weed species composition. This fact is the basis for selective weed control, which will be discussed in Chapter 9. *Selectivity* means that some plants in the crop-weed association are killed, while others are not. Thus, when certain herbicides are used, a change toward more tolerant weed species is often observed. For example, a tendency for grass weeds to be favored over broadleaf weeds usually follows repeated annual applications of 2,4-D, and other similar herbicides in cereal production (Figure 8.2).

An extreme shift in weed species composition can result in herbicide resistance caused by the selection of resistant plants in a species that was formerly susceptible to a particular chemical. Many instances of herbicide resistance

TABLE 8.1 Comparison of Douglas Fir and Red Alder Stand Development with and without a Competition Release Treatment at the Toledo Cutoff Unit, Oregon[a]

Stand Age (years)	*Treatment or Activity*	*Species*	*Mean Stand DBH (in.)*	*Stocking or (Removals) (trees/acre)*	*Basal Area (ft²/acre)*	*Merchantable Volume (ft²/acre)*
Treated Area						
0	Site prep and natural seeding	Douglas-fir	—[b]	—	—	—
6	Competition Release	Douglas-fir	—	400	—	—
33	Measurement	Douglas-fir	10.3	200	132	—
45	Stand exam	Douglas-fir	14.0	187	199	7,173
	Commercial thin	Douglas-fir	12.6	(114)	99	3,572
	Residual stand	Douglas-fir	15.9	72	100	3,601
60	Final yield	Douglas-fir	20.6	(72)	166	7,182
	Total yield	Douglas-fir	—	(186)	265	10,754
Untreated Area						
0	Site prep and natural seeding	Douglas-fir	—	—	—	—
33	Measurement	Red alder	9.3	272	127	—
55	Final yield	Red alder	13.6	(165)	166	6,680
	Total yield	Red alder	—	(165)	166	6,680

Source: Walstad and Kuch, 1987.

[a]The released area develops into a Douglas fir stand, whereas the untreated area becomes a red alder stand.

[b]Dash (—) indicates that no data were available for that treatment and stand age.

have been documented as a result of repeated herbicide applications for weed control (Tables 3.4 and 3.5). The first examples involved the triazine class of herbicides (Radosevich and Holt, 1982; LeBaron and Gressel, 1982; Chapter 3), but many other examples involving other herbicides and plant species also exist.

FIGURE 8.2 Grass weeds present in a cereal field previously sprayed with 2,4-D. In this case, grass weeds are favored over broadleaved plants by the herbicide eventually, after repeated annual applications, causing a shift in the weed species composition. (Photograph provided by A.P. Appleby, Oregon State University.)

Influence of Weed Control on Other Organisms

Regardless of the method used, weed control suppresses, removes, or destroys vegetation, which results in modification of the environment and habitat of other organisms (crop pests, non-pests, and beneficials) (Figure 8.3). Because weeds, like crops, are primary producers (Chapter 1), weed control must be recognized as a component of management programs that target crop pests but sometimes affect other organisms (Figure 1.14).

There is little reason to maintain a level of plant species diversity in a crop field if the objective is only to control weeds. In this case, the presence of several weed species might actually decrease the level of weed control achieved and, therefore, crop yields. However, if insect or disease management also is a consideration, the presence or absence of certain weed species can be a significant factor in the success of those programs. Altieri et al. (1977) observed that a diversity of plant species reduced the magnitude of insect attacks in several cropping systems. The increased level of species diversity in those experiments either decreased the incidence of phytophagus insects or increased the numbers of beneficial insects, which then lowered the pest insect population numbers (Table 8.2). Vegetation diversity also can lead to increased insect attack in some cases, however, since weeds may provide a food source or habitat for insect pests

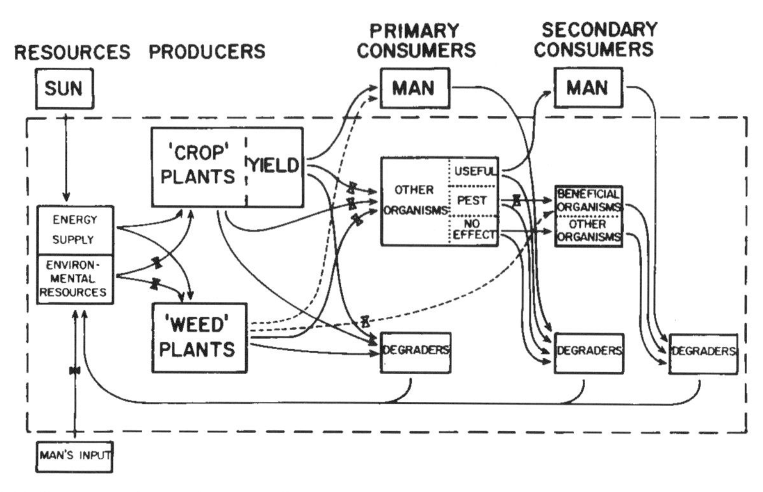

FIGURE 8.3 Diagram depicting the position of weeds in the flow of energy/resources in the agroecosystem. Solid "valves" indicate direct effects on primary producers (plants); open "valves" indicate indirect effects on consumers (non-plants). (From Norris, 1982.)

TABLE 8.2 Effect of Weeds on Specialized Herbivores

Arthropod Species	*Plant Attacked/Weeds Present*	*Change in Population*[a]	*Year*
Acarina			
Tetranychidae			
Eotetranychus sexmaculatus (Riley)	Citrus/unspecified weeds	–	8 years
E. willametei (Ewing)	Grape/Johnsongrass	–	1965
Eutetranychus banksi (McG.)	Citrus/unspecified weeds	0	7 years
Metatetranychis citri (McG.)	Citrus/pasture grasses	–	?
Panonychus citri (McG.)	Citrus/unspecified weeds	–	10 years
Tenuipalpidae			
Brevipalpus phoenicis (Geijskes)	Citrus/ unspecified weeds	+	5 years
Ertophyidae			
Phyllocoptruta oleivora (Ashm.)	Citrus/unspecified weeds	0	15 years
	Oranges/*Crotalaria* spp.	0	1942
		0	1943

TABLE 8.2 Continued

Arthropod Species	*Plant Attacked/Weeds Present*	*Change in Population*[a]	*Year*
Thysanoptera			
Thripidae			
Megalurothrips sjostedti (Tryb.)	*Vigna unguiculata*/corn, weeds	0	1978
Hemiptera			
Pentatomidae			
Antestiopus intricata	Coffee (arabica)/ undescribed weeds	–	1938
Coreidae			
Ambylpelta cocophaga (China)	Coconut palm/viney undergrowth	–	?
	Coconut palm/*Paspalum conjugatum, Mimosa predica, Triumfetta bartrami*	0	?
Homoptera			
Cicadellidae			
Empoasca fabae (Harris)	Alfalfa/grass weeds	–	1981
Empoasca kraemeri (Ross and Moore)	Dry beans/described weeds	–	1975
	Dry beans/grass weeds	–	1975
	Dry beans/broadleaf weeds	–	1975
Empoasca sasakii (Dwor.)	Ladino clover/orchard grass	–	1974
		–	1985
		–	1976
Scaphytopius acutus (Say)	Peach/*Dactylis glomerata*	0	1980
Erythroneura elegantula (Osb.)	Grapes/blackberry	–	?
Aphididae			
Brevicoryne brassicae (L.)	Brussels sprout/described weeds	–	1968
	Brussels sprout/*Spergula arvensis*	–	1977
	Brussels sprout/undescribed weeds	–	1965
		–	1966
		–	1967

(continued)

TABLE 8.2 Continued

Arthropod Species	*Plant Attacked/Weeds Present*	*Change in Population*[a]	*Year*
Aphididae			
Brevicoryne brassicae (L.)	Cabbage/white clover, two vars.	–	1981
	Cabbage/creeping bentgrass	–	1981
	Cabbage/chewings fescue	–	1981
	Collards/described weeds	–	1978
	Cole crops, five vars./weeds	–	1957
		–	1958
Myzus persicae (Sulz.)	Brussels sprout/described weeds	–	1968
	Brussels sprout/undescribed weeds	–	1965
		–	1966
		–	1967
	Collards/described weeds	–	1978
Aphis gossypii (Glover)	Cotton/undescribed weeds	0	?
Aphis fabae (Scopoli)	Brussels sprout/undescribed weeds	–	1965
		–	1966
		–	1967
Chromaphis juglandicola (Kalt.)	Walnut/ground cover	–	1964
		–	1965
Aleyrodidae			
Aleyrodes brassicae (Wlk.)	Brussels sprout/undescribed weeds	–	1966
		–	1965
		–	1967
Diaspidae			
Chrysomphalus aonidum (L.)	Citrus/weeds	–	3 years
	Citrus/weeds	–	9 years
		0	2 years
		+	4 years
Lepidosaphes gloveri (Pack)	Citrus/weeds	+	4 years
		0	1 year

TABLE 8.2 Continued

Arthropod Species	*Plant Attacked/Weeds Present*	*Change in Population*[a]	*Year*
Diaspidae			
Lepidosaphes beckii (Newm.)	Citrus/weeds	–	11 years
		0	4 years
Parlatoria pergandii (Comst.)	Citrus/weeds	–	13 years
		0	1 year
Aonidiella citrina (Coq.)	Citrus/weeds	–	4 years
Quadraspidiotus perniciosus (Comst.)	Orchards/*Phacelia* cover crop	–	?
Coleoptera			
Scarabeidae			
Oryctes rhineoceros (L.)	Oil palm/*Pueraria* sp. and *Flemingia* sp.	–	?
	Oil palm/ferns, grasses, creepers	–	?
Chalcosoma atlas (L.)	Oil palm/ferns, grasses, creepers	–	?
Curculionidae			
Anthonomus grandis (Boh.)	Cotton/*Ambrosia trifida*	–	?
Chrysomelidae			
Acalymma vittatum (F.)	*Cucurbita maxima*/undescribed weeds	–	1981
Phyllotreta cruciferae (Goeze)	Collards/undescribed weeds	–	1974
	Collards/fava bean, weeds	–	1981
	Collards/golden rod	–	1980
	Cole crops, five vars./ undescribed weeds	–	1957
		–	1958
	Cabbage/white clover, two vars.	–	1981
		–	1982
	Cabbage/creeping bentgrass	–	1981
	Cabbage/*Poa pratense*	–	1982
	Cabbage/chewings fescue	–	1981
		–	1982

(continued)

TABLE 8.2 Continued

Arthropod Species	*Plant Attacked/Weeds Present*	*Change in Population*[a]	*Year*
Chrysomelidae			
Phyllotreta cruciferae (Goeze)	Broccoli/white clover	–	1982
Phyllotreta striolata (F.)	Collards/undescribed weeds	–	1974
	Cole crops, five vars./ undescribed weeds	–	1957
		–	1958
Psylloides punctulata (Melsheimer)	Cole crops, five vars./ undescribed weeds	–	1957
		–	1958
Coccinellidae			
Epilachna varivestis (Mulsant)	Dry bean/described weeds	–	1981
	Dry bean/*Brassica kaber*	–	1981
	Soybean/described weeds	–	1981
		–	1982
	Soybean/broadleaf weeds	–	1981
		–	1982
	Soybean/grass weeds	–	1981
		–	1982
Lepidoptera			
Pieridae			
Pieris rapae (L.)	Brussels sprout/described weeds	–	1968
	Brussels sprout/white clover	–	1970
	Brussels sprout/undescribed weeds	–	1965
		–	1966
		–	1967
	Brussels sprout/*Spergula arvensis*	0	1977
		0	1978
	Cabbage/white clover, two vars.	–	1981
		–	1982
	Cabbage/creeping bentgrass	0	1981
	Cabbage/chewings fescue	0	1981

TABLE 8.2 Continued

Arthropod Species	*Plant Attacked/Weeds Present*	*Change in Population*[a]	*Year*
Lepidoptera			
Pieridae			
Pieris rapae (L.)	Collards/undescribed weeds	+	1974
	Broccoli/white clover	0	1982
Nocutidae			
Heliothis zea (Boddie)	Corn/selected weeds	0	1978
		0	1979
	Corn/undescribed weeds	0	1978
		0	1979
Spodoptera frugiperda (J.E. Smith)	Corn/undescribed weeds	–	1978
		–	1979
	Corn/selected weeds	–	1978
		–	1979
Mamestra brassicae (L.)	Brussels sprout/*Spergula arvensis*	–	1978
		–	1977
	Brussels sprout/undescribed weeds	–	1965
		–	1966
		–	1967
Plutellidae			
Plutella xylostella (L.)	Brussels sprout/undescribed weeds	0	1965
		0	1966
		0	1967
Pyralidae			
Evergestis forficalis (L.)	Brussels sprout/undescribed weeds	0	1966
		–	1965
		–	1967
	Brussels sprout/*Spergula arvensis*	–	1978
Tortricidae			
Cydia pomenella (L.)	Apple/weeds	–	?
Cydia molesta (Busck.)	Peach/described weeds	–	?

(continued)

TABLE 8.2 Continued

Arthropod Species	Plant Attacked/Weeds Present	Change in Population[a]	Year
Lepidoptera			
Lasiocampidae			
Malacosoma americanum (F.)	Apple/described weeds	–	?
Diptera			
Anthomyiidae			
Delia brassicae (Wiedemann)	Brussels sprout/*Spergula arvensis*	–	1977
		–	1978
	Cabbage/clover	–	1973
		–	1974
		–	1975
		–	1976

Source: Adapted from Andow, 1988.

[a]The change in population refers to population densities in the weedy system compared with the weed-free system. Year of the study is also given.

that damage crops (Table 8.2). In those cases, weed control has been shown to decrease the incidence of insect pests in numerous crops. Depending upon the situation, weed species can act directly as alternative food sources for phytophagus insects, provide a food source for phytophagus insects on which beneficial insects feed, or modify the microenvironment allowing survival of pests or beneficial natural enemies during adverse conditions (Norris, 1982; see discussion of facilitation in Chapter 7).

The role of weed species diversity on pathogen and nematode populations is unclear. It is known that maintaining large areas of a single crop or tree species can result in widespread disease. However, within a field, evidence also suggests that weed control leads to less disease and nematode problems. Unfortunately, there is little information indicating whether changes in abundance of diseases or nematodes result from either maintaining weed diversity or practicing weed control.

Because of the lack of clear answers about the role of weed diversity in overall pest management, land managers and farmers are faced with a potential dilemma. Should all weeds be controlled or should some of them be left to assist in the management of other pest organisms? To answer this question, the costs and benefits of weed control must be weighed against potential costs and benefits derived from alternative strategies to control other pest populations. It may be possible for weeds to be managed in such a way as to maintain crop yields and sustain beneficial populations of other organisms. However, such management tactics must be examined carefully for each cropping system, since any increase in plant diversity will most likely complicate weed control efforts.

TOOLS USED FOR WEED CONTROL

Many tools are used to disrupt, suppress, or eliminate vegetation. They fall into the general categories of physical methods (e.g., fire or manual and mechanical implements), cultural practices, organisms, and chemicals. The tools for weed control are most often used alone to uproot, cut, or otherwise kill vegetation that is not wanted or needed. However, most tools also can be used in combination to develop longer-term, more effective, or more selective tactics of weed control than can be achieved if they are used independently.

Existing vegetation usually must be removed from a field before it is suitable to grow crops. This initial process of vegetation removal is accomplished by plowing in agriculture and site preparation in forestry and range management. The actual process may involve slashing and burning the vegetation, tilling the soil, flooding, or some other method to both remove vegetation and prepare the field for planting. Soon after the crop is planted, other tools can be employed to remove weeds from around the crop plants. This later suppression of vegetation is accomplished by various implements or organisms, such as hoes, saws, knives, mowers, animals, herbicides, or some forms of tillage. The purpose of these practices is to maintain the crop in an environment that is conducive to optimal plant growth.

Effectiveness is often the primary criterion for selecting tools for weed control. However, considerations other than just biological effectiveness also enter into the tool selection process. These include a tool's potential to disrupt soil stability or cause erosion and its impact on air and water quality, as well as issues related to economics, worker safety, and public health.

PHYSICAL METHODS OF WEED CONTROL

Physical methods of weed control are any technique that uproots, buries, cuts, smothers, or burns vegetation. They consist of hand pulling and hoeing, fire, flame, tillage, mowing and shredding, chaining and dredging, flooding, and mulches and solarization.

Hand Pulling and Hoeing

Hand pulling and hoeing are the oldest and most primitive forms of weed control. However, it is estimated that over 70 percent of the world's farmers, mostly in developing countries, still use hoes or other manual implements to cultivate their cropland. A variety of hand implements have been developed for the removal of weedy vegetation (Figure 8.4). These tools range from rather primitive devices to sophisticated implements. Hoeing and pulling are most effective on annual or simple perennial plants that are not able to sprout from roots or other vegetative organs.

Even though manual methods of weed control have declined in developed nations, hand hoeing is still practiced in certain high-value crops or when other types of weed suppression are not possible. For example, hoeing is still practiced in crops for which selective herbicides have not been developed or where the area to be weeded is so small that it is not accessible for most equipment. Hand hoeing and pulling also can be effective and economical for "rouging" the few individual weeds that escape other control measures or that infest a field for the first time. These few individual plants have the potential to replenish the soil seed reservoir of a field if they are left unattended (Chapter 4).

Fire

Fire is another tool that has been available to humans for centuries for the manipulation of vegetation. It is still used extensively to remove vegetation and crop residues in agriculture (Figure 8.5), and to prepare forest lands for regeneration (Figure 8.6) after clear-cut logging. The burning of agricultural fields to remove residues of the previous crop can suppress pathogens, insects, or weeds that might occur in the new crop. In addition, fire has been used to manage fuel

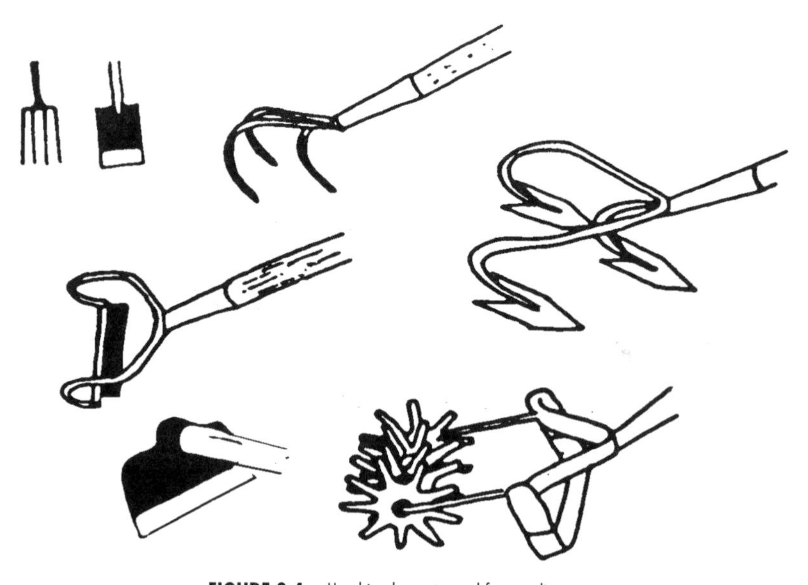

FIGURE 8.4 Hand implements used for weeding.

A

B

FIGURE 8.5 Use of fire in agriculture. (a) Slash and burn agriculture in the tropics and (b) field burning in the Central Valley, California. (Photographs provided by L.C. Burrill and S.R. Radosevich, respectively, Oregon State University.)

FIGURE 8.6 Using prescribed fire to prepare a forest site for tree planting. The helicopter is used to rapidly ignite the perimeter of the site. (Photograph by E. Cole, Oregon State University.)

breaks in vast shrublands that are prone to catastrophic wildfire (Figure 8.7). Broadcast burning is also an accepted and effective method used periodically to increase rangeland productivity by stimulating growth of certain fire-adapted grass species. Fire is sometimes used to remove weeds and other residue from along roadsides, canal banks, ditches, and vacant areas (Figure 8.8).

Flame

For selective weed control in certain crops, heat can be applied with a tractor-drawn flamer (Figure 8.9). The method of selective flaming is to direct heat toward the ground and avoid injuring the crop. Flaming is used in crops in which the growing points (meristems) are either beneath the soil and protected or the crop is relatively tall and woody. In either case, the crop plants withstand the heat of the burner, whereas small succulent weeds do not. The principle behind flaming is to control the intensity and exposure of the heat, causing cell rupture, rather than combustion of plant material. The effects of flaming may not be apparent until several hours after treatment. This weed control technique has been used successfully in alfalfa, cotton, sugarcane, soybean, and peppermint.

FIGURE 8.7 A controlled burn used to convert shrublands to grasslands. The technique is also used to maintain fuel breaks in dense shrubs or chaparral. (Photograph by S.R. Radosevich, Oregon State University.)

FIGURE 8.8 Fire used to remove vegetation along a ditch used for irrigation. (Photograph provided by W.B. McHenry, University of California, Davis.)

FIGURE 8.9 A tractor-drawn flamer used to control weeds in alfalfa and peppermint. (Photograph by R.F. Norris, University of California, Davis.)

Tillage (Cultivation)

Tillage is disturbance of the soil. A major benefit of tillage is prevention and suppression of weeds. Tillage suppresses weeds by breaking, cutting, or tearing them from the soil, thus exposing the vegetation to desiccation, and by smothering them. Repeated tillage may also deplete weeds from fields by diminishing seed or vegetative propagules in the soil, providing that "escaped" plants are not allowed to reproduce. Repeated tillage may also exhaust the carbohydrate reserves of perennial weeds, thus suppressing them.

Tillage is important to crop production for reasons other than weed control. Some of these other reasons for tillage are

- seed bed preparation
- burial of crop residues
- control of plant pathogens, insects, and rodents
- temporary improvement of soil physical conditions
- improvement of surface conditions for rainfall reception
- altered surface roughness
- incorporation of fertilizers, herbicides, or soil amendments
- moderation of soil surface microenvironment

Weed seed near the soil surface usually are not injured by tillage, but effective control of weed seedlings can result if the tillage is timed properly. Weed mortality is most likely to result from tillage when the plants are small. Annual, biennial, and simple perennial weeds are susceptible to tillage practices that sever shoots from the root or uproot the entire plant. Only the shoots of creeping and woody perennials are killed by tillage.

Best results from tillage occur when the soil is dry and the weather is hot. These conditions allow the optimum opportunity for weeds to desiccate and die following soil disturbance. Substantial control of even perennial vegetative reproductive organs (rhizomes, stolons, etc.) can result when tillage is performed under hot, dry conditions. For example, nearly 60 percent reduction in Johnsongrass (*Sorghum halepense*) ramets occurred from a single properly timed tillage in contrast to an untilled treatment (Radosevich et al., 1975; Table 8.3). In that experiment, which was conducted in California where the summers are hot and dry, the July tillage allowed sufficient time for desiccation of severed rhizomes to occur before the fall rains.

SUPPRESSION OF SEEDLING WEEDS BY TILLAGE

Covering small weeds with soil smothers them by disrupting the reception of light necessary for plant life. This is an especially useful technique for controlling small weeds that occur within rows of crop plants. Covering small seedling weeds with soil can be accomplished by a variety of manual and mechanical tools, such as rolling cultivators, that are designed for that purpose.

Repeated tillage can reduce the level of weed seed present in the soil because a portion of the seed in the soil will usually germinate when the proper environmental stimuli (light, moisture, temperature) are provided. Tillage brings weed seed to the soil surface where germination is most likely to take place and then subsequent tillage kills them. A seed bank can be depleted in this manner, providing the germinants are not allowed to mature and reproduce. However, if the disturbance is not frequent enough to prevent weeds from producing seed, annual weeds can respond to the reduced densities caused by tillage and increase both size and seed output (Chapter 5). Frequent cultivations tend to disfavor the long-term presence of perennial weed species.

TABLE 8.3 Reduction in Johnsongrass (*Sorghum halepense*) Ramet Survival from Tillage[a]

Not Tilled	*Tilled on July 16, 1973*	*Tilled on September 20, 1973*
10.7	4.3	7.1

Source: Radosevich et al., 1975.

[a]Density (ramet/m^2) determined July 22, 1974.

Field experiments have demonstrated that two- to five-fold increases in germination of many dicotyledonous and some grass weeds result from tillage operations performed during the day as compared to identical tillages performed at night (Figure 8.10) (Scopel et al., 1994). These authors also demonstrated that light acting through the photoreceptor phytochrome is the environmental signal that allows seed to detect such disturbances to the soil. The enhancement of weed seed germination (Figure 8.10) is due to light that penetrates into the soil during the actual tillage operation. These experiments indicate that if tillage operations are performed at night or with light-proof implements, substantial reductions in weed germination and abundance result.

SUPPRESSION OF PERENNIAL WEEDS BY TILLAGE

Perennial and biennial weeds can be severely suppressed by repeated tillage of fields that are infested by them. Repeated tillage results in a process called *carbohydrate starvation.* Following each tillage operation, new shoots emerge from the root or rhizome system of perennial weeds. The emergence of new shoots requires energy in the form of stored carbohydrates. The objective of repeated tillage is to deplete the carbohydrate reserves in the underground storage organs, eventually causing starvation and death of the plants. Cultivations every 10 to 14 days after emergence of new vegetative growth, for at least one growing season, is necessary for maximum carbohydrate starvation of herbaceous perennial weeds (Figure 8.11). Perennial weeds actually may be stimulated by cultivation if it is not done frequently enough to cause carbohydrate starvation of the underground storage system.

MECHANICAL EQUIPMENT USED FOR TILLAGE

In modern agriculture, mechanical equipment is typically used before and after crop planting to control weeds. The tools used for mechanical tillage are either *primary* or *secondary* tillage implements. Primary tillage implements are used to break and loosen the soil from 10 to 90 cm in depth. Usually, primary tillage is done only at the beginning of each growing season. Secondary tillage equipment is used to disturb the soil to 10 cm or less. These implements are used more often than those for primary tillage operations. Often several cultivations are necessary during a growing season to prepare the soil for planting, ensure optimum crop development, and control weeds. The type and size of tillage implement depends on the crop, soil, acreage, and power available to operate the equipment. In all cases the power to drive the implements is provided by animals or engines, rather than by humans.

Many of the methods and tools for mechanical tillage have been in widespread use for decades in the United States. Excellent discussions of the methods and tools used for mechanical tillage are provided by Ross and Lembi (1985), Aldrich (1984), and others. Thus, we relied heavily upon those texts for the following examination of mechanical tillage tools and practices. The reader

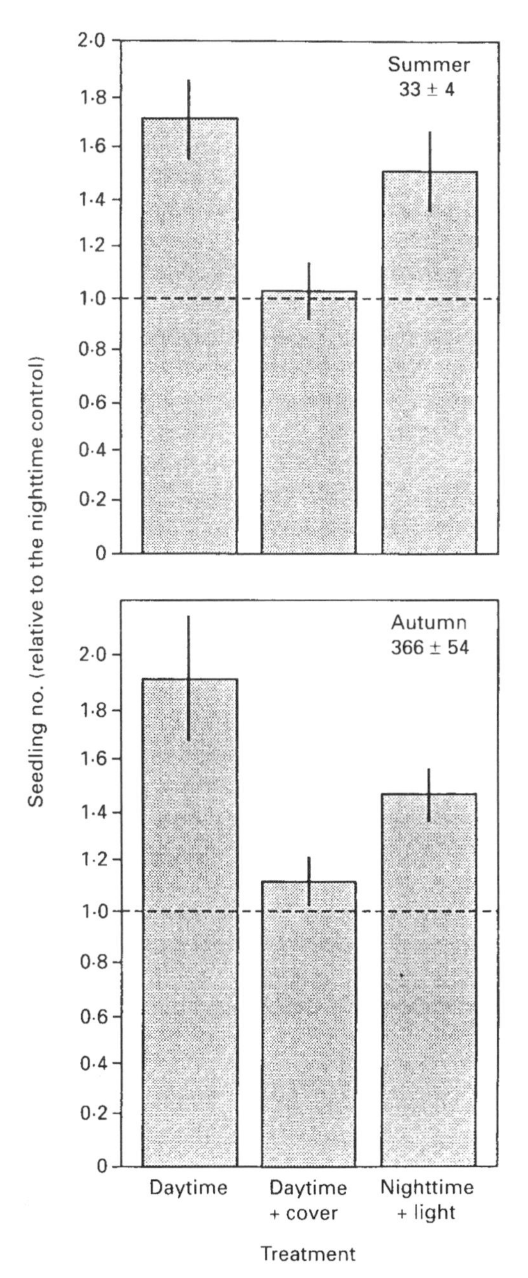

FIGURE 8.10 Effect of manipulating the light conditions during cultivation on the emergence of dicotyledonous weed seedlings compared with a nighttime (no light) control. Absolute densities after nighttime cultivations are indicated at the top of each panel in plants per m^2. Seedling counts were performed three weeks after cultivations. Species were (summer) *Amaranthus retroflexus, Solanum nigrum,* and *S. sarrachoides*; (autumn) *Lamium purpureum, Cerastium vulgatum, Veronica* sp., and *Stellaria media.* (From Scopel et al., 1994.)

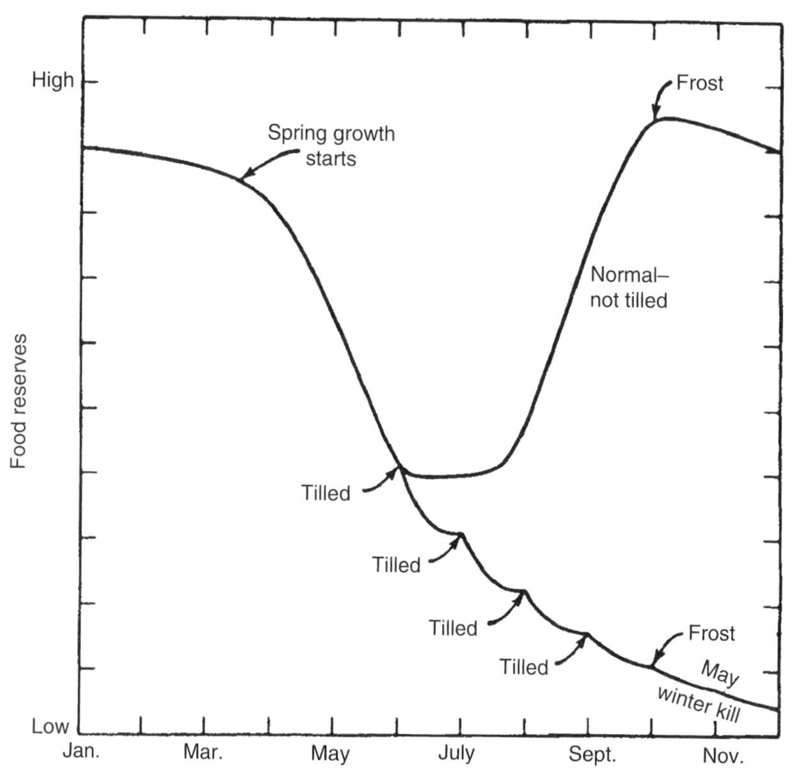

FIGURE 8.11 The principle of carbohydrate starvation. In perennial weeds, shoots are repeatedly removed by tillage, thus reducing the carbohydrate reserves of the plant until eventual death occurs. (Modified from Ashton and Monaco, 1991.)

is also referred to those texts for a more complete and comprehensive review of this subject.

Implements used for primary tillage. The most common tool used for primary tillage is the plow. Plows range from primitive animal-drawn tools (Figure 8.12) to modern implements. Initial soil preparation and crop residue removal, rather than weed control, is often the objective of plowing. Plowing implements include the moldboard, disk, chisel, and sweep plows.

The *moldboard plow* (Figure 8.13) was developed and perfected in the early 1800s (Ross and Lembi, 1985). The tool "cuts" the soil from the side and bottom of a furrow, causing the soil to become inverted. In the process of plowing, weeds and crop residues are covered over with soil. Crop residues and sod may sometimes impede the moldboard plow by clogging the implement while it is in operation. Coulters or rolling disks are sometimes attached to the front of the implement to cut crop residues or sod, and thereby assist plowing. Often, several plows are hooked together on a common frame (Figure 8.13). This tan-

FIGURE 8.12 Animal-drawn plows. (Photographs by L.C. Burrill, Oregon State University.)

dem arrangement makes the plow wider, with the result that fewer passes need to be made across a field.

The *disk plow* (Figure 8.14) has circular blades that are 50 to 90 cm in diameter, and are spaced about 35 cm apart. The blades or disks are tilted slightly, at approximately a 20-degree angle. Each disk blade is mounted on its own axle. The implement cuts and turns the soil profile to a depth of 10 to 30 cm. Soil inversion is less complete than with the moldboard plow, and more plant residues remain on the soil surface (Ross and Lembi, 1985). The implement

FIGURE 8.13 Moldboard plow. (Photograph by S.R. Radosevich, Oregon State University.)

FIGURE 8.14 Disk plow. (Photograph by A.P. Appleby, Oregon State University.)

also is more adaptable to a variety of soil conditions and has a lower energy requirement than the moldboard plow (Ross and Lembi, 1985).

The *chisel plow* (Figure 8.15) was first developed about 25 years ago for cereal production in the midwestern and western United States. Since its development, it has undergone rapid acceptance as a replacement for the moldboard plow in many cropping systems. Ross and Lembi (1985) indicate that nearly half the corn and soybean acreage in the midwestern U.S. is plowed with this implement. The implement consists of a heavy frame to which curved spring-like shanks are attached. The shanks are about 30 cm apart along the frame. A chisel point approximately 10 cm in width is attached to the bottom of each shank. The tool is operated at about 30 cm in soil depth. Soil inversion does not occur with this tool; instead, a tearing and loosening action results. Considerable plant residue may remain on the soil surface, ranging from 30 to 75 percent depending upon the design and operation of the implement (Ross and Lembi, 1985). The use of the chisel plow impedes soil erosion by incorporation of plant residues into the soil and by clod formation.

Sweep plows (Figure 8.16) are used when the amount of crop residue is minimal and it is desirable to leave it on the soil surface. The size of these implements varies according to the width of the sweeps. Some sweeps are large, 2 m or more, while others are relatively small, 30 to 50 cm. The sweeps generally are operated at a shallow soil depth of 10 to 15 cm. Often sweep plows, chisel plows, and to some extent disk plows are used in areas where wind erosion and moisture loss are major considerations for crop production. They are effective tools for weed removal that also restrict soil erosion by residue maintenance and clod or ridge formation (Ross and Lembi, 1985).

FIGURE 8.15 Chisel plow. (From Ross and Lembi, 1985.)

FIGURE 8.16 Sweep plow. (Photograph by A.P. Appleby, Oregon State University.)

The *brushrake* or brushblade is an implement sometimes used by foresters to uproot shrubs and to prepare flat to moderately sloping sites for tree planting (Figure 8.17). The implement consists of a frame on which sturdy tines are spaced approximately 30 cm apart. The entire implement then is attached to the blade of a crawler-type tractor. As the tractor moves, the tool is "raked" through the soil, uprooting shrubs and small hardwood trees that are subsequently piled and burned. This tool and heavy-duty *rangeland disk harrows* (Figure 8.18) are the primary implements available to foresters and range managers for primary tillage during site preparation.

Implements used for secondary tillage. Secondary tillage operations are used to prepare the soil for planting by leveling, firming, and preparing a seedbed. Weeds are also controlled effectively with secondary tillage implements. The tools to accomplish secondary tillage are disk harrows, rolling cultivators, spring-tooth harrows, spike-tooth and other harrows, rod weeders, and power-driven tillers.

The *disk harrow* (Figure 8.19) is a widely used implement for secondary tillage and field preparation in agriculture. The size of the tool and the arrangement of the disk blades vary according to the uses of the implement. The blades of most units are from 45 to 65 cm in diameter, concave, and attached to a common axle. Some disk harrows consist of a single set or "gang" of blades. These implements are used in orchard, vineyard, and other horticultural crops. For row crops, two gangs of blades are set in tandem to form a V-shape. The action of the disk blades moves and inverts the soil, breaks clods,

FIGURE 8.17 Brushrake (brushblade). (Photograph by R.G. Wagner, Oregon State University.)

FIGURE 8.18 Rangeland disk harrow. (Photograph by S.R. Radosevich, Oregon State University.)

FIGURE 8.19 Agricultural disk harrow. (Photograph by A.P. Appleby, Oregon State University.)

and chops weeds and crop residues. *Mulch treaders* are similar to disk harrows, except that the circular blades of the disk are replaced with coulters or spiked blades. A major use of treaders is in a stubble-mulch system, where residue maintenance is important (Ross and Lembi, 1985).

Rolling cultivators (Figures 8.1 and 8.20) consist of a series of pronged wheels that are mounted along a common axle. The implement is pulled over the soil in such a way that the points or prongs strike both the soil and vegetation. Rolling cultivators are used for shallow tillage, crust breaking, and for uprooting small weeds. The ability to kill weeds is improved by creating a slight angle to the wheels.

The *spring-tooth harrows* (Figure 8.21) consist of a frame to which several series of spring shanks or tines are attached. The shanks are at least 20 cm apart to avoid plant residue build-up during cultivation. Variously shaped *sweeps, chisels*, or *points* are sometimes attached to each shank. In the western United States spring-tooth harrows are used for seedbed preparation for dryland cereals, or for fallow-ground maintenance. Many *field cultivators* are similar to the spring-tooth harrow, except the shanks are less flexible. These implements uproot weeds and stir and level the soil, but their effect on residue incorporation is limited.

Harrows (Figure 8.22) are implements with fixed, rigid spikes or coiled tines that are dragged over the soil to uproot small weeds and smooth the soil surface. Harrows are sometimes used behind disks or other cultivation equipment to assist in seedbed preparation. Harrowing over a field can remove a high percentage of weeds, when it is done prior to planting.

FIGURE 8.20 Rolling cultivator. (Photograph by Karin Thalinphant, Oregon State University.)

FIGURE 8.21 Spring-tooth harrow. (Photograph by A.P. Appleby, Oregon State University.)

FIGURE 8.22 Several variations of spike-toothed harrows. (Photograph by A.P. Appleby, Oregon State University.)

FIGURE 8.22 Continued

Rod weeders are implements that consist of a rod about 2 cm in diameter (Figure 8.23) that rotates as it is moved through the soil. The rod is placed at a 2- to 5-cm depth below the soil surface and rotates in the direction opposite to that in which the tractor is moving. The action of the rod lifts, uproots, or breaks off plants from the soil. The implement is usually 10 to 12 m wide, and supports four to five rods. They are best used in soil free of stones. Rod weeders are used almost entirely on fallow ground, often in stubble mulch systems, to conserve soil moisture by suppressing weed growth (Ross and Lembi, 1985).

Rotary tillers are operated by using the power take-off unit on a tractor. These implements mix and pulverize soil. Rotary tillers also are known as rototillers, bed mulchers, and rotovators (Figure 8.24). Power tillers operate by the cutting action of many horizontal "knives" that are mounted along a horizontal shaft. Each knife blade cuts a vertical slice of soil as the shaft revolves. The implement disturbs the soil to a depth of 5 to 10 cm. Rotary tillers are often used for vegetable production or other small-seeded crops where uniformity in planting, irrigation, or herbicide incorporation into the soil is desired.

Tools used for cultivation between crop rows. Various *sweeps, knives* and *listers* attached to cultivator shanks can be used to make furrows, cut weeds, or throw soil into the crop row to smother seedling weeds growing there. These tools usually attach to a drawbar mounted on, or pulled behind, a tractor (Figure 8.25). Names such as V-sweep, cultivator knives, shovels, sweeps, duckfoot, chisels, and listers are used to describe these various cultivating devices (Ross and Lembi,

FIGURE 8.23 Rod weeder. (Photograph by L.C. Burrill, Oregon State University.)

FIGURE 8.24 Rotary tiller. (Photograph by Floyd Colbert, California Polytechnic State University, San Luis Obispo.)

A

B

FIGURE 8.25 Various (A) sweeps, (B) knives, and (C) chisels for row crop cultivation. (Photographs by R.F. Norris, University of California, Davis.)

(*continued*)

C

FIGURE 8.25 Continued

1985). Some of these attachment tools are used only to cut or uproot weeds, others move soil for smothering small seedlings, and still others accomplish both activities. *Rotary cultivators* are somewhat like the rolling cultivator described earlier. However, the pronged wheels of the rotary cultivator usually are spaced or arranged in groups (gangs) along a frame that is pulled behind a tractor (Figure 8.26). These implements both uproot seedling weeds and throw soil to the center of the crop row, smothering small weeds that are present there. Rotary cultivators are often used for the first several cultivations in row crops.

Cultivation in orchards and vineyards presents special tillage problems because the crop plants are large and spaced widely apart in rows. Vegetation between rows may be disked, but weeds within the rows must be removed with minimal injury to the trees or vines. An implement used in orchards and vineyards is the *spring-hoe weeder*. This tool uproots weeds on the berm, a raised row on which trees or vines are planted. Caster wheels attached to the implement reduce damage to the vine or tree trunks. The *French plow* (French hoe) or *grape hoe* is also used in vineyards to remove weeds from within the row. This implement has a "trip" mechanism that prevents it from contacting the vine as cultivation proceeds (Figure 8.27).

PROBLEMS WITH TILLAGE

The influence of tillage on weed seed in the soil is uncertain. In some instances tillage may decrease weed seed abundance, while it may maintain seed abun-

FIGURE 8.26 Rotary cultivator. (Photograph by S.R. Radosevich, Oregon State University.)

dance in other situations. Stirring weed seed in the soil profile often enhances germination and assists in depleting the reservoir of dormant seed, if control measures occur before weeds reproduce. Seed burial by tillage may prolong the time that seed of some weed species exist in the soil, increasing their responsiveness to light stimuli and susceptibility to predation and decay (Chapter 4). It also is possible that repeated tillage favors some weed species or biotypes through selection pressure (Chapter 3).

FIGURE 8.27 Grape hoe. (Photograph by the Green-hoe Company, Portland, New York.)

Tillage assists the dispersal of some weeds. Roots and rhizomes of herbaceous perennial weeds can be spread from field to field with tillage implements. In addition, infrequent tillage enhances perennial weeds by stimulating sprouting from severed vegetative organs. Tillage also can damage crop roots. However, such injury can be minimized if late-season and deep cultivations are avoided. Tillage exposes soil to weathering that can make wind and water erosion possible. However, equipment has been developed and practices can be used that minimize soil erosion. Compacted soils also have been attributed to mechanical tillage, especially when equipment operations are performed on wet soils.

ALTERNATIVES TO CONVENTIONAL TILLAGE

Several alternatives to conventional tillage have been developed to aid in soil conservation, reduce energy requirements, and decrease expenses. These systems, as described by Ross and Lembi (1985), are as follows.

Conservation tillage. Any tillage system that reduces the loss of soil or water when compared to a non-ridged clean tillage system. This tillage system generally leaves a layer of crop residue on the soil surface that slows soil erosion and conserves water.

Minimum tillage. The minimum amount of tillage required for crop production or for meeting the tillage requirements under existing soil and climatic conditions. This system means eliminating excess tillage operations.

Reduced tillage. Use of primary tillage in conjunction with special planting techniques to reduce or eliminate secondary tillage operations.

No tillage. A 2- to 5-cm-wide seedbed is prepared with a special coulter or disk attached ahead of the planter. The crop seed is planted directly through the crop residue and no further tillage is required. In no-till systems, weed control is accomplished by chemical methods.

Ridge tillage. A form of conservation tillage that appears to overcome many of the soil microenvironmental, soil compaction, and weed control problems associated with other conventional and untilled systems. In the spring, the tops of the ridges are tilled for planting (Figure 8.28). This removes crop residues from the ridge tops and disturbs the soil enough to create a seedbed. Soil on the ridges also is generally warmer than that between the ridges or in fields without ridges. Warmer soils facilitate crop emergence. Tilling only the tops of the ridges disturbs fewer weed seed, reducing weed germination. Erosion is slowed because soil and crop residues between the rows are not disturbed. Weeds that emerge later in the growing season tend to be between the ridges. Cultivations can easily control these weeds and less soil compaction occurs in the crop rows, thereby further enhancing crop growth and water infiltration (National Research Council, 1989).

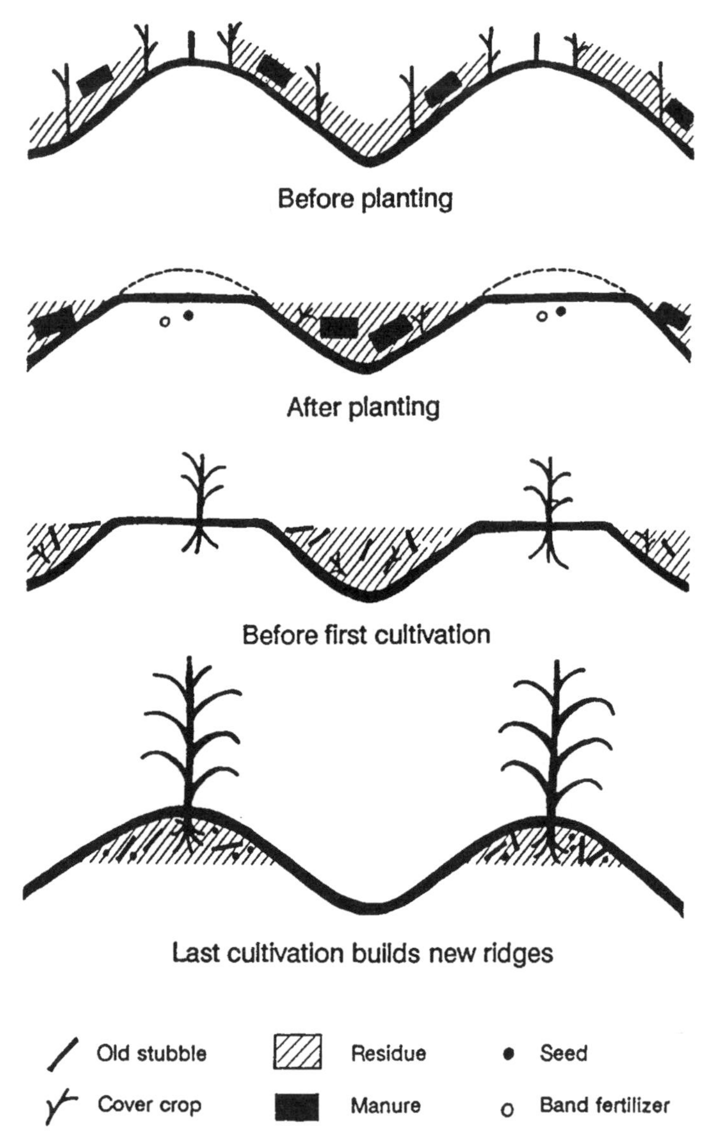

FIGURE 8.28 Ridge tillage advantages in alternative production systems. The planter tills 5 to 10 cm of soil in a 15-cm band on top of the ridges. Seeds are planted on top of the ridges and the soil from the ridges is mixed with crop residue between the ridges. Soil on the ridges is generally warmer than soil in flat fields or between ridges. Warm soil facilitates crop germination. The presence of crop residues between the ridges also reduces soil erosion and increases moisture retention. Mechanical cultivation during the growing season helps control weeds, reduces the need for herbicides, and rebuilds the ridges for the next season. (From: Dick Thompson, in National Research Council, 1989.)

Mowing and Shredding

Mowing is used to control weeds by cutting or shredding their foliage. Mowing is usually accomplished to facilitate other management activities by removal of weed growth, such as to suppress weeds temporarily in orchards and vineyards to assist harvesting operations. Mowing weeds also can reduce water use in orchard and vine crops, since weeds without foliage cannot use soil moisture rapidly. In some situations, it is desirable to decrease the amount of vegetation that is already present without killing it. Powerline rights-of-way, some roadsides, ditchbanks, abandoned cropland, or vacant lots are areas where weeds are mowed periodically. Mowing of herbaceous vegetation should occur before the plants set seed in order to discourage weed seed dissemination. Since mowed plants, even annuals, usually regenerate new shoots, frequent mowing is often required to prevent seed production.

Mowing can suppress some perennial weeds through carbohydrate starvation, discussed earlier in this chapter (Figure 8.11). Similar to tillage, frequent mowing stimulates new branch development, which eventually depletes the plants' carbohydrate reserves if it is done often enough. Mowing every few weeks for at least one or two growing seasons is usually necessary to suppress herbaceous perennial vegetation satisfactorily in this way. At no time during the growing season can the weeds be allowed to replenish their underground supply of carbohydrates if this system of weed control is to be effective.

Mowing alters the stature of some weeds. A common morphological response of plants that have been mowed is to regenerate several new shoots from below the cut. Therefore, repeated mowing can change the appearance of a weed from a single-stemmed, tall and upright form, to a plant with multiple branches that are relatively prostrate. Weeds that occur in frequently mowed crops, such as alfalfa or turf, often assume this prostrate appearance. Even low-growing prostrate weeds produce seed, so weed seed abundance in the soil probably will increase when mowing is practiced for weed control.

Often, mowing is used in conjunction with a cover crop to control weeds. This practice is usually more successful in suppressing weed growth than mowing alone. Cover crops planted between rows of trees in orchards and rows of vines in vineyards and maintained by mowing are examples of how vegetation in combination with mowing can be used for weed control (see section on living mulches later in this chapter). Other crops, such as alfalfa, pastures, and turf are mowed as a normal production practice, which results in the suppression of some weeds as well.

EQUIPMENT USED FOR WEED MOWING

There are four basic types of mowers: the sickle-bar mower, rotary mower, flail, and reel mower (Ross and Lembi, 1985). The *sickle-bar mower* is a tool used for cutting alfalfa and other types of hay. It consists of a cutting bar with attached guards and a knife (sickle) that is driven back and forth in a horizontal direction. The motion of the sickle through the guards cuts the plants. This tool is also used to cut tall weeds in pastures and along roadsides.

The *rotary mower* is an implement that consists of a horizontal blade attached to a perpendicular revolving shaft. It is used for mowing weeds in orchards (Figure 8.29a), along roadsides, or in vacant areas. A heavy-duty machine (Figure 8.29b) also has been developed for cutting brush and trees from under powerlines, along other rights-of-way, and in certain forest situations. The *flail* is similar in operation to the rotary mower. It consists of rods, blades, or chains attached to a revolving shaft. *Reel mowers* are used for management of turf. This tool consists of several spiral-like blades that rotate against a stationary cutting blade. Although an important tool for turf, the reel mower is used sparingly for weed control.

TOOLS USED FOR CUTTING AND MOWING WEEDS BY HAND

Various kinds of *hand sickles, scythes*, and *machetes* are used to cut weeds and other vegetation. In many areas of the world, these tools are still in common use for the suppression of herbaceous vegetation. Often, manual methods are the only practical means to manage vegetation in small, close-cornered areas, where economics or topography do not allow more elaborate tools, or where other methods of weed control have been restricted.

Manual cutting is an important way to alter the composition of forest stands by removal of cull, unmerchantable, or weed trees. This practice also can be used to remove shrubs that are in proximity to commercially desirable trees. The gasoline-powered chain saw is the major tool for accomplishing these activities (Figure 8.30). Most hardwood tree species and some conifers sprout after the stem and foliage have been removed. However, the amount of sprouting may be diminished significantly by cutting trees at certain times of the growing season. For example, the sprouting ability of red alder (*Alnus rubra*) is reduced when cutting is performed in July and August, rather than at other times of the year.

Chaining and Dredging

Chaining involves the use of a heavy chain, similar to that used to anchor ships, which is dragged between two tractors (Figure 8.31). In some cases, a metal blade is welded across each link of the chain. As the chain "rolls" between the two tractors, shrub stems are crushed and some plants are uprooted. This procedure is used to prepare shrublands or chaparral for rangeland improvement. *Dredges* are used primarily to remove submerged and emersed aquatic weeds from canals and rivers. Chaining also is used for aquatic weed control as a means of "dragging" canals to tear loose aquatic plants growing there.

A

B

FIGURE 8.29 Rotary mowers used for vegetation control in (A) orchards and (B) forestry. (Photographs by C.L Elmore, University of California, Davis, and S.R. Radosevich, Oregon State University, respectively.)

FIGURE 8.30 Brush cutting using a chain saw. (Photograph by R.G. Wagner, Oregon State University.)

Flooding

Flooding is used in some regions to control established herbaceous perennial weeds. It has been used successfully to control Johnsongrass (*Sorghum halepense*), Russian knapweed (*Centaurea repens*), hoary cress (*Cardaria draba*), and silverleaf nightshade (*Solanum elaeagnifolium*). Complete submergence for one to two months during the summer is necessary to kill these species. Water depths of 15 to 25 cm are necessary so that the weeds cannot extend their foliage above the water surface. Flooding for only a few weeks rarely has an adverse effect on vegetative reproductive organs of weeds or weed seed buried in the soil.

In areas suitable for rice production, rotation to this crop permits both crop production and control of some perennial weeds. However, annual weeds, such as barnyardgrass (*Echinochloa crus-galli*) and sprangle top (*Leptochloa* spp.), are associates of rice culture and are not controlled well by flooding. Both weed species have been suppressed in rice by maintaining high water levels (15 to 20 cm). However, this practice is not desirable because deep water is also detrimental to rice at certain stages of development.

Mulches and Solarization

The purpose of *mulches* is to exclude light from germinating plants. The exclusion of light inhibits photosynthesis, causing the plants to die. Commonly used

A

B

FIGURE 8.31 (A) Use of an anchor chain for conversion of shrublands to grasslands and (B) to uproot aquatic weeds from an irrigation canal. (Photographs by S.R. Radosevich, Oregon State University, and W.B. McHenry, University of California, Davis, respectively.)

mulches are straw, manure, grass clippings, sawdust, rice hulls or other crop residues, paper, and plastic. Recently, artificial mulches made of woven plastics have become available for use. These mulches exclude particular wavelengths of light, usually in the photosynthetically active region of the light spectrum, but allow water, nutrients, and air to penetrate into the soil. Mulches are most effective for controlling small annual weeds, but larger plants and some perennials can also be suppressed by using mulches. Crops in which mulches are used are strawberries, pineapple, sugarcane, some vegetable crops, and home gardens. Organic food growers often rely heavily on mulches for weed control.

Soil solarization involves covering moist, tilled soil with clear plastic to kill imbibed weed seed. The plastic sheets are left covering the soil surface for about four weeks. Long periods of full solar radiation, which cause elevated soil temperatures, are needed for best results. It is uncertain whether high temperature or other factors increase mortality of weed seed. Some experiments suggest that soil microflora and microfauna are also injured or killed by solarization.

CULTURAL METHODS OF WEED CONTROL

Cultural methods of weed suppression often occur during the normal process of crop production. These practices include weed prevention, crop rotation, crop competition, mulches and cover crops, and harvesting operations.

Weed Prevention

The *prevention* of a weed problem is usually easier and less costly than control or eradication attempts that follow weed introductions, because weeds are most tenacious and difficult to control after they become established. If weeds are allowed to develop a reservoir of seed or buds, they usually will be present in the field for many years, even decades. McHenry and Norris (1972) suggest the following measures to prevent the introduction of weeds into non-inhabited fields:

- Use "clean" (weedless) crop seed for planting.
- Use manure only after thorough fermentation to kill weed seed.
- Clean harvesters and tillage implements before moving to non-weed-infested areas.
- Avoid transportation and use of soil from weed-infested areas.
- Inspect nursery stock or transplants for seed and vegetative propagules of weeds.
- Remove weeds that are near irrigation ditches, fence rows, rights-of-way, and other non-crop land.

- Prevent reproduction of weeds.
- Use weed seed screens to filter irrigation water.
- Restrict livestock movement into non-weed-infested areas.

Other practices used to prevent and avoid potential weed problems at the state, regional, or national level are *weed laws, seed laws*, and *quarantines*.

The *Federal Noxious Weed Act* was enacted in the United States in 1975. This law prohibits entry of weeds into the U.S. by providing crop inspection for weed seed at ports of entry. The law also allows establishment of quarantines, and provides for the control or eradication of weeds that are new or restricted in distribution. Other local, county, and state *weed laws* have been enacted so that property owners or public agencies must maintain a program of weed prevention or control on their lands. These weed laws permit the formation of *weed control districts* at the local level. It is the obligation of the district, through the activities of a superintendent, to diminish or restrict the occurrence of certain noxious weeds within its jurisdiction. The success of such laws depends upon the level of funding available, the knowledge of the superintendent about weed control measures, and the cooperation of the public and private landowners to see that weed suppression programs are implemented.

Seed laws are used primarily to assure the purity of crop seed and to restrict the dissemination of weed seed across political boundaries. In the United States, every state has a seed law that generally conforms to the statutes of the *Federal Seed Act*. Most states define two categories of noxious weeds, usually primary and secondary, that must be excluded from crop seed. No crop seed containing any primary noxious weed seed may be sold under this law, while only a small percentage (approximately 0.25%) of secondary noxious weeds may be present in crop seed.

Quarantines are enacted to isolate and prevent the dissemination of noxious weeds within a defined area or region. However, very few quarantines have been enacted against weed species. A notable example of a successful quarantine is one established for witchweed (*Striga asiatica*) containment in portions of North and South Carolina (Figure 8.32). The strict regulation of farm material (farm products, residue material, etc.), in combination with other weed control methods, has restricted this weed to the quarantine area.

Crop Rotation

Crop rotation is the practice of growing different crops on the same land from year to year. It is an important method of weed control in many annual and short-lived perennial crops. Certain weed species often are associated with particular crops (e.g., barnyardgrass in rice, mustard in cereals, dodder in alfalfa, foxtail in corn, etc.). Therefore, populations of such weeds usually will increase whenever the crop is grown on the same ground continuously for several years.

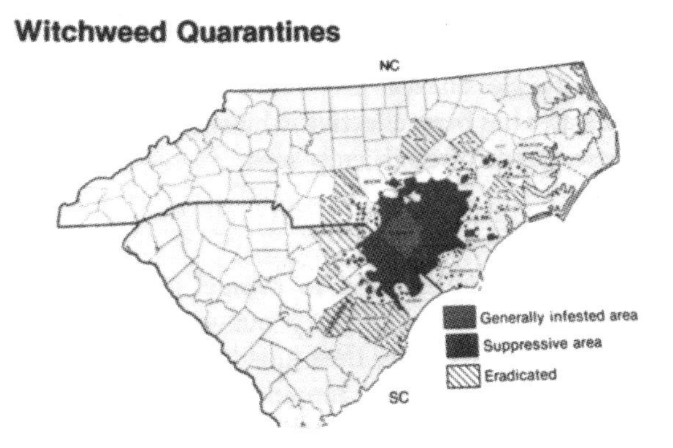

FIGURE 8.32 Areas of North and South Carolina under quarantine for witchweed. (From R. Eplee, Witchweed Methods Development Center, Whiteville, North Carolina; Ross and Lembi, 1985.)

This increase in weed abundance happens because the same environmental or cultural conditions that favor crop production also tend to favor the weeds.

Weed associations with crops may be discouraged by growing in sequence crops that have sharply contrasting growth and cultural requirements. This practice discourages the development of weed populations that are well adapted to the growing conditions of any particular crop. Crop rotation also permits the use of different herbicides or other tools to select against weed populations that might be herbicide resistant or tolerant.

The rotation of a solid-seeded crop, like alfalfa or cereals, to a row crop, like tomato, cotton, corn, or soybean, often allows a concomitant shift in weed control practices because of the differing cultural techniques necessary to grow crops in narrow versus wide rows. The rotation from crop production to fallow (no crop) also permits weed control measures that are not possible when a crop is always present. During the fallow period, use of a different form of tillage implement or control method may be possible to reduce weed abundance. Anderson (1977, 1996) notes that crop rotations, in addition to providing weed control, often improve crop yields and quality by enhancing soil conditions and disease or insect control.

Crop Competition

Crop yields often depend on the amount, size, and proximity of weeds present after crop emergence. Weed vigor is similarly influenced by crop abundance, size, and proximity. Therefore, cultural practices that shift the balance of competition toward the crop usually will disfavor weed occurrence and improve crop yields. Factors that improve *crop competitiveness* include:

- selection of well-adapted crop varieties
- optimum planting date
- optimum planting arrangement (row spacing)
- soil amendments, such as fertility and lime
- proper water management
- use of "smother" crops

Any practices that provide vigorous uniform crop establishment usually will assist in reducing weed prevalence. Numerous examples exist in which poor crop development allowed increased weed growth. Poor performance of the crop may be genetic in origin (selection of the wrong cultivar) or caused by an array of cultural and environmental factors. The choice of planting date also influences the level of crop competition and necessity for weed control. For example, alfalfa in California may be planted either in autumn or spring. When seeded in the fall, seedling alfalfa plants are exposed to several months of cool, wet weather that slows their growth. In contrast, winter annual weeds grow vigorously under those environmental conditions, making chemical weed control necessary in fall-planted alfalfa stands. Alfalfa planted in the spring grows more quickly than that planted in the fall, thus it can be more competitive with weeds.

Other cultural practices such as row width can improve crop competitiveness. Rodgers et al. (1976) observed that decreasing row widths improved the competitiveness of cotton. When cotton was grown in rows 105 cm apart, 14 weeks of weed control were required to prevent yield losses of cotton. However, the weed control period was reduced to 10 and 6 weeks when the row widths were decreased to 77 and 52 cm, respectively. These required "weed-free" periods corresponded to the time it took for cotton to develop a closed canopy. As crop plants grow and increase in leaf area throughout the growing season, many of them become highly competitive to weed seedlings below them. Thus, in some crops, control measures that inhibit weeds soon after crop emergence can provide apparent season-long weed suppression. It is this observation that suggests the existence of critical time periods for weed control following crop emergence (Chapter 5).

Cultural practices that provide adequate soil fertility and water availability are necessary to ensure good vigorous crop growth. Poor irrigation or fertility practices create crop stress, which may favor weed occupation, abundance, or competitiveness. Excess water or fertility also may disfavor crop growth and favor the occurrence of water-tolerant or nitrophilous weeds.

Some crops can suppress weed growth significantly through an ability to grow fast or because they are planted at high density. Crops with these characteristics are called *smother crops*. They include foxtail millet, buckwheat, rye, sorghum, sudangrass, sweetclover, sunflower, barley, soybean for hay, cowpea, clover, and silage corn (Ross and Lembi, 1985). Often, these crops are solid seeded or planted in very closely spaced rows. They also may be used in rotations or mixtures with other crops.

Living Mulches and Cover Crops

In *living mulch* systems, two crops are grown simultaneously, although one of them is usually more economically important. Although the term "living mulch" has a recent origin, the concept does not. For example, intercropping of corn and legumes was studied in the 1930s and 1940s as a means of improving production efficiency of both crops. The current living mulch concept, however, involves growing row crops in an established plant cover or "green mulch." In this way, the occurrence of bare soil is minimized and weed seed germination is reduced. Cropping systems presently under study that use another crop species as a green mulch include corn, soybean, several vegetable crops, and dry beans. *Cover crops* are similar to living mulches, except that cover crops usually are perennial grass or legume mixes grown between vine or tree crop rows.

The benefits of cover crop and living mulch systems are reduced soil erosion, stabilization of soil organic matter layers, improved soil structure, reduced weed abundance and competition, and diminished soil compaction. In addition, some cover crops and living mulches may be harvested for forage or may help reduce pest problems (Altieri et al., 1977). Legume mulches and covers supply nitrogen to the crop. However, substantial losses in crop yield can result from living mulches if competition between the "mulch" and crop plants is allowed to develop. Minimal competition has been found in soybean-winter rye and cabbage-fescue systems, but nearly all other living mulch systems require some form of "mulch" suppression by either chemical or mechanical means. Little competitive effects between cover crop species and vine or tree crops are observed when cover crops are used appropriately.

Harvesting

Although not considered a method of weed control, harvesting can provide a certain level of weed removal or suppression, especially in short-statured herbaceous perennial crops. For example, it is common to harvest alfalfa several times during a growing season. The timing of harvest operations, relative to water availability and germination characteristics of weed species, can substantially improve weed control throughout the entire growing season. For example, summer annual weeds present in alfalfa at the time of first cutting can be reduced by timing harvests to "shade-out seedlings" that germinate as the crop canopy develops. The frequency of grazing on rangelands and pastures has a similar effect on weed abundance and distribution in those production systems.

BIOLOGICAL CONTROL: USING NATURAL ENEMIES TO SUPPRESS WEEDS

The crop environment consists of many other organisms, including weeds, pathogens, insects, and animals. Some of these organisms utilize specific plant species as a food source. Many weeds were introduced into new regions without these associated "natural enemies." For this reason, weeds often grow as solid extensive stands in new areas of introduction, whereas in their native area they exist as scattered patches or clumps, which is due to the feeding activities of natural enemies. The lack of these organisms following weed introduction allows the weed population to increase rapidly to levels that eventually can conflict with human interests.

Biological weed control is the use of living organisms to lower the population level or competitive ability of a weed species so that it is no longer an economic problem (Figure 8.33). Numerous examples of successful biological weed control that utilizes one or more natural enemies are demonstrated in Figure 8.34. Rosenthal et al. (1989) indicate that this method of weed control differs from other methods in several ways:

- It does not necessarily kill the weed outright; instead, often only the competitive or reproductive ability of the affected plant is reduced.
- It may be slow acting, often requiring years to achieve acceptable control levels. Thus, biological control should not be attempted when the destruction of the weed is needed immediately.
- It is relatively inexpensive, especially in contrast to the high costs of developing herbicides.
- Biological control is selective. Control agents must be host specific for the weed they affect to prevent damage to crops or native plants. This selectivity is an advantage when control is directed at only one weed species. However, it is a disadvantage when several weed species must be suppressed at the same time.
- Because of the high host specificity, biological control agents should not cause harmful side effects.
- Biological weed control often is permanent. The object of biological control is not to eradicate the weed. Ideally, some of the weed population should always be present to maintain a population of the natural enemy. Such control is permanent once the weed and natural enemy populations are in equilibrium (Figure 8.33).

Procedures for Developing Biological Control

The first step in developing a biological control program is to determine the suitability of the weed problem for biological control measures. Economic losses caused by the weed plus the costs of control measures must exceed the

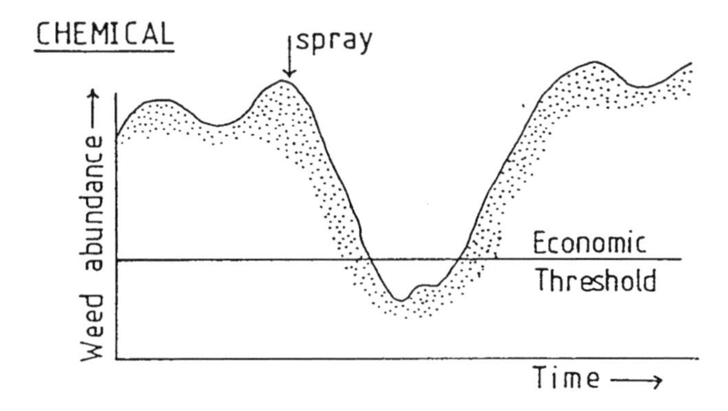

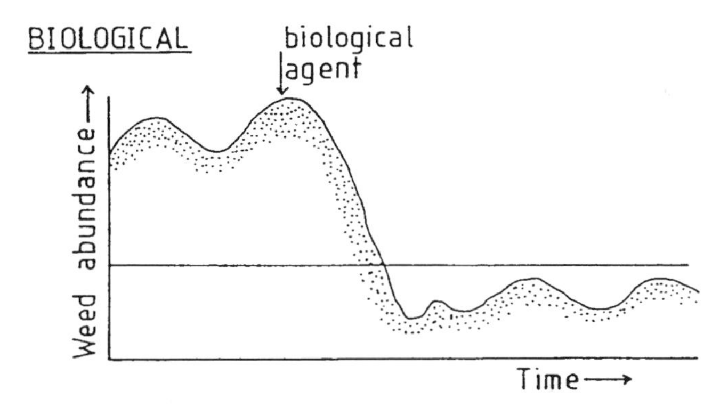

FIGURE 8.33 Diagram showing the weed suppression over time from biological and chemical control methods. Note that successful biological control results in an equilibrium of weed abundance below an economic threshold through time. (From Adkins, 1995.)

cost of the biological control project. Generally, biological control has been most useful when (1) current control measures are inadequate, (2) land values are low, and (3) no closely related crops or other plants of economic/ecological importance are present within the region of weed infestation.

Next, natural enemies of the weed species must be surveyed in both native and naturalized locations of the weed. The biological control agent must be damaging to the weed and be able to survive in the area of introduction. Thus, careful study of the distribution and feeding behavior of the potential biological control agent is needed. Host specificity in feeding, development, and reproduction must be demonstrated by the potential biological control agent.

When selecting target weeds and potential organisms for host specificity testing in the United States, the Federal Working Group for Biological Control of Weeds (WGBCW) must be consulted. This organization is composed of representatives of the U.S. Department of Agriculture, U.S. Department of the Interior,

A

B

FIGURE 8.34 Examples of biological control using (A) *Chrysolina* beetles for St. Johnswort control, (B) Cinnabar moth larvae on tansy ragwort, (C) talapia fish for aquatic weed control, and (D) "feeder" geese for grass suppression. (Photographs courtesy of W.B. McHenry and C.L. Elmore, University of California, Davis.)

C

D

FIGURE 8.34 Continued

Environmental Protection Agency, and U.S. Army Corps of Engineers. The recommendations of the WGBCW are made to the Animal and Plant Health Inspection Service (APHIS) of the U.S. Department of Agriculture. APHIS gives the final approval for importation, testing, and release of biological control agents in the United States. Testing of biological control agents involves experiments for host specificity in quarantine facilities and eventual release in the field.

Rosenthal et al. (1989) suggest several methods for implementing a biological control program for weeds. The implementation method used most is the "*classical*" *approach*. This method relies on the utilization of exotic herbivores or pathogens with sufficiently narrow host ranges. It is the procedure just described above. These species are usually sought in areas where the weed-herbivore or weed-pathogen association co-evolved. The classical approach has been most effective against naturalized weeds as well as perennial weeds commonly found in infrequently disturbed habitats.

Another useful method of biological control is *augmentation*. With augmentation, the biological control agent is collected or mass reared, then periodically released to control weeds. This approach is particularly suitable for areas in which the natural enemy is unable to survive adverse climatic conditions or when its population is insufficient to maintain acceptable control. Augmentation depends upon the ease of collection, rearing, transport, and distribution of the biological control agent.

Grazing

Grazing is, perhaps, the oldest and most common form of biological weed control. It can be accomplished using a wide array of animals that eat vegetation, including large ruminants and ungulates, birds, insects, and fish (Figure 8.34). For example, geese are sometimes used to control grass weeds growing in peppermint and orchards. Sheep are used at times to suppress herbaceous weeds in fast-growing, established alfalfa stands, while blackberry can be controlled effectively by goat grazing. Certain species of fish also have been used to suppress aquatic weeds in canals and lakes.

Mycoherbicides

Plant pathogens have been used effectively to control weeds in augmentative-type biological control programs because plant pathogens are easily and cheaply cultured on artificial media, whereas insects and other biological control agents are not. Furthermore, pathogens may be applied to fields using the same techniques and devices as used for herbicides (Chapter 9). An organism used in this manner has been termed a *bio-herbicide*, or if the organism is a fungus, a *mycoherbicide*.

The use of microorganisms in this manner is new and presents an exciting opportunity for the expansion of weed control technology. *Colletotrichum gloeosporioides* f.sp. *aeschynomene* is currently being tested as a mycoherbicide for the control of northern jointvetch (*Aeschynomene virginica*). Candidates for mycoherbicides must produce large amounts of easily collected inoculum and be

- easily cultured in the laboratory
- highly virulent to the weed
- selective to desirable plant species
- safe to humans and animals

The development of microorganisms or the toxic substance they sometimes produce is discussed further in Chapter 7.

Allelopathy

Allelopathy is any harmful, direct or indirect, effect of one plant on another through the production of chemicals that enter the environment (Chapter 7). Allelopathy has emerged as an intriguing method of using plants or plant residues to control weeds. Many *smother crops* may be allelopathic to other species or themselves (autotoxic). Therefore, the use of smother crops may be a form of biological control that uses plants.

It is uncertain whether toxins produced by allelopathic species arise from the plants themselves or the decay of plant residues. However, the formation of toxic substances by microbial decay of plant residues may be an important mechanism for controlling seedling weeds with natural mulches. The use of allelopathic plants, or substances isolated from them and produced transgenetically, may become an important form of weed control in the future.

CHEMICAL CONTROL

Chemicals, like other methods of weed control, have been used for centuries to suppress or remove weeds. Crafts (1975) indicates that solutions of sodium nitrate, ammonium sulfate, iron sulfate, and sulfuric acid were all effective treatments for weed control in cereal crops by 1900. Early nonselective uses of other chemicals, often industrial by-products or salt, had been used prior to that time to kill weeds in non-cropland areas. The use of chemicals for weed control expanded rapidly during the middle portion of this century, following the discovery of several synthetic organic substances that killed or suppressed vegetation. Herbicidal oils and dinitrophenolic compounds were among the first organic chemicals to be used for weed control. However, it was the discovery of the plant growth suppressing ability of 2,4-D (2,4-dichlorophenoxyacetic acid) during World War II that led to the expansive growth of chemical tools for weed control (also see pages 33 through 34).

Herbicides

Herbicides are synthetic chemicals used to kill or suppress unwanted vegetation. Herbicides now lead all other pesticide groups in total acreage treated, amount produced, and total value from sales (National Research Council, 1989). Ross and Lembi (1985) list the following reasons for the overwhelming success of herbicides as tools for weed control:

- Herbicides allow the control of weeds where cultivation is difficult, for example, within and between narrowly spaced crop rows.
- Herbicides reduce the number of tillage operations needed for crop establishment. The amount of tillage reduction may be only a few operations or entire reliance on chemical weed suppression, such as in no-tillage systems.
- Controlling weeds with chemicals often permits earlier planting, since some tillage operations can be eliminated.
- Herbicides have reduced the amount of human effort expended for hand and mechanical weeding. In crops where herbicides are available, the costs associated with weeding often can be reduced substantially. These cost reductions may be directly or indirectly associated with reduced managerial requirements related to the hiring, overseeing, or housing of labor.
- Weeds that cannot be controlled economically by tillage or other means often can be controlled effectively and efficiently by herbicides.
- Land removed from production because of perennial weeds often can be rehabilitated using herbicides.
- Production losses associated with the fallowing of land because of severe weed abundance can be reclaimed through herbicide use.
- Herbicides allow greater flexibility in the choice of management systems. Less reliance on crop rotation patterns, tillage implements and timings, and fallow periods allows greater selection of crops and management options.
- Mechanical damage to crops can be reduced.

PROBLEMS WITH HERBICIDES

It should be realized, however, that herbicides are only one type of tool available for weed control. Because of the effectiveness of herbicides, there is sometimes a tendency among growers, land managers, and their advisors to expect that any weed problem can be controlled effectively by chemical means. Such an attitude can lead to more expensive or less-effective weed suppression because other, proven methods such as prevention and sanitation, tillage, crop competition, and rotation are overlooked as viable options.

Other potential problems associated with herbicide use are (1) injury to nontarget vegetation, (2) crop injury, (3) residues in soil or water, (4) toxicity to other nontarget organisms, and (5) concerns for human health and safety. The increased legal and regulatory requirements for herbicide application and worker safety are other concerns associated with the use of some herbicides. In many cases, these problems or disadvantages can be overcome by proper selection, storage, handling, transportation, and application of the chemical.

TIMING AND USES OF HERBICIDES

Herbicides often are applied at various times during a growing season depending upon differences in crop and weed development, or stage of growth. Specific terms are used to describe these differences in herbicide application time (Figure 8.35). These terms also explain how herbicides are used and categorized.

Preplant. Applications are made to soil before the crop is planted. They are typically made before or during seedbed preparation and before the crop is sown.

Preemergence. Applications are made to the soil after the crop is sown, but before emergence of the crop or weeds.

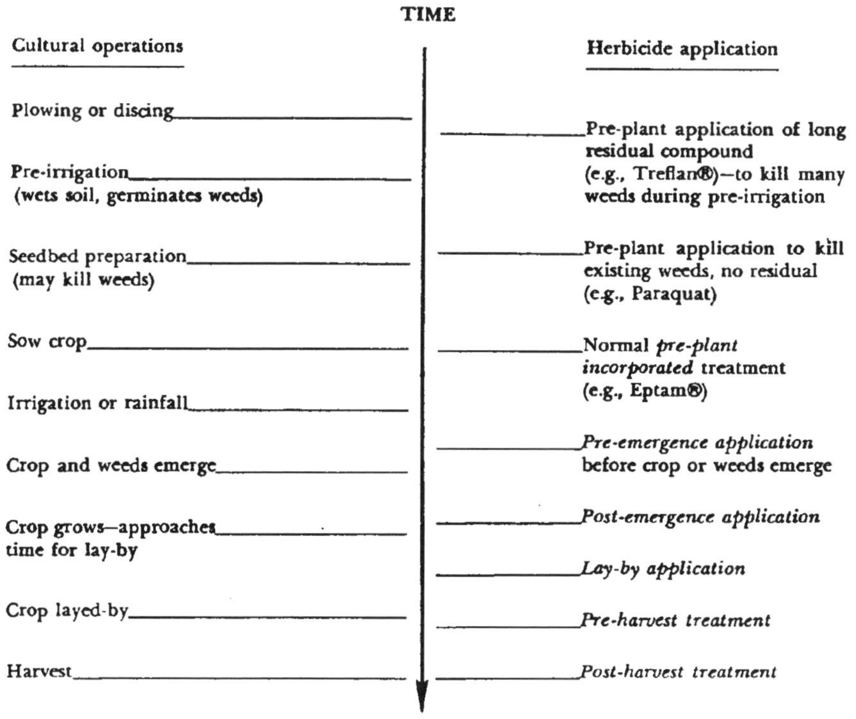

FIGURE 8.35 Diagram showing the sequence of herbicide application, cultural practices, and stages of crop and weed growth. (From McHenry and Norris, 1972).

Postemergence. Treatments are applied to both crop and weeds after they have germinated and emerged from the soil. Usually this term implies an application of the herbicide early in development of the plants. However, other postemergence applications at later stages of development also are possible. They include

Lay-by. A herbicide application made to row crops that is the last equipment operation in the field until harvest.

Pre-harvest. An application of herbicide made prior to harvest usually to desiccate the crop foliage and remove weeds that might interfere with harvesting operations.

Post-harvest. A herbicide application that is made to control weeds after harvest, but which is not strictly part of the weed control program for the next crop.

Figure 8.35 demonstrates how the sequence of herbicide applications fits with cultivations and stage-of-growth of weeds and crop plants.

Because of the importance of herbicides to modern weed control, various aspects of herbicide use will be developed in the next chapters. These topics include basic aspects of herbicide chemistry, herbicide properties and modes-of-action, selectivity, application and calibration, and safety and regulatory aspects of herbicide use.

PRINCIPLES FOR WEED MANAGEMENT SYSTEMS

Every weed management program is only part of a total crop production system. Therefore, any combination of environmental manipulation and cultural techniques to reduce weed abundance must be compatible with other farm, forest, or range management objectives. In this regard, interactions between soil, fertility, insect and disease pests, and direct weed control procedures may be especially important. For example, weeds may indicate a loss of soil nutrients, and manipulations of weed abundance may only exacerbate a deteriorating soil condition (Forcella and Harvey, 1983; Ghersa et al., 1994a). Furthermore, the manipulation of weed abundance, in addition to affecting crop-weed competition, also may influence the performance of natural enemies of crop pests (Table 8.2), sometimes resulting in decreased crop yields. Altieri and Liebman (1988) suggest that the major contributions that weed biologists can make to weed management systems are to: (1) determine the ecological factors that govern weed abundance, (2) discern conditions and times when weeds are most vulnerable to management tactics, (3) provide information to predict accurately the response of weeds to various controls, and (4) elucidate the functional links between environment, weeds, crops, and other species. They believe that such understanding will contribute greatly to improved management of weeds in agricultural (Figure 8.36) and other ecosystems.

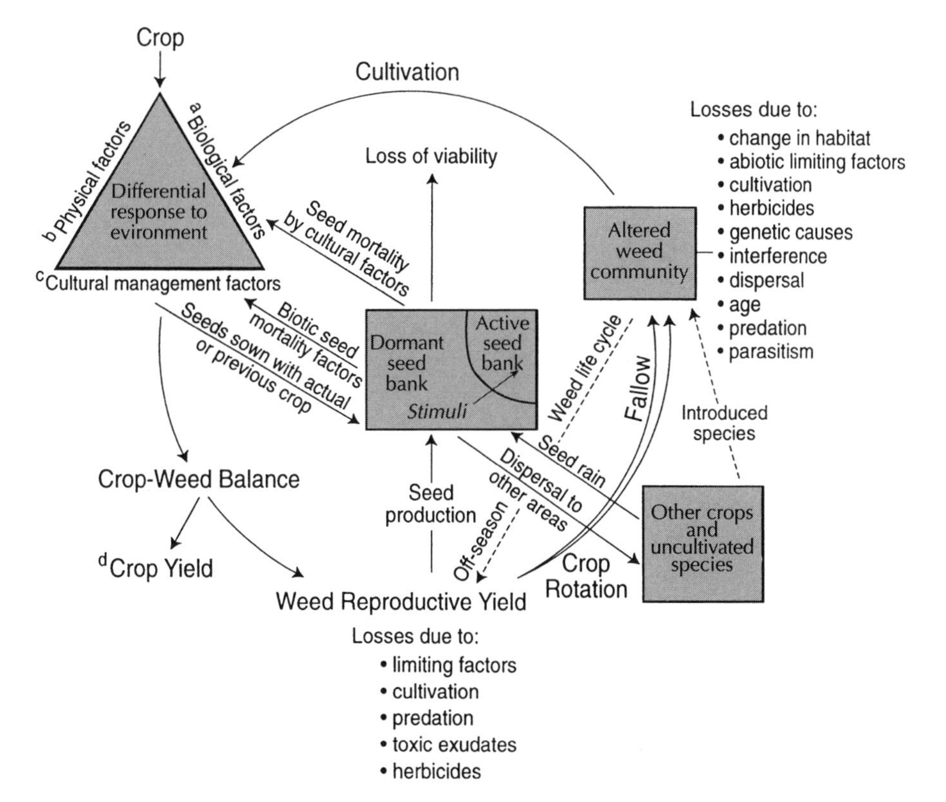

FIGURE 8.36 Conceptual model of a weed community cycle in an annual cropping system. [a]Weed community composition, density, and temporal duration; crop species, density, spatial distribution, and duration; allelopathy; insect herbivores and plant pathogens. [b]Soil moisture level, light intensity, temperature, rainfall, and soil fertility. [c]Irrigation, tillage, chemical herbicides, fertilizer use, cropping pattern, planting time, crop spacing, rotation, and mulching. [d]$y = 1/(a + bx)$ where y = crop weight (g/m^2), x = population density of specific weed (plants/m^2), a = reciprocal of crop weight in a weed free plot and b = measure of factors influencing competition. Stable equilibrium reached where $\frac{\text{No. of crop seeds produced}}{\text{No. of weed seeds produced}} = 1$. (Modified from Altieri and Liebman, 1988.)

Realistically, however, it is not possible to know everything about every (or any) weed species in order to develop effective weed management strategies. Zimdahl (1993) recognizes this limitation and suggests that development of weed management systems should include the following components:

- incorporation of ecological principles
- use of plant interference and crop-weed competition
- incorporation of economic and damage thresholds
- integration of several weed control techniques including selective herbicides
- frequent supervision of weed management by a professional weed manager employed to develop a program for each crop-weed situation

He further suggests, after Beck (1990), that effective weed managment should include accurate mapping of weed infestations, prioritization of weed problems, selection of appropriate control tactics, integration of tools used for direct control, systematic implementation of management over time and space, a record keeping and evaluation process, and continued vigilance against further weed infestation. Altieri and Liebman (1988), on the other hand, believe that practical weed suppression in most agroecosystems may require knowledge of only a few biological features and key functional relationships of the weed in question, which may necessitate changes in only a few key management decisions. However, a broad knowledge base may be needed in most cases in order to ascertain which biological features are most critical to effective control (Table 8.4). Below is a list of principles based on one first posed by Altieri and Liebman (1988) that should be considered when designing integrated weed management systems.

- Crop monocultures seldom use all the environmental resources available for plant growth. The resulting ecological niches, therefore, are susceptible to invasion by weeds and should be protected.
- Weed populations are either active, as photosynthesizing plants, or dormant, usually seeds. Thus, the seed bank as well as the above-ground vegetation should be considered when determining weed abundance.

TABLE 8.4 Nonchemical Methods of Managing Weeds and the Ecological Principles upon Which Each is Based

Ecological Principle	Weed Control Practice
Reduce inputs to and increase outputs from soil seed bank	Prevention, soil solarization, weed control before seed set
Allow crop earlier space (resource) capture	Early cultivation, using crop transplants, choice of plant date
Reduce weed growth and thus space capture	Cultivation, mowing, mulching
Maximize crop growth and adaptability	Choice of crop variety, early planting
Minimize intraspecific competition of crop, maximize crop space capture	Choice of seeding rate, choice of row spacing
Maximize competitive effects of crop on weed	Planting smother or cover crops
Modify environment to render weeds less well adapted	Rotation of crops, rotation of control methods
Maximize efficiency of resource utilization by crops	Intercropping

Source: Altieri and Liebman, 1988. Modified after Holt, J.S., University of California, Riverside, unpublished data, 1986.

- The reproductive capacity and seed survival of weeds determine the composition and abundance of the succeeding weed community in a cropping system. In intensive cropping systems, the type of weed community and its abundance are a direct product of the crop and how it is managed. In developing countries, where mixed cropping is traditionally practiced in contrast to more intensive forms of crop production, the complex cropping patterns appear to disfavor weeds. Additionally, in less intensive cropping systems than those presently practiced in many developed nations, weed management appears to have less relevance than overall cropping practices in determining the weed community composition.
- The cropping pattern can be a powerful agent in reducing weed densities by preemption of environmental resources to crop plants. Weed control tactics, while usually reducing the abundance of weeds, can shift the composition, density, and spatial distribution of weeds in fields. These shifts in the spatial or temporal dynamics of weed communities and populations ultimately may increase weed abundance and, therefore, crop competition over the long term. Although shifts in species composition often appear inevitable, the practice of growing crops with divergent life cycles and using rotational sequences of crops and weed control tactics are effective strategies to prevent any one weed species from becoming dominant. These practices reduce selection pressure on weed communities and populations.
- Single measurements of weed density are usually not adequate to determine weed impacts. Most crops have thresholds of weed tolerance, expressed as density, amount of biomass, or period of time before significant economic loss in crop yields results. However, these thresholds vary among cropping systems, weed species, and environmental constraints. The thresholds also may have to be adjusted when criteria other than crop loss are used to determine the economic, ecological, and social effects of weeds.
- Suppressed crop growth cannot always be explained by crop-weed competition. At times, weeds may simply indicate a deteriorated soil or resource base. Allelopathy also may be a mechanism through which weeds affect plant growth, and vice versa. In addition, crops and weeds may coexist without crop yields being reduced economically and beneficial effects are possible from some crop-weed associations.
- Weed populations and communities are regulated by a combination of factors (e.g., predation, parasitism, environmental stress, interference, and direct-control methods), and at several stages of the weed's life cycle. Thus, control of weeds can be achieved through direct or indirect means. During direct control the weed plant is physically or chemically removed, while indirect controls rely on biological functions of the crop, the weed, and their associates. Both strategies have been successful, although indirect regulation may be most appropriate for complex cropping systems.

Unfortunately, at this point there are few, if any, fully integrated weed management systems, and each developing system must adapt to the local realities of environment, economics, and production practices. It is also quite likely that direct controls, especially tillage and herbicides, will remain as important components of most weed management systems for the next several years and perhaps decades. However, the sole reliance on such tools should diminish as better understanding of environment and biology are incorporated into these weed management systems.

SUMMARY

Vegetation management is the fostering of beneficial vegetation and the suppression of undesirable plants. Weed control is often a component of vegetation management. Physical, cultural, biological, and chemical methods have been used to control weeds. Weed control practices influence plant communities by direct reductions of plant density and by alteration of species composition. Reductions in plant density by weed control procedures usually are obvious. However, shifts in weed species composition as a result of weed control are more subtle. All tools used to control weeds alter the species composition of crop-weed stands to some extent. Herbicide resistance is an extreme example of a shift in weed species composition following repeated use of the same weed control chemical.

Weed control methods remove, suppress, or destroy vegetation. This activity results in the modification or disruption of the habitat of other organisms, both pests and non-pests. The presence of weeds also influences the habitats of other organisms. Thus, it is important for weeds and weed control to be considered as a component of management programs that involve other beneficial and non-beneficial organisms.

Many tools are available for the physical disruption, suppression, or elimination of vegetation. These tools or techniques are fire and flame; manual pulling, hoeing, and cutting; and various mechanical implements. Tillage is a principal means of seedbed preparation and weed control in agriculture and many tools have been developed to accomplish tillage effectively. However, some problems exist with conventional tillage systems, and other alternative systems that reduce or eliminate tillage have been devised. Chaining, dredging, flooding, and artificial mulches are other physical methods of weed control. These tools are used effectively in certain circumstances and locations to control weeds.

Cultural practices used to control weeds are prevention, crop rotation, crop competition, living mulches and cover crops, and harvesting. In general, any practices that favor crop growth also will disfavor weed abundance, unless the crop is grown repeatedly for a number of years. Quarantines and weed laws also represent preventive methods of weed control.

Biological control utilizes natural enemies to suppress weed species. There are defined procedures that must be met before new natural enemies can be introduced into the United States for biological weed control. There have been numerous successful natural enemy introductions that have effectively controlled specific weed species. Both allelopathy and mycoherbicides may become important biological tools for weed suppression in the future.

Chemicals that are used to suppress weeds are called herbicides. Herbicides are the most used form of pesticide in the United States. Although herbicides are not problem-free, there are many reasons why they are such a popular form of weed control.

Chapter

9

Herbicide Use and Application

Herbicides are chemicals used to suppress or kill unwanted vegetation. They are used to reduce weeds in cropland, forest plantations, rangeland, and many other situations, such as roadsides and rights-of-way where weed growth is sometimes a problem. Herbicides have become a major technological tool for agriculture and are responsible, at least in part, for significant increases in crop production during the last quarter of a century. The USDA estimates that herbicides represent over 80 percent of all pesticides used in the United States. For example, approximately 225,000 tons of herbicide were used in 1984 (see also p. 36 and p. 388). Because of the importance of these chemicals to modern agriculture, the next two chapters describe more about them, their uses, characteristics, and environmental fate.

HERBICIDE DEVELOPMENT AS COMMERCIAL PRODUCTS

Few herbicides, if any, are initially synthesized solely for their plant killing properties. Rather, most manufacturers prepare and "screen" numerous chemical structures for a variety of purposes, including potential herbicidal activity. It is likely that a single chemical manufacturer will synthesize and test thousands of potential herbicides in a single year. It is during this primary synthesis and screening phase that a chemical is identified as a potential herbicide.

Herbicide development, following discovery, is a process of systematic chemical modification and examination for improved biological activity. It is an empirical procedure, based on both experimentation and experience, in which chemists systematically add various substituent groups to the parent compound. Each of these "new" chemicals is also tested to find the material with greatest biological potential for plant susceptibility. Further laboratory, greenhouse, and field experiments (Figure 9.1) are then conducted with the most promising materials to determine plant selectivity, soil persistence, and other physical and biological characteristics that influence the fate of the chemical in the environment.

The cost to develop a single new herbicide ranges from 20 to 50 million dollars, which is a substantial amount of money even for a large corporation. It also takes from six to twelve years for the research and regulatory processes to be completed. The primary agency responsible for the regulation of herbicide development in the United States is the Environmental Protection Agency (EPA), which enforces federal laws requiring pesticides to be effective and safe. Herbicides must kill unwanted vegetation but not injure crops. They also must not enter the food chain or cause adverse effects to the environment. The necessary data to meet federal requirements enforced by the EPA involves experiments on toxicology, biology, chemistry, and biochemical degradation of the chemical. In addition, the effects of the chemical on air and water quality, soil microorganisms, wildlife, and fish must be determined by the pesticide manufacturer.

FIGURE 9.1 Screening chemicals in the field for herbicidal properties. (Photograph by A.P. Appleby, Oregon State University.)

Laws Governing Herbicide Registration and Use

In the United States, there are two laws that provide the authority to regulate pesticide development and use. These laws are the Federal Insecticide, Fungicide and Rodenticide Act (FIFRA), and some portions of the Food, Drug, and Cosmetic Act (FDCA).

FIFRA was first enacted in 1947 and was rewritten in 1972. This law enforces the concept that the benefits from pesticide use must be in balance with concerns about public health and environmental impacts. It provides for registration and cancellation of pesticides, creates a classification system for pesticides based on toxicity, and allows states to regulate pesticide use in a manner consistent with federal regulations. The FDCA requires the establishment of tolerances for pesticides in food, feed, fiber, and water.

Pesticides, including herbicides, cannot be distributed or sold in the United States unless they are registered with the EPA. Pesticides are classified by the EPA as being for either *general* or *restricted* use. The criteria for classification as a restricted-use pesticide are

- danger or impairment of public health
- hazard to farm workers
- hazard to domestic animals and crops
- damage to subsequent crops by persistent residues in the soil

In addition, hazard to surface and ground water supplies is an important criterion for herbicide regulation and restriction. Many herbicides are not as toxic as some other types of pesticides, such as insecticides. However, certain herbicides are very toxic and should be used with extreme caution. Furthermore, laboratory tests with animals indicate that some herbicides may be toxic following chronic exposure for several months or years.

Pesticide uses, needs, and environmental concerns often vary among the states in the United States. For this reason, the regulatory aspects of herbicide use are influenced strongly by state laws. All states have pesticide worker-safety and restricted materials regulations, which specify safe worker practices for individuals who handle or apply pesticides. These regulations are implemented to reduce the risk of pesticide exposure to people. It is the employers' responsibility to provide a safe working environment for their employees and to see that they are following safe practices as defined by the law. In most states, it is the Department of Food and Agriculture or a similar agency that has the responsibility for pesticide regulation and worker safety.

INFORMATION ON THE HERBICIDE LABEL

The label printed on a herbicide container is considered to be a legal document that specifies how the material should be used to ensure its safety and effectiveness. All labels must show clearly the following information:

- product trade name
- name of the registrant (usually the manufacturer of the product)
- net weight or measure of the contents
- EPA registration number
- registration number of the formulation plant or factory
- an ingredients statement containing the name and percentage of the active ingredient of the product
- percentage of inert ingredients
- use classification, that is, general or restricted
- a warning or precautionary statement

Toxicity categories of herbicides. Warning and precautionary statements on the pesticide label are concerned with human toxicity and environmental, physical, and chemical hazards associated with each material (Maddy et al., 1989). A *toxicity category* is assigned to every pesticide based on levels of hazard indicators. Each toxicity category and its indicator is shown in Table 9.1. The signal word

TABLE 9.1 Toxicity Categories and Hazard Indicators of Pesticides[a]

	Toxicity Categories			
Hazard Indicators	*I*	*II*	*III*	*IV*
Oral LD_{50}	Up to and including 50 mg/kg	From 50 thru 500 mg/kg	From 500 thru 5000 mg/kg	Greater than 5000 mg/kg
Inhalation LC_{50}	Up to and including 0.2 mg/liter	From 0.2 thru 2 mg/liter	From 2 thru 20 mg/liter	Greater than 20 mg/liter
Dermal LD_{50}	Up to and including 200 mg/kg	From 200 thru 2000 mg/kg	From 2000 thru 20,000 mg/kg	Greater than 20,000 mg/kg
Eye effects	Corrosive, corneal opacity; not reversible within 7 days	Corneal opacity: reversible within 7 days; irritation persisting for 7 days	No corneal opacity; irritation reversible within 7 days	No irritation
Skin effects	Corrosive	Severe irritation at 72 hours	Moderate irritation at 72 hours	Mild or slight irritation at 72 hours

Source: Maddy et al., 1989.

[a]LD_{50}, dosage required to kill 50% of a test population; LC_{50}, concentration required to kill 50% of a test population.

danger is required for a pesticide meeting any criterion for Toxicity Category I. Toxic Category II materials require the signal word *warning*, while pesticides in Categories III and IV use the word *caution*. Both Federal and State laws require that pesticides be used in accordance with the instructions printed on the label.

CHEMICAL PROPERTIES OF HERBICIDES THAT AFFECT USE

Most herbicides are organic chemicals, primarily made up of carbon and hydrogen atoms. The carbon atoms of organic molecules bind together to form "chains." The simplest organic compound is methane. It is composed of a single carbon atom that is bonded to four hydrogen atoms (CH_4). If a hydrogen atom in methane is replaced by another carbon, ethane (C_2H_6), a two-carbon chain, is formed. Long chains of interconnected carbon atoms can be made in this way. The chains may be straight, branched, or cyclic.

Organic compounds composed of only carbon and hydrogen, such as those described above, are called *hydrocarbons*. Hydrocarbons are *saturated* when all available bonds are occupied by an atom of carbon or hydrogen. *Unsaturation* occurs when two carbon atoms share more than one bond. There may be only a few or many double and triple bonds in an organic molecule. Acetylene (C_2H_2) and benzene (C_6H_6) are examples of unsaturated hydrocarbons. Benzene is a common constituent of many herbicides. Organic chemicals that are arranged in an unsaturated ring configuration (e.g., benzene) are also called *aromatic* hydrocarbons.

Only a few elements other than carbon and hydrogen are found in organic compounds, including herbicides. These elements are oxygen, nitrogen, sulfur, phosphorus, and the halogens (chlorine, fluorine, iodine, and bromine). Organic chemicals having other atoms than carbon as part of their ring structure are called *heterocyclic* hydrocarbons. Many herbicides are heterocyclic compounds. In addition, most herbicides contain at least one halogen atom (salt-forming elements) as part of their molecular structure. Alcohols (R–OH), organic acids (R–COOH), and esters (R–O–R) are forms of organic compounds that dramatically influence chemical and physical properties. These structures tend to be highly reactive and influence such properties of herbicides as water solubility, electrical charge, and potential to vaporize.

Chemical Structure

Each herbicide has inherent chemical properties that influence its ability to kill plants. The biologically active portion of a commercially manufactured herbicide is called the *active ingredient*. This is the fundamental molecular composition and configuration of the herbicide. In addition to biological activity, chemical and physical properties of the herbicide can determine the method of application and use. For example, the active ingredient of 2,4-D is the acid form of that herbicide. However, the herbicide is rarely sold or applied in that form because it does not penetrate leaves or kill plants as well as other forms of the chemical. The chemical also is difficult to formulate (see pp. 404–407) and therefore is more expensive than other forms of 2,4-D.

An active ingredient can be altered slightly by chemical processes, such as esterification, which may improve biological activity, alter the method of application, or influence the herbicide's fate in the environment. Herbicides that are derived from alcohols, phenols, and organic acids often are more soluble in water than those that are not. In contrast, ester forms of herbicides are relatively more soluble in oil or organic solvents and have a tendency to produce vapors. The loss of herbicide as vapor is called *volatility*, which is related to the vapor pressure of the herbicide (see p. 404).

Organic salts of herbicides may also be formed during the manufacturing process. The acid form and two organic salts of 2,4-D are shown below. These forms of 2,4-D are rather unlikely to volatilize, but are soluble in water. The ester forms of 2,4-D are soluble in organic solvents and are more likely to vaporize than the organic salts. However, the size of the ester linkage to the parent 2,4-D acid molecule also influences the degree of volatility of the chemical The isobutyl ester is highly volatile, whereas the butoxyethyl ester of 2,4-D has much lower volatility characteristics.

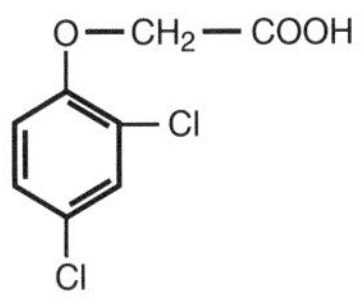

acid of 2, 4-D

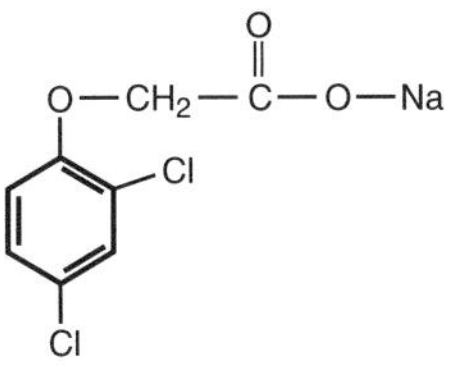

sodium salt of 2, 4-D

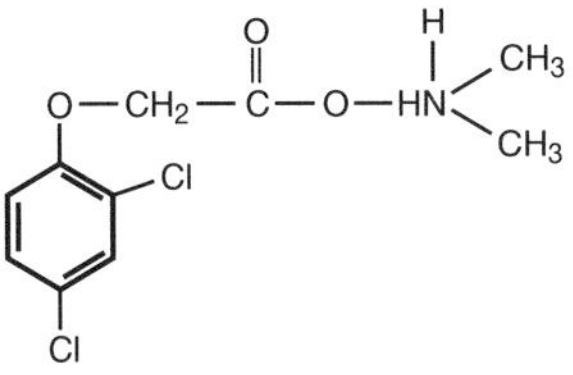

dimethylamine salt of 2, 4-D

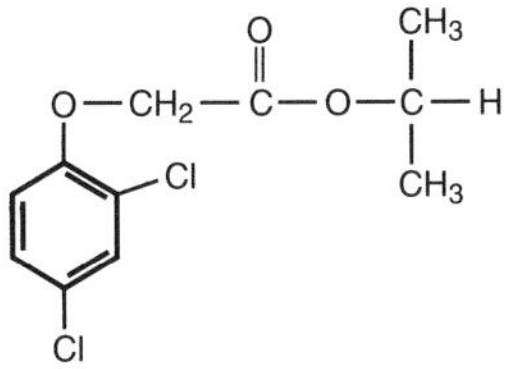

isopropyl ester of 2, 4-D

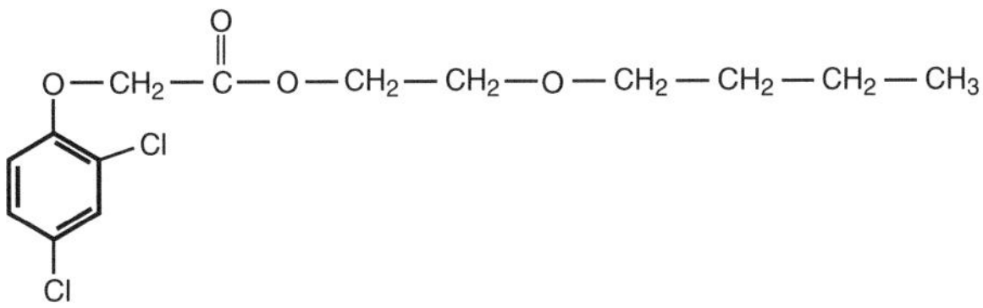

butoxyethyl ester of 2, 4-D

As just demonstrated for 2,4-D, it is possible that rather small changes in chemical form can alter significantly the chemical properties, uses, and effectiveness of herbicides. The chemical and physical properties of some common organic herbicides are presented in Table 9.2.

TABLE 9.2 A Summary of Information about Some Herbicides

Common Name	Oral LD_{50}[a] (mg/kg)	Toxicity class[b]	Leaching class[c]	Water solubility[d]	Volatility[e]	Site of Uptake	Soil Persistence[f]
			Ureas				
Chloroxuron	3700	III	1	2.7	Low	Roots	3 months
Diuron	3400	III	2	42	None	Roots and foliage	6 months
Fenuron	6400	IV	3	2400	None	Roots	3 months
Tebuthiuron	644	III	2	2300	None	Roots	12–15 months
			Triazines				
Atrazine	3080	III	3	34	Low	Roots, some foliage	6 months
Propazine	>5000	IV	3	8.6	None	Roots	6 months
Simazine	>5000	IV	3	3.5	None	Roots (little if any foliage)	6 months
Prometon	2980	III	3	750	Low	Roots and foliage	6 months to many years
Prometryn	3150	III	2	48	Low	Roots and foliage	2 months
Metribuzin	1090	III	3–4	1220	Low	Roots and foliage	2 months
			Uracils—Pyrimidines				
Bromacil	5200	IV	4	815	None	Roots	6 months
Terbacil	>5000	IV	3	710	None	Roots	6 months
			Acylanilides				
Propanil	1384	III	2	50,000	None	Foliage	1–3 days
			Pyridazinones				
Pyrazon/ chloridazon	3600	III	2	400	Low	Roots	1–2 months
			Bis-carbamates				
Phenme-dipham	8000	IV	1	10	None	Foliage	1 month
Desmedipham	10,250	IV	1	7	None	Foliage	1 month

TABLE 9.2 Continued

Common Name	*Oral LD_{50}[a] (mg/kg)*	*Toxicity class[b]*	*Leaching class[c]*	*Water solubility[d]*	*Volatility[e]*	*Site of Uptake*	*Soil Persistence[f]*
			Dinitroanilines or Toluidines				
Benefin	10,000	IV	1	1	Low	Roots	>6 months
Dinitramine	3000	III	1	1	Low	Roots	2–3 months
Ethalfluralin	>10,000	IV	1	0.3	Moderate	Roots	2–3 months
Fluchloralin	1550	III	1	1	Moderate	Roots	12 months
Isopropalin	5000	IV	1	<1	Low	Roots	6 months
Oryzalin	>10,000	IV	2	2	Low	Roots	2 months
Pendi-methalin	1250	III	1	0.5	Moderate	Primarily roots	~4 months incorporated
Profluralin	2200	IV	1	0.1	High	Roots	6 months
Trifluralin	<10,000	III	1	1	High	Roots	3 months

Source: Adapted from Zimdahl, 1993.

[a]Values are acute oral toxicity given in milligrams/kilogram of body weight for technical material (95% pure) for adult male rats.

[b]Toxicity category; see Table 9.1 this text.

[c]Leachability divided into five classes: class 1, immobile —> class 5 very mobile.

[d]Values are in ppm for unformulated molecules.

[e]Only none (insignificant) low, moderate, and high are used.

[f]Values are approximations that will be changed by soil, rainfall, temperature, and cropping conditions and represent how long phytotoxic activity will remain in soil.

Water Solubility and Polarity

If a chemical is soluble in water, a solution forms when the two substances are mixed. The solvent action of water is based on the ability of water molecules to form hydrogen bonds and dipole-dipole interactions with other molecules and ions. Many chemicals, such as alcohols, organic acids, phosphates, nitrates, chlorates, ammonium compounds, and sugar, are held in solution with water by hydrogen bonding. Water also is electrically asymmetrical or polar, since the centers of positive and negative charge are located at different molecular points (Figure 9.2). Other types of chemicals that also are polar readily dissolve in water due to such dipole-dipole interactions.

As a general rule, polar substances dissolve in other polar substances. Ionizable salts, such as table salt (NaCl), dissolve in water in this way. Herbicides that are produced as salts are quite water-soluble and are usually formulated to be applied

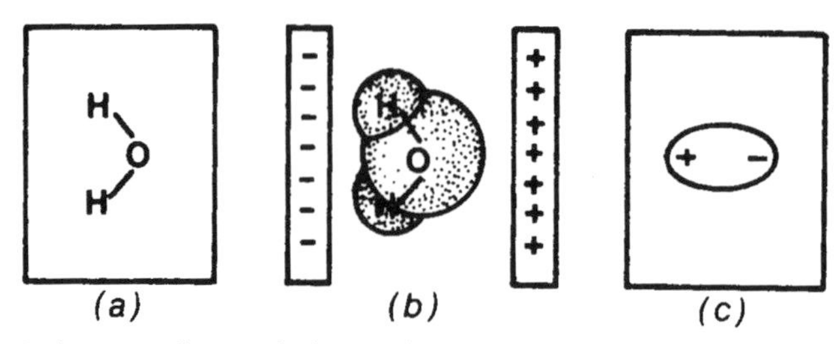

FIGURE 9.2 Dipolar structure of water molecule. (a) Hydrogen atoms are positively charged (though not ionized in the ordinary sense) and the oxygen is negatively charged. (b) Orientation of water molecules in electrical field. (c) Simple diagram of a polar molecule, such as water, showing positive and negative portions. (From Slabaugh and Parsons, 1966, p. 158.)

in water. In contrast, nonpolar substances are practically insoluble in water. For example, oil is a nonpolar solvent and does not mix well with water because of differences in polarity. Thus, herbicides that are soluble in nonpolar, oil-like solvents are not very soluble in water. The solubilities of various herbicides in water and other solvents are listed in Table 9.2.

Water is the major substance used to disperse (that is, spray) herbicides over a field. Therefore, the water solubility of a herbicide determines, to some extent, the type of product that is formulated and how it is applied. Water solubility is also important because it influences herbicide movement in the soil profile (see p. 411 and Chapter 10).

Volatility

Volatility occurs when a chemical changes from the solid or liquid state of matter to the gaseous state. The tendency of chemicals to volatilize is determined by their vapor pressure, which is measured and expressed in mg of mercury (Hg). Herbicides with low vapor pressures (e.g. 10^{-7} mm Hg) are relatively nonvolatile, while those with high vapor pressures (e.g. 10^{-3} mm Hg) volatilize readily (Table 9.2). Both chemical form and formulation influence the ability of herbicides to volatilize.

Formulations

The active ingredient of many herbicides is unsuitable to use as a commercial product. It usually must be refined by the manufacturer prior to sale and use. The final product, or *formulation*, contains the active ingredient of the herbicide and any "inert" ingredients, such as solvents, emulsifiers, diluents, and so on, that enhance the marketability or biological activity of the chemical (see pp. 428–430). Herbicides usually are formulated for ease of transportation and application in water, but other types of carriers include diesel oil, fertilizer solutions, and dry material for granule applications. Formulations must be

- compatible with the carrier, usually water
- convenient and safe to handle
- capable of being applied accurately and uniformly

Some herbicides are formulated as a number of different products, all containing the same active ingredient. These products are developed to enhance the particular chemical or physical properties of the herbicide, improve weed control or crop selectivity, reduce animal toxicity, or provide an economic advantage to the manufacturer. Large differences in effectiveness, rate of application, hazard, or cost often exist among such herbicide formulations.

Anderson (1977, 1996), Ross and Lembi (1985), and Zimdahl (1993) provide excellent discussions of the types and uses of herbicide formulations.

FORMULATIONS FOR USE AS LIQUIDS

Formulations that can be used as liquid sprays include *water-soluble powders* and *liquids, emulsifiable concentrates, wettable powders*, and *water-dispersible liquids* and *granules*. Each of these types of formulation is briefly described below.

Water-soluble powders (SP). Herbicides formulated as water-soluble powders are dry, fine solids that dissolve readily into water. In many cases, the formulation is the water-soluble salt of the active ingredient. A solution usually forms in the spray tank but a spray solution additive (adjuvant, see pp. 428–430) is sometimes required to increase herbicidal effectiveness. These formulations are polar, form solutions in water, and require little agitation in the sprayer tank once they dissolve. They do not mix well with oil carriers.

Water-soluble liquids (S or SL). Herbicides formulated as soluble liquids are similar to water-soluble powders, except they are already in liquid form. Like water-soluble powders, they completely dissolve in water. Many water-soluble liquid formulations contain a wetting agent as an inert ingredient.

Emulsifiable concentrate (E or EC). The active ingredient of herbicides formulated as emulsifiable concentrates usually is nonpolar and relatively insoluble in water. Solubility in organic nonpolar solvents is possible, however. In order to mix the herbicide with water for field application, the herbicide active ingredient is dissolved first in an appropriate organic solvent. An emulsifier then is added to the solution. Emulsifiers are substances that aid the suspension of one liquid in another (see p. 428). They are used most often to disperse oil-like materials into water. Emulsifiers are unique substances because they possess both polar and nonpolar properties within the same molecule (Figure 9.3). The emulsifier molecules orient at the water-organic solvent interface (Figure 9.3). It

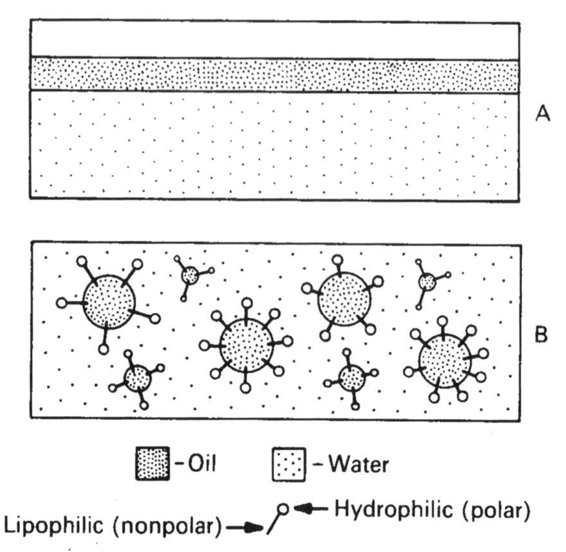

FIGURE 9.3 (A) Water and oil without emulsifier. (B) Emulsifiers link oil and water particles, enabling oil droplets to become suspended in water. (From Ross and Lembi, 1985.)

is the combination of herbicide active ingredient, solvent, and emulsifier that constitutes the emulsifiable concentrate formulation.

When the EC formulation is added to water in a sprayer, the emulsifier also orients with the water, which prevents the oil-like herbicide solution from forming large drops that would separate from the water carrier (Figure 9.3). A stable liquid-in-liquid suspension, called an emulsion, is formed when the herbicide product is added to water and agitated prior to application. Emulsions usually are milky in appearance. The section on adjuvants (pp. 428–430) in this chapter also discusses this type of herbicide formulation.

Wettable powders (W or WP). The active ingredient of herbicides formulated as wettable powders is nearly insoluble in water and may not be soluble in organic solvents. The solid active ingredient is ground to a fine powder. Wettable powders suspend readily in water because a dispersing agent (p. 428) is added to the formulation. Usually, wettable powder formulations consist of 50 to 80 percent active ingredient. Agitation is required to keep the small herbicide particles suspended in the water of a spray tank.

Water-dispersible liquids (WDL) or flowables (F). Water-dispersible liquids are similar to wettable powders since the active ingredient of the herbicide consists of a finely ground, nearly water-insoluble powder. In this case, the active ingredient is suspended as a highly concentrated liquid system. The particle size of water-dispersible liquid formulations is usually less than in wettable powders, so less vigorous agitation in the sprayer tank is required to avoid precipitate formation.

Water-dispersible granules (WDG). Herbicides of this formulation are dry mixtures of finely ground active ingredient, dispersing agents, and diluents (dry, inert carriers) that are mixed and dried to form granules. The granules are added directly to the spray tank. This type of formulation pours readily from its container and disperses into water without clumping. At this point, the formulation becomes suspended in the water carrier, which must be agitated to avoid settling of the herbicide to the bottom of the sprayer tank. WDG formulations are generally easier to measure and pour and produce less dust than do wettable powders.

DRY HERBICIDE FORMULATIONS

Some herbicide active ingredients are formulated dry as granules or pellets for direct application without dilution in water. These formulations also have low concentrations of active ingredient and are less hazardous than other formulations for this reason. Because of their low concentration, dry formulations are usually more expensive than other formulations when their price is based on cost per unit of herbicide active ingredient.

Granules (G). Solid herbicide formulations that are not powders and are less than 10 mm^3 in size are granules (Ross and Lembi, 1985). Usually, only 2 to 20 percent of the formulation is the active ingredient. The inert, dry carrier of granular formulations is made from a variety of organic and inorganic substances, such as clay colloids, starch polymers, ground plant residues, or dry fertilizers. Penetration of the herbicide through foliage to the soil surface is often enhanced by granular applications. In addition, loss of the active ingredient by volatilization may be diminished by formulating some herbicides as granules.

Pellets (P). Pellets are dry formulations larger than granules that are used for "spot" applications to control isolated herbaceous or woody plants. They also are used to penetrate plant canopies or establish a grid pattern of herbicide application, a practice sometimes used with soil-active herbicides to remove undesirable vegetation from rangelands.

HERBICIDE CLASSIFICATION

There are approximately 150 herbicide active ingredients. Most of these basic forms of herbicides are further refined and formulated, creating several hundred commercial products. Because of the number of herbicides available, it is necessary to distinguish among them somehow. Herbicides most often are classified according to similarities in (1) *chemical structure*, (2) *use*, and (3) *effects on plants*. Herbicides also are classified according to *toxicity* or hazard level, as discussed earlier. Thorough discussions of herbicide classification are provided by Ross and Lembi (1985), Zimdahl (1993), and several other texts on weed control.

Classification Based on Chemical Structure

Classification systems based on chemical structure catalog herbicides by *chemical similarity*. This is the classification system used in the Weed Science Society of America (WSSA) *Herbicide Handbook* (WSSA, 1994), which provides a brief description of the various herbicides that are used in the United States. However, new herbicide development and registration are a continual process, so even that survey is likely to be both incomplete and outdated. The primary use, formulations, water solubility, and acute oral toxicity for each herbicide is usually provided by this method of classification. In addition, the principal manner in which the herbicides of each chemical group suppress plants is described.

HERBICIDE NOMENCLATURE

Every herbicide is named in three ways. Since herbicides are chemicals, each active ingredient has a *chemical name* to describe its chemical structure. The chemical constituents that make up the herbicide active ingredient can be determined and similarities to other chemicals can be found in this way. Herbicides also are commercial products. Therefore, each herbicide has a *trade name* given to it by its manufacturer that distinguishes it from other products and assists in its sale. However, some herbicides may be manufactured by several companies and each gives its product a different trade name. To avoid confusion, herbicides also are provided a *common name* by the Weed Science Society of America. This common name refers to all herbicide products that have the same active ingredient. The herbicides in the WSSA *Herbicide Handbook* are organized according to common name, although the chemical name of each herbicide is also provided.

Classification Based on Use

When herbicides are classified based on how they are used, they are first characterized as being either selective or nonselective (Figure 9.4). *Selective* herbicides are chemicals that suppress or kill certain weeds without significantly injuring an associated crop or other desirable plant species. Usually some weed species also are not injured by selective herbicides. *Nonselective* herbicides, in contrast, suppress whatever vegetation is treated and are used with the intention that none of the plants survive. Many herbicides occur in both categories, however, because differential phytotoxicity among plants (selectivity) is not an absolute characteristic of the chemical (McHenry and Norris, 1972). Rather, selectivity depends upon the rate of herbicide applied, method of application, and many other plant and environmental factors (Ashton and Monaco, 1991). Herbicide selectivity is an important principle of modern weed control and will be discussed in much greater depth later. A partial list of herbicides based on how they are used is provided in Table 9.3.

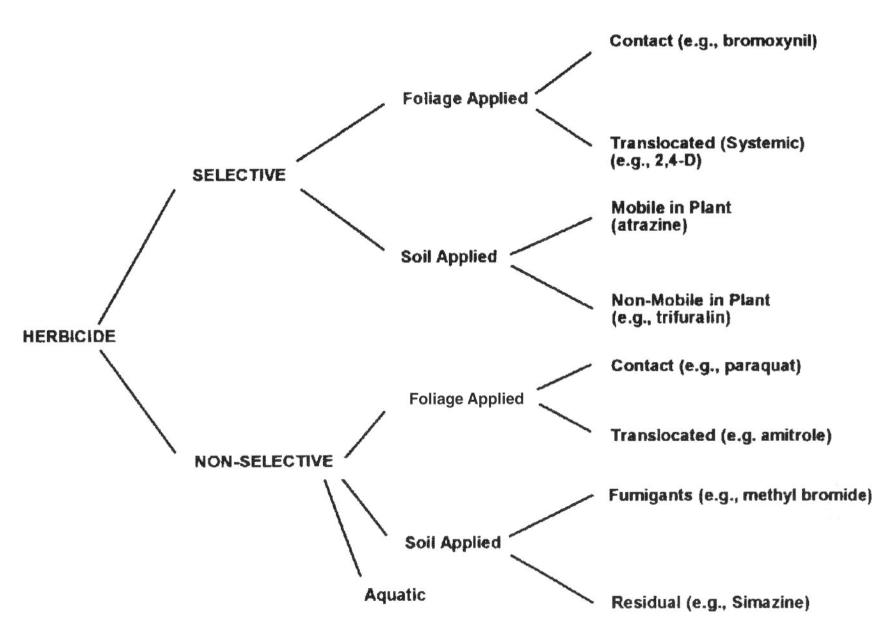

FIGURE 9.4 Herbicide classification based on use. (From McHenry and Norris, 1972.)

TABLE 9.3 A Partial List of Herbicides Based on How They Are Used

Common Name	*Trade Name*	*Common Name*	*Trade Name*
	Aquatic Herbicides		
acrolein	Acrolein	copper sulfate	several
copper chelate	Komeen	diquat	Diquat
copper chelate	Cutrine	endothall	Hydrothol, Aquathol
copper sulfate	K-lox		
	Contact Herbicides		
ammonium sulfamate	Ammate X	endothall	several
bentazon	Basagran	fluazifop	Fusilade
bifenox	Modown	oxyfluorfen	Goal
bromoxynil	Brominal, Buctril	paraquat	Paraquat, Gramoxone
cacodylic acid	several	phenmedipham	Betanal
chloroxuron	Tenoran	propanil	Stam
desmedipham	Betanex	sethoxydim	Poast
diclofop	Hoelon	difenzoquat	Avenge
dinoseb	several	sulfuric acid	several
diquat	Diquat, Rezlone	weed oils	several

(continued)

TABLE 9.3 Continued

Common Name	*Trade Name*	*Common Name*	*Trade Name*
	Foliar Translocated Herbicides		
amitrole	several	DSMA	several
asulam	Asulox	fosamine	Krenite
barban	Carbyne	glyphosate	Roundup, Rodeo
2,4-D	several	MCPA	several
2,4-DB	Butoxone, Butyrac	mecoprop	several
2,4,5-T	several	MSMA	several
dalapon	Dowpon	picloram	Tordon, Amdon
dicamba	Banvel	silvex	several
dichlorprop	several	triclopyr	Garlon
	Soil-applied Herbicides		
alachlor	Lasso	hexazinone	Velpar
ametryne	Evik	linuron	Lorox
atrazine	several	metham	Vapam
benefin	Balan	methyl bromide	several
bensulide	Betasan, Prefar	metolachlor	Dual
borax	several	metribuzin	Lexone, Sencor
bromacil	Hyvar X	molinate	Ordram
butylate	Sutan	napropamide	Devrinol
chloramben	Amiben	naptalam	Alanap
chloroxuron	Tenoran	nitralin	Planavin
chlorpropham	Furloe	norflurazon	Solicam, Zorial
chlorsulfuron	Glean	oryzalin	Surflan
cyanazine	Bladex	oxadiazon	Ronstar
cycloate	Ro-Neet	oxyfluorfen	Goal
DCPA	Dacthal	pebulate	Tillam
diallate	Avadex	pendimethalin	Prowl
dichlobenil	Casoron	prometon	Pramitol
diethatyl	Antor	prometryn	Caparol
diphenamid	Enide	pronamide	Kerb
diuron	several	propachlor	Ramrod
EPTC	Eptam	propazine	Milogard
ethalfluralin	Sonalan	propham	Chem Hoe
ethofumesate	Nortron	pyrazon	Pyramin
fenac	Fenac	siduron	Tupersan
fluometuron	Cotoran	simazine	several

TABLE 9.3 Continued

Common Name	*Trade Name*	*Common Name*	*Trade Name*
	Soil-applied Herbicides		
sodium chlorate	several	terbutryn	Igran
sulfmeturon-methyl	Oust	thiobencarb	Bolero
TCA	several	triallate	Far-go
tebuthiuron	Spike	trifluralin	Treflan
terbacil	Sinbar	vernolate	Vernam

Source: Gowgani et al., 1989.

SOIL-APPLIED HERBICIDES

These herbicides (Figure 9.4, Table 9.3) are applied before planting, before crop or weed emergence, or after the plants emerge. These times of herbicide application are referred to as *preplant, preemergence*, or *postemergence,* respectively. Soil applied herbicides must be moved into the soil profile by water or mechanical incorporation to be effective since they are usually taken up by plant roots, underground structures, or seed. Movement in soil is an important factor that influences herbicide persistence and fate and is discussed in Chapter 10. The phytological activity of soil-applied herbicides depends on the degree of inherent plant tolerance, the location of the herbicide in the soil, and depth of plant roots. Some soil-applied herbicides are applied as bands, either over or between crop rows, to enhance selectivity and decrease costs of application.

FOLIAGE-APPLIED HERBICIDES

These chemicals (Figure 9.4) injure plants when applied to the leaves or stems. Some herbicides, either by their rapid action or limited movement, injure only the portion of the plant actually touched or contacted by the chemical or spray solution and are called *contact* herbicides. Herbicides in this category are usually applied to foliage (McHenry and Norris, 1972; Table 9.3). Paraquat, bromoxynil, and dinoseb are examples of foliage-applied contact herbicides. In some cases, herbicides may be directed away from crops or applied in shields to minimize foliage exposure to these chemicals.

Some soil-applied and many foliage-applied herbicides move or translocate in treated plants. Thus, *translocated* (Table 9.3, Figure 9.4) or *systemic* herbicides move in the plant after application. Herbicides of this type often effectively suppress root, rhizome, or shoot growth at a considerable distance from the point of application, that is, either the soil (roots) or the foliage. Herbicide uptake, translocation, and mechanism of action are considered in Chapter 10.

SOIL RESIDUAL HERBICIDES

These herbicides are chemicals applied to the soil that may be used selectively at some rates and in certain conditions, but that at higher rates suppress plant growth for several months to years (e.g. some soil-applied herbicides in Table 9.3). These herbicides were called soil sterilants in the past. This nomenclature is discouraged now since the herbicides do not "sterilize" the soil, but kill plant seedlings for a prolonged period of time. Most of the herbicides in this category translocate to some degree in germinating seedlings.

SOIL FUMIGANTS

Herbicides that are always nonselective are the soil-applied fumigants and certain chemicals used for aquatic weed control. Soil fumigants (Figure 9.4) are gasses applied to the soil that kill all vegetative plant growth. These herbicides are usually applied prior to crop planting and weed emergence. Small germinating weeds are most susceptible to the treatment, while dormant seed often tolerate the chemicals. The length of time soil remains weed-free following soil fumigation depends upon

- chemical used and amount applied
- soil type
- soil water status
- extent of weed dissemination from adjacent areas

Soil fumigation is usually expensive and is used only to prepare the soil for high-value crops. Also, due to the relatively high toxicity level or adverse environmental effects of these chemicals, many are being re-examined for possible cancellation of registration. For example, the use of methyl bromide, a commonly used soil fumigant, is being discontinued because of its adverse impact on atmospheric ozone.

AQUATIC HERBICIDES

Most herbicides used for aquatic weed control are considered to be nonselective (Figure 9.4) in the aquatic environment. These materials are applied either directly to the water or to the soil in canals, ponds, and lakes (Table 9.3).

Classification According to Biological Effect

Grouping herbicides according to the way they kill or suppress plants is another method of classification. Herbicides are identified as

- plant hormone regulators
- cell division inhibitors
- photosynthesis inhibitors

- cell membrane disrupters
- inhibitors of general cell metabolism

This method of classification requires that the cause of plant injury for specific herbicides be known. Although knowledge about specific biochemical changes that result from herbicide use is sometimes incomplete, enough information is usually available about specific herbicides to place them into broad categories of cellular dysfunction (Chapter 10) and symptomology (Ross and Lembi, 1985; Zimdahl, 1993).

HERBICIDE SYMPTOMS AND SELECTIVITY

Since herbicides alter the ability of plants to grow, various structural features of plants change following exposure to these chemicals. These visible changes in plant structure or morphology are the *symptoms* of herbicide effects. Herbicides are often selective or applied in a selective manner, meaning that some plants are injured by the chemical while others are not. Herbicide selectivity is the foundation of chemical weed control in most crops and it is the basis for most herbicide uses in agriculture.

Symptoms of Herbicide Injury in Plants

The symptoms of herbicide exposure include abnormal cell development and division, epinasty (twisting), leaf chlorosis and necrosis, albinism, altered geotropic and phototropic responses, and reduced formation of cuticle and waxes.

ABNORMAL TISSUES AND TWISTED PLANTS

Epinasty is the bending or twisting of stems and leaves. This symptom is most characteristic of herbicides such as 2,4-D, dicamba, and picloram that influence hormonal regulation in plants (Figure 9.5) (Ashton and Crafts, 1981). Increased tillering of shoots or callus formation on roots is sometimes a response to low rates of these herbicides. Other "formative" symptoms occur during development and include thickened coleoptiles or leaves, multiple shoot formation at internodes, reduced internode length, and abnormal seedling development. Herbicides of the chlorinated aliphatic acid, chloroacetamide, dinitroaniline, nitrile, and thiocarbamate groups cause such formative symptoms in treated plants (Ashton and Crafts, 1981, Table 9.2). Epinasty and formative symptoms are caused by abnormal cell division, cell enlargement, and tissue differentiation as a result of herbicide exposure.

FIGURE 9.5 Epinasty occuring in a mustard plant as a result of herbicide treatment. (Photograph by A.P. Appley, Oregon State University.)

DISRUPTION OF CELL DIVISION

The process of cell multiplication (mitosis) is inhibited by many herbicides, including chemicals belonging to the carbamate, thiocarbamate, and dinitroaniline groups (Ashton and Crafts, 1981; Zimdahl, 1993; Table 9.2). The symptoms of these herbicides range from suppressed root or shoot development in whole plants (Figure 9.6) to aberrant and multinucleate cells. Although the general symptoms of mitotic disruption are similar for a large number of herbicides, detailed microscopic examinations have shown important differences in the mechanism of action (Chapter 10) among specific chemicals. The primary site of toxic action of mitotic inhibitors is located in meristematic regions of plants, such as forming root tips or buds.

LEAF CHLOROSIS, NECROSIS, AND ALBINISM

Leaf *chlorosis* is a common symptom of herbicides that inhibit photosynthesis (Ashton and Crafts, 1981; Zimdahl, 1993). Chlorosis is the bleached yellow appearance of leaves following the degradation of chlorophyll in treated plants. Some herbicides cause chlorosis along the leaf veins, while chlorosis between veins is the symptom of other herbicides (Figure 9.7). *Necrosis* is the most advanced or extreme case of chlorosis. It often takes several days to weeks for chlorosis and finally necrosis to develop following treatment. Herbicides that produce these symptoms in plants are usually members of the urea, uracil, and triazine groups (Table 9.2).

FIGURE 9.6 Symptoms of mitotic disruption in (*top*) pigweed and (*bottom*) a grass. (Photographs by C.L. Elmore, University of California, Davis, and A.P. Appleby, Oregon State University.)

FIGURE 9.7 Inter-veinal (*top*) and marginal (*bottom*) chlorosis. (Photographs by C.L. Elmore, University of California, Davis, and A.P. Appleby, Oregon State University, respectively.)

Other herbicides, such as glyphosate, DSMA, MSMA, dalapon, and some phenoxy-type herbicides, can cause chlorosis. However, their symptoms are caused by a lack of chlorophyll, which makes other leaf pigments more obvious following herbicide application. *Albinism* results when new foliage is devoid of chlorophyll (Figure 9.8). Amitrole is a herbicide that causes this symptom of herbicide exposure. A few herbicides kill plant foliage so rapidly that only necrosis results from herbicide treatment (Figure 9.9). A general disruption and dysfunction of cell membranes is responsible for such rapid and dramatic symptoms. Contact-type herbicides, such as paraquat, diquat, bromoxynil, dinoseb, endothall, and nitrofen, are responsible for this type of herbicide injury (Ashton and Crafts, 1981; Table 9.3).

ALTERED GEOTROPIC AND PHOTOTROPIC RESPONSES

Geotropism is the ability of plants to orient and grow downward in response to gravity. The ability of plants to grow toward light is *phototropism*. Naptalam and 2,3,6-TBA have been reported to alter these responses (Ashton and Crafts, 1981).

REDUCED LEAF WAXES

Cuticular and epicuticular waxes are complex structures that cover the foliage of plants. The primary role of these structures is to restrict water loss from the plant. Herbicides in the thiocarbamate and aliphatic acid groups inhibit epicuticular wax formation, in addition to causing other morphological symptoms already mentioned.

FIGURE 9.8 Albinism as a result of amitrole application. (Photograph by A.P. Appleby, Oregon State University.)

FIGURE 9.9 Severe necrosis in foliage following herbicide application. (Photograph by A.P. Appleby, Oregon State University.)

Herbicide Selectivity

Selectivity usually depends on the degree of plant tolerance to the herbicide. Some plants are inherently tolerant to certain chemicals, while others have evolved herbicide resistance after repeated exposure to a chemical. Tolerant and resistant plants usually degrade or metabolize the chemical to nonphytotoxic substances. In some cases of resistance, the herbicide does not affect the site of toxic action in treated plants. Although tolerance and resistance are common, herbicide selectivity among plants is often conditional; that is, it depends on the rate and timing of herbicide application, placement of the herbicide relative to location and stature of crop and weeds, and numerous other plant and environmental factors that influence herbicide performance. Some of the factors that influence herbicide selectivity are

- plant tolerance to the herbicide
- herbicide rate (dosage)
- time of application
- stage of weed and crop development
- weather patterns
- variation in microenvironment or microtopography
- variation in resource level
- soil type and pH

Weed scientists have made substantial efforts to understand and take advantage of the selectivity patterns of herbicides. Many of the principles of how herbicides are used to attain selective chemical weed control have been reviewed by others (Ashton and Harvey, 1971; Ashton and Monaco, 1991; Ross and Lembi, 1985). These principles involve the role of plant morphology and physiology, chemical properties, and environmental factors in the differential susceptibility of plants to herbicides.

PLANT FACTORS

Plant factors that influence the way weeds and crops respond to herbicides are *genetic inheritance, age, growth rate, morphology*, and *physiological and biochemical processes*. The most effective use of herbicides results when these factors are considered.

Genetic inheritance. Plant species within a genus often respond to herbicides in a similar manner. However, responses to herbicides by plants in different genera often vary. The reason is that plants with similar taxonomic traits often have similar genetic and enzymatic components. Thus, crops and weeds that belong to the same genera usually are susceptible to the same herbicides. For example, herbicides that do not injure tomato also fail to control nightshade weed species because the crop and weeds are members of the same taxonomic family and have a similar biochemical make-up. This rule-of-thumb is not general, however, because varieties of many crops are known to respond differentially to herbicides.

Plant age and growth rate. Young plants or weed seedlings are usually killed more easily than large mature vegetation. In addition, some preemergence herbicides that suppress seed germination often are not effective when used to control larger, better-established plants. Plants that grow rapidly generally are more susceptible to herbicides than are plants that grow slowly.

Morphology. The morphology or growth habit of plants can determine the degree of sensitivity to some herbicides. Morphological differences in root structure, location of growing points, and leaf properties between crops and weeds can determine the selectivity pattern of some herbicides. Annual weeds in a perennial crop usually can be controlled by herbicides because of their differences in root distribution and structure (Figure 9.10). For example, perennial crops such as alfalfa can recover from moderate herbicide injury to foliage, whereas annual weeds, because of their size and shallow root system, will be killed by the same herbicide application (Ashton and Harvey, 1971).

The meristematic regions of most grasses, such as cereal crops and grass weeds, are located at the base of the plant or even below the soil surface (Figure 9.11). The growing points are protected from herbicide exposure by the foliage

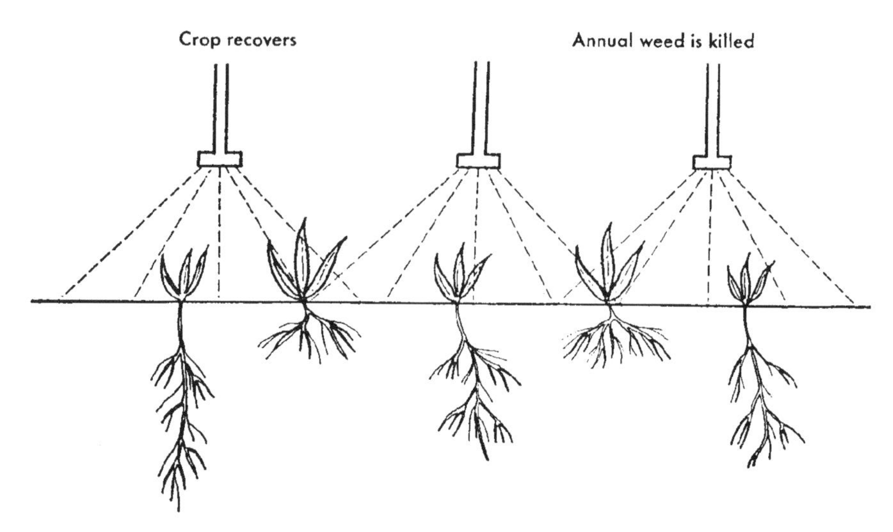

FIGURE 9.10 Selectivity based on difference in root size or structure. Plant roots 1,3,5 (*left to right*) are deep and extensive and therefore recover from surface sprays of herbicide. Weed roots 2 and 4 are close to the soil surface and easily killed by such sprays. (From Ashton and Harvey, 1971.)

or soil that surrounds them. Thus, herbicide that contacts only foliage may injure some leaves but will not influence severely the ability of the plant to grow. In contrast, most broadleaved plants have their growing points exposed at shoot tips and leaf axils (Figure 9.11). For this reason, these plants are more susceptible than grasses to foliage-applied herbicides, especially contact herbicides.

Leaf properties of some plants can impart insensitivity or tolerance to certain herbicides, while other plants are controlled effectively. Spray droplets do not adhere well to the surfaces of narrow, upright, waxy leaves that characterize many monocot crops such as cereals and onion (Figure 9.12). Thus, spray droplets do not adequately cover such leaves following herbicide application. The effect of the herbicide is therefore reduced. In contrast, broadleaved plants have wide leaves that are usually horizontal to the main stem. Leaves of broadleaved plants intercept more spray than leaves of grasses and spray droplets spread more evenly over broadleaved foliage (Figure 9.12). Herbicide effectiveness is best when spray interception and coverage are greatest.

Physiological and biochemical processes. The physiology of a plant influences the ability of a herbicide to enter it following application. This process is called *absorption*. The extent of herbicide movement in a plant (*translocation*) after it has been absorbed is also a physiological process. Both absorption and translocation are important processes governing herbicide activity, and are considered in much more detail in Chapter 10. Herbicide absorption and

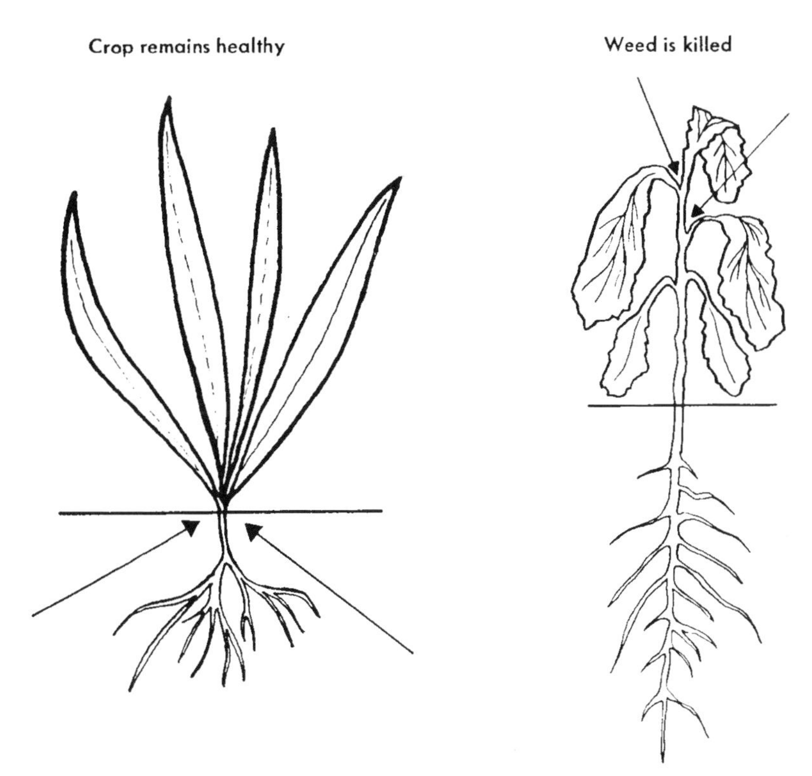

FIGURE 9.11 Selectivity based on location of growing points. *Left,* growing points (arrows) are protected from sprays. *Right,* growing points are exposed. (From Ashton and Harvey, 1971.)

translocation vary markedly among plant species. Generally, species of plants that absorb and translocate herbicides most readily are killed by them.

Biochemical and biophysical processes are also important plant factors determining herbicide selectivity. *Adsorption* can be responsible for differential herbicide susceptibility among plant species. During this process, a herbicide is bound so tightly by cellular constituents (usually cell walls) that it cannot be translocated readily and is therefore inactivated (Figure 9.13). Membrane stability is another biochemical/biophysical process that results in herbicide selectivity among plants. In this case, the cell membranes of tolerant plants can withstand the disruptive action of the herbicide (Figure 9.14). The ability of carrot to withstand the toxicity of certain oils is an example of this form of herbicide selectivity.

Enzyme inactivation, herbicide activation, and herbicide inactivation are biochemical processes that can occur in plants in response to herbicide treatment. Since all plants cannot perform these processes equally well, they also form the basis for herbicide selectivity. *Enzyme inactivation* occurs when a her-

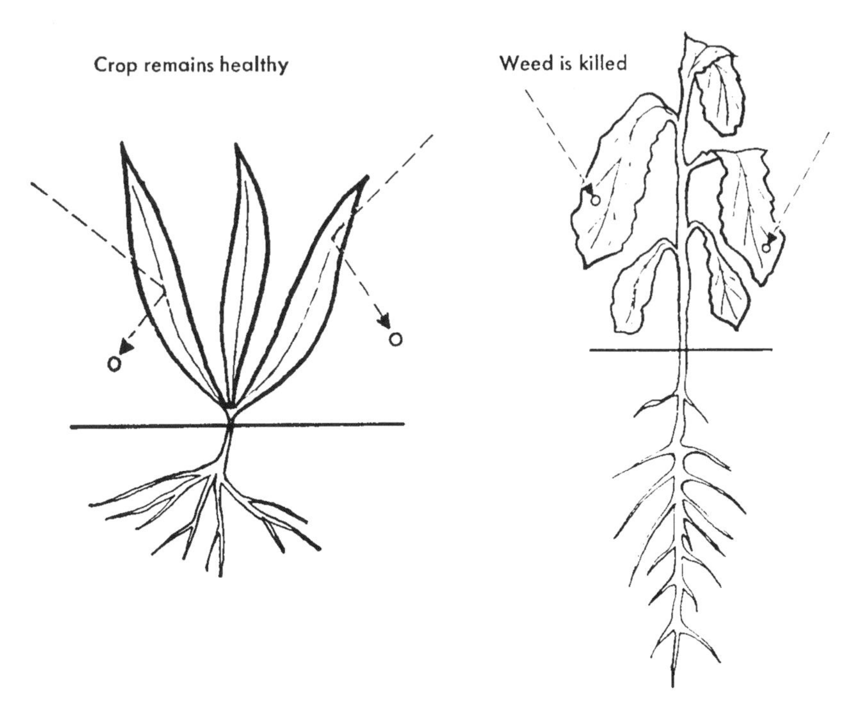

FIGURE 9.12 Selectivity based on leaf properties. *Left*, on narrow, upright leaves spray bounces off and does not affect the plant. *Right*, on wide, horizontal leaves spray sticks and kills the plant. (From Ashton and Harvey, 1971.)

bicide reduces the activity of a particular enzyme in a plant species (Figure 9.15). Thus, some plants are killed, but others are not. *Herbicide activation* results when a nontoxic chemical is transformed in a plant into a herbicide. An example of this process is the transformation of 2,4-DB into 2,4-D in susceptible plants, but not in tolerant ones (Chapter 10). *Herbicide inactivation* occurs

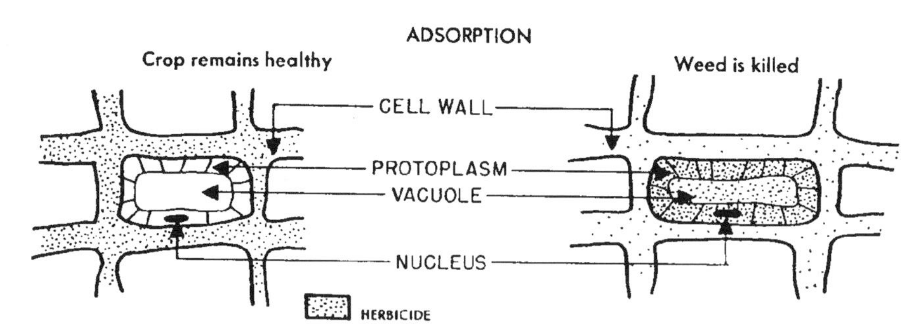

FIGURE 9.13 Selectivity based on physiological binding (adsorption in plants). *Left*, herbicide is adsorbed by cell walls and is prevented from reaching the cytoplasm of a treated plant. *Right*, herbicide is not adsorbed by the cell wall and reaches the cytoplasm. (From Ashton and Harvey, 1971.)

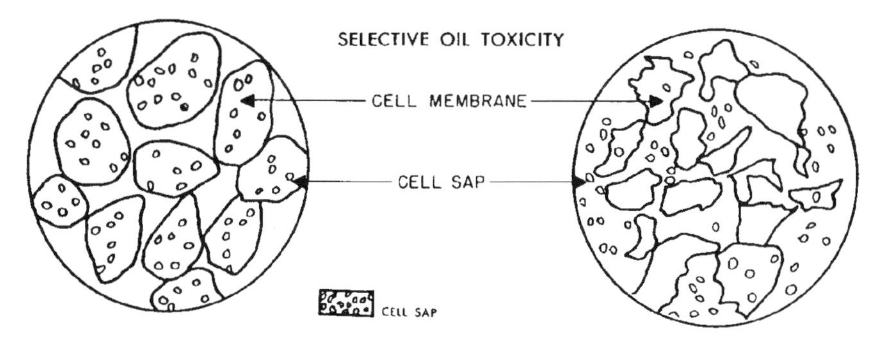

FIGURE 9.14 Selectivity based on membrane integrity. *Left,* carrot cell membranes are resistant to selective oil. Membranes stay intact, keeping cell sap inside. *Right,* weed cell membranes are damaged by selective oil, allowing cell sap to leak into intercellular spaces. (From Ashton and Harvey, 1971.)

when a herbicide is degraded in a treated plant to nontoxic materials. As we will see in Chapter 10, there are many examples of this form of differential sensitivity of plants to herbicides.

CHEMICAL FACTORS

The *structure and form* of the herbicides themselves can influence the tolerance of plants to them. Even small changes in molecular configuration of a herbicide can modify its chemical properties and also its effects on plants. Differences between two herbicides, benefin and trifluralin, offer an example of this type of herbicide selectivity. The only difference between the two herbicides is that a ($-CH_2-$) group is moved from one side of the molecule to the other (Ashton and Monaco, 1991). However, benefin will control many weeds without harming lettuce, while trifluralin kills lettuce even at low rates.

The *formulation* of a herbicide is also an important consideration for herbicide selectivity. Herbicides are formulated in a number of different ways to improve transportation, storage, application, or marketing. Herbicide formula-

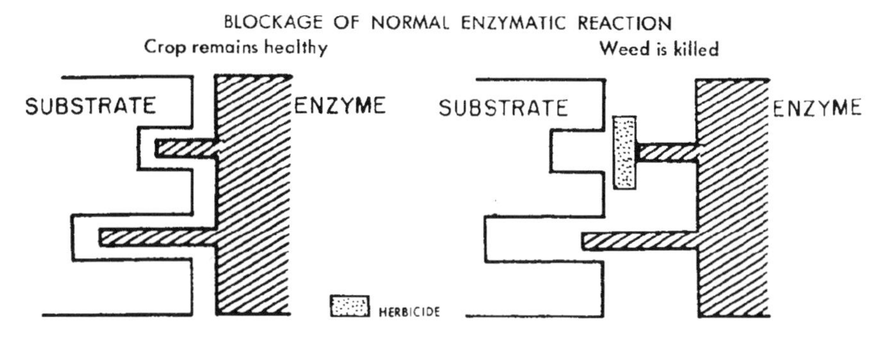

FIGURE 9.15 Selectivity based on enzyme inactivation. *Left,* herbicide does not interfere with enzyme reaction and metabolism. *Right,* herbicide alters the structure and attachment of the enzyme and upsets metabolic processes. (From Ashton and Harvey, 1971.)

tion also may enhance herbicide selectivity by increasing toxicity in susceptible plants or decreasing activity in tolerant ones. The use of granular formulations, emulsifiers, or surfactants are examples of herbicide formulations used to improve selectivity.

A herbicide may also be directed away from susceptible plants, such as a crop, which imparts a type of selectivity. The use of *shields*, *wipers*, and *directed sprays* (Figure 9.16) are examples of this positional type of herbicide selectivity (Ashton and Harvey, 1971; Ashton and Monaco, 1991).

ENVIRONMENTAL FACTORS

Factors of the environment that influence herbicide selectivity are *soil type, rainfall and irrigation patterns*, and *temperature* (Ashton and Harvey, 1971). Soil type and the amount of precipitation or irrigation determine the location of herbicides in the soil profile. In general, herbicides tend to move more readily in sandy soil than in clay and in wet soil than in dry (Chapter 10). Temperature and soil moisture also determine the rate of herbicide degradation in the soil and the rate of plant growth. Warmer temperatures and greater soil moisture generally promote more rapid microbial and chemical degradation of herbicides than do cooler, drier conditions. Warm temperatures and moist soils also promote more rapid plant growth and thus more rapid onset of herbicidal injury.

Some soil-applied herbicides that are not biochemically selective in crops may be functionally selective by their placement in the soil profile (Figure 9.17). This type of selectivity requires differential rooting habits between the crop and weeds and an understanding of the factors that influence vertical herbicide movement in the soil (Ashton and Monaco, 1991). The placement of herbicides in the soil relative to the roots of crops and weeds is an important principle of herbicide selectivity. The factors that influence herbicide movement in soil are considered later (pp. 438–443) and in Chapter 10.

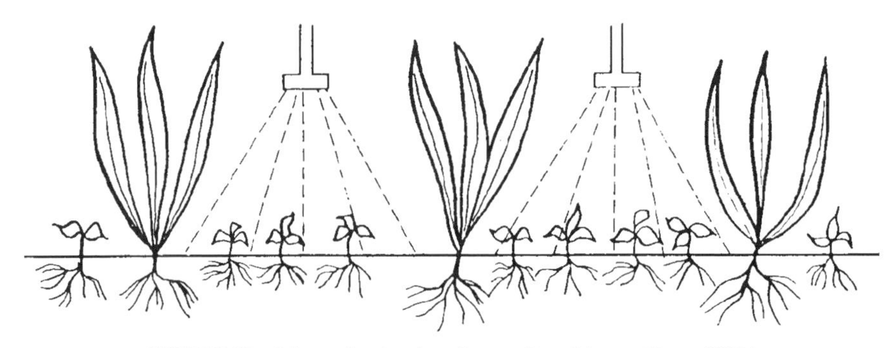

FIGURE 9.16 Selectivity based on directed sprays. (From: Ashton and Harvey, 1971.)

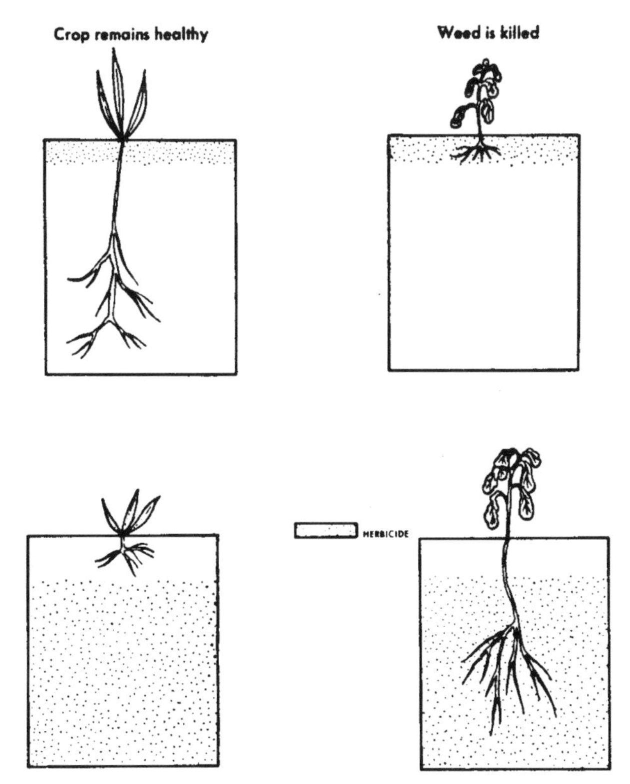

FIGURE 9.17 Selectivity based on herbicide placement in the soil. *Top,* deep-rooted crop (*left*) is not affected by herbicide that remains near the soil surface. Shallow-rooted weed is killed by herbicide that stays near the surface. *Bottom,* shallow-rooted crop (*left*) remains alive if the herbicide moves below its rooting zone. Deep-rooted plants are killed when the herbicide is leached into the deeper zones of the soil. (From Ashton and Harvey, 1971.)

HERBICIDE APPLICATION

As discussed thoroughly by Ross and Lembi (1985) and others, herbicides must be applied accurately and uniformly to an area of land or foliage to be effective, because too much of the chemical may damage crop plants, while too little will not provide acceptable weed control. It is also necessary for the herbicide to arrive at the targeted area and not be displaced by drift, volatility, leaching, or runoff (Chapter 10). Damage to susceptible crops or undesirable—even toxic—residues in food, feed, water, or soil may result if excessive herbicide displacement occurs. Improved accuracy can be achieved by proper calibration and operation of herbicide application equipment (Gebhart and McWhorter, 1987).

Both ground and aerial applications of herbicides are used in agriculture, forestry, and range management. Frequently, ground applications are made by tractor-drawn sprayers, but hand applications of herbicides are also common in some locations and weed control situations. Aircraft are also used to apply herbicides, but special precautions and equipment are necessary when herbicides are applied by aircraft.

Proper Herbicide Rate

Herbicides are used within a specified range of dosages or rates. The rate of herbicide usually is expressed as the amount of chemical per unit of ground area to be covered. Common units of herbicide rates are pounds per acre or kilograms per hectare. In addition, the rate of herbicide applied usually is expressed in terms of chemical active ingredient as well as the amount of formulated product.

The reason to calibrate herbicide application equipment is to assure that the chemical needed for optimal weed control is spread uniformly over the specified area. The procedure for calibration is similar for both ground and aircraft applications (see Anderson, 1977, 1996; Ross and Lembi, 1985; or Zimdahl, 1993 for specific examples). However, some special calibration techniques are necessary for aircraft because of the speed and extent of area covered by this kind of equipment.

Proper Herbicide Distribution

The uniformity obtained from a herbicide application depends on several factors, including topography of the land, type and quality of equipment, skill of the operator, and certain weather conditions, especially wind and temperature. Unfortunately, it is often impossible to determine the degree of uniformity until after a herbicide has been applied. Strips of injured crop plants or uninjured weeds in a treated field indicate poor uniformity of application. If poor application is suspected, both the equipment and its operation should be examined to determine where improvements can be made.

Herbicide applicators should avoid overlapping spray swaths or slowing the speed of application. Both of these situations increase the amount of herbicide applied. Turning a sprayer in the field can cause an application problem because the operator may reduce speed or turn while the sprayer is in operation. Turning while the sprayer is in operation causes a high dose of herbicide to be applied on the inside of each turn and less at the outside turning edge. It is best to apply one or two swath widths around the perimeter of the field being treated to avoid this source of uneven herbicide application.

Herbicides are always applied as a mixture with some other material, such as water, oil, or dry diluent. The *carrier* is used to dilute and disperse the herbicide over the field.

LIQUID CARRIERS

The most widely used liquid carrier for herbicide applications is water. Water is the preferred carrier for spray systems because it is abundant and inexpensive (Ross and Lembi, 1985). The volume of the carrier in which a herbicide is applied can influence the uniformity achieved. With herbicides applied as liquids, the water volume needed for proper application usually is determined by the chemical characteristics of the herbicide. However, the density of the vegetative canopy being treated can also influence the uniformity of application and control obtained from some foliage-active herbicides. The amount of coverage (uniformity) generally increases with increased carrier (water) volume, especially when the vegetation is very dense. Nonetheless, some herbicides require less complete coverage than others to be effective and may even perform best at low carrier volume. The volume of carrier usually is less important when herbicide applications are made to the soil surface, rather than to foliage. Most applications are made within the range of 5–50 gallons of herbicide-water mix per acre.

Other types of liquid carriers include oils and liquid fertilizers. Diesel oil is the most common oil carrier. It is often mixed with 2,4-D or other herbicides to enhance penetration into the bark or waxy foliage of plants. Except for dormant applications to the bark of trees, however, it is rare to use diesel oil as the only carrier in the spray mixture. For example, 2,4-D usually is mixed with oil at a ratio of 1 part oil to 10 parts water. A unique manner in which diesel oil, some herbicides, and water are mixed is as an *invert emulsion*. This type of carrier becomes very thick and mayonnaise-like. Penetration into bark and leaves is sometimes improved by this procedure. However, drift control is the main reason for its use.

In some instances, nonphytotoxic or phytobland oils are used to enhance effectiveness of postemergence herbicides. These oils have a saturated ring structure or are long-chain hydrocarbons. The nonphytotoxic oils facilitate entry of the herbicide into the leaf tissue of weeds. Because foliage is nonpolar and relatively water repellent (Chapter 10), the oil coats the leaves and allows greater retention of the herbicide on the foliage. Nonphytotoxic oils have been used as carriers for atrazine applications for weed control in corn and sorghum.

SOLID CARRIERS

Solid carriers include certain types of clays, vermiculite, plant residues, starch polymers, and some types of dry fertilizers. These materials are usually mixed with the herbicide during formulation and both are applied directly to the field. Further dilution during application is not required. The choice of solid carrier often is determined by the herbicide manufacturer.

The formulation of herbicides with dry fertilizers represents an interesting form of carrier system. Often the herbicide is impregnated into or coated over the fertilizer. This meth[illegible] been used for many years in the "weed and feed" lawn prod[illegible]ts. Redu[illegible] application result from this practice, since both weed control and fertility can be accomplished with a single treatment. Only certain types of fertilizers can be used effectively with herbicides. Coated nitrate and urea are not used for this purpose. Best coverage of the soil surface with dry herbicide formulations, such as granules and pellets, is achieved with high volumes of the dry carrier.

GASEOUS CARRIERS

Air is the most common form of gaseous carrier used for herbicide applications. However, air usually is not mixed with the herbicide, but is used as a propellant for a herbicide-liquid carrier system. Forced air is used to apply herbicide-water mixtures as small drops or a mist. These mist blowers deliver the herbicide solution or suspension in a stream of air to the plants being treated. Often low volumes of the herbicide-water mixture can be applied in this way, resulting in low application costs.

Compressed air, carbon dioxide, and nitrogen also have been used as delivery systems for herbicide-water mixtures. However, few of these systems have been developed for commercial use. Forced air driven equipment has been developed for the application of granular and pellet formulations of herbicides.

ADJUVANTS

An adjuvant is a material that is mixed with a spray solution or suspension to improve the performance, handling, or application of herbicides. Adjuvants are inert chemicals that are classified according to their use, rather than chemical or physical properties (Ross and Lembi, 1985). For many herbicide products, the adjuvants are formulated with the active ingredient at the time of manufacture (see pp. 404–407). At other times, it is desirable to add a specific material, such as a surfactant, to improve or enhance the performance of the formulated product. Caution should be used when selecting additional adjuvants for use with herbicide products because a loss of selectivity can result.

Terms often used to describe adjuvants include *activator, additive, dispersing agent, emulsifier, spreader, sticker, surfactant, thickener*, and *wetting agent*. Each of these terms has a specific meaning:

Activator. Material used to enhance the activity of a herbicide, often by altering the pH of the spray mixture

Additive. Material added to a spray solution or suspension, that is, another term for adjuvant

Dispersing agent. Substance that enhances the dispersal of a powder in a solid-liquid suspension

Emulsifier. Material that aids in the suspension of fine drops of one liquid in another, for example, oil in water

Spreader. Substance added to spray solutions or suspensions that allows more thorough contact between spray droplets and a leaf surface (same as a wetting agent)

Sticker. Substance used to increase the amount of spray deposit remaining on a plant surface following application

Surfactant. Material that accentuates the emulsifying, spreading, or wetting properties of a spray solution at a surface or interface, that is, a surface-active agent

Thickener. Material used to reduce the number of fine droplets produced by nozzles during herbicide application

Wetting agent. Same as spreader

Surfactants are a special class of adjuvant that is used to enhance herbicide effectiveness. They usually are mixed with aqueous solutions or suspensions of herbicides for postemergence weed control. The term *surfactant* is a contraction for surface-active-agent since these materials cause a reduction of surface (gas-liquid) and interfacial (liquid-liquid and solid-liquid) tensions between substances. How surfactants enhance herbicidal action is not understood completely. However, herbicide entry into treated plants is aided by surfactants and phytotoxic responses to many foliage active herbicides are increased by them.

The surface active nature of surfactants means they can orient at interfaces as a result of their chemical structure. Surfactants have both lipophilic (oil soluble) and hydrophilic (water soluble) portions on the same molecule. Surfactant molecules are similar to emulsifiers (which is a type of surfactant) in this respect (Figure 9.3). After application, surfactant molecules orient themselves between the leaf and spray droplet, modify the interfacial tension between them, and provide more intimate contact of the spray droplet with the leaf surface. However, the improved wetting of leaf surfaces does not correspond completely with increased phytotoxicity of herbicide-surfactant-water mixtures. For example, maximum leaf wetting occurs in the range of 0.01 to 0.1 percent surfactant in the spray mixture, while maximum phytotoxicity usually results in the surfactant concentration range of 0.2 to 0.5 percent. Thus, much more surfactant is needed to enhance the phytotoxicity of herbicides than simply to improve leaf wetting by the spray mixture (McHenry and Norris, 1972).

Surfactants are classified according to their electrical charge, or tendency to ionize at the hydrophilic portion of the molecule. They are classified broadly as *anionic* (negatively charged), *cationic* (positively charged), or *nonionic* (neutral). Most surfactants used with herbicide emulsions and suspensions are nonionic because they are relatively unaffected by hard water and are compatible with many types of herbicides. Anionic and cationic surfactants are used much less than nonionic surfactants because the reactive nature of their electrical charge sometimes causes them to bind to other molecules and precipitate from the spray solution or suspension. Surfactants generally are believed to intensify the activity of herbicides by

- creating uniform spreading or wetting on leaf surfaces
- increasing spray droplet retention
- improving spray droplet and leaf surface contact

- solubilizing nonpolar plant substances
- causing enzymatic denaturation or membrane dysfunction

Herbicide Application Equipment

SPRAYERS

The components of herbicide sprayers are similar for both ground and aircraft applications. Common features of a herbicide sprayer are: *tank, agitation device, pump, pressure regulator, hoses*, and *nozzles* (see Ross and Lembi, 1985; Ashton and Monaco, 1991 for details). These basic components and their arrangement are shown schematically in Figure 9.18. Ideally, the components of a sprayer should be resistant to rust, abrasion, and corrosion. The sprayer usually is attached directly to a tractor or mounted on a trailer that is pulled.

Tank. Sprayer tanks are made from a variety of materials, for example, stainless steel, fiberglass, or reinforced plastic. The size of the tank depends on the number of nozzles, size of field or area to be sprayed, volume per area of spray

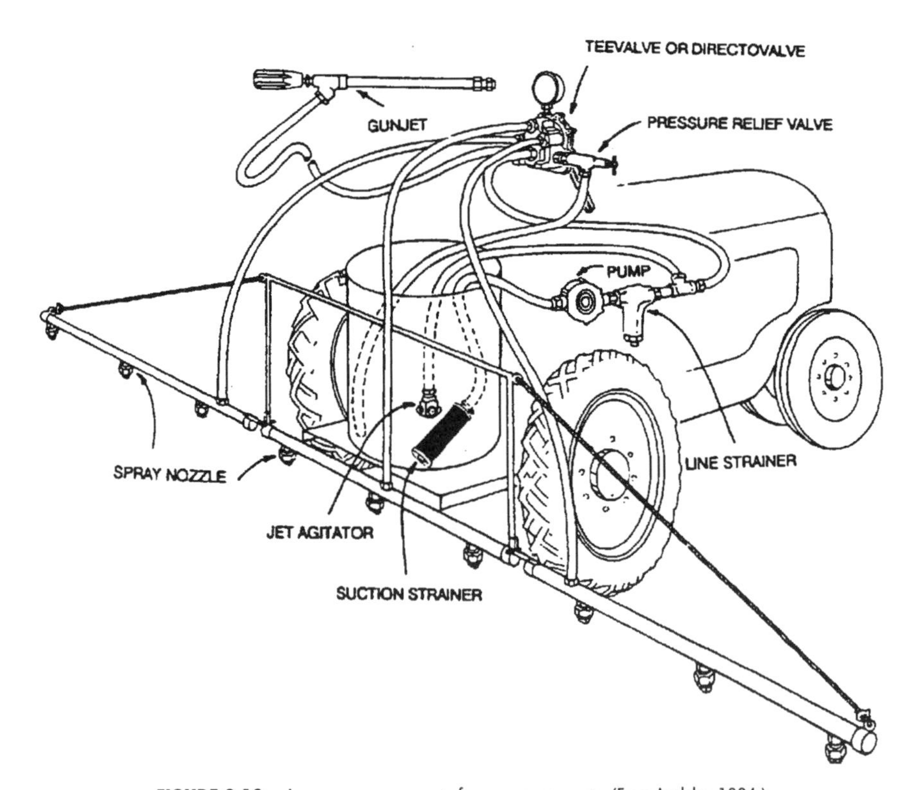

FIGURE 9.18 A common arrangement of sprayer components. (From Appleby, 1994.)

mixture applied, and type of carriage system (i.e., ground, aircraft, etc.) used. Appleby (1994) indicates that the ideal tank for a sprayer should be

- resistant to rust and corrosion
- formed to allow proper agitation of the spray liquid
- strong enough to withstand rough use
- relatively lightweight
- fitted with splash-proof filler opening and removable coarse screen filter
- easily drained and cleaned
- low cost

Agitation devices. Thorough and continuous agitation of the spray mixture is necessary for uniform herbicide application. If the herbicide is not agitated sufficiently in the spray tank, it may settle out (precipitate), causing less herbicide to be applied to the field. Two types of agitation devices often are used in sprayers: mechanical and hydraulic. The type and amount of agitation required depends on the shape of the tank and formulation of the herbicide used. Usually mechanical agitation by paddles on a shaft is preferred when the tank bottom is not round. Herbicides formulated as emulsions usually can be agitated satisfactorily by hydraulic means, but mechanical agitation is necessary for proper mixing of most wettable powder formulations.

Pumps. All sprayers must have a pump to force the spray mixture through the sprayer system and out the nozzles. The size of pump is determined by the volume of liquid moved through the sprayer system, volume per area output desired, and speed of operation. There are several types of pumps used for ground applications of herbicides. These are the piston, centrifugal, gear, roller-impeller, and diaphragm pumps (Ross and Lembi, 1985; Appleby, 1994). Other types of pumps are used for manual sprayers and aircraft application equipment, which are discussed in later sections.

Pressure regulator and pressure gauges. A critical component of all sprayers is a pressure regulator. It provides the following functions (Appleby, 1994):

- controls the sprayer system pressure and spray output through the nozzles
- protects hoses, seals and other sprayer components from excess pressure
- routes excess spray liquid into a bypass line and back into the tank

Hoses, line filters, and cut-off valves. Hoses connect the various components of the sprayer. They should be able to withstand the pressure or vacuum generated by the pump, and be non-corrosive to the chemicals used in the sprayer. Hose size is determined by the maximum pressure at which the sprayer is expected to operate. Most sprayers contain a line filter or strainer to remove debris from the spray mixture that might damage the pressure regulator or clog nozzles. A cut-off valve allows the operator to control the liquid flow through the sprayer and various sections of the sprayer boom.

Nozzles. A nozzle is an atomizing device used to create and disperse droplets in a specific orientation, forming a spray pattern. A complete nozzle assembly consists of a body, screen, and tip or orifice plate (Figure 9.19). Since there are a variety of conditions for spraying herbicides, nozzles are manufactured with a wide array of tips that vary in orifice size and angle (Appleby, 1994). The manufacturers of sprayer nozzles supply information concerning the volume of output and angle for each tip. Nozzles are made from many different substances, such as brass, aluminum, stainless steel, and so on.

Sprayers may have only a single nozzle, as for some band applications. It is more usual, however, for several nozzles to be arranged systematically in a horizontal direction along a metal bar to form a *boom*. Perhaps the most common type of spray nozzle is that emitting a *fan* pattern. When viewed from the side, the spray pattern from such a nozzle is flat because the orifice shape is elliptical. The pattern also tapers, so that less herbicide is dispersed at its edges. To compensate for this effect, the nozzles are arranged so that the spray patterns of adjacent nozzles overlap by 30–50 percent (Figure 9.20). Usually, the nozzles are spaced about 18–24 inches apart and the angle of each nozzle is between 65–80 degrees. Nozzle height (i.e., height of the boom) is critical to attain proper overlap of the spray patterns.

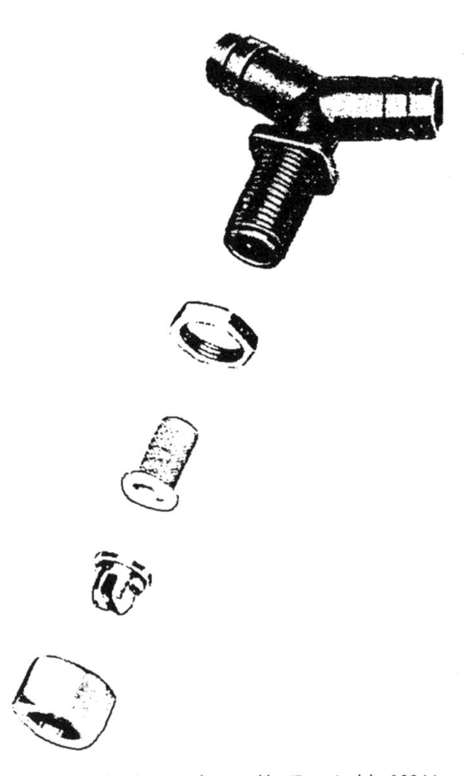

FIGURE 9.19 Nozzle assembly. (From Appleby,1994.)

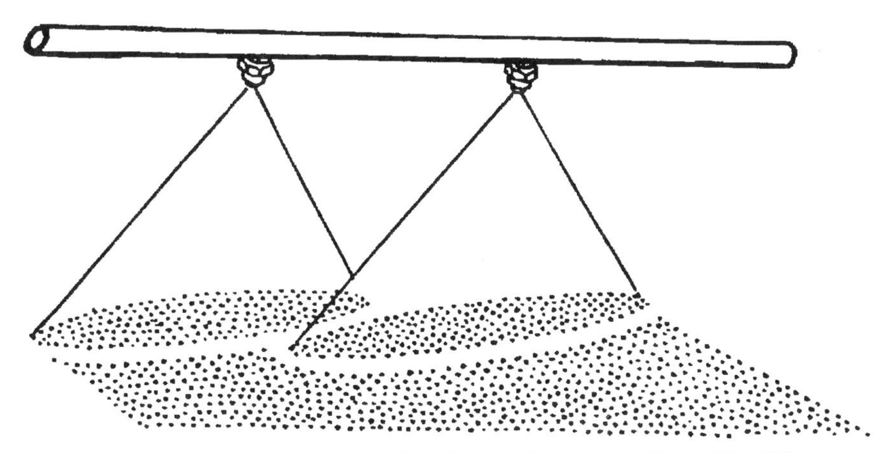

FIGURE 9.20 Nozzle arrangement on a boom showing overlapping patterns. (From Appleby, 1994.)

A special type of flat-pattern nozzle, an *even flat fan*, is manufactured to provide even spray distribution at the edges of the pattern. It is often used as a single nozzle for band applications of herbicides that are used in certain row crops, orchards, and vineyards. The *flood nozzle* has a wider pattern of spray dispersal, usually 115–147 degrees, and a larger droplet size than flat fan nozzles. The potential for drift is reduced with this nozzle type, although the distribution of the herbicide usually is less even than with fan-type nozzles. *Spinning nozzles* are devices that create relatively uniform droplet sizes of the spray liquid by using centrifugal force. The spray liquid is either gravity fed or hydraulically fed into the nozzle, and droplets are formed by the circular rotation of a notched plate or cone. The size of the droplets varies depending upon the speed of rotation, with the smallest droplets being formed by fast plate rotation. The advantage of spinning nozzles is that large and rather uniform droplets can be delivered to the target, thus reducing the potential for herbicide drift.

EQUIPMENT FOR SPECIAL APPLICATIONS

There are several types of equipment that have been developed for special applications of herbicides by ground vehicles. This equipment allows efficient herbicide application, minimal chemical loss, enhancement of weed control, or improved crop selectivity.

Opto-electronic weed detection. One way to minimize herbicide use and exposure is to apply the chemical only where it is intended, that is, directly onto weeds. Computerized weed detection systems attached to sprayers now allow automatic discrimination between weeds, crops, and non-plant objects. It is estimated that such systems can reduce herbicide usage by 70 percent or more in orchards and vineyards. This new technology is based on the fact that growing plants have different spectral reflectance in the visible and near infrared light range that can be detected and used to discriminate among plant species and background using high-precision opto-electronics and microprocessors.

Recirculating sprayers. Spray mixture that is not intercepted by vegetation is caught and reused with a recirculating sprayer (Figure 9.21). The herbicide is sprayed as a stream across the crop row, then the unused herbicide is intercepted by a catching device and returned to the sprayer tank. Weeds must be several inches taller than the crop for proper equipment use and for herbicide selectivity to result. The advantage of this application tool is that herbicide that does not contact vegetation does not fall to the soil surface or onto the crop, which reduces injury to susceptible crops, diminishes deposition of the chemical on the soil, and improves the use efficiency of the herbicide.

Wipe-on devices. Another type of equipment used to apply herbicides also takes advantage of the height difference between tall weeds and the crop. The herbicide is applied through the wicking action of the applicator to tall-growing weeds (Figure 9.22). These applicators include the rope wick, wedge wick, sponge wick, and carpet roller devices (Stroud and Kempen, 1989). The cost of this equipment is relatively inexpensive and the amount of actual herbicide-carrier mixture applied can be quite low (0.25 to 0.5 gallons per acre).

Granule spreaders. Application of herbicides as granules requires equipment designed to spread the dry formulated chemical uniformly. Granular applications often are advantageous for drift control or for penetration of dense vegetation canopies (Appleby, 1994). Relatively few herbicides are applied as granules, except in lawn or ornamental treatments, or in treatments that include both herbicides and fertilizers.

FIGURE 9.21 Recirculating sprayer. (From Ross and Lembi, 1985.)

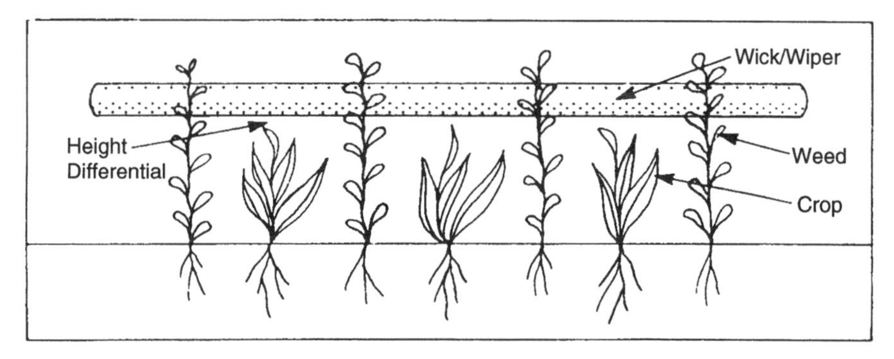

FIGURE 9.22 Application of herbicide through a wick/wiper requires a sufficient height differential between crop and weeds to allow the herbicide to contact the weed without damaging the crop. (From Stroud and Kempen, 1989.)

SPRAYERS FOR MANUAL APPLICATIONS

Many herbicide applications are accomplished using the operator for both power and mobility. Sprayers developed for manual operation are used frequently in developing nations (Fisher and Deutch, 1985) and in some forest and rangeland situations where access by ground or air equipment is limited (Figure 9.23). These sprayers are also used on small farms and by homeowners. The components of pressurized "backpack" sprayers are similar to those of other sprayers (Figure 9.18), except they are smaller. Spray can be delivered by using a lance or wand, or a hand-held boom of nozzles.

FIGURE 9.23 A manual (backpack) application of herbicide in a forest tree plantation. (Photograph by R.G. Wagner, Oregon State University.)

LOK sprayers. The lever-operated knapsack (LOK) sprayer is a versatile tool that can be used to apply herbicides nearly anywhere a person can walk (Figure 9.23). It is useful for applying large or small amounts of spray material as strips, spots, or broadcast applications. In addition, LOK sprayers are relatively cheap, requiring little capital investment. Therefore, they are ideal tools for use on small farms and in developing countries.

The basic components of a LOK sprayer (Figure 9.24) are similar to other hydraulic energy sprayers that are attached to various vehicles for energy and mobility (Figure 9.18). The components consist of a *tank*, which is usually five gallons or less, a *lever action pump*, and a *spray delivery system*. When being used, the sprayer is positioned on the operator's back and secured with carry-

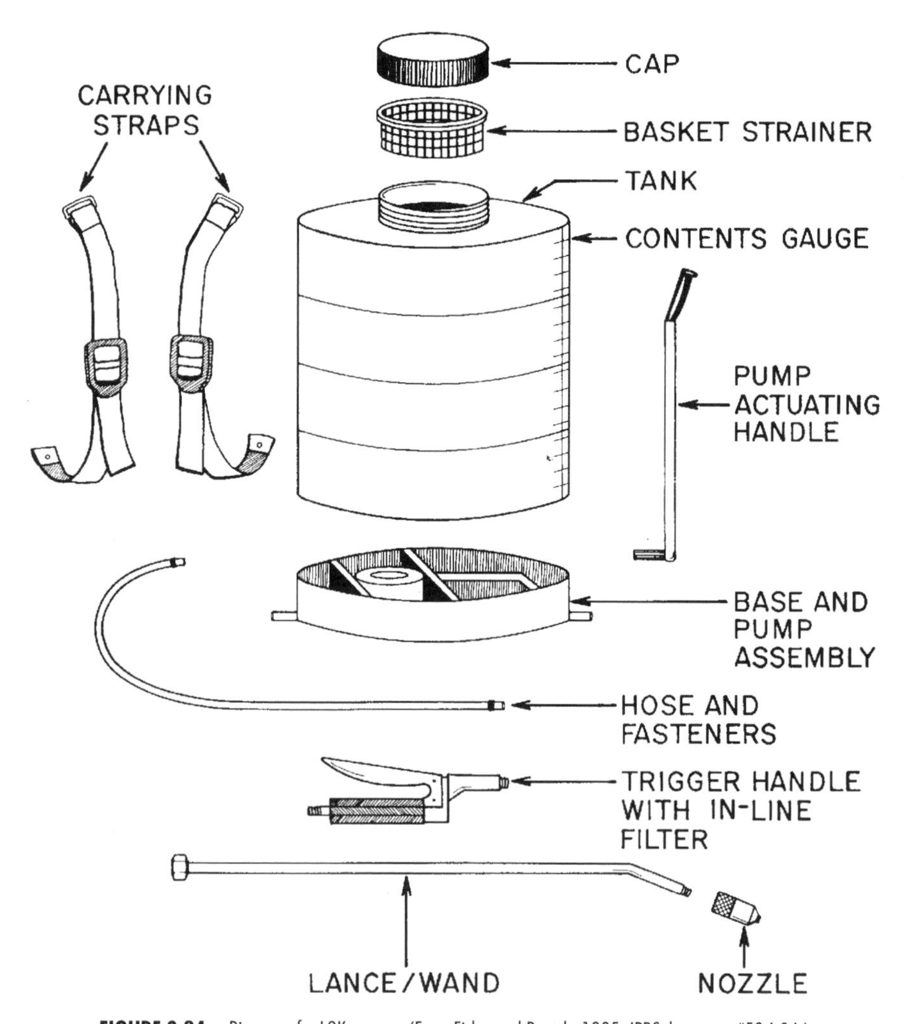

FIGURE 9.24 Diagram of a LOK sprayer. (From Fisher and Deutch, 1985. IPPC document #53-A-84.)

ing straps. A lever-actuated pump handle is moved in a vertical (up and down) direction by the operator (Fisher and Deutch, 1985). This action supplies hydraulic pressure to the system. The hand not doing the pumping holds a lance, wand, or boom of nozzles for spray delivery. A control valve or trigger allows the pressurized spray to exit through the appropriate delivery system.

Compressed gas sprayers. The lever-actuated pump in some sprayers can be replaced by a cylinder of compressed gas or by an air pump. As with LOK sprayers, the spray liquid can be delivered through a single-nozzle wand or through a boom.

Mist blowers. Mist blowers are another class of herbicide application devices that use air flow to disperse the spray liquid. Although mist blowers can be quite large and needed to be moved by tractors during ground applications, others are small and can be used for manual "backpack" operations. This tool uses a gasoline-powered engine to generate an airflow into which the spray liquid is metered, usually by gravity. The air stream then atomizes the liquid into droplets that are dispersed through a wand to the targeted vegetation. This tool is often used on relatively inaccessible areas for brush suppression.

AIRCRAFT APPLICATIONS

Herbicides represent approximately 25 percent of all pesticide applications made by aircraft in the United States (Akesson and Walton, 1989). The major advantage of aerial applications over those made using ground vehicles or manual operations is the ability of aircraft to cover large areas rapidly. The ability to cover muddy ground and very large vegetation is also an advantage of aircraft applications. Most agricultural applications of herbicides by air are made with fixed-wing aircraft. Airplanes with a load capacity of about 1.5 to 2.0 tons are the most common aircraft used. Helicopters are used less extensively for agricultural applications than fixed-wing aircraft. However, helicopter applications are prevalent for many forestry and rangeland uses where terrain often limits the access of airplanes (Figure 9.25).

Granule applications with aircraft. A common device used for aerial application of granular herbicides is the "ramair" spreader (Akesson and Walton, 1989). With this device, air from the airplane propeller enters the spreader, picks up the material, and discharges it. Vanes inside the spreader channel the granules toward the vortex created by the aircraft wings. The swath of herbicide applied is approximately equal to the wingspan of the aircraft. The height of the aircraft also influences the swath width, but height usually is maintained at 20 or 30 feet above the ground. As the rate of herbicide discharge increases, the material tends to concentrate at the center of the swath, so careful metering of the herbicide is required during application.

Helicopters usually fly at speeds approximately half that of fixed-wing aircraft. The ramair spreader will not function properly at such low air speeds, so

FIGURE 9.25 Aerial application of herbicide by helicopter. (Photograph by R.G. Wagner, Oregon State University.)

another device is used for herbicide dispersal. A rotary-disc spreader (Figure 9.26) is hung from the helicopter for this purpose. The pattern of herbicide dispersal from this device is equivalent to that from the ramair spreader.

Liquid applications with aircraft. Herbicide applications by aircraft are usually made as liquid solutions, emulsions, or suspensions. The sprayer tank, which ranges in capacity from 200 to 500 gallons, is mounted over the wings of the aircraft, and the sprayer pump is powered by a small propeller-like "windmill." The spray material discharges through a valve between the pump and boom. When the valve is shut, a venturi device creates a vacuum in the system that closes check-valves in each nozzle and eliminates leakage. Helicopter-mounted sprayers are similar to those used for fixed-wing applications, except that power for the pump is supplied by an electric or hydraulic motor.

Methods to Enhance Herbicide Effectiveness

MECHANICAL INCORPORATION INTO SOIL

During or shortly after application, herbicides often are physically mixed into the upper portions of the soil profile. This "*incorporation*" of the chemical into the soil reduces volatility from the soil surface, decreases photodecomposition, and provides uniform distribution throughout the seed germination zone of the soil profile. It is possible to incorporate some herbicides by rainfall or sprinkler

FIGURE 9.26 Granule herbicide being dispensed from a helicopter. (Photograph by M. Newton, Oregon State University.)

irrigation. However, mechanically mixing the chemical into the soil is a desirable practice when

- the herbicide is volatile
- the herbicide has poor solubility in water
- rainfall patterns are not dependable

Cultivation equipment that is attached directly to the sprayer or follows soon after the herbicide application often is used for herbicide incorporation (Figure 9.27). Mechanically mixing herbicides into the soil has been accomplished by power incorporation, disk harrows, and the use of ground-driven rolling cultivators (see Chapter 8). Each tool has advantages as well as disadvantages that must be considered in relation to the chemical properties of the herbicide and soil characteristics of the field. Generally, a firm, well-formed seedbed without large clods enhances mechanical incorporation of herbicides.

The optimum method of mixing herbicides with soil is by *power incorporation*. The herbicide is mixed uniformly and to a specific depth within the row or treatment band (Figure 9.27 [top]). Often, the power incorporater is attached to the tractor-sprayer unit or follows closely behind it during application. Herbicide mixing into soil following broadcast applications can also be accomplished with a *disk harrow* (Figure 9.27 [bottom]). This method of incorporation is economical and usually effective. However, the soil condition and extent of overlapping swaths can affect uniformity of herbicide distribution with disk harrows. The depth of herbicide incorporation into the soil profile will also vary

FIGURE 9.27 Herbicide incorporation with a rotary tiller (*top*) and disk harrow (*bottom*). (Photograph courtesy of F.O. Colbert, California Polytechnic State University, San Luis Obispo.)

according to the type of equipment used. Some herbicides can be mixed effectively with the soil by a *spring-tooth* or *spike harrow* but this form of incorporation usually does not provide enough mixing for use with most soil-applied herbicides. *Rolling cultivators* can be useful incorporation devices if only shallow mixing of the herbicide into the soil is necessary. The presence of clods severely limits this method of incorporation and its use usually is restricted to non-volatile herbicides.

SPRINKLER IRRIGATION AND RAINFALL INCORPORATION

Leaching (Chapter 10) is used to incorporate certain herbicides into the rooting zone of seedling weeds. The water solubility, charge, vapor pressure, and ability of the chemical to adsorb to soil constituents all influence the uniformity and depth of incorporation obtained by this method (pages 485–488). Soil properties, especially the amount of soil organic matter, affect the way herbicide is moved in the soil profile by water. Least movement occurs in soils with high organic matter content. The amount of water received or applied also influences the depth of the chemical in the soil profile. Too much water can move the herbicide deep into the soil profile, while too little water may cause the herbicide to remain on the soil surface. Wind also may disturb the pattern of herbicide incorporation if poor distribution of water from sprinklers occurs. In spite of these considerations, incorporation of herbicides with water is effective for many soil-applied herbicides. It also is not expensive, especially if rainfall is used as the method of incorporation.

APPLICATION IN IRRIGATION WATER

The application of herbicides in flood, furrow, or sprinkler irrigation water has the advantage of incorporating the herbicide as it is being applied. However, the irrigation system itself must be monitored frequently to assure even distribution of the water and herbicide.

Applications in surface-flow irrigation systems. A constant head siphon device (Figure 9.28) is necessary for application of herbicides by flood, furrow, or basin irrigation systems. The device is attached to the herbicide container and allows uniform metering of the chemical into the irrigation system. The length of time required for application varies during irrigations. For this reason, it is best to monitor the system throughout the irrigation cycle so that changes in application rate do not occur.

Applications by sprinkler and drip irrigation. An injection system is necessary to apply herbicides by sprinkler irrigation. Often the tank containing the chemical is attached to the suction side of the centrifugal pump of the irrigation system. The herbicide is then drawn into the water during irrigation. The amount of herbicide entering the water is regulated by an orifice or valve that is placed between the tank and irrigation pump. A pressurized herbicide injection system

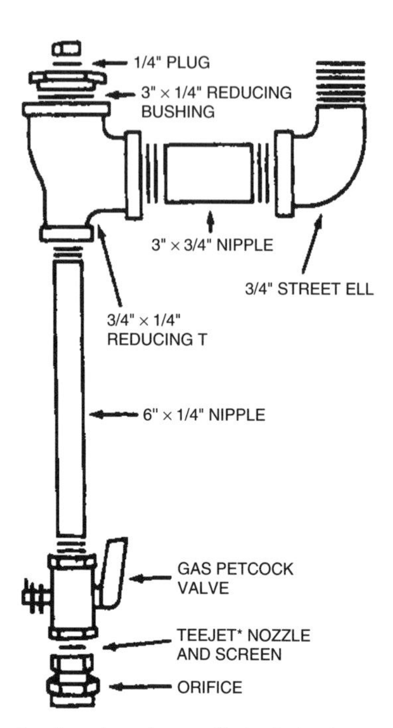

FIGURE 9.28 A constant head siphon device for application of herbicide directly into irrigation water. (From Crop Protection Chemicals Reference 1986.)

sometimes is attached to irrigation systems to meter the flow of chemical into the water. In this case, the herbicide tank is pressurized with carbon dioxide or compressed air cylinders. Wettable powder formulation of herbicides must be mixed with water and diluted before injection into sprinkler irrigation systems. Safety or vacuum-relief valves are an important component of herbicide injection into irrigation water. These valves prevent back-flow of the herbicide into the water supply source.

SOIL FUMIGATION

Soil fumigants are organic chemicals that have high vapor pressure and low solubility in water. The objective of soil fumigation is to place the liquid chemical into the soil where it can volatilize and disperse as a gas throughout the available soil pores. A physical seal placed over the treated soil is necessary to prevent rapid loss of the chemical into the air. Depending upon the chemical used to fumigate the soil, a barrier can be formed by packing or firming the soil surface, applying sprinkler irrigation, or placing a plastic sheet over the treated area. The degree of chemical volatility and level of toxicity determine the method of barrier formation, which is often the critical step for soil fumigation.

In addition to the presence of a physical barrier to chemical volatility in the air, the effectiveness of soil fumigation also depends on soil temperature, mois-

FIGURE 9.29 A device to inject herbicide directly into soil as a concentrated band. (Photograph by C.L. Elmore, University of California, Davis.)

ture, and organic matter content; dosage delivered; and time of exposure. In general, warm soil temperatures, adequate soil moisture that softens weed seed coats and allows chemical diffusion, and well-decomposed organic matter will facilitate optimal weed suppression by soil fumigation.

DIRECT INJECTION INTO SOIL

Certain volatile herbicides can be injected into the soil as a concentrated band or line of chemical. The herbicide diffuses through the soil air spaces, inhibiting germination of seed that are contacted. Other herbicides have been injected below the soil surface with a blade (Figure 9.29). In this case, a concentrated layer of the herbicide is applied that inhibits sprouting of shoots of perennial weeds, such as field bindweed.

SUMMARY

Herbicides are chemicals that kill or suppress the growth of plants. The discovery, manufacture, and use of herbicides represents one of the major technological advances of modern agriculture. Most herbicides are discovered by the systematic synthesis and testing of chemicals for biological activity. The U.S. Environmental Protection Agency is the primary organization responsible for the development and regulation of herbicides in the United States. The Federal Insecticide, Fungicide, and Rodenticide Act (FIFRA) and the Food,

Drug and Cosmetic Act (FDCA) are two major laws that regulate herbicide use in this country.

Important properties of herbicides include their chemical form, water solubility, and vapor pressure. Most herbicides are modified or formulated before sale and use as commercial products. The formulation of a herbicide contains the active ingredient of the chemical and various inert ingredients, such as solvents and emulsifiers. These "inert ingredients" usually improve the use or phytological activity of the product. Herbicides may be applied dry, or as granules or pellets, but they are usually applied as a liquid spray.

Herbicides are classified according to similarities in chemical structure, use, and the biological process they inhibit. Plants often exhibit characteristic symptoms of cell disruption, hormone imbalance, leaf chlorosis, necrosis, or albinism following herbicide application. These symptoms reflect the inhibition of a vital plant process in exposed plants. Most herbicides are selective, meaning that some plants are injured by them while others are not. However, selectivity is relative and depends on the rate of herbicide applied and numerous other plant, chemical, and environmental factors. The plant factors that influence herbicide selectivity are genetic inheritance, age, growth rate, morphology, physiology, and certain biochemical/biophysical processes. Herbicide structure or configuration, formulation, and placement also influence differential herbicide sensitivity among treated plants. Environmental factors that affect the movement of herbicides in the soil also influence selectivity patterns among crop and weed species.

Herbicides are most effective as tools for weed control when they are applied accurately and uniformly. The use of properly designed, calibrated, and operated equipment assures both accuracy and uniformity of application. Sprayers and dry formulation spreaders are the most common types of application equipment for herbicide use. Sprayers are composed of a tank, agitation device, pump, pressure regulator, hoses, and nozzles. A method of movement and power also is a necessary component for all sprayers. Some sprayers are pulled by tractors or mounted directly on them. Other sprayers, such as the LOK sprayer, use humans as a source of both power and locomotion. Manual applications of herbicides are most appropriate for use on small farms, in developing nations, and for homeowners.

Sprayers and granule spreaders also may be attached to aircraft for rapid and widespread application of herbicides. However, special calibration and application techniques are necessary to avoid drift. Recirculation sprayers and wipe-on devices represent innovations in herbicide application that reduce chemical usage. Often, herbicides require mixing into the soil profile to control weeds effectively. Incorporation can be accomplished by a variety of mechanical tools, precipitation, or irrigation. In some cases, herbicides may be applied directly with irrigation water or injected into the soil. Fumigation with volatile herbicides requires special techniques to assure herbicidal effectiveness and avoid displacement to the atmosphere.

Chapter

10

Action and Fate of Herbicides

HERBICIDE ACTION

The ability of herbicides to suppress plant function depends on many biological and environmental factors. The *mode of action* of a herbicide includes all of the physiological, anatomical, morphological, and biochemical processes that result in phytotoxicity. These include *absorption* (uptake), *transport, metabolism*, and *biochemical inhibition*. The mode of action also includes all the events that eventually lead to plant death following herbicide application. Thus, environmental factors that influence the *fate*, *persistence*, and *availability* of herbicides for uptake are components of the mode of action. The specific biochemical process or site of inhibition in treated plants that results in growth suppression or death is termed the *mechanism of action*. Both the mode and mechanism of action of many herbicides are known. Ashton and Crafts (1981), Duke (1985), and Devine et al. (1993) provide complete reviews of both of these important processes.

Herbicide Uptake by Plants

Herbicides enter plants through shoots (leaves, stems, buds), roots, other belowground organs, and seed. The process of herbicide entry into treated plants is called *absorption*, which involves contact, penetration, and movement of the chemical into the plant. *Transport* is herbicide movement in the plant after absorption.

FOLIAR ABSORPTION

Many herbicides are applied to plant surfaces as foliar sprays (Chapter 9). Leaves are the primary means of herbicide entry through shoots, although herbicide absorption can also occur through other aerial organs. For example, substantial amounts of some herbicides are absorbed through stems or emerging coleoptiles. Foliar herbicide absorption includes the following three steps:

- retention of spray droplets on a leaf surface
- penetration of the herbicide into plant cells
- movement into the cytoplasm of the plant cell

Leaf anatomy. Figure 10.1 (Taiz and Zeiger, 1991) shows the various tissues that must be entered for herbicidal activity to occur. The *epidermis* and asso-

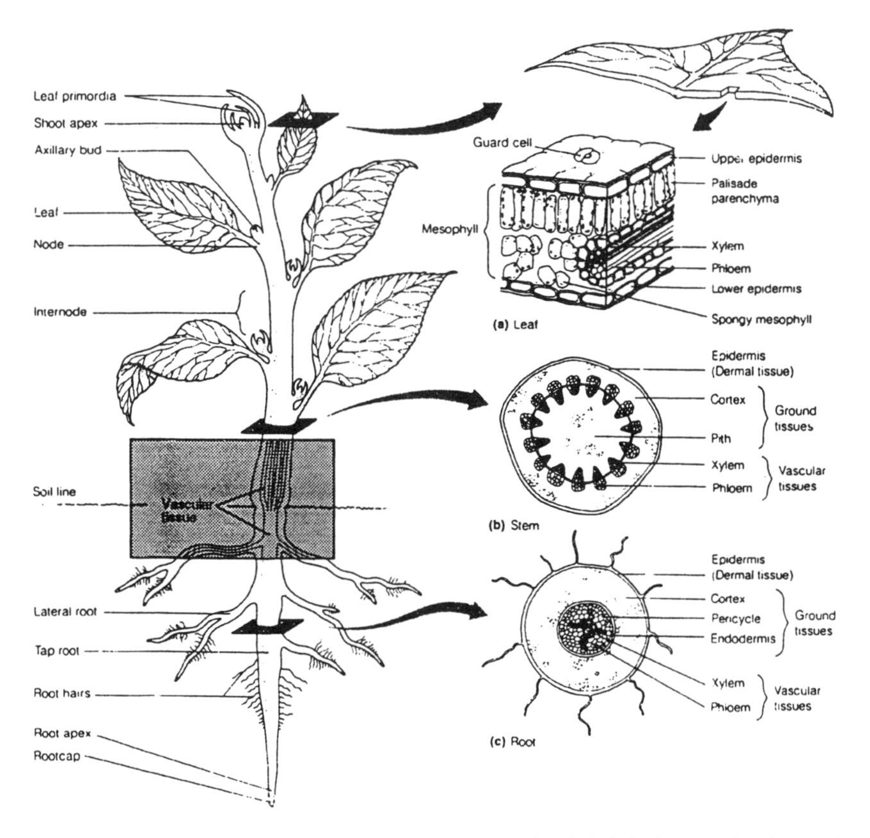

FIGURE 10.1 Diagram of the body of a seed plant, showing cross sections of (a) the leaf, (b) the stem, and (c) the root. (From Taiz and Zeiger, 1991.)

ciated *cuticle* is a single layer of cells that covers the entire above ground plant surface. This layer protects the shoot from excessive moisture loss and is the primary barrier to the absorption of herbicides into plant foliage. *Epicuticular waxes,* which form the outermost covering of the shoot, are a complex of lipoidal and crystalline-like substances (alkanes, alcohols, paraffins, glycerides, and sterols) that form a water-repellent barrier between epidermal cells and the air. Environmental conditions influence the amount and form of epicuticular waxes produced by the plant, with greatest wax deposits occurring when there is high light, moderate temperature, and low relative humidity. Although the cuticle completely covers the leaf surface, the amount of cuticle present varies among plant species and even among locations on the same leaf. Leaf age and environmental conditions during leaf development affect the amount of cuticle present on individual leaf surfaces (Hess, 1985).

Trichomes are plant hairs located on leaf surfaces (Figure 10.2a,b). They assist in absorption of water, reduce transpiration, and protect the leaf from abrasion. Densely packed trichomes retard wetting. However, trichomes sometimes increase wetting, retention, and penetration of herbicides by physically holding small spray droplets on the leaf surface. Trichomes also may be preferential sites for uptake for certain herbicides.

The cuticle overlies and merges with the epidermal cell walls (Figures 10.2 and 10.3). It is composed of intergrading layers of pectin, cutin, and waxes. The external portions of the cuticle grade into the cutin and pectins at the cuticle-cell wall interface. Often, microfibrils of cellulose and pectin strands exist throughout the cuticule (Figure 10.3). There is a gradual transition in polarity of the cuticle from the outer wax-like surface to the inner pectin and cell wall interfaces (Figure 10.3). Epicuticular and cuticular waxes are nonpolar and hydrophobic (oil-loving), forming a water-repellent barrier on the leaf surface. The internal surface of the cuticle (cutin, pectins, cellulose) is polar and hydrophilic (water-loving). This transition in polarity from the leaf surface to the cell wall is an important aspect of the cuticle that influences how herbicides are absorbed (Ashton and Crafts, 1981; Hess, 1985; Devine et al., 1993).

Stomata are small pores or openings into the internal structure of leaves surrounded by guard cells (Figures 10.1a and 10.2c,d) that regulate the flow of gases between the leaf and the atmosphere. Stomata regulate carbon dioxide uptake by leaves. The carbon dioxide is used in photosynthesis, which gives off oxygen as a product. The loss of water vapor through transpiration is also regulated by stomata. Regulation of gas exchange is a critical function that assures normal growth and development of plants. Penetration through stomata is minimal for most herbicides (Ashton and Crafts, 1981; Hess, 1985).

Beneath the epidermis of leaves is the *mesophyll*, composed of parenchyma cells (Figure 10.1a) where most leaf photosynthesis occurs. Veins in leaves are found in the mesophyll and consist of both xylem and phloem. These two components of the plant vascular system conduct water and inorganic salts (xylem) and carbohydrates and other substances (phloem) throughout the plant.

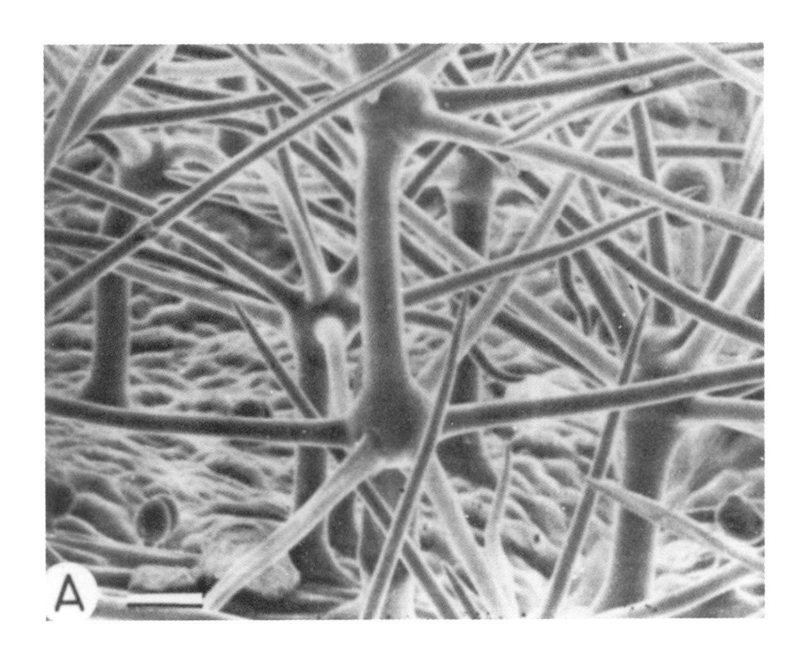

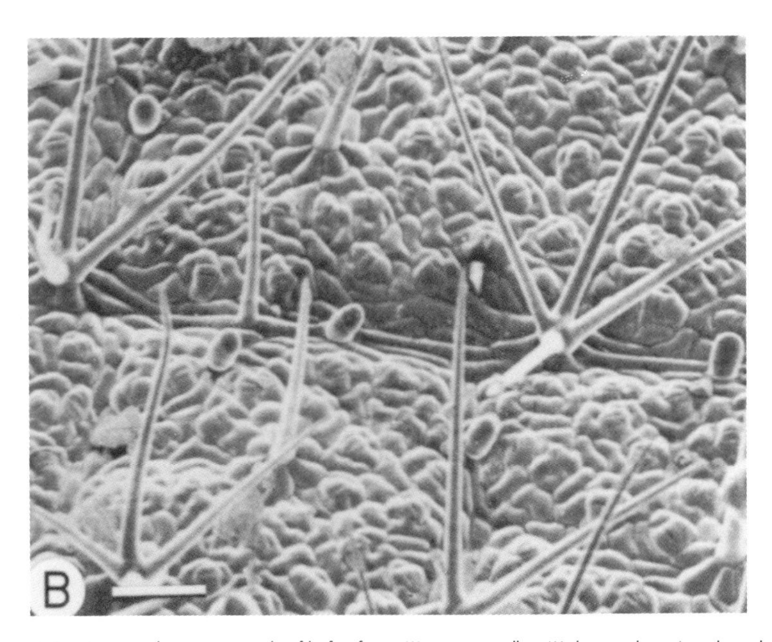

FIGURE 10.2 Scanning electron micrographs of leaf surfaces. (A) common mullein (*Verbascum thapsus*), scale marker = 50 microns. (B) Lower epidermis of velvetleaf (*Abutilon theophrasti*), scale marker = 100 microns. (C) Cabbage (*Brassica oleracea* var. *capitata*), scale marker = 25 microns. (D) Sugarbeet (*Beta vulgaris*), scale marker = 25 microns. (Micrographs by F.D. Hess, in Hess, 1985.)

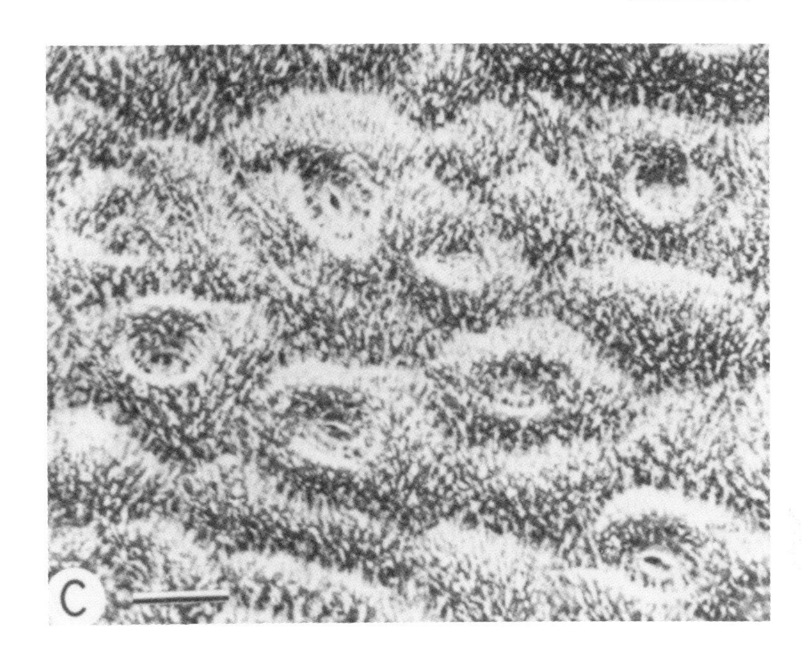
C

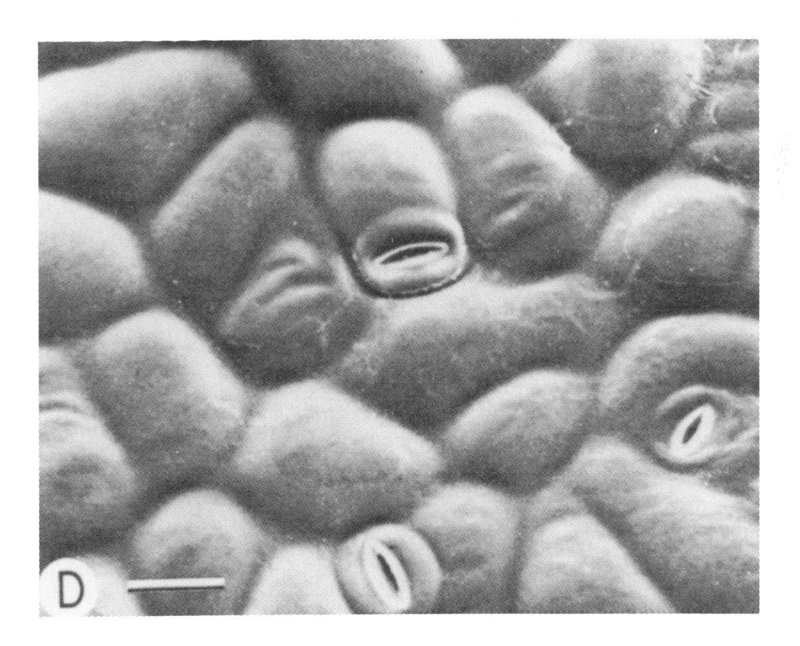
D

Pathways for foliar absorption. A herbicide moving from a leaf surface to the cytoplasm of cells first encounters epicuticular waxes, then cuticle, pectic layers, cell walls, and finally the plasmalemma (outer membrane) of the cell (Figure 10.3). These substances create gradients in polarity and water solubility that all chemicals must traverse to enter plant leaves. Two pathways have been proposed to explain the absorption of both polar and nonpolar herbicides into plant foliage (Ashton and Crafts, 1981; Hull et al., 1982). These pathways are the *aqueous* and *lipoidal* routes of herbicide entry, respectively (Figure 10.4).

The cutin and pectinaceous strands of the cuticle are believed to constitute the *aqueous route* for herbicide absorption into leaves (Figures 10.3 and 10.4). After entry into cracks, punctures, or fissures in the epicuticular waxes on the leaf surface, the herbicide moves internally along the relatively polar components of the cell wall (Figure 10.4). Since water-soluble herbicides generally do not penetrate leaf surfaces easily, the aqueous route is enhanced by a hydrated atmosphere that causes expansion of the distance between epicuticular and cuticular wax plates and makes herbicide movement through the cuticle easier. Maleic hydrazide and paraquat are herbicides that enter plant leaves via the aqueous route. The *lipoidal pathway* is thought to consist of the nonpolar portions of the cuticle, primarily embedded waxes (Figure 10.4). Pores, which function as channels for wax movement outward to the leaf surface, also may be part of the lipoidal route. However, the existence and role of such pores for herbicide movement through the cuticle are uncertain. Herbicides that are soluble in oil or oil-like solvents readily penetrate the cuticle by the lipoidal route of uptake. Sethoxydim, dinoseb, and 2,4-D (ester form) are examples of foliage-active herbicides that enter plants in this way.

Methods to improve foliar absorption. Many aspects of leaf surfaces change with leaf age and plant development. These factors may either restrict or improve herbicide retention and subsequent penetration of foliage (Hess, 1985,1987). Surface waxes generally increase with leaf age, but cracks and fissures in the cuticle also increase as leaves grow old. Weathering and abrasion

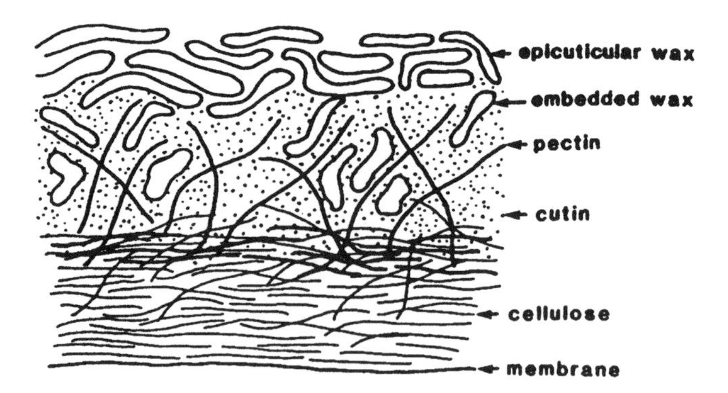

FIGURE 10.3 Diagrammatic representation of the location of components of plant cuticles. (From Hess, 1985.)

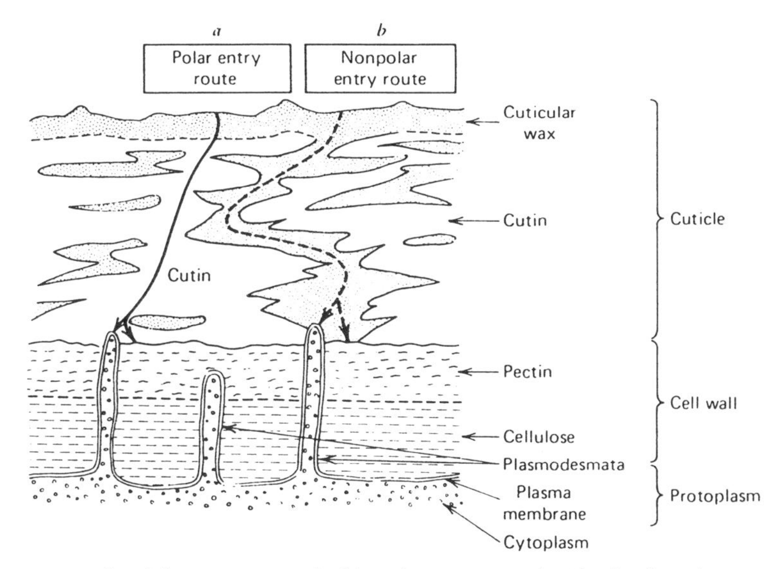

FIGURE 10.4 Hypothetical diagram representing the foliage absorption aspects of cuticle-cell wall-protoplasm structure. Hypothetical routes of entry for (a) polar and (b) nonpolar herbicides. (From Ashton and Crafts, 1981.)

decrease the quantity and uniformity of surface waxes, which may improve herbicide absorption. The topography of leaf surfaces also can influence foliar absorption of herbicides. For example, leaves with a predominant vein structure usually retain more herbicide than leaves with smooth surfaces (Chapter 9).

Foliar absorption of herbicides can be enhanced by modifying the carrier to improve leaf surface wetting and retention. Herbicides applied in oils or oil-based carriers generally are retained and penetrate leaf surfaces better than herbicides applied only in water. Surfactants and wetting agents improve herbicide absorption by decreasing the surface tension between the spray droplets and the leaf surface (Chapter 9). The volume of the herbicide solution or suspension being sprayed can also influence absorption. The carrier volume should be sufficient to cover the plant canopy without causing excess spray "run-off" from treated foliage, which reduces herbicide availability. Some herbicides (e.g., glyphosate) are most effective when applied in lower (5 gallons per minute [gpm]) than higher (30 gpm) carrier volumes for reasons not fully understood. Spray droplet size can also influence the amount of herbicide absorbed by foliage. Greatest absorption results when spray droplets are small. However, spray droplets should be 100 microns or larger to reduce drift to adjacent crops or property.

Although little can be done to alter the environment during herbicide application, such factors as moisture (rainfall and relative humidity), temperature, and light intensity can influence the absorption of foliage-applied herbicides. Generally, herbicides are absorbed and kill plants most effectively when conditions are humid, temperatures are hot, and light intensities are high.

ENTRY THROUGH PLANT STEMS

The direct application of herbicides to plant stems is rare, except to control woody plants, generally trees and shrubs. However, the stems of herbaceous crops and weeds usually are exposed to herbicides that are applied primarily to leaves. Herbicide applications directed to soil at the base of plants may also result in some stem exposure.

Penetration of herbicides through the bark of trees or shrubs presents a much more difficult problem than penetration into foliage or herbaceous stems. Bark is a suberized covering of cork cells that represents a formidable barrier to herbicide penetration. When bark is uniform, without cracks or fissures, aqueous sprays of herbicides usually are ineffective. An oil carrier is usually required for adequate herbicide absorption through woody stems. Stem treatments for shrub and tree control are made directly to the bark as basal sprays, through cuts in the stem or trunk (frill treatments), or directly to the cut-surface of stumps.

ABSORPTION FROM SOIL

Many herbicides are applied directly to the soil, while others eventually arrive there by the washing action of rainfall or sprinkler irrigation on plant foliage. Roots are the primary absorbing organ for herbicides present in soil, although absorption by other subterranean plant organs has been observed.

Root anatomy. Plants have two types of root systems, *taproot* and *fibrous*, which are remarkably similar in anatomy despite their difference in external structure. Although the root anatomy of all plants is not the same, the most common arrangement has been described in the following manner. The most internal portion of a root is composed of the *vascular cylinder*, which contains the two conducting tissues, xylem and phloem, arranged in the distinctive manner shown in Figure 10.5. Cells of the phloem function in the long-range transport of food material, mostly sugars, which are produced by photosynthesis in leaves. Xylem elements, either *tracheids* or *vessel elements*, are elongated cells that form thick secondary walls. These cells die after the walls are formed, producing a continuous system of xylem in which water, nutrients, and other substances are transported upward through the plant (Weir et al., 1982; Salisbury and Ross, 1992).

The layer of cells immediately next to the vascular cylinder in roots is the *pericycle*. Cell division occurs in this tissue and lateral roots are produced when given the proper physiological signal. External to the pericycle is the *cortex* (Figure 10.5). This tissue, composed primarily of parenchyma cells, stores carbohydrates and has a large amount of air space. The innermost row of cortex cells, next to the pericycle, form the root *endodermis*, a specialized tissue that regulates the flow of material into the vascular cylinder (Figure 10.5). Such regulation is possible because of the special modified walls of endodermis cells, a region called the *Casparian strip*. The Casparian strip is perpendicular to the

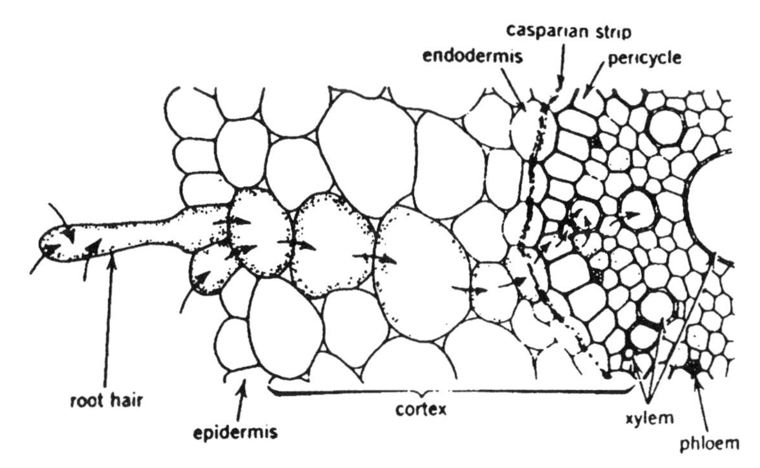

FIGURE 10.5 Transverse section of a root showing pathway of water uptake (arrows). (Reproduced with permission from Esau, 1965.)

outside root surface and impermeable to water, nutrient ions, and most other substances. Therefore, everything entering or leaving the vascular cylinder of a root must pass through the cell membranes of the endodermis.

The outermost layer of root cells form the *epidermis* (Figure 10.5), which can produce *root hairs*. In most plants, root hairs function for only a few days, so they are constantly dying and being produced. Root hairs dramatically increase the surface area of the rooting zone and hence the absorbing capacity of a root system.

Pathways of herbicide absorption into roots. Herbicides enter roots by three possible routes: *apoplast, symplast*, and *apoplast-symplast* (Devine et al., 1993). Collectively, all living portions of a plant form the symplast, which is the interconnected, continuous, living protoplasm of plants. The apoplast comprises all the nonliving plant tissues and any spaces between cells. Phloem and other living cells are the major components of the symplast, whereas xylem, intercellular spaces, and cell walls form the apoplast. The *apoplastic route* allows herbicides to enter roots freely until they encounter the Casparian strip of the root endodermis. After passing through the endodermis, they then enter the xylem of the vascular cylinder. The *symplastic route* involves initial herbicide entry through the cell walls of root hairs and subsequent movement into the cytoplasm of the epidermis and cortex. The herbicide then passes through the cytoplasm of the endodermis, avoiding the Casparian strip, and enters the phloem of the vascular cylinder by means of interconnecting protoplasmic strands (Figure 10.5). The *apoplast-symplast route* is identical to the symplastic route, except that after passing through the endodermis the herbicide may reenter cell walls and then enter the xylem of the vascular cylinder (Figure 10.5). Some herbicides are restricted to only one route of entry. However, it is more likely that most herbicides enter plant roots through more than a single pathway (Devine et al., 1993). The route or routes of entry are determined by the particular chemical and physical properties of each herbicide.

ABSORPTION BY SEED AND COLEOPTILES

Absorption by seed and young shoots of plants is important for herbicides that inhibit germination and seedling development. There are numerous examples of herbicides that are absorbed primarily by seed or germinating seedlings (Chapter 9). In the majority of cases, herbicide entry into seed occurs in a passive manner with the water necessary for germination (Anderson, 1977, 1996; Aldrich, 1984). However, certain volatile herbicides and some soil fumigants enter dry seed as a gas.

Some soil-applied herbicides are absorbed primarily through young shoots or coleoptiles of emerging plants when they contact herbicide-treated soil. Absorption occurs with the soil water or as a gas, if the herbicide is volatile. The activity of EPTC, trifluralin, alachlor, and several other herbicides is dependent on shoot absorption as weed seedlings emerge through soil treated with herbicides (Hess, 1987), although their site of action may be in a different portion of the plant.

Herbicide Movement in Plants

Regardless of whether a herbicide enters a plant through leaves, stems, roots, or seed, it must eventually reach a specific location (site) to suppress physiological function. The movement of a systemic herbicide from its point of entry to its site of action in the plant usually occurs by means of *translocation* in the phloem or with water movement in the xylem. Although most herbicides move better in one tissue than the other, all herbicides are capable of movement in both xylem and phloem (Devine et al., 1993). Herbicides often are grouped in four major ways, depending on the manner and distance traveled in exposed plants. Herbicides may

have limited mobility in treated plants,
move primarily in the symplast, especially the phloem,
move primarily in the apoplast, especially the xylem,
move in both the symplast and apoplast.

HERBICIDES WITH LIMITED MOBILITY

Foliage-applied herbicides that kill plant tissue quickly after treatment usually do not move well in plants. These are *contact* herbicides according to the classification scheme described in Chapter 9. The movement of some soil-applied herbicides from the root surface into root cells is another form of short-distance transport (Hess, 1985, 1987). Trifluralin and other dinitroaniline herbicides only move a few cells in developing roots but effectively inhibit plant growth even though long-distance transport does not occur.

HERBICIDES THAT MOVE IN THE SYMPLAST

Herbicides that move in the symplast are applied primarily to foliage (Figure 10.6). After absorption, the chemical moves from cell to cell in treated leaves through interconnecting strands of cytoplasm until the phloem is reached. Once

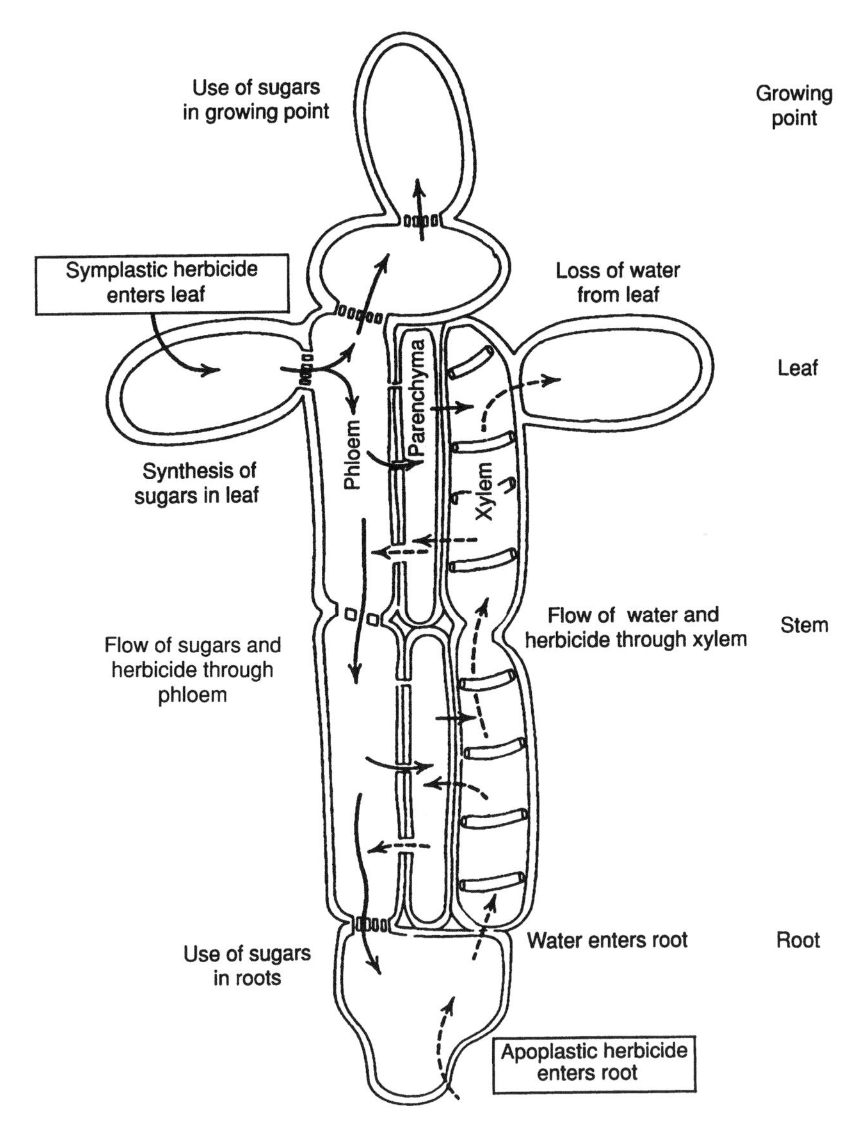

FIGURE 10.6 Diagram representing routes of translocation of herbicides in plants. (Adapted from Bonner and Galson, 1952. Copyright 1952, W.H. Freeman and Company; in Ashton and Crafts, 1981.)

in the phloem, herbicides in this group move throughout the plant in association with sugars produced during photosynthesis. Phloem-mobile herbicides accumulate where sugars are being used or stored. This form of translocation is called *source-to-sink* movement (Ashton and Crafts, 1981; Hess, 1985) because it occurs in accordance with osmotic pressure gradients established between "source" and "sink" tissues in plants. Source tissues are those that have an excess of carbohydrate available, some of which is available to sink tissues for growth, metabolism or storage. The driving force for source-to-sink transport is the difference in tugor pressure between cells that have available sugars (high tugor) and cells that utilize sugars (low tugor). The actual mechanism of sugar (usually sucrose) entry into phloem is not completely understood. However, it is believed that sucrose is concentrated in the phloem conducting elements by either an active loading process or direct transfer from mesophyll cells. The high osmotic potential of this concentrated solution causes a simultaneous influx of water and the resultant turgor pressure forces flow toward tissues of low turgor pressure. Herbicides are generally considered to move passively in the direction of solute flow in the phloem, although the actual path taken by herbicide molecules to reach the phloem is not well understood (Devine et al., 1993).

Herbicides used to kill perennial herbaceous and woody plants usually are most effective when applied to foliage at a time when photosynthates are being moved readily to belowground storage or reproductive organs. Triclopyr and 2,4-D are herbicides that kill plants in this way. Movement of symplast-mobile herbicides into underground organs can take several hours or even days. Regrowth of underground shoots often results if aboveground plant parts are killed too rapidly, and before adequate herbicide translocation has occurred.

HERBICIDES THAT MOVE IN THE APOPLAST

Herbicides that move in the apoplast are absorbed primarily through roots (Figure 10.6). Some herbicides of this group also are absorbed through the foliage of treated plants, but the distance transported is usually less than when they are absorbed by roots (Ashton and Crafts, 1981; Devine et al., 1993). Thus, these herbicides are generally applied to the soil. Examples of apoplastically transported herbicides are members of the triazine, uracil, and substituted urea groups. These herbicides enter roots with soil water and move throughout the plant in the xylem. Transport occurs because of the water potential gradient that develops as water is absorbed by roots and transpired by leaves. The herbicides that move in the apoplast accumulate in the veins and margins of treated leaves (Figure 9.7).

Herbicide transport in either xylem or phloem is always with the water in those tissues. However, much more water moves through the xylem than through the phloem. Water also moves most rapidly in xylem. Herbicides, such as atrazine, that easily diffuse across cellular membranes can enter the xylem from nearby cells and be swept into the transpiration stream. Herbicides of this type actually can exist in both symplast and apoplast, but movement in the xylem is more likely.

HERBICIDES THAT MOVE IN BOTH SYMPLAST AND APOPLAST

Most herbicides are not restricted to the symplast or apoplast but can move in both types of tissue (Devine et al., 1993). Herbicides will move predominantly in either xylem or phloem depending on the relative ease with which these chemicals cross cell membranes, such as the plasmalemma. The tendency or ability of herbicides to move in the apoplast or enter the symplast depends in part on pH gradients that develop across membranes of the cell. The chemical characteristics of each herbicide in relation to cellular pH also influence whether the herbicide moves predominantly in the apoplast or symplast. Amitrole and glyphosate are examples of two herbicides that move readily in both the symplast and apoplast.

Transport across cell membranes. Herbicides must enter the living cytoplasm of plant cells in order to cause biochemical dysfunction, which means that at some point every herbicide must cross the plasmalemma (cell membrane) into a plant cell. The plasmalemma and other cell membranes are fluid, lipid bilayers with proteins embedded in them (Figure 10.7). Since many herbicides are

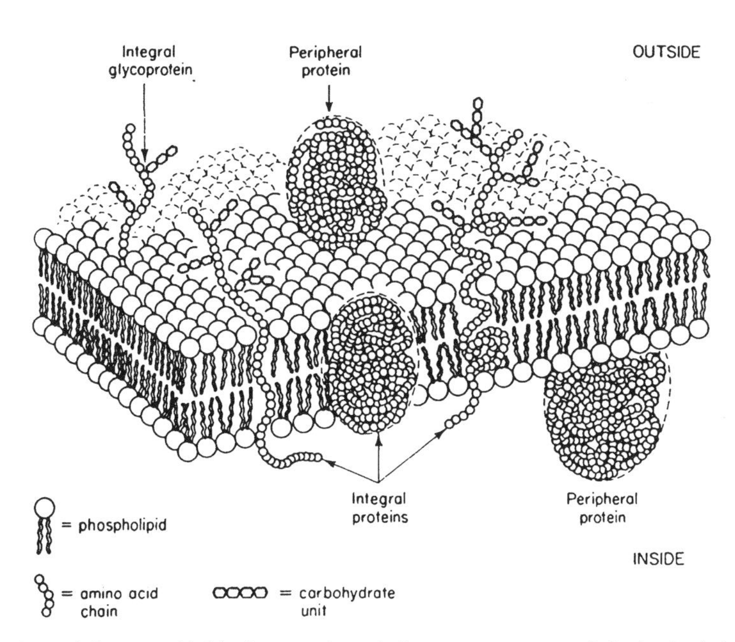

FIGURE 10.7 Fluid mosaic model of the plasma membrane of cells. Proteins are asymmetrically distributed in the lipid bi-layer. Carbohydrate units may be attached to protein or lipid but only on the outer surface. The tonoplast structure is similar except glycoproteins may not be present. (From *Biology of the Cell*, 2d edition, by Stephan Wolfe. Copyright 1981, Wadsworth, Inc. Reprinted by permission of Wadsworth Publishing Company; in Balke, 1985.)

quite lipophilic, they are often associated with membranes and membrane function. The plasmalemma regulates the entry of molecules, including some herbicides, into plant cells (Figure 10.8) through *active transport* using proton (H^+) gradients across the membrane as the driving force. The proton gradient is established by the relative difference in pH and charge between the sides of the

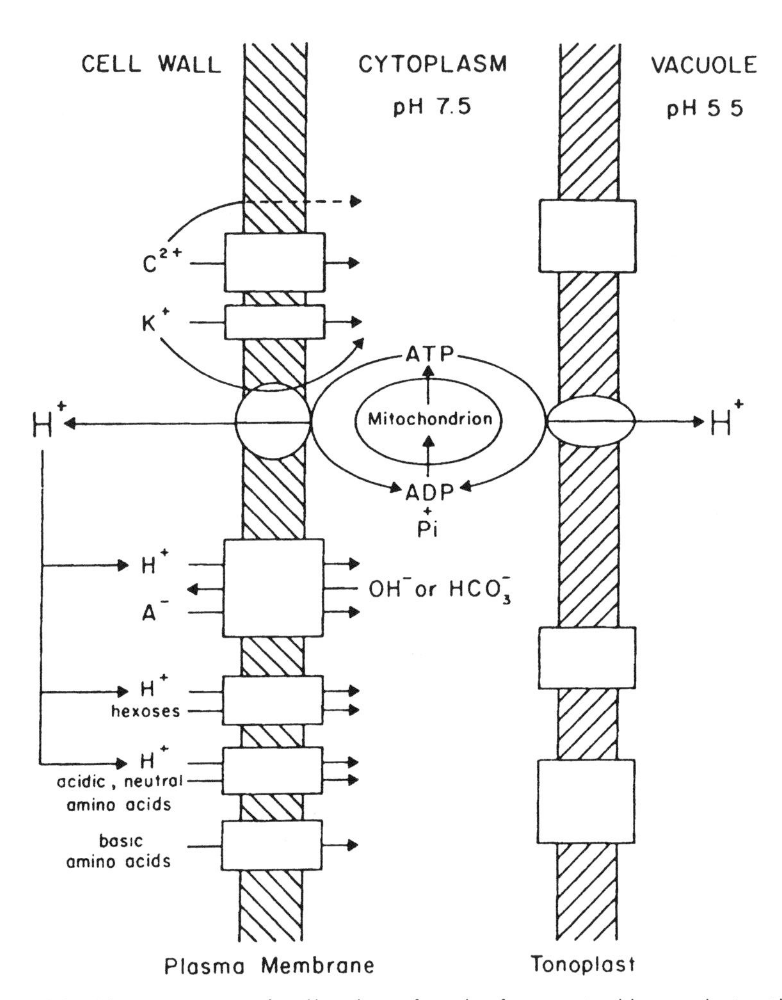

FIGURE 10.8 Schematic representation of possible mechanisms for coupling of energy to mineral, hexose, and amino acid absorption into plant cells. ATP produced in mitochondria is used by the plasma membrane and tonoplast to pump H^+ out of the cytoplasm. The resultant H^+ gradient and electrogenic transmembrane potential provide the driving force for absorption of various solutes. K^+ absorption across the plasma membrane may be mediated by the ATPase or a separate carrier protein. Divalent cation (C^{2+}) absorption may be via a carrier or by passive diffusion (represented by broken line). Monovalent anion (A^-) absorption may be by co-transport with H^+ or by counter-transport with OH^- (or HCO_3^-). Hexoses and acidic or neutral amino acids move by co-transport on separate carriers, whereas basic amino acids move without an accompanying H^+ or OH^- (like K^+). Similar carriers are believed to transport solutes at the tonoplast. (From Balke, 1985; in Duke 1985. Copyright 1985, CRC Press Inc., Boca Raton, Florida. With permission.)

membrane. Differences in cell pH range between 5 and 8, with the highest pH being on the inner side of the membrane (Balke, 1985). If herbicide molecules exist as acids, pH can affect herbicide movement by altering their lipid solubility through ionization. Herbicide movement through the plasmalemma by active transport occurs via carrier proteins (Devine et al., 1993).

Most herbicide molecules are believed to enter plant cells passively or by diffusion (Hess, 1985; Devine et al., 1993). Passive movement across the plasmalemma by hydrophilic substances is related to molecular size of the herbicide and its oil/water partition coefficient, which can be influenced by cellular pH. Regardless of molecular size, herbicides that are hydrophobic and soluble in organic solvents dissolve readily into or through cell membranes. As herbicides are transported in either xylem or phloem, some of the chemical moves into adjacent cells by diffusion across the plasmalemma. Cell dysfunction results when the chemical accumulates to a sufficient concentration in those cells.

Herbicide Transformation in Exposed Plants

The chemical structure of a herbicide can be changed while it is in a plant. Most of these biotransformations result in *deactivation*, or reduced phytotoxicity, of the herbicide. However, *activation* or enhanced phytotoxicity can result from some biotransformations. In many cases, the transformation results from chemical degradation, although additions and substitutions of other molecules to the herbicide are also possible. The fate of herbicides in exposed plants has been the focus of exhaustive research and therefore cannot be discussed completely here. Several excellent reviews are available, however, and they should be consulted for more information concerning this topic (Ashton and Crafts, 1981; Kearney and Kaufman, 1988, Shimabukuro, 1985).

PHASES AND TYPES OF HERBICIDE TRANSFORMATIONS

If not killed by a herbicide, exposed plants usually detoxify the chemical according to a three-step process proposed by Shimabukuro and colleagues (1982, 1985) (Table 10.1). Phase I reactions generally detoxify or activate the herbicide and predispose the herbicide to the transformations of Phase II. Conjugations with other molecules occur during this second phase and usually result in complete loss of herbicidal activity. Herbicides are subsequently metabolized further or incorporated into insoluble plant residues during Phase III reactions. Insoluble plant residues are the end product of herbicide biotransformations because plants have no method for excretion (Table 10.1).

Plants alter the molecular configuration of herbicides through an array of chemical reactions—in particular, oxidation, hydroxylation, hydrolysis, conjugation, and ring cleavage. Plant species vary markedly in their ability to biotransform herbicides and most possess several alternative pathways for herbicide metabolism. Most of these transformation reactions are catalyzed by specific enzymes.

TABLE 10.1 Generalized Summary of Pesticide Behavior and Fate in Plants

Characteristics	*Initial Properties*	*Phase I*	*Phase II*	*Phase III*
Reactions	Pesticide ⟶	Oxidation; reduction; hydrolysis ⟶	Conjugation ⟶	Secondary conjugation or incorporation into biopolymers (insoluble residues)
Solubility	Lipophilic	Amphophilic	Hydrophilic	Hydrophilic and insoluble
Transport	Selective mobility	Modified or reduced mobility	Limited or immobile	Immobile
Phytotoxicity	Toxic	Modified or less toxic	Greatly reduced or nontoxic	Nontoxic

Source: Shimabukuro et al., 1982.

Oxidation. Oxidation reactions of herbicides in exposed plants are common and result in a less phytotoxic substance than the herbicide active ingredient. Examples include hydroxylation of bentazon (Figure 10.9), *N*-dealkylation of atrazine (Figure 10.10), and demethylation of monuron (Figure 10.11). Each of these examples is a Phase I reaction. An activation reaction that occurs in some plants is the β-oxidation of 2,4-DB (Figure 10.12). In this reaction, the relatively inactive 2,4-DB is transformed in susceptible plants to the more toxic 2,4-D. Molecules with an even number of carbon atoms in the side chain are degraded, two carbons at a time, into 2,4-D, whereas molecules with an odd number of carbon atoms in the side chain are eventually degraded by the same process to 2,4-dichlorophenol, which is nontoxic. This latter reaction is a form of deactivation.

Hydroxylation. The primary structure of many herbicides consists of an aromatic carbon ring (Chapter 9). The hydroxylation (Phase I) of the ring of some herbicides is an important mechanism of detoxification. An example of ring hydroxylation of atrazine that results in hydroxy-atrazine (nonphytotoxic) is shown in Figure 10.13. Other examples are the ring hydroxylation of 2,4-D and dicamba.

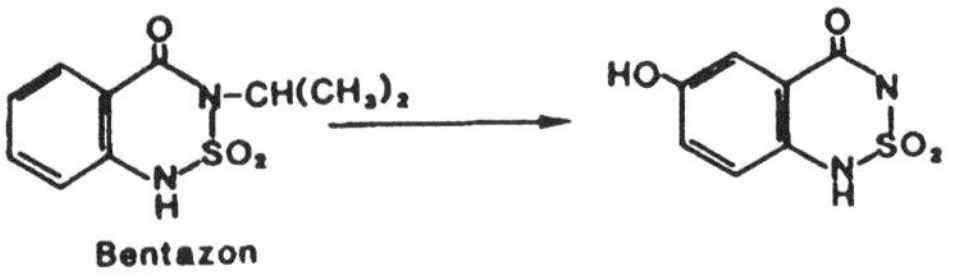

FIGURE 10.9 Aryl hydroxylation of bentazon. (From Shimabukuro, 1985.)

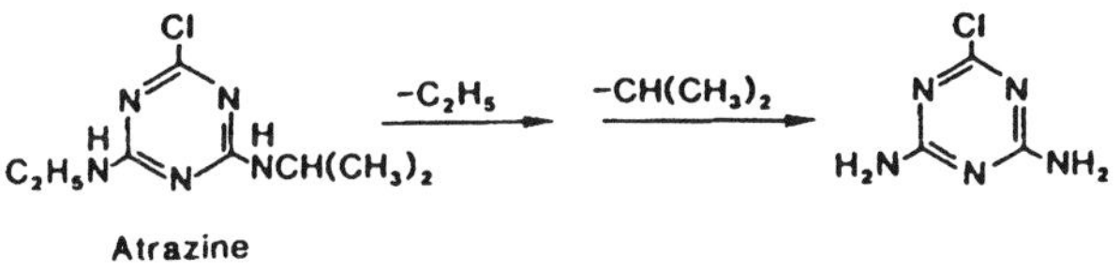

FIGURE 10.10 *N*-dealkylation of atrazine. (From Shimabukuro, 1985.)

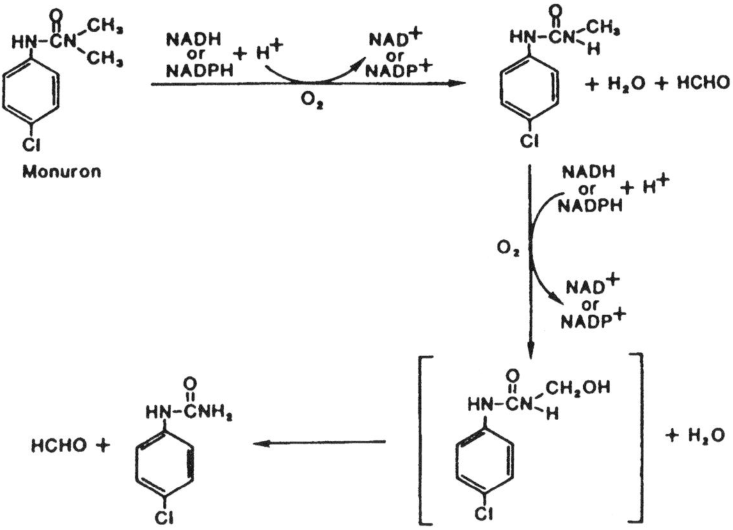

FIGURE 10.11 The mechanism for mfo-catalyzed *N*-demethylation of monuron. (From Shimabukuro, 1985.)

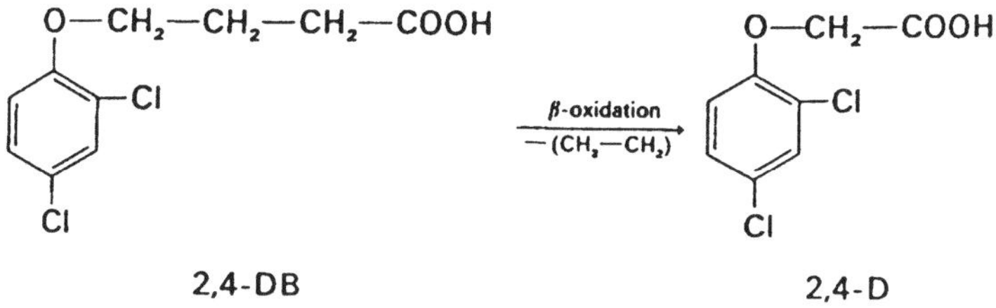

FIGURE 10.12 Beta-oxidation of 2,4-DB to 2,4-D. (From Ross and Lembi, 1985.)

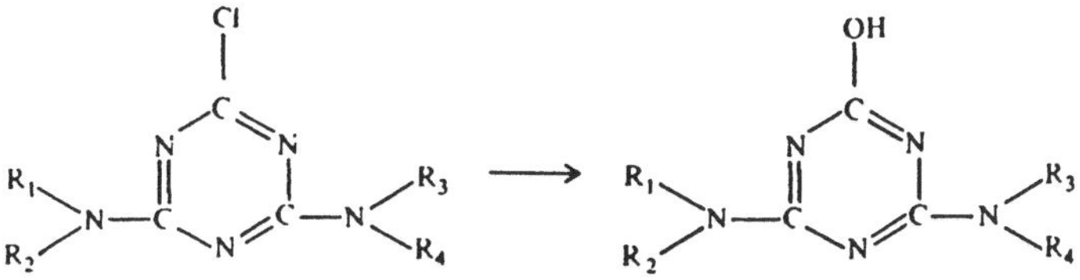

FIGURE 10.13 Hydroxylation of a 2-chlorotriazine molecule accompanied by dechlorination. (From Ashton and Crafts, 1981.)

Hydrolysis. Hydrolysis (Phase I) is an important selective mechanism for herbicides in the carbamate, thiocarbamate, substituted urea, sulfonylurea, and triazine groups of chemicals. Hydrolysis may result in either activation or deactivation of the herbicides. Shimabukuro (1985) indicates that propanil (Figure 10.14) is hydrolytically degraded (deactivation) by an aryl-acylamidase. This enzymatic reaction results in differential tolerance between rice (tolerant) and barnyardgrass (susceptible) to propanil. The hydrolysis of benzoylprop-ethyl (Figure 10.14), chlorfenprop-methyl, and dichlofop-methyl to their herbicidally active acids (activation) is the basis for activity and selectivity among grass species.

Conjugation. Conjugation is the reaction of a herbicide with another substance in the plant, resulting in a new compound of higher molecular weight. Generally, conjugation is a process in which plants convert herbicide molecules into less toxic and more water-soluble metabolites (Phase II, Table 10.1). Conjugation reactions also function in translocation of some herbicides. Conjugation of herbicides with glucose happens readily in most plant species. Such glycoside formation is an important mechanism for symplastic translocation. The conjugation of herbicides with amino acids, such as glutamic and aspartic acid, are also well-known biological reactions. These conjugation reactions have little effect on the ultimate activity of herbicides, although they too are important for translocation.

Conjugation with glutathione, a polypeptide, is a major mechanism for herbicide detoxification in plants. Glutathione conjugation with atrazine, propachlor, alachlor, metolachlor, and fluorodifen have all been characterized (Figure 10.15; Shimabukuro 1985). Up to 90 percent of the propachlor in corn can be converted to nonphytotoxic water-soluble metabolites by glutathione conjugation. Both enzymatic (e.g., atrazine) and nonenzymatic (e.g., EPTC) conjugation of herbicides with glutathione are possible.

Ring cleavage. Some evidence exists for the partial cleavage of the aromatic ring found in most organic herbicides. However, the process appears to be

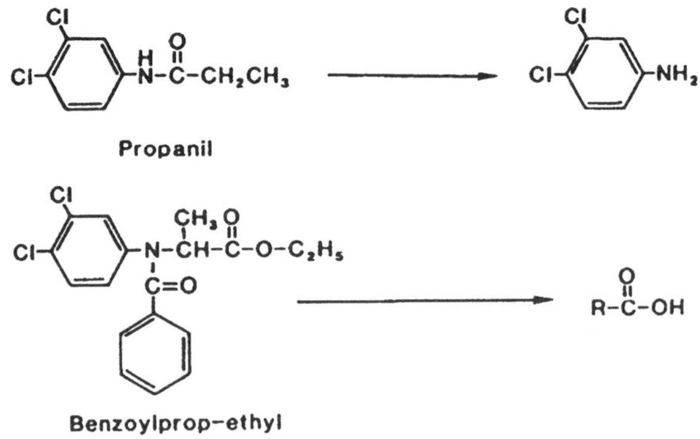

Propanil

Benzoylprop-ethyl

FIGURE 10.14 Hydrolytic metabolism of propanil and benzoylprop-ethyl. (From Shimabukuro, 1985.)

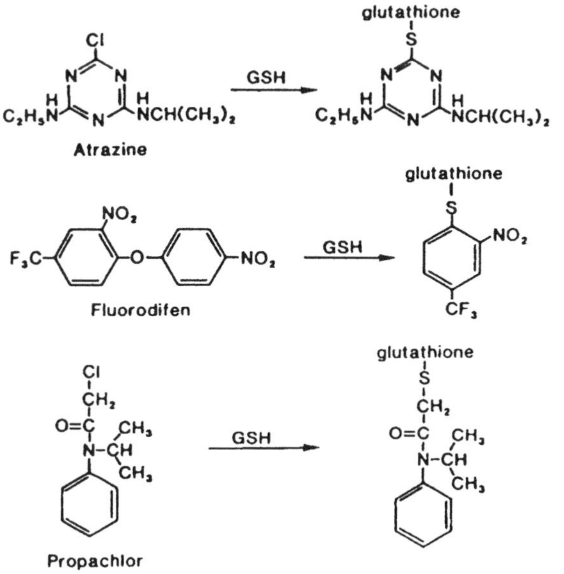

FIGURE 10.15 Glutathione conjugates of atrazine, fluorodifen, and propachlor. (From Shimabukuro, 1985.)

slow and relatively unimportant in higher plants. It is more likely for the ring to be incorporated into insoluble, nonphytotoxic residues (Phase III, Table 10.1) following biological substitution or removal of the various substitutions on it. There is little evidence for the complete degradation of any herbicide ring structure to CO_2 at a biologically important rate.

HERBICIDE SAFENERS

Safeners or *protectants*, when used in combination with a specific herbicide, improve the tolerance of crops to the chemical. For example, *N,N*-diallyl-2,2-dichloracetamide, a safener, is mixed with some thiocarbamate herbicides to improve the tolerance of corn. Naphthalic anhydride is used as a seed protectant to reduce corn injury from thiocarbamate herbicides. The tolerance of sorghum to metolachlor has been improved by using α-[(1,3-dioxalan-2-ylmethoxy)iminol] benzene acetonitrile as a safening agent. Increased glutathione conjugation is the primary mechanism for the improved herbicide tolerance of certain plants that have been treated with safeners (Devine et al., 1993).

Biochemical Processes Inhibited by Herbicides: Mechanisms of Action

After a herbicide has entered the plant and reached a specific site in plant cells, it inhibits a biochemical process. The specific biochemical process inhibited often depends on the chemistry of the herbicide and, in some cases, the plant species involved. Both the site and biochemical reaction inhibited are known for many herbicide groups. For others, only the effect of the herbicide, rather than the cause of the response, is known. Most herbicides fall into five categories, based on the cellular site or biochemical process inhibited by them. These processes/sites are *photosynthesis, respiration, cell division, nucleic acid metabolism* and *protein synthesis*, and *membrane function* (Ashton and Crafts, 1981; Devine et al., 1993).

PHOTOSYNTHESIS

Photosynthesis is the conversion of light energy, water, and carbon dioxide into the chemical energy of carbohydrates. As discussed in Chapter 6, the light-requiring reactions of photophosphorylation ("light reactions") occur in pigment-containing chloroplast membranes called *thylakoids* found in the chloroplasts of plant cells (Figure 10.16). Photons of light are absorbed by chlorophyll molecules in the thylakoids, which drives a flow of electrons from water to $NADP^+$ (Figure 10.17 and 6.4). Light absorption and electron flow of photosynthesis occur in two photosystems in the light reactions (Figure 10.17 and 6.4). Each photosystem contains a special form of chlorophyll, called P_{680} in Photosystem II and P_{700} in Photosystem I, accessary pigments, and electron-carrying proteins, which function together as a "reaction center." Each reaction center pigment donates an electron, following photon absorption, to its respective electron acceptor, Q (PSII) and ferredoxin (PSI; Figure 10.17 and 6.4). The

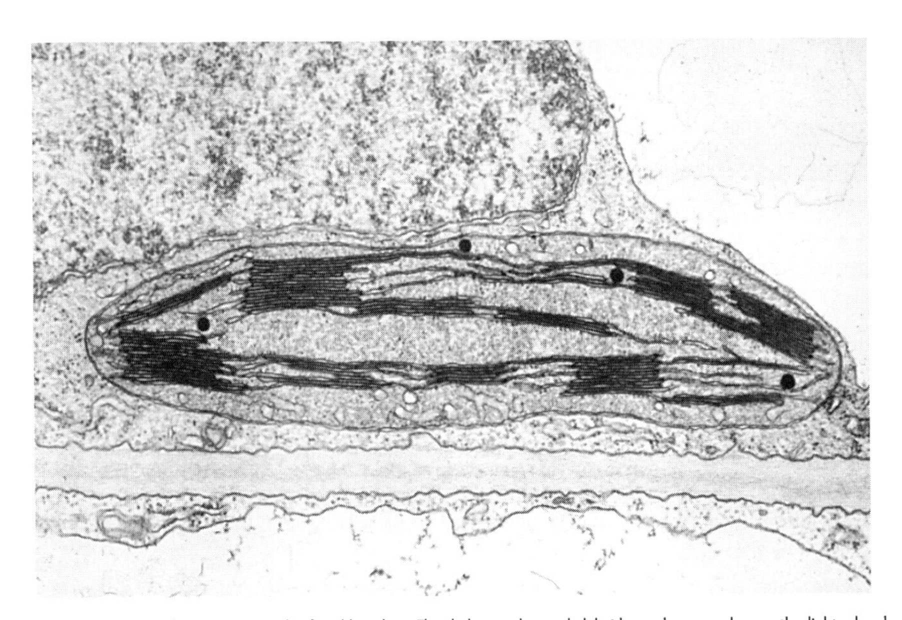

FIGURE 10.16 Electron micrograph of a chloroplast. The dark strands are thylakoid membranes, whereas the light-colored matrix is the stroma of the chloroplast. (Micrograph by W. Mast, University of California, Davis.)

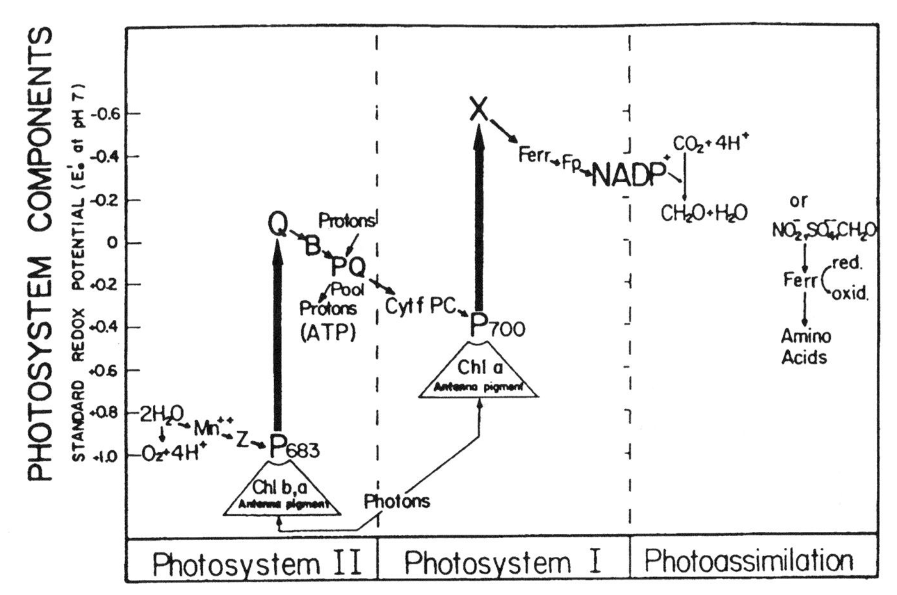

FIGURE 10.17 Components of photosynthesis schematically presented as a two-photosystem electron-transport pathway coupled to photoassimilation of CO_2, NO_2^-, and SO_4^{2-}. The CO_2 assimilation and electron-transport pathway are presented stoichiometrically. No stoichiometry is implied in the electron-transport components as drawn. For example, PQ is a pool within a leaf chloroplast that may be ten-fold larger than other electron-transport components, or Photosystem I may be present at two- to three-fold excess over Photosystem II in a leaf chloroplast. The abbreviations used are Z, an unidentified component donating electrons to P_{683}; P_{683}, the reaction center chlorophyll of Photosystem II; Q, the unidentified acceptor of electrons from P_{683}; B, the protein probably bound to a quinone; PQ, plastocyanin; P_{700}, the reaction center chlorophyll of Photosystem I; X, the acceptor center for electrons from P_{700}, which is likely an iron-sulfur center; Ferr, ferredoxin; and Fp, flavoprotein ferredoxin NADP+ reductase. (From Black, 1985.)

electrons are then transferred through a series of oxidation-reduction (redox) reactions that involve plastoquinone (PQ) and various cytochromes to the final acceptor NADP+.

Electron flow results from the "splitting" (photolysis) of water, a process known as the Hill reaction. Electron flow through the reaction centers and redox components generates oxygen and energy used to produce ATP, and eventually reduced carbohydrate (Figure 10.17). Electron flow through Photosystems II and I, as depicted in Figure 10.17 and 6.4, has been termed the *Z-scheme* of photosynthesis to describe the movement of electrons along a gradient of redox potential. Herbicides bind to and inhibit various chemical components of the photosynthetic light reactions [see Black (1985) or Devine et al. (1993) for specific herbicides and a more in-depth discussion].

In the Calvin cycle ("dark reactions") which takes place in the liquid *stroma* of the chloroplast (Figure 10.16), CO_2 is fixed ultimately into six carbon sugars using the high-energy products (NADPH and ATP) formed during photophosphorylation. There are three primary methods of synthesizing CO_2 into sugars, but two, the C_3 and C_4 pathways, are predominant in most crops and weeds. These differences in the dark reactions are important physiological distinctions among plant species and are discussed in more detail in Chapter 6. However, most herbicides have limited impact on the dark reactions.

Electron-transport inhibitors. The greatest number of herbicides function by blocking the electron-transport system of photosynthesis. For example, nearly fifty percent of all commercial herbicides block electron transport between the primary acceptor in PSII (Q) and P_{700} (Figure 10.17) (Black, 1985). Many of these herbicides bind to a protein, often called D-1, located between Q and plastoquinone in the electron-transport chain. The binding of herbicide molecules to this protein prevents the further acceptance and transfer of electrons to other components of the sequence, effectively inhibiting photosynthesis. Examples of electron-transport inhibitors are atrazine, diuron, bentazon, and many other herbicides.

Although the inhibition of photosynthesis eventually prevents the production of carbohydrate in the dark reactions, the action of electron-transport inhibiting herbicides is too rapid to be explained by plant death by starvation. Light is necessary for symptoms (Figure 9.7) and death to result from these herbicides, which also suggests that inhibition of the dark reactions is not their mechanism of action. Rather, chlorosis and necrosis result from the photooxidation of chlorophyll and other chloroplast pigments that occurs when electrons no longer move through the electron-transport chain because of the herbicide (Ashton and Crafts, 1981).

Electron-transport diverters. Other herbicides attain biological activity by diverting electrons from the electron-transport system and subsequently forming compounds that destroy membrane integrity. These herbicides, represented by paraquat and diquat, intercept electrons in PSI near the primary electron receptor P_{700} (Figure 10.17). With the addition of the intercepted electrons, the herbicide molecules become chemical *free radicals*, which are highly reactive, unstable substances that transfer electrons to molecular oxygen, forming superoxide in plant cells (Black, 1985). Subsequent reactions result in the formation of hydrogen peroxide and hydroxyl radicals, which cause cell membrane damage and eventually plant death (Devine et al., 1993).

Inhibitors of photosynthetic pigments. Some herbicides inhibit the direct formation of pigments other than chlorophyll. The non-photosynthetic pigments affected by some herbicides (amitrole, fluridone, and norflurazon) are the carotenoids, which protect chlorophyll from photooxidation (Ashton and Crafts, 1981). Plant growth following treatment with these herbicides is usually white (albinism, Figure 9.8) or light yellow in color, which suggests that direct disruption of the light reactions is not the cause of herbicidal injury. Treated plants often require several weeks to die, indicating that inhibited carbohydrate production may be the cause of plant mortality.

RESPIRATION

During respiration, energy obtained from oxidation of carbohydrates, proteins, and lipids in plant cells is converted into the phosphate-bond energy of adenosine triphosphate (ATP; Figure 10.18). These complex molecules, predominantly carbohydrates, first are enzymatically converted to acetyl-coA. Acetyl-coA is further metabolized to CO_2 by enzymes located in the *mitochondria* in a process known

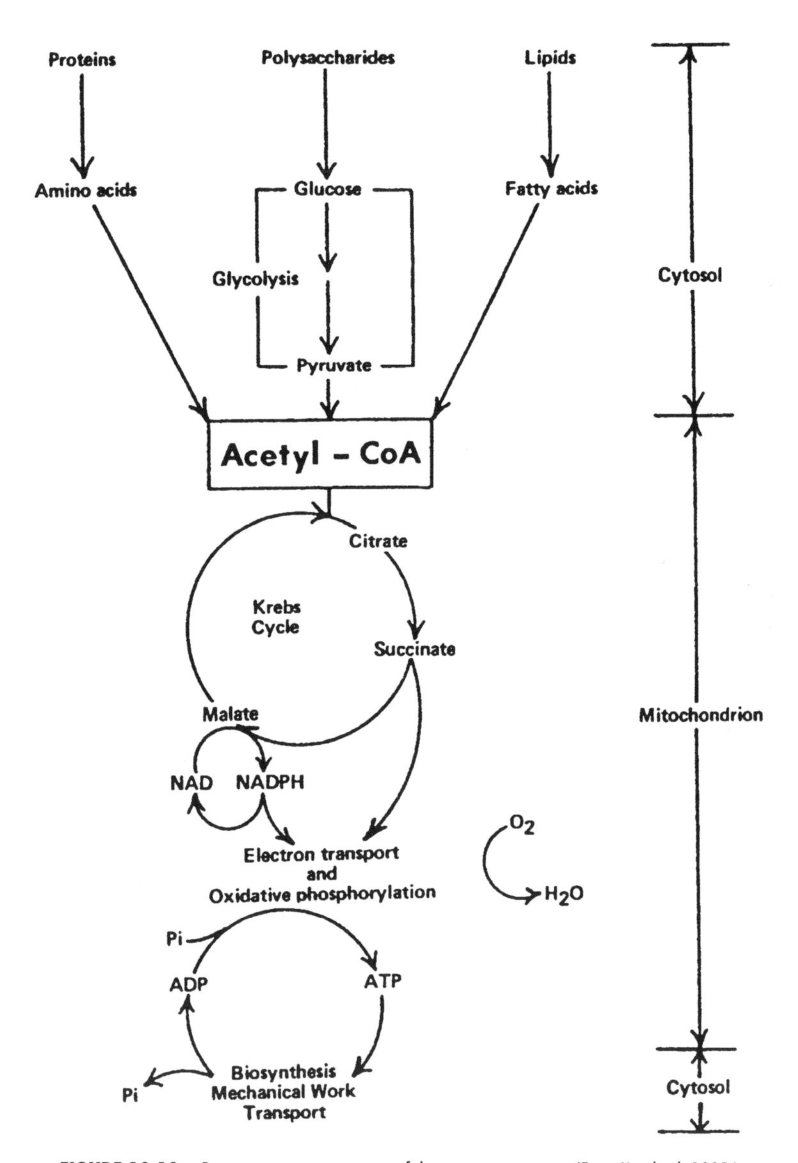

FIGURE 10.18 Diagrammatic representation of the respiratory process. (From Moreland, 1985.)

as the citric acid or Krebs cycle. Electrons and protons generated from the organic acids in the Krebs cycle are then transferred through a series of oxidation-reduction reactions in the inner membrane of the mitochondria (Figure 10.18). Simultaneously, as electrons move through the electron-transport system, electron energy is transformed into the chemical energy of ATP. This reaction involves the esterification of inorganic phosphate (P_i) with adenosine diphosphate (ADP). In this process, called *oxidative phosphorylation* (Figure 10.18), oxygen is reduced to water by the electron-transport (cytochrome) system and ATP is produced. ATP is the most important molecule in all living organisms since it provides the energy necessary to drive all other biosynthetic, mechanical, and transport activities of cells. Many herbicides interfere with the production or use of ATP (Moreland, 1985); however, such interference may suppress plant growth only indirectly (Devine et al., 1993).

Electron-transport inhibitors. Electron-transport inhibitors are herbicides that interrupt electron flow in the respiratory electron-transport system (Figure 10.18). These herbicides combine with one of the protein carriers in the system and thus prevent subsequent oxidation-reduction reactions. When electron flow is interrupted, the associated (coupled) reactions of phosphorylation also are inhibited. Moreland (1985) has reviewed extensively herbicides that disrupt respiratory processes and indicates that most herbicides that affect electron transport also affect phosphorylation (production of ATP). According to Devine et al. (1993), however, no current herbicides have mitochondrial respiration as their primary site of action since they are much more effective as inhibitors at some other site.

Oxidative phosphorylation uncouplers. Like photophosphorylation in chloroplasts, oxidative phosphorylation in mitochondria can be inhibited either through direct effects on ATP synthesis or by uncoupling, that is, by dissipation of the proton gradient that drives phosphorylation. Many of the herbicides listed by Moreland (1985) that inhibit electron transport also inhibit phosphorylation as a result, while other herbicides inhibit or uncouple phosphorylation directly. However, lower concentrations of some herbicides are required to inhibit phosphorylation than to inhibit electron transport. Therefore, it is assumed that inhibiting ATP formation in mitochrondria is the primary mechanism by which those herbicides inhibit respiration (Ashton and Crafts, 1981; Moreland, 1985). Bromoxynil and dinoseb are examples of phosphorylation uncouplers.

CELL DIVISION

Cell division in plants consists of mitosis and cytokinesis and is the first step in tissue differentiation and plant growth (Weir et al., 1982). *Mitosis* is the process by which the cell nucleus divides into two identical "daughter " nucleii such that each new cell has identical genetic components. There are four stages of mitosis—*prophase, metaphase, anaphase*, and *telophase*. *Interphase*, the phase between successive mitotic events, is not strictly a stage of mitosis since it is the period of preparation for cell division. Division of the cytoplasm following mitosis is called *cytokinesis*. The steps of cell division are outlined below.

1. *Interphase*: Preparation for cell division
   DNA replication
2. *Mitosis*: Identical daughter nucleii produced
   Prophase—chromosomes condense and become distinct
   Metaphase—chromosomes align on equatorial plane
   Anaphase—chromatids (duplicate chromosomes) separate
   Telophase—daughter nucleii form
3. *Cytokinesis*: Different parcels of cytoplasm formed

An important structure involved during cell division is the *spindle apparatus* (Figure 10.19). This structure forms in interphase and becomes pronounced

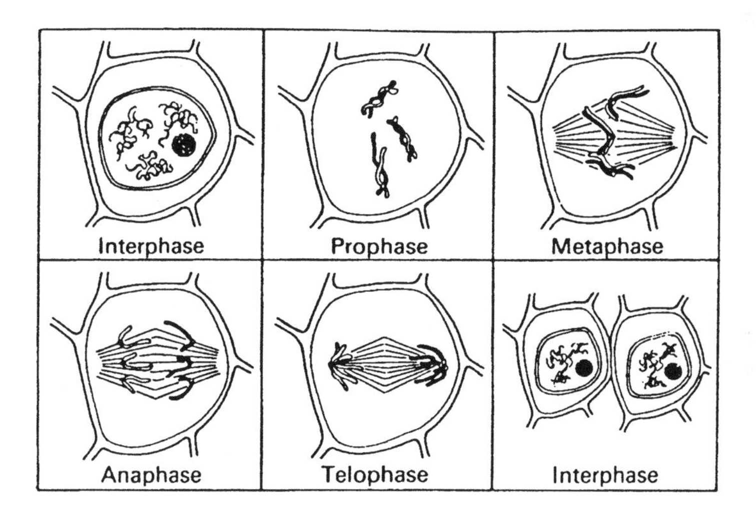

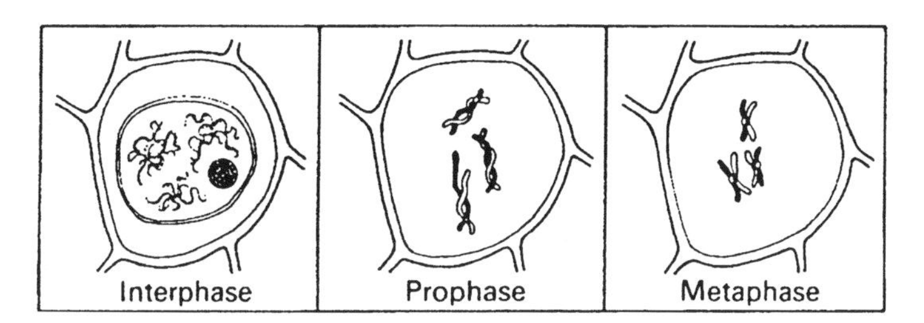

FIGURE 10.19 *Top two rows*, Diagram of normal mitotic sequence; *bottom row*, effect of dinitroaniline herbicides on mitotic sequence in root cells. (By F.D. Hess, in Ross and Lembi, 1985.)

during metaphase, anaphase, and telophase. Spindle apparatii are composed of filamentous protein structures called *microtubules*, which are made of polymerized *tubulin*. The microtubules are the functional elements of mitosis because they determine the plane at which cells divide and subsequent organization of new cells. The dinitroaniline herbicides and pronamide prevent spindle apparatus formation in root cells (Figure 9.6). Prophase is normal following treatment with these herbicides but the spindle apparatus is absent because the herbicides bind to tubulin, preventing formation of normal microtubules. The chromosomes are unable to begin metaphase and the chromatids cannot migrate to their respective poles at anaphase (Figure 10.20). Without the proper distribution of nuclear material (DNA) into daughter nucleii, new cell formation is abnormal (Figure 10.19 and 10.20) and mitosis and plant growth cease (Hess, 1987; Bartels, 1985).

Propham and chlorpropham, carbamate herbicides, also cause the spindle apparatus to become dysfunctional. The chromosomes of plants treated with these herbicides accumulate in various parts of the cell, resulting in abnormal multinucleate cells that cannot continue mitosis (Hess, 1987; Devine et al., 1993). Apparently, these herbicides affect the microtubule organizing center, but do not prevent polymerization of tubulin into microtubules.

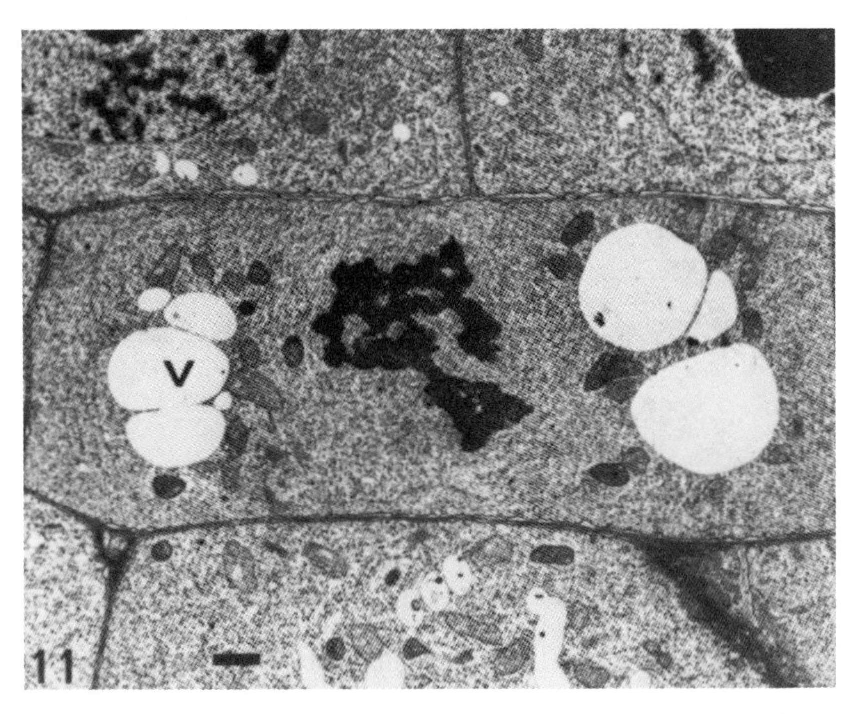

FIGURE 10.20 Nucleus of cotton root cell treated with trifluralin. Arrested metaphase division occurs as a result of microtubule disappearance in late prophase. Scale marker equals 1 micron. V symbolizes vacuole. (From Hess and Bayer, 1974. With permission of The Company of Biologists Ltd.)

There are several herbicides to which a precise mechanism of action cannot be attributed. These herbicides include members of the phenoxy acetic acid group (e.g., 2,4-D) and the thiocarbamate group (e.g., EPTC), in addition to bensulide, diphenamid, napropamide, siduron, and the substituted amides. In general, these herbicides inhibit development (probably cell division) of seedling roots and shoots (Ross and Lembi, 1985).

NUCLEIC ACID METABOLISM AND PROTEIN SYNTHESIS

Nucleic acid (DNA and RNA) formation and protein synthesis are aspects of cell function that are closely linked to cell division. These processes involve the transfer of genetic information to form functional and structural proteins in plant cells. They begin with the replication of DNA, transcription of RNA, and finally translation of the genetic information into proteins. It is obvious that any herbicide that alters or interacts significantly with the reactions of these processes will affect the growth and development of treated plants (Ashton and Crafts, 1981). However, Devine et al. (1993) point out that few, if any, herbicides are known to affect either nucleic acid or protein synthesis directly, although reports of indirect effects of herbicides on these processes are common.

For example, exposing a susceptible plant to a herbicide such as 2,4-D enhances its RNA polymerase activity, causing greater RNA and protein synthesis, followed by massive cell proliferation in primarily meristematic tissues. Because of this extensive and distorted cell proliferation, food reserves of treated plants are used up rapidly, causing death (Hanson and Slife, 1969; Ashton and Monoco, 1991). Other herbicides may affect metabolism by inhibiting formation of a precursor to protein synthesis. For example, inhibition of the biosynthesis of aromatic amino acids, which are components of proteins, is the mechanism of action of glyphosate (Kearney and Kaufman, 1988), and sulfonylurea herbicides bind tightly to acetolactate synthase (ALS), which inhibits the synthesis of branched-chained amino acids (Beyer et al., 1988).

MEMBRANE FUNCTION

All plant membranes, even though involved in different cellular functions, have a remarkably similar structure, composed of a lipid bilayer in which proteins are imbedded as discussed earlier (Figures 10.7 and 10.8). Membranes are involved in nearly every process that occurs in cell biology. For example, electron transport and phosphorylation—which occur in respiration and in the light reactions of photosynthesis, as well as protein synthesis—all occur in association with cellular or organelle membranes. In addition, the plasmalemma and tonoplast (outer membrane of vacuoles) are important regulators of solute transport into and out of cells and organelles (Balke, 1985). Since membranes are so critical to the function of plant cells, it is not surprising that any herbicide that perturbs membrane integrity will disrupt normal plant development.

brane disruption. The symptoms of membrane disruption are severe wilting and foliage desiccation from cell leakage. Herbicides that create cell leakage are grouped into three categories. These include herbicides that

- are oil soluble
- form free radicals in living cells
- uncouple oxidative phosphorylation

Each of these categories of herbicides has been discussed in other sections of this chapter. Herbicides that are oil soluble or oils themselves solubilize the plasmalemma or tonoplast, allowing cellular contents to leak into the intercellular spaces of leaf tissue. Bipyridilium and nitro-substituted diphenylether herbicides form free radicals by diverting electrons from photosynthetic (Figure 10.17) and mitochondrial (Figure 10.18) electron-transport systems. Subsequent membrane dysfunction results from the formation of hydrogen peroxide or similar substances toxic to cell membranes. Dinitrophenol herbicides also disrupt the functioning of electron transport in membranes but exert their effect by uncoupling oxidative phosphorylation (Figure 10.18).

Disruption of lipid synthesis. Lipids are essential to the integrity and function of cell membranes (Figures 10.7 and 10.8) and the proper function of enzymes. For example, there are mitochondrial, plastidic, and cytoplasmic lipids that are synthesized through several different independent pathways (Balke, 1985). Simply suppressing or inhibiting such a pathway alters the relative proportions of lipid present in a membrane, which could have dramatic effects on membrane function. Several substituted pyridazinones, cyclohexanedione, and aryloxyphenoxy herbicides are known to alter lipid content or composition of higher plants (Ashton and Monaco, 1991).

FATE OF HERBICIDES IN THE ENVIRONMENT

The occurrence and persistence (fate) of herbicides in the environment has become a serious concern for farmers, foresters, scientists, and the general public. Figure 10.21 demonstrates the diversity and interrelationship of environmental processes that lead to herbicide movement, detoxification, degradation, or persistence. In order for herbicides to be effective, they must persist long enough to kill weeds. However, if persistence is too long, herbicide injury to non-target plants, undesirable residues in crops, or contamination of various components of the environment (air, soil, water, organisms) may result. The length of time herbicides persist in a field following application is determined by

- displacement (movement) to other environmental compartments
- adsorption to soil
- decomposition (degradation)

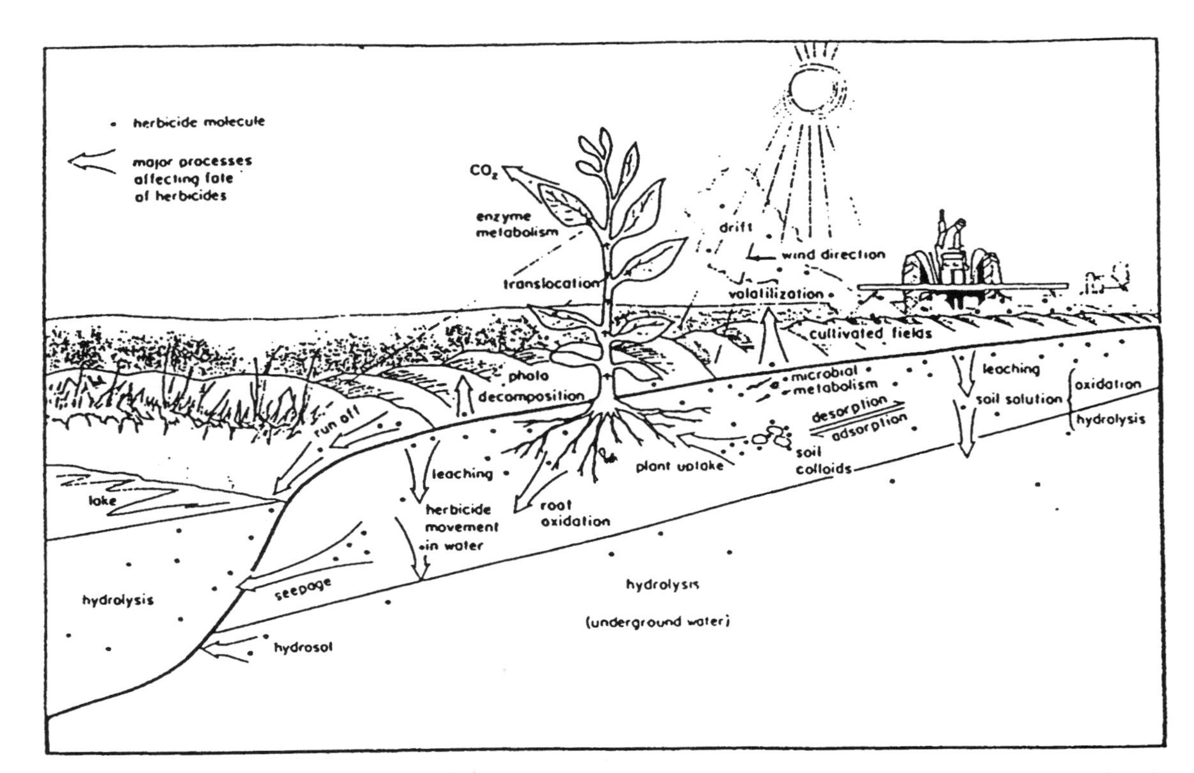

FIGURE 10.21 Diagram of the interrelations of processes that lead to displacement, adsorption, or decomposition of herbicides in the environment. (From Akobundu, 1987.)

Knowledge about environmental processes is necessary in order to understand the patterns of herbicide persistence. This same information can also be used to avoid displacement of herbicides from their intended site of application or to improve their decomposition. Minimal environmental contamination results when the processes that regulate herbicide fate are understood, because then application methods or management techniques can usually be devised that will reduce herbicide impacts to most environmental compartments.

Herbicide Movement in the Environment

Movement of a herbicide from an intended site of application or through the soil profile was once considered to be a loss of the chemical from the environment. However, this assumption is not necessarily true if a proportion of the herbicide applied simply moved to another environmental compartment (air, soil, surface water, ground water) from the treated field. Chemical displacement can be in any direction, vertical or lateral, in the soil, across the soil, into plants, or into the atmosphere (Figure 10.21). Movement into the atmosphere is determined by herbicide vapor gradients and air circulation patterns. Movement on and in the soil is determined primarily by the flow of water.

Sometimes, herbicide movement is desirable and several techniques have been developed that use movement to improve selectivity (Chapter 9). At other times, herbicide movement results in the chemical being where it is not intended. Unnecessary or unexpected environmental contamination resulting from herbicide use can be avoided if the processes of herbicide displacement, adsorption, and decomposition are understood.

HERBICIDE MOVEMENT IN AIR

Drift is the physical displacement of a herbicide in the form of particles or droplets from the intended target during application. Since most herbicides must move through the air in order to reach target vegetation or the soil surface, the opportunity for drift is always present. Drift is of concern because it may result in herbicide injury to plants not intended to be controlled, herbicide residues in adjacent tolerant crops, or contamination of land or surface water. Drift is an important way in which herbicides move in the environment.

Methods to reduce drift. As discussed in Chapter 9, herbicides are most often applied as a mixture with water and dispensed as a spray of droplets. Most of the droplets fall rapidly to the targeted vegetation or soil. However, some droplets do not. It is these droplets, usually of very small size, that result in drift (Table 10.2). Figure 10.22 shows the pattern of spray coverage for a typical aircraft application. The greatest potential for drift exists when droplets are small, usually less than 100 microns in diameter. Fine droplets can remain in the air for a long time and travel for an indefinite distance. A major way to reduce

TABLE 10.2 Spray Droplet Size and Its Effect on Drift

				Evaporating Water[a]	
Droplet Diameter	*Steady State Fall Rate*	*Time to Fall 3 m in Still Air*	*Drift distance in 3 m Fall with 5 km/h Wind*	*Lifetime of Droplet*	*Distance Droplet Falls in Lifetime*
(μm)	(m/s)	(s)	(m)	(s)	(cm)
5	0.00076	3960	4816	0.04	<2.5
10	0.003	1020	1372	0.16	<2.5
20	0.012	230	338	0.64	<2.5
50	0.075	40	54	3.5	27
100	0.279	11	15	14	384
150	0.46	8.5	8	36	1200
200	0.721	5.4	5	56	4263
500	2.139	1.6	2	400	15,250
1000	4.0	1.1	1	1620	>15,250

Source: Bode, 1987.

[a]Air temperature 30°C; R.H. = 50%.

drift of herbicides is to apply them with equipment systems or nozzle types that produce relatively large droplets. The microfoil boom is a special application device that is used on aircraft to reduce drift by producing droplets of very large size. Reducing operation pressure, spraying with materials of high viscosity, adjusting nozzle spacing and boom height, and using hoods or shields are other ways to decrease drift.

During ground application, less herbicide will be lost by drift if nozzles are close to the soil surface or vegetation being sprayed. When herbicides are applied inside shields, the potential for drift is substantially reduced. Wind and temperature inversions also influence the potential for herbicides to drift dur-

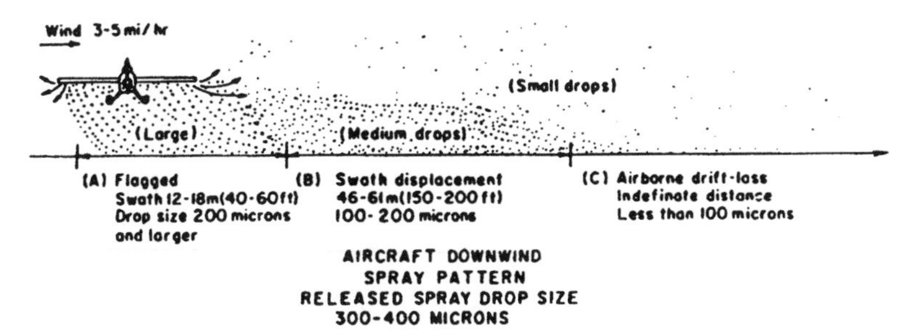

FIGURE 10.22 Effect of drop size and wind condition on aircraft swath displacement. (From Akesson and Walton, 1989.)

ing and after application. For this reason, herbicides are normally applied when the wind speed is less than 5 mph and during the morning when temperature inversions are least likely. In addition, spraying downwind results in less drift than spraying upwind or against a cross-wind. Some herbicide formulations are less likely to drift than others. For example, granule and pellet formulations will not drift as far as small droplets. Some herbicide applicators mix drift retardant chemicals (see adjuvants, Chapter 9) in the spray solution/suspension to create large droplets that fall to the target area rapidly. Such spray additives can reduce coverage of the herbicide on the foliage of treated plants, however, and sometimes diminish the effectiveness of weed control by some herbicides.

Use of buffer zones. Injury to susceptible crops or sensitive areas often can be avoided by precautionary measures before and during herbicide applications. For example, sensitive areas, such as stream beds and riparian areas in forest plantations and range environments, are protected by wide buffer zones that are adjacent to the area being treated (Figure 10.23). Buffer zones vary in size according to the herbicide being applied, method of application, and local ordinances and regulations. Aerial applications of herbicides generally require wider buffer zones than other methods. Similarly, if susceptible crops are adjacent to a field to be treated with a herbicide, it is wise to leave an untreated strip of land between the two fields. This procedure reduces the likelihood of drift reaching the susceptible crop. If a field can be sprayed before a susceptible crop in an adjacent field emerges, damage often can be avoided.

FIGURE 10.23 Buffer zones along stream near adjacent forest plantations. (Photograph by David Hibbs, Oregon State University.)

Volatility. *Volatilization* is the change of a solid or liquid into a gas. All chemicals, including herbicides, can volatilize (vaporize) depending upon the vapor pressure of the chemical and temperature. As seen in Chapter 9, some herbicides volatilize readily, while others volatilize very little, if at all, under normal environmental conditions. Because volatility is an important source of herbicide displacement from treated soil and vegetation, herbicide labels and precautions should be followed closely to reduce air contamination and injury to nearby susceptible vegetation. Usually, volatile soil-applied herbicides are mechanically mixed with soil to minimize losses to the environment as vapors (Chapter 9). Non-incorporated volatile herbicides are lost more rapidly from wet soil, so incorporation must proceed more rapidly than with dry soil.

Incorporation of certain herbicides, such as the thiocarbamates, into the soil enhances adsorption and thus diminishes displacement via volatility. This practice also reduces the potential for vertical and lateral herbicide displacement in the soil with water. Incorporation of both volatile and nonvolatile herbicides into the soil often improves weed control, as well. There are several methods that can be used to mix or incorporate herbicides into the soil mechanically (Chapter 9).

Herbicides in Soil

Herbicides are often applied directly to the soil or mixed with it for weed control. Other herbicides eventually settle on the soil during or after herbicide application. Herbicide runoff from treated foliage and the decomposition of herbicide-injured plants are other sources of herbicide entry into the soil system. Since most herbicides either are applied to the soil for weed control, or eventually arrive there, herbicide interactions within the soil system are a very important aspect of herbicide persistence and fate in the environment.

CHARACTERISTICS OF SOIL

Soil is the substance on the surface of the earth in which plants grow. One of its components is the basic mineral parent material that results from the weathering of rock. Soil also contains water, gases, organic matter and numerous types of living organisms. Any parcel of land, wherever it is located, is probably composed of several different *soil types* that vary in terms of physical structure, texture, profile, organic matter content, and fertility. All soils have four basic phases: solid, liquid, gaseous, and biological. Soil also is a dynamic system and therefore subject to the entry and loss of substances like herbicides, which can associate and interact with all four phases.

Solid phase. Solids are the mineral and organic components of soil that make up its bulk or mass. This phase ranges from large rocks and gravel to particles of microscopic dimensions (Table 10.3). The amounts of sand, silt, clay, and organic matter determine the soil's texture and type. As the size of soil particles

TABLE 10.3 The Size, Number, and Surface Area of Soil Particles

Particle Type	*Diameter (mm)*	*Number of Particles/g**	*Surface Area (sq cm/g)*
Very coarse sand	2.00–1.00	90	11
Coarse sand	1.00–0.50	720	23
Medium sand	0.50–0.25	5,700	45
Fine sand	0.25–0.10	46,000	91
Very fine sand	0.10–0.05	722,000	227
Silt	0.05–0.002	5,776,000	454
Clay	below 0.002	90,260,858,000	8,000,000

Source: Millar, Turk, and Foth, 1965.

*Assumed to have spherical shapes, based on maximum diameter of the particle type.

declines, both the number of particles and soil surface area increase dramatically. Thus, soils with high levels of silt and clay have much more surface area exposed than soils with a large amount of sand (Table 10.3). A soil texture triangle that shows relative amounts of sand, silt, and clay often is used to categorize soils according to the size and amount of solid particles they contain.

Organic matter, humus, is a component of soil that is formed by the decay of plant material. It provides much of the basic structure and fertility to a soil and must be maintained continually for adequate soil tilth and productivity in agriculture. Most soils that are predominantly mineral or inorganic have organic matter contents of 0.5 to 5 percent. A few soils have very low organic matter contents, only 0.25 percent or less. Organic soils, developed from old lakes or marshes where plant growth was abundant but biomass decomposition was minimal, have much higher organic matter contents than mineral soils.

The most chemically active components of soil are humus and clays because of their large surface area and charge. These two soil components are less than one micron in size, which is small enough that they act like colloids (Table 10.3). Soil colloids have the following characteristics:

- a very high surface to volume ratio
- capacity to adsorb other materials
- ability to act as catalysts for chemical reactions

Inorganic colloids, typically clays, are crystalline and have a definite size and shape. In contrast, organic colloids (humus) are amorphous, without constant form, size, or shape. Although both types of colloids are usually present in soils, the organic colloids are the most reactive.

Humus. Unlike clays, there is no definite structure or chemical identity to humus. It is an amorphous, shapeless, nearly odorless substance formed from the decomposition of organic matter. It is about 75 percent ligno-protein. Humus has a very high cation exchange capacity (Table 10.4). Humus also helps form soil aggregates that are necessary for good soil tilth and porosity. The continued formation and replenishment of humus by decay of plant residues provide for storage and release of soil nutrients, especially nitrogen, phosphorus, sulfur, and potassium.

Clays. There are two general types of clays: alumina silicates and hydrous oxides (iron and aluminum hydroxides). The alumina silicate clays are found in temperate regions and are categorized into three general classes: *kaolinite, montmorillonite*, and *hydrous mica.* Kaolinite is the most common clay. It is found in highly weathered soils, such as those of the southeastern United States. Montmorillonite and hydrous mica are found most often in the prairie and arid soils of the midwestern and western United States. Hydrous oxide clays are found primarily in tropical and subtropical regions.

The structure of the various types of clay is similar, consisting of thin, laminated sheets of either silica ($2SiO_2 \cdot H_2O$) or alumina ($Al_2O_3 \cdot 3H_2O$). Some of these sheets are flat while others are hexagonal in shape. Some are frayed at the ends; others are fibrous. As seen in Figures 10.24 and 10.25, the manner in which the silica and alumina sheets are arranged and bonded determines the type of clay that is formed. Hydrous oxide clays are formed from the intense weathering conditions of tropical and subtropical climates. The silicon in silicate clays is removed under such conditions, leaving iron and aluminum oxides as the soil colloids. Such soils are red or yellow and do not erode easily.

TABLE 10.4 The Cation-Exchange Capacity and Surface Area of Some Soil Constituents

Soil Constituent	*Cation-Exchange Capacity (meq/100 g)*	*Surface Area (m^2/g)*
Organic matter	200–400	500–800
Vermiculite	100–150	600–800
Montmorillonite	80–150	600–800
Dioctahedral vermiculite	10–150	50–800
Illite	10–40	65–100
Chlorite	10–40	25–40
Kaolinite	3–15	7–30
Oxides and hydroxides	2–6	100–800

Source: Reprinted with permission from Bailey and White, 1964. Copyright by the American Chemical Society.

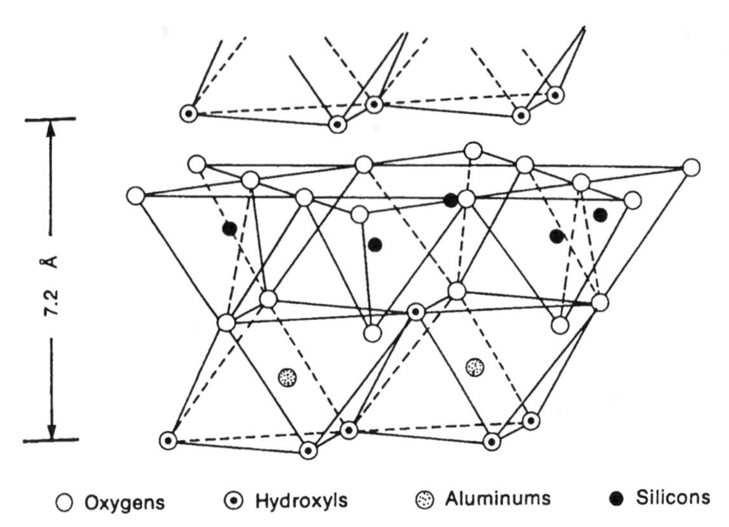

FIGURE 10.24 Structure of kaolinite. (In Anderson, 1977; From C.A. Black, *Soil-Plant Relationships* [New York: John Wiley & Sons, 1968].)

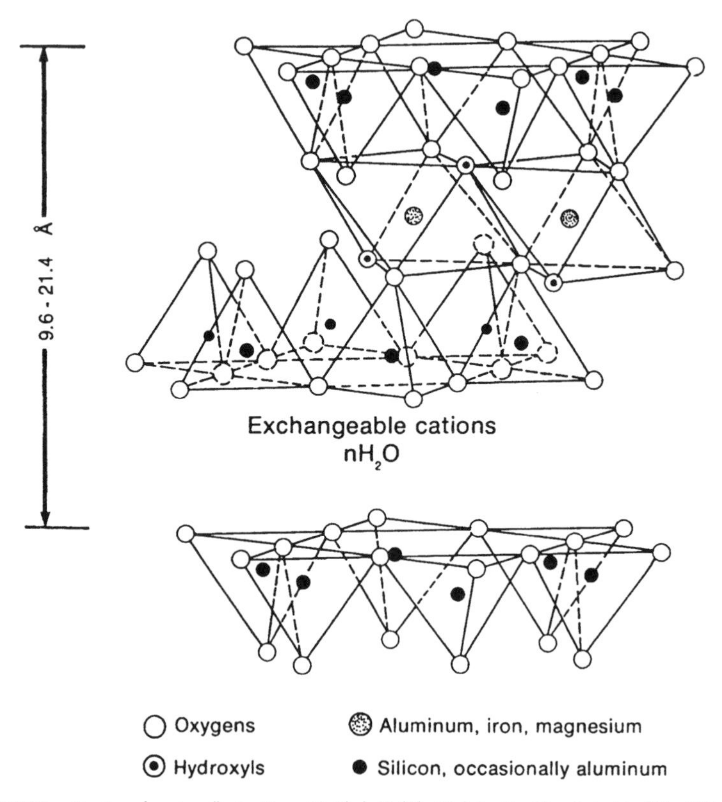

FIGURE 10.25 Structure of montmorillonite. (From C.A. Black, *Soil-Plant Relationships* [New York: John Wiley & Sons, 1968].)

Chemical properties of clay and humus. The overall electrical charge of soil is always negative because both clay and humus maintain net negative charges. Positively charged ions (cations), such as K^+, NH_4^+, Mg^{++}, and so on, are attracted and bind to these colloidal components of the soil. The sum total of exchangeable cations that a soil can bind is called its *cation exchange capacity* (CEC). Organic matter has a much higher cation exchange capacity than most clays (Table 10.4). For this reason, soils with high amounts of organic matter in them tend to be the most reactive and fertile. The ability of soil colloids to attract and bind cations and other substances is called *adsorption.* Although some anions (negatively charged ions or substances) also can be adsorbed by soil, anions usually are repelled by soil colloids, remain in the soil solution, and tend to move with it. The downward movement of substances with the soil solution is called *leaching.*

The pH of soil is also determined by the colloidal fraction, that is, the clay and humus, of a soil type. Soils that are leached extensively by rainfall have had most cations, except H^+, Al^{+++} and Fe^{+++} removed. High concentrations of H^+ result in acid soils with low soil pH levels (pH of 5.0 or less). Soils of arid regions are high in pH, because relatively mobile cations, such as Na^+, which are bases, are not extensively leached. Levels of pH that exceed 8.0 are often found when such conditions prevail and alkaline soils result. Soil adsorptive capacity, as expressed by CEC, organic matter content, and pH influence the ability of herbicides to bind and move in soil. These topics will be addressed more fully later.

Liquid phase. Water makes up the liquid phase of soil. It is essential for plant life since soil water must supply the large demands of evaporation and transpiration. The amount of water held by a soil is its *water-holding capacity* (Figure 10.26). This is determined by the number and size of voids or pore spaces that exist among the soil solids. The *field capacity* of soil (Figure 10.26) is the percentage of water remaining after the soil is saturated and then allowed to drain freely.

The liquid phase of soil has free, bound and hygroscopic water. *Free water* is associated primarily with large pore spaces, whereas *bound water* is held tightly to soil colloids and moves only by capillary action. *Hygroscopic water* binds so tightly to soil particles that it is unavailable for plant use. Many nutrients (ions and other substances) are dissolved in soil water. A dynamic equilibrium exists between materials adsorbed onto soil colloids and those dissolved in the soil solution. This balance can influence markedly the amount of nutrients or herbicide available for absorption by plants. Most substances present in the soil enter plant roots only as a component of the soil solution.

Gaseous phase. All of the gases found in the atmosphere are found in the air spaces of soil. Thus, a waterlogged, saturated soil has no gases associated with it (Figure 10.26), except those which are dissolved in the soil water. Soil at field capacity, in contrast, has both water and gases in pore spaces. The difference

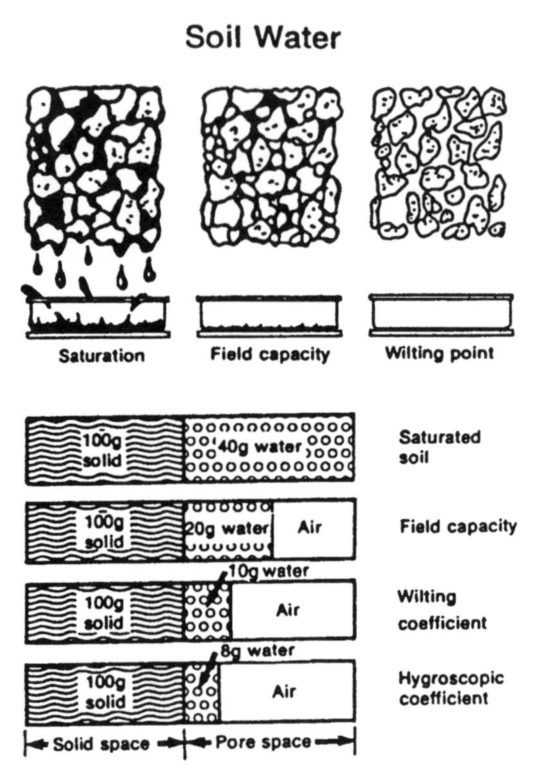

FIGURE 10.26 Diagrams showing the volumes of solids, water, and air in a well-granulated silt loam soil at different moisture levels. (From Buckman and Brady, *The Nature and Properties of Soils* [New York: Macmillan, 1969], p. 163. Copyright 1969 by Macmillan Publishing Co. Inc.)

between saturation and field capacity is the amount of pore space available for gases. Oxygen is the most essential soil gas because proper root function cannot occur without an adequate supply of it.

Biological phase. In addition to the inert chemical system described above, soil is a biological matrix of living organisms. The biological system of soil is made up of many species of animals, such as rodents, insects (eggs, larvae, and adults), and worms. Seed, plant roots, and other underground plant structures also are components of the living system of soil. Most soils contain a reservoir of weed seed and buds (Chapter 4). However, the principal and most numerous component of the soil biota are microorganisms. Soil-borne microorganisms are bacteria, fungi, actinomycetes, protozoa, slime molds, and algae.

Certain species of soil microorganisms are responsible for symbiotic nitrogen fixation, while some fungi (mycorrhizae) form associations with roots to enhance nutrient uptake (Chapter 7). Still other microorganisms are plant pathogens and are the cause of plant diseases. Soil microorganisms are largely responsible for the decomposition and decay of plant residues and other sources of organic matter. They also play a major role in nutrient cycling. Microorganisms also decompose herbicides and virtually any other natural or synthetic organic substance in soil. The degradation of herbicides by soil microorganisms is an important way in which chemicals are removed from soil, which decreases herbicide longevity and persistence.

HERBICIDE ADSORPTION TO SOIL

Adsorption is the adhesion of chemicals to the surfaces of solids. In soil, adsorption of herbicides is a colloidal process that involves the negatively charged particles of clay and organic matter (Figure 10.27). *Desorption* is the release of herbicide molecules into the soil solution. Through adsorption and desorption an equilibrium forms that regulates the amount of herbicide on soil colloids and in the soil solution (Figure 10.27). The amount of herbicide adsorbed to the soil depends upon the amount of clay, organic matter, and moisture present in each soil type, and also upon the ionization properties of each herbicide.

Adsorption of herbicides to soil colloids is the most important process affecting herbicide availability to plants and its persistence in the soil. Virtually all herbicides are adsorbed to some extent. Adsorption also influences the amount and rate of microbial degradation of herbicides in soil. When adsorbed, herbi-

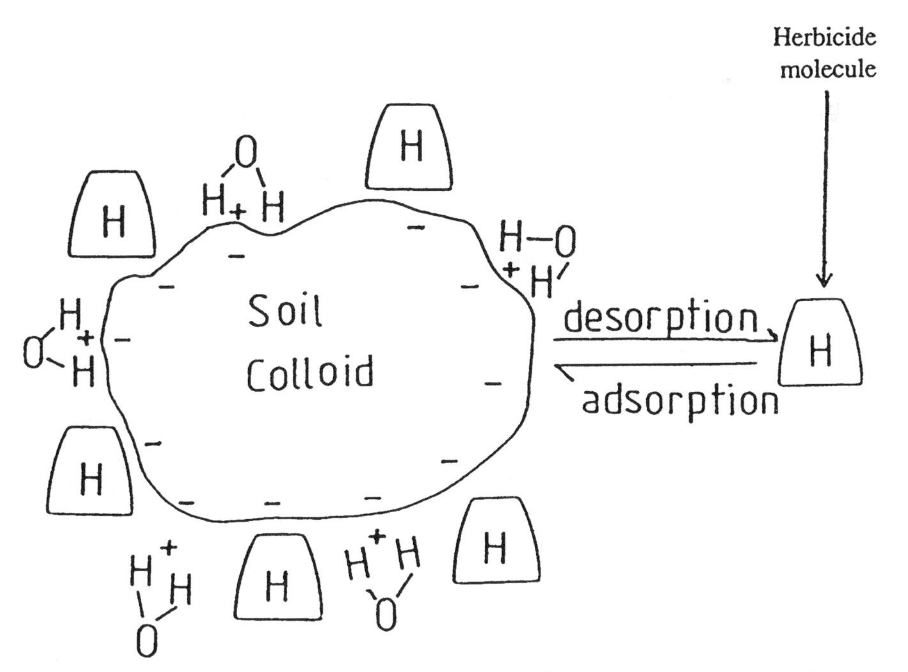

FIGURE 10.27 Schematic diagram of adsorption of molecules on soil colloids. Water molecules can compete with the herbicide molecules for adsorption sites on colloids. (From Adkins, 1995.)

cides are not available for plant uptake, lateral or vertical movement (leaching), or degradative processes. Herbicides first must be desorbed from soil particles in order for other soil processes, such as those just mentioned, to happen.

Influence of soil colloids on adsorption. Most herbicides are adsorbed by either clay or humus particles. However, humus contains many more adsorptive sites than clay (Table 10.4). Usually, only a small percentage of humus, one percent or less, is enough to affect the adsorptive properties of a soil. Because of the sensitivity of herbicide adsorption to the level of soil organic matter, both plant injury and herbicide persistence can be affected significantly by this factor. Herbicide recommendations and manufacturers' labels usually caution users concerning the activity and persistence of herbicides when applied to certain soils. Often the basis for such precautions is the amount of clay and organic matter present in the soil.

Sometimes activated carbon is added to soil to reduce the phytotoxicity of a herbicide to a susceptible crop. Activated carbon decreases herbicide availability by providing a large number of sites for adsorption. Once a herbicide is adsorbed by activated carbon, it is not easily desorbed from it. Activated carbon sometimes is applied as a narrow band over seeds of susceptible crop plants to increase herbicide tolerance (Figure 10.28). Activated carbon is sometimes mixed in the soil profile to reduce the amount of herbicide available to a susceptible crop in a rotation sequence.

FIGURE 10.28 Bands of activated charcoal applied over crop seed rows to improve herbicide selectivity. (Photograph by S.R. Radosevich, Oregon State University.)

Influence of water on adsorption. Water effectively interferes with adsorption by displacing herbicide ions or molecules from soil colloids. A thin film of water can surround soil colloids in wet soils, thereby making herbicide adsorption difficult. Herbicide that is not adsorbed will either remain in the soil solution, where leaching, degradation, or absorption by plants will occur, or volatilize into the air.

The equilibrium between the herbicide adsorbed to soil colloids and that in the soil solution (Figure 10.27) is also influenced by the water solubility and vapor pressure of the chemicals (Chapter 9). Herbicides with poor water solubility often adsorb more readily to soil than highly soluble chemicals. Usually, within a class of herbicides (Chapter 9) adsorption is inversely proportional to water solubility. This means that within a herbicide class, herbicides with low water solubility usually adsorb to soil best. However, water solubility is not a good indicator of adsorptive ability if the chemistry among herbicides being compared varies widely.

Volatile chemicals, such as the thiocarbamate herbicides, are easily displaced from soil colloids by soil moisture. These herbicides are prone to loss from the soil as vapors, unless special precautions are taken to enhance adsorption. Generally, adsorption of thiocarbamate herbicides is enhanced by incorporating (mixing) them into dry soil (see discussion above). Herbicide persistence in the soil is increased in this case because the volatile herbicides are not displaced to the atmosphere. Weed control is also enhanced by incorporating thiocarbamate herbicides into dry soil, for the same reason. As soil is wetted by precipitation or irrigation, these herbicides are desorbed and made available for plant uptake.

Influence of chemical charge on adsorption. Most herbicides act in the soil as either weak acids or bases (Figure 10.29). The degree that various herbicides ionize depends on the soil pH and the ionization constants of each chemical. Since the colloidal component of soil is negatively charged, herbicides that form a positive ion when dissolved in water (cations) are adsorbed readily to soil. Herbicides that form cationic molecules (e.g., paraquat) are bound so tightly to soil that they are unavailable for plant uptake. Under acid conditions (soil pH of 5 or 6), many herbicides undergo *protonation*, which makes neutral and anionic chemicals relatively more attractive to the negatively charged soil colloids. Under neutral or alkaline conditions (pH 7 and above), the opposite reaction occurs. Generally, herbicides will be repelled by colloids in alkaline soils, making them relatively more available for plant uptake or movement with soil water.

Over the pH range of most agricultural soils, many herbicides behave as though neutral in charge. Adsorption of herbicides with neutral charge is determined by chemical properties other than charge, such as water solubility, vapor pressure, or molecular size and shape.

HERBICIDE MOVEMENT WITH WATER

Lateral displacement of herbicides from soil can occur with surface runoff water (Figure 10.21). However, many agricultural fields are flat or have only

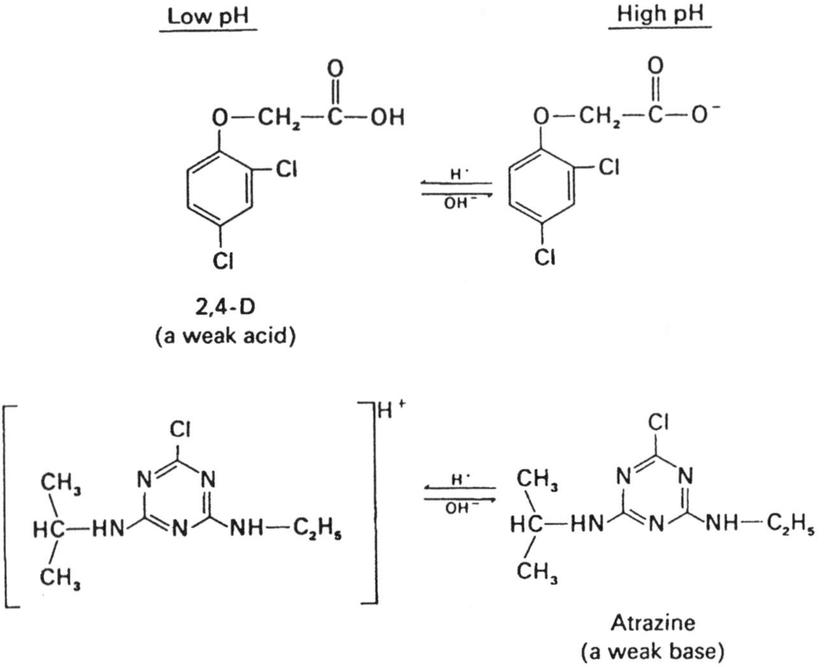

Low pH

High pH

2,4-D
(a weak acid)

Atrazine
(a weak base)

FIGURE 10.29 Effect of pH on the ionization properties of weak acids and bases. The proton on atrazine at low pH probably resonates among N atoms on the ring and side chain. (From Cruz et al., 1968, in Ross and Lembi, 1985.)

small inclines that do not favor lateral displacement of soil-applied chemicals. Because of herbicide adsorption to soil colloids, lateral movement on steeper slopes is also difficult. However, some lateral displacement of herbicides adsorbed to soil particles has been demonstrated when precipitation patterns are intense and movement of soil particles results.

In forest and rangeland situations, some lateral movement of herbicides with surface water can occur if the chemical is applied over a stream bed. Therefore, care must be taken during herbicide applications to assure that herbicides are not applied near active or dry streams. Stream beds and riparian zones are protected from herbicide exposure by regulations that specify the size of untreated buffer strips adjacent to them.

The *vertical displacement* (usually downward) with water of substances in the soil is called *leaching* (Figure 10.21). Actually, herbicide leaching can be in any direction in the soil profile, depending on where the chemical is placed and the direction of water flow. In fields that are irrigated by ditches or rills, lateral and sometimes even upward leaching of herbicides through soil may result in loss of crop selectivity and poor weed control. Upward water movement concentrates the herbicide in the crop row, at the soil surface, or along the ditch sides.

Downward leaching of herbicides in the soil profile is most usual because of the percolating action of water from rain or irrigation through the soil profile.

Soil incorporation of herbicides with water depends on leaching and is often used to achieve the desired placement (Figure 10.30). The most important factor that influences herbicide leaching is adsorption (Figure 10.27). Herbicide that is adsorbed is not present in the soil solution and therefore cannot leach until desorption occurs. Leaching is also dependent on the water solubility of the herbicide. Generally, herbicides that are relatively insoluble in water leach poorly and remain near the soil surface. For many such herbicides, the chemical tends to concentrate in the soil profile within a few centimeters of the soil surface (Figure 10.30).

The depth and amount of herbicide leached vertically is influenced not only by the adsorptive capacity of the soil but also by the amount and long-term duration of percolating water from irrigation and rainfall. The accumulation of herbicides deep in the soil profile or in ground water also will depend on the tendency of herbicides to degrade in the soil environment. Deep leaching was sometimes used in the past to "remove" herbicide residues from the rooting zone of susceptible crops or as a means of increasing crop tolerance to herbicides used in prior rotations. This practice is now discouraged.

Residues of herbicides in ground water supplies have been found in some regions where these chemicals are used repeatedly and extensively. There currently is substantial debate concerning the toxicity, cause, and importance of such residues in ground water. The residues probably arise because adequate time has not been allowed between herbicide applications for soil adsorption and degradative processes to work sufficiently. Decreasing the reliance of farmers on a single herbicide, greater use of crop rotations, and use of different her-

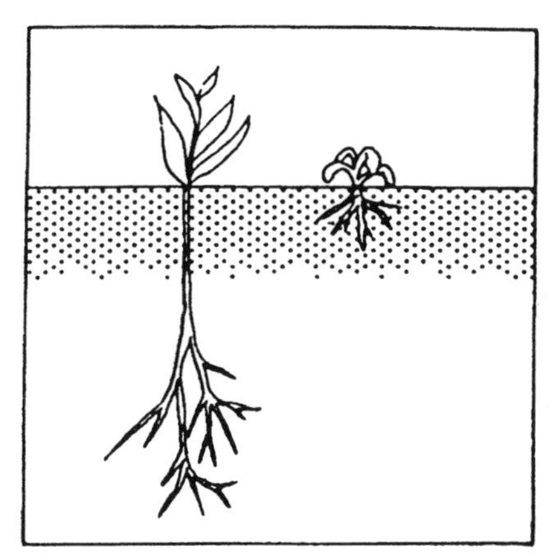

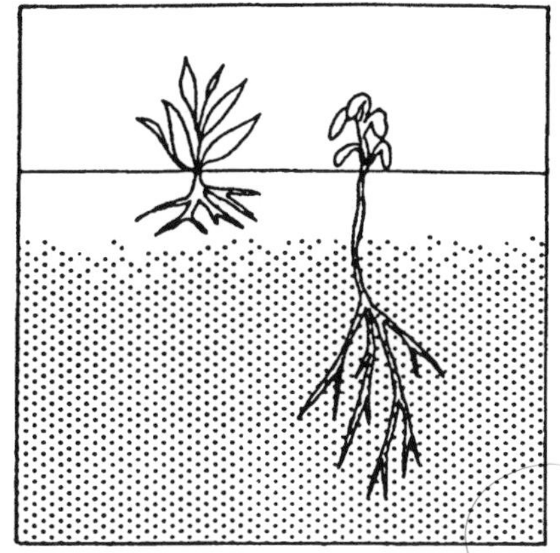

FIGURE 10.30 The position of the herbicide in the soil profile influences control of shallow- or deep-rooted plants. When herbicide remains near the surface (*left*), the deep-rooted plant is not affected, while the shallow-rooted plant is killed. When the herbicide is leached deeper in the soil profile (*right*), the shallow-rooted plant is not affected, while the deep-rooted plant is controlled. (From Ashton and Monaco, 1991.) Herbicide residue that is leached too deeply may also contaminate ground water.

bicides or weed control techniques probably will reduce the incidence of herbicide residues in ground water.

A model of economic and environmental risk from herbicides in water. An ecological-economic modeling system (Figure 10.31) was developed by the U.S. Environmental Protection Agency to compare the risks and benefits of possible bans on atrazine and other triazine herbicides that are used widely for corn and sorghum production (EPA, 1993). These herbicides were chosen for the analysis because they are the most detected group of pesticides in surface and ground water in the United States. Current use of atrazine in the midwestern United States constitutes almost 12 percent of the total agricultural usage of herbicides in the nation. About 42 million pounds of atrazine (active ingredient) and 63 million pounds of all triazines are used each year.

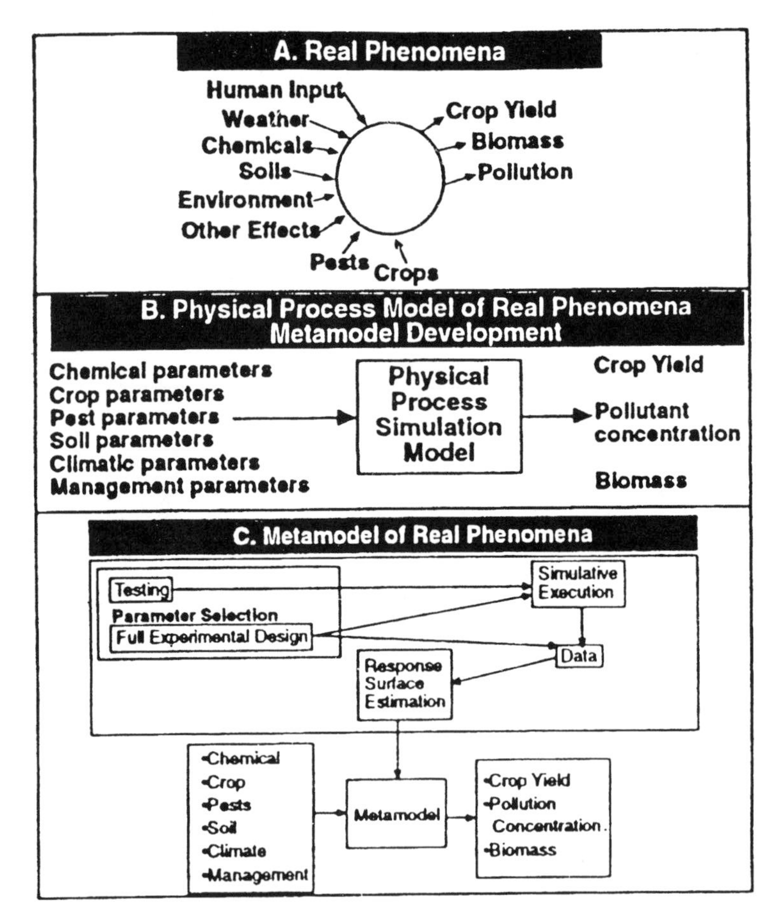

FIGURE 10.31 Process of megamodel development for a Comprehensive Environmental Economic Policy Evaluation System (CEEPES) of atrazine and other triazine herbicides. (From EPA, 1993.)

The peak and average chemical concentrations found in surface and ground waters were transformed in the EPA's model into a unitless measure of risk called exposure value (EPA, 1993). By using a benchmark for environmental hazard, such as Maximum Contamination Level for long-term health exposures and ten-day Health Advisories for short-term exposure from drinking water, the EPA calculated the exposure values for each herbicide as

$$\text{Exposure Value} = \frac{\text{predicted concentration}}{\text{environmental benchmark concentration}}$$

This procedure normalized concentration levels and allowed a comparison of risks over a range of herbicides (Table 10.5). For example, the first row in Table 10.5 shows that atrazine concentration levels exceeded the short-term benchmark of 100 ppb (the drinking water Health Advisory level for short-term exposure) in 43, 43, and 8 percent of the soils in the study region when the crop was grown under conventional tillage, reduced till, and no-till systems, respectively. In the study, the exposure of aquatic vegetation to herbicides, an indicator of ecological impact, was also high. The majority of the exposures exceeded the aquatic benchmarks that have been either proposed as standards or derived from existing EPA guidelines, often by a factor of 20.

Corn yields were estimated to decrease by 2.8 and 4.1 percent under an atrazine and all triazine ban, respectively (EPA, 1993). These yield decreases, after accounting for higher commodity prices, were projected to cause an average annual decrease of $365 million (atrazine ban) and $526 million (all triazine ban) in national economic welfare, with most of the burden falling on producers in the "Corn Belt." The EPA determined that overall herbicide use would not be likely to decrease under an atrazine ban; instead, total triazine use was projected to increase by 27 percent. The use of alternative triazine and non-triazine herbicides from substitute weed control practices was projected to increase, as would the surface and ground water concentrations of most of those herbicides. Thus, the environment would not necessarily be better off with an atrazine ban, though exposures to that herbicide in surface and ground water would decline and be of less concern in terms of health effects. A ban on all triazine herbicides was therefore considered preferable to a ban on atrazine when the importance of human health was considered (Figure 10.32) in relation to economic risk. In other words, an atrazine ban would buy little in the overall hazard index (Figure 10.32), whereas a 50 percent increase in cost for a triazine ban would lead to about half the environmental/health hazard.

The analysis found that there would be a major shift to and widespread use of the sulfonylurea herbicides under a triazine ban, with their concentrations being well below human health benchmarks. But, EPA stressed that uncertainties still exist with these chemicals regarding selection for weed resistance and hazards to aquatic and non-target terrestrial vegetation, even from the extremely low concentrations expected in surface water or atmospheric drift. The loss of triazine herbicides could also lead to greater conventional tillage for

TABLE 10.5 Exposure Distribution in Surface Water for the Three Scenarios

	Percent of Soils with Concentrations Exceeding EPA Benchmarks		
	Conventional Tillage	*Reduced Tillage*	*No-tillage*
	Baseline		
Atrazine	43.10	42.87	8.08
Atrazine <1.5 lb	14.75	15.80	2.38
Dicamba	18.14	0.00	0.00
Cyanazine	24.91	0.92	0.00
Bentazon	2.39	45.90	0.00
Metolachlor	3.01	8.91	0.00
Alachlor	20.83	32.65	0.00
Simazine	86.95	67.66	40.12
Propachlor	6.65	38.44	—
Atrz_Slow_Decay	28.31	26.06	4.26
Atrz_Fast_Decay	11.73	0.01	1.13
	Atrazine Ban		
Dicamba	35.08	1.21	0.00
Cyanazine	37.57	90.15	0.00
Bentazon	67.74	0.00	—
Metolachlor	0.06	26.93	12.63
Alachlor	4.70	31.61	18.28
Simazine	96.68	42.57	40.05
	Triazine Ban		
Dicamba	26.66	5.60	0.00
Bentazon	51.41	84.96	—
Metolachlor	0.00	13.59	0.00
Alachlor	0.67	0.00	0.00

Source: EPA, 1993.

weed control. More widespread use of conventional tillage would, in turn, lead to greater soil erosion and water-quality problems from sedimentation.

Herbicide Decomposition in the Environment

The final factor that accounts for herbicide persistence is *decomposition* or *degradation*. This is the process of destruction of the original herbicide mole-

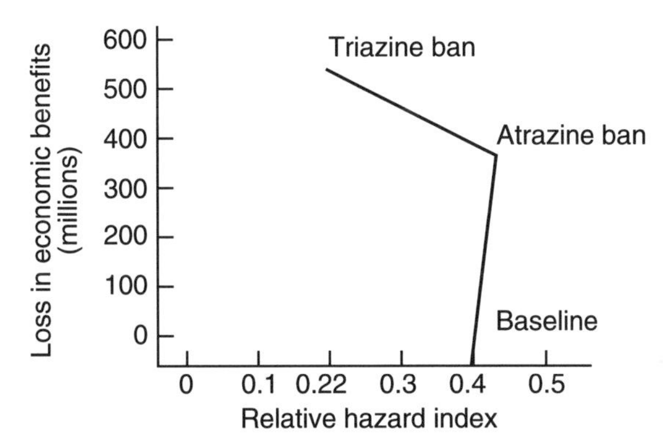

FIGURE 10.32 Plot of relative hazard index (using summed exposure values) against loss in economic benefit. (From EPA, 1993.)

cule and usually loss of herbicidal activity. After degradation, parts of the original herbicide structure remain as different molecules. These "breakdown" products ultimately may be decomposed further to simple organic molecules, but more complex break-down products may be incorporated into organic residues. Often, degradative processes occur in the soil. However, they are not restricted to soil and may occur in water, air, plants, microbes, and animals.

Persistence or degradation curves exist for most herbicides. These curves usually document actual destruction of a particular herbicide in soil, water, or air through time (Figure 10.33). Herbicide persistence is expressed as the *half-life* or time required to degrade fifty percent of the applied material. However, the half-life of a herbicide is not absolute, because it depends on the soil type, temperature, and the concentration of the herbicide applied. Thus, herbicide decomposition, like many other aspects of herbicide science, is relative to a number of other factors. Most organic herbicides are quite degradable, so they do not accumulate appreciably in the soil. They have half-lives that range from only a few days to several months (Figure 10.33). Even relatively long-lived soil-applied herbicides, such as atrazine, usually do not persist sufficiently in the soil to build up or cause plant injury beyond one or two growing seasons. However, small concentrations or residues of some herbicides may persist in soil or water for a long time, especially if they are used repeatedly (Figure 10.34).

PHOTOCHEMICAL DECOMPOSITION

Some herbicides, when exposed to sunlight, undergo photochemical reactions that result in degradation of the chemical molecules. However, photolysis is a unique feature of each herbicide. Some herbicides are photochemically decomposed with relative ease, while others do not degrade well in sunlight. Photochemical degradation is an important process of herbicide loss from surfaces, such as soil and leaves. It can be reduced or prevented by incorporating herbicides into the soil soon after application. Soil incorporation of herbicides shields the chemical from the ultraviolet wavelengths from the sun, which are the most destructive.

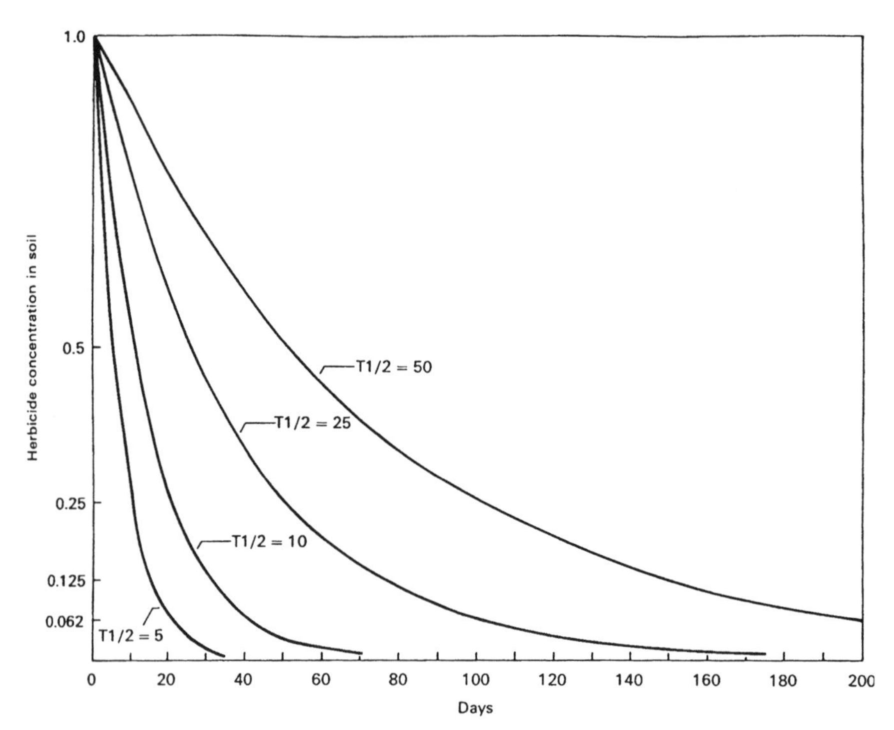

FIGURE 10.33 Persistence curves for herbicides with half-lives of 5, 10, 25, and 50 days. Actual half-life curves for 2,4-D and dicamba in Oklahoma soils approximate the 5- and 25-day curves, respectively. (From Altom and Stritzke, 1973, in Ross and Lembi, 1985.)

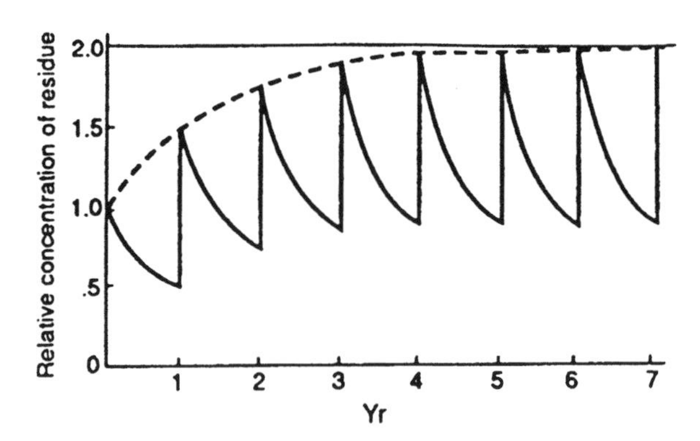

FIGURE 10.34 Residue pattern for a single annual herbicide application and a half-life of one year. (Reproduced with permission of American Chemical Society; copyright 1976.)

CHEMICAL DECOMPOSITION

Herbicides can be decomposed or degraded in soil by purely chemical means. However, it often is difficult to distinguish between chemical decomposition and other forms of herbicide loss, such as microbial degradation. Chemical decomposition of herbicides may result from the following types of reactions: *oxidation-reduction*, *hydrolysis*, and the *formation of water-insoluble salts and chemical complexes*. These reactions are unique for each herbicide. Herbicide decomposition has been the subject of thorough reviews by Audus (1964, 1976), Kearney and Kaufman (1988), and Shimabukuro (1985).

MICROBIAL DECOMPOSITION

Many types of organisms can degrade herbicides. However, microorganisms are the primary agents responsible for the degradation of herbicides in the soil system. Organic herbicides are subject to a wide array of soil microorganisms that utilize these compounds as a source of carbon, nutrients, and energy. In the process of using herbicides as a food source, the microorganisms also destroy the herbicidal properties of the chemical and break the material down into simpler substances. A portion of the herbicide molecule may be incorporated into the microorganism itself. This organic matter is, in turn, decomposed as the microbes die and are recycled in the soil.

Microorganisms degrade herbicides through biological reactions that are the result of specific enzymes. These enzymes often are secreted by the microorganism to break down complex organic products. The following degradative reactions are known to result in microbiological decomposition of herbicides. Similar biochemical reactions also occur in plants, as already discussed. These reactions result in alterations of the parent molecule and include (Anderson, 1977, 1996):

- dehalogenation—removal of chlorine, bromine, or other halogen atoms
- dealkylation—removal of organic side chains
- hydrolysis—removal of amides or esters
- beta-oxidation—cleavage into carbon units of two
- ring hydroxylation—addition of hydroxyl (–OH) groups to the aromatic ring structure
- ring cleavage—breaking the structure of the aromatic ring
- reduction—usually refers to the addition of hydrogen to NO_2 groups under anaerobic conditions

Numerous factors influence microbiological herbicide degradation in the soil. These factors include soil conditions, such as moisture, temperature, aeration, pH, and organic matter content. Generally, if any of these conditional factors of the soil is reduced, the rate of herbicide decomposition also is diminished. Other factors that influence decomposition are the dose and structure of the herbicide, adsorption to soil colloids, and the composition and density of the microbe population.

The concentration or dose of herbicide applied to the soil is important in degradation because degradation takes longer if a lot of the chemical is present. The structure of a herbicide also has a pronounced effect on the ability of microbes to degrade it. For example, the difficulty of degrading a herbicide usually increases as the amount of chlorine or other halogen atoms on the molecule increases. Short-chain ester or alkyl groups are easier to cleave from parent herbicide molecules than are long-chain constituents. Herbicides that are adsorbed strongly to soil colloids are unavailable for either degradation or plant uptake. For this reason, cationic herbicides, such as paraquat and diquat, persist almost indefinitely when adsorbed to soil particles.

Soil enrichment. The influence of soil microbe population dynamics on herbicide degradation and persistence is an interesting phenomenon. Because the microbial process of herbicide degradation is enzymatic, the population of microbes often must adjust to the presence of the chemical. This time required for adjustment results in a period when little herbicide degradation is observed. During this lag period, the microbes are adapting to the new carbon source and multiplying as they utilize the material. Degradation proceeds faster as more microbes become available to use the herbicide. Following this second rapid phase of decomposition, the herbicide concentration is low, and the microbial population diminishes accordingly. In the event that a subsequent application of the herbicide is made, the lag time and time period of rapid degradation are reduced. The phenomenon of a reduced lag time and more rapid degradation following subsequent herbicide applications to soil is called *soil enrichment.* This occurs because a population of microbes already adapted to use the herbicide is present in the soil. However, enrichment is not a universal occurrence among herbicide groups, being most noted in the phenoxy, thiocarbamate, and certain other herbicide classes.

Enrichment is important because it can reduce the length of time that a herbicide exists intact in the soil. Some herbicide manufacturers supply a microbial retardant with their herbicides to increase the longevity of their products for weed control. Although enrichment is often viewed as a problem, it may be possible to use enrichment in the future to diminish herbicide residues in the soil or dispose of unwanted pesticide wastes.

Effects of herbicides on soil microflora and fauna. The microflora and fauna in soil are large and diverse. It is, therefore, not surprising that some herbicides should affect certain species of microbes and soil-borne animals. Generally, negative effects of herbicides are reversible, meaning that the population level or composition of species is decreased for a while but subsequently improves. Beneficial organisms known to be affected negatively by specific herbicides include nitrogen-fixing bacteria (*Rhizobium*) and some mycorrhizal fungi (*Rhizophagus*). In addition, the incidence of some plant pathogens has increased as a result of herbicide use (Chapter 7).

SUMMARY

Herbicides interfere with vital plant processes. The ability of these chemicals to kill weeds depends on many factors of plant anatomy, morphology, and physiology. The mode of action of a herbicide includes all of the plant and sometimes environmental factors necessary to disrupt plant growth. Mechanism of action refers to the specific biochemical process inhibited by a herbicide in a plant. Herbicides must enter exposed plants to be effective. The process of herbicide penetration and entry is called absorption. Absorption of herbicides occurs primarily through leaves or roots, although herbicides also enter plants via stems and various subterranean organs. Two pathways, the lipoidal and aqueous routes, are proposed for foliar absorption of herbicides by plants. These pathways refer to the ability of chemicals to penetrate the cuticle of treated plants in different ways. All substances that enter roots must pass through the cytoplasm of the endodermis. Three pathways of absorption have been postulated for root absorption of herbicides: apoplastic, symplastic, and apoplastic-symplastic.

Transport is the movement of herbicides in plants. Herbicides are grouped into those with limited mobility, those that move in either the symplast or apoplast primarily, and those that move in both apoplast and symplast. Source-to-sink movement is an important way herbicides move in the plant symplast. The ability of herbicides to move readily in either the symplast or apoplast probably depends on cellular pH gradients and subsequent diffusion across cell membranes. Herbicides that enter plant cells often are transformed into other substances, a process which may either deactivate or activate the herbicide. Herbicide metabolism in treated plants is an important basis for differential selectivity.

Most herbicides inhibit photosynthesis. The predominant way herbicides interfere with this important plant process is by inhibiting or diverting electron transport in the light reactions. Photosynthetic pigments also may be inhibited or destroyed by some herbicides. Respiration, cell division, protein synthesis, and membrane function are other plant processes affected by herbicides.

The fate of herbicides in the environment is a concern of many segments of society. The soil acts as an important buffer governing the persistence and fate of most herbicides in the environment. As long as the soil system remains healthy, possible adverse effects from herbicides in the environment probably can be minimized. Soil is the mineral substance on the surface of the earth in which plants grow. It has four phases: a solid, liquid, gaseous, and biological. Herbicides either are applied directly to the soil or eventually arrive there by run-off or in plant residues. Herbicide persistence and thus environmental fate are both influenced by the soil processes of adsorption, movement, and decomposition. Herbicides are adsorbed most readily by dry soil. Adsorption also is affected by the chemical charge and water solubility of each herbicide. The amount of organic matter and clay in soil also influences herbicide adsorption dramatically. Herbicides that are adsorbed to soil colloids are not readily available for absorption by plants, movement in the environment, or degradation.

Movement of herbicides in the environment results from drift, volatility, and lateral and vertical displacement in soil by water (leaching). Herbicide displacement from chemical drift can be avoided or reduced by using appropriate equipment or spray additives that produce large spray droplets. Buffer areas around sensitive areas also are an effective way to minimize the impacts of spray drift. Herbicide leaching in the soil is an important mechanism for incorporation and activity. It also may result in herbicide contamination to ground water supplies if appropriate care is not taken to prevent it.

Herbicide decomposition can occur by photochemical, chemical, or microbiological means. Degradation by microbes is probably the most important mechanism for herbicide decomposition in soil. The soil microflora may become enriched, which enhances microbial degradation of certain herbicides. Some herbicides may also adversely affect beneficial microflora and fauna, or enhance temporarily the incidence of plant diseases.

Chapter

11

Weed Control in a Social Context

In earlier chapters we discussed the nature of weediness, many of the biological principles that influence weed abundance and the technology of weed control. Most of the previous discussion has focused on weeds and weed control procedures at the field, tree stand, or perhaps farm level. However, weeds may extend much farther than individual fields, and the benefits and costs of weed control may extend much further than to individual farmers, foresters, and land managers. For example, consumers of agricultural and forest products may benefit from lower-priced food or more abundant lumber, and users of natural resource areas may benefit from greater access to recreational areas as a result of weed control procedures. Many of these same people also may have legitimate concerns about the presence of chemical residues in food or water, public safety, soil erosion, or other impacts that weed control techniques might have on them or their environment. Others may be concerned about the overall vitality of an industry or profession as new technologies are introduced and others are regulated. All of these issues extend beyond the aims of individuals to the needs, wants, and expectations of society in general. These societal issues are the subject of this chapter.

SOCIETAL AIMS AND INDIVIDUAL OBJECTIVES

Auld et al. (1987) indicate that differences in societal and individual goals arise when some benefits or costs of an activity occur outside the domain of the individual decision maker. These *externalities* result in consequences to others that would not normally be taken into account in making private decisions. In relation to weed control, an external benefit arises if a farmer controls a readily dispersed weed on his farm, thereby reducing its incidence on neighboring farms. On the other hand, if one farmer fails to control that weed species, then other farmers may be subjected to an external risk of infestation and may incur an additional cost for weed control or crop yield loss. Similarly, a weed control tactic may be employed by a farmer that has no immediate or obvious impact on the land, other than to control weeds. If, however, many farmers use that tactic, the impact on water quality or public safety may be substantial, with the additional costs being borne primarily by consumers or other users of the land resource. A primary reason why externalities are important to Weed Science is that weeds often occur over vast areas of farm, forest, or rangeland, and have the propensity to spread easily over large areas. The same can be said for some techniques used to control weeds.

Weeds in a Regional Context

There are many examples of the widespread distribution of weeds. One of the earliest examples is reported in Hitchcock and Clothier (1898), who describe the distribution of native and introduced weeds in Kansas as the prairie was being developed for agriculture (Figure 11.1). A similar study was done by Manson (1932), who described the occurrence of wild oat (*Avena fatua*) throughout several provinces of central Canada (Figure 11.2). These studies are augmented by more recent descriptions of widespread infestations of weed species, for example, leafy spurge (*Euphorbia esula*), purple loosestrife (*Lythrum salicaria*), downy brome (*Bromus tectorum*), and Paterson's curse (*Echium plantagineum*) (Auld et al., 1987). The proclivity to disperse widely is a common characteristic of many weed species (Chapters 1 and 4), and any harmful organism that is spreading or has the capacity to spread poses a threat to areas that are uninfested. Such organisms pose a real or potential cost to owners, users, and other species on that land. Thus, a spreading species represents a problem to more people than just those whose land it currently occupies. Such situations make a strong case for legislation (weed laws), quarantine districts, or other governmental interventions to reduce or slow the spread of some weed species. In addition, governmental organizations whose objectives include weed suppression may have less budgetary constraints than individual farmers, ranchers, or land managers (Auld et al. 1987), which justifies a regional context in which to view weeds.

Market-Driven Considerations

Weed control can have both widespread economic benefits and costs. However, private profits gained from weed control may not always or fully reflect community gains or losses (net benefits) from those practices. Auld et al. (1987)

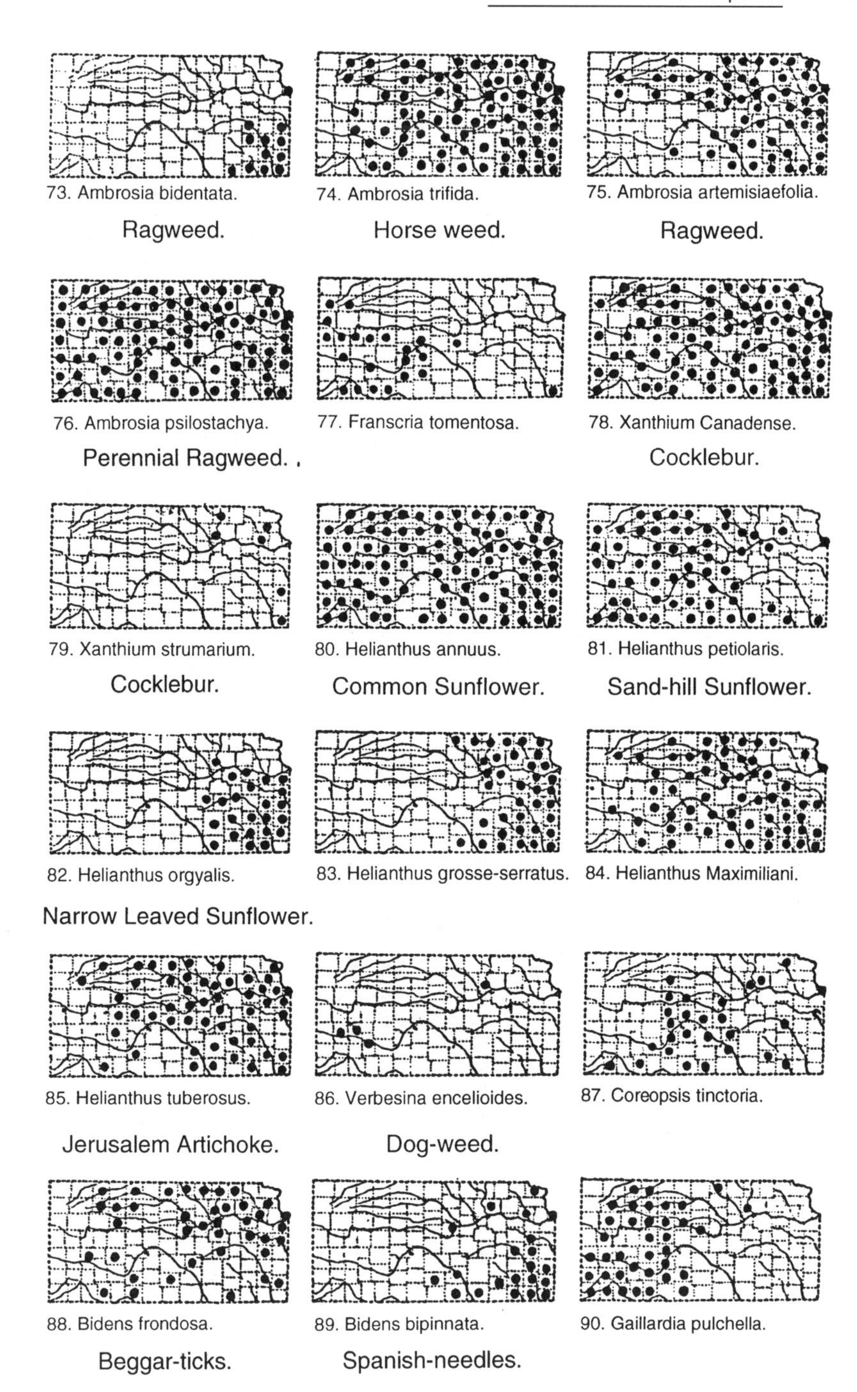

FIGURE 11.1 Distribution of weeds in counties in Kansas in 1898. (From Hitchcock and Clothier, 1898.)

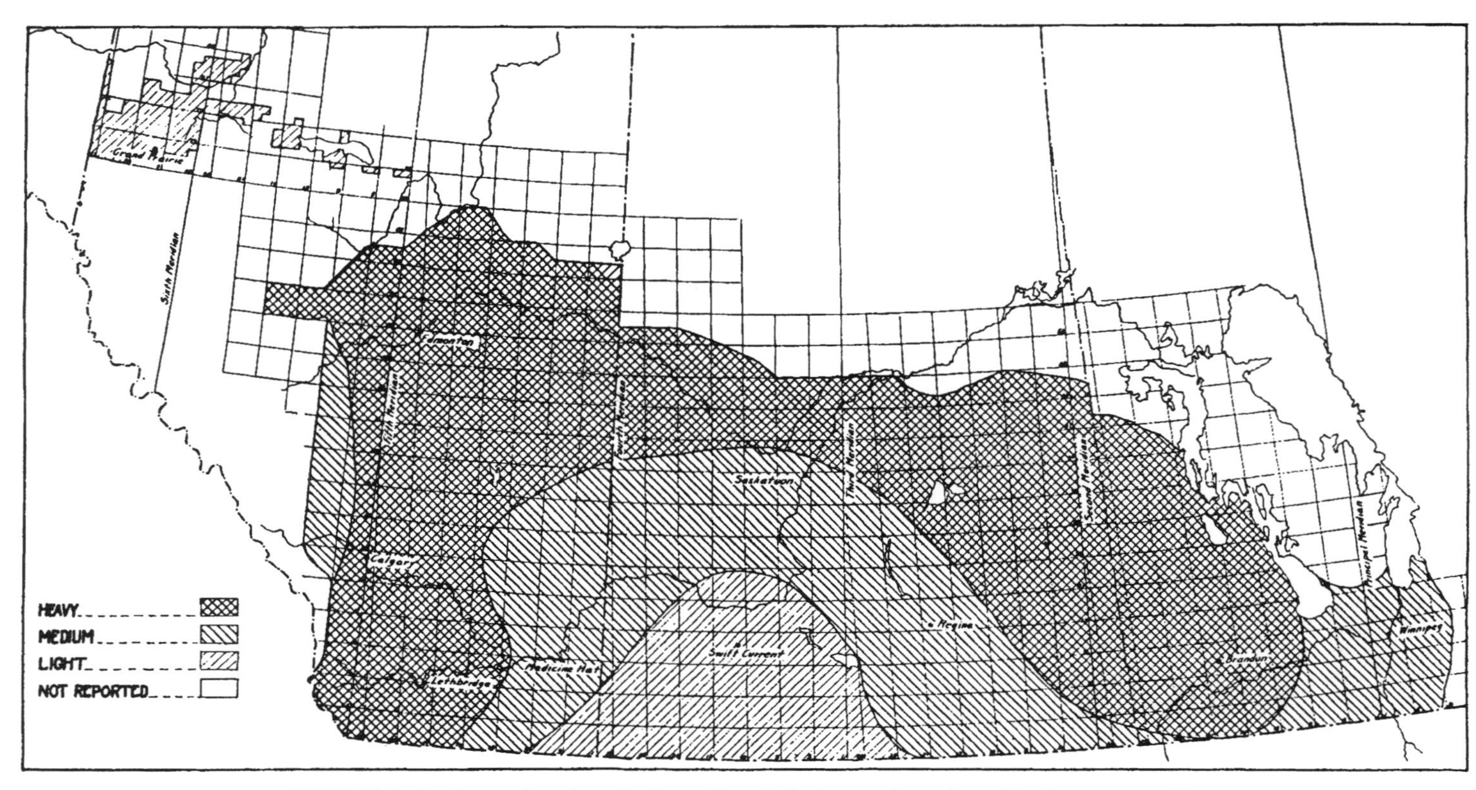

FIGURE 11.2 Distribution and prevalence of wild oat in Alberta, Saskatchewan, and Manitoba in 1931. (From Manson, 1932.)

identify two main reasons why the private profits gained by agriculturists, timber companies, or ranchers from weed control may not fully reflect social net benefits from such procedures. According to Auld et al.,

- Even within a fully efficient system of markets, some of the gains from weed control are likely to be distributed to consumers or purchasers of commodities in places where these commodities register an increase in yield or quality following weed treatment. Market competition may cause the price of such commodities to fall and this may benefit the consumer or purchaser of them, as well as those who supply tools to accomplish the weed control activity. Thus, to assess the overall benefits of weed control in this context, industry-wide gains to consumers or purchasers as well as suppliers (farmers and their suppliers) must be considered.
- Some costs or benefits of economic activities involving weed control may not be taken into account in the market system. For example, herbicide drift from one property may damage crops or other attributes on another property, and yet the herbicide user may pay no compensation. In addition, widespread herbicide use or tillage for weed control may result in reduced water quality, litigation, environmental assessments, and remediation. In all such cases, the actual costs to society are much greater than those borne by the user.

These types of market factors and externality costs need to be included when calculating the societal value of weed control. This subject has been considered in detail by Auld et al. (1987) and will be explored briefly later in this chapter.

SOCIAL CONFLICTS AND THEIR RESOLUTION

Conflicts about weeds and weed control arise when the positive effects of a control measure are felt by one group of people and the negative effects are felt by another group. Such conflicts sometimes occur between people who make their living in rural areas (farmers or natural resource managers), the people or professions that provide tools or production information to them, and others who do not live or make their living from the agriculture/natural resource sector. These divergent populations are sometimes separated by distance and other cultural attributes into rural and urban populations, with the latter being described by the former as "the public." Until the last two decades, the public left farmers, foresters, and land managers alone to make their own decisions about how to grow crops, produce wood, or manage grazing lands. These decisions included those about weed control and weed control procedures.

The situation now is radically different, with virtually every sector of agriculture, forestry and range management influenced by public views. For example, Kimmins (1991), speaking about forestry, states that

> Thanks almost entirely to the dedicated efforts of environmentalists, there has been a dramatic awakening of public opinion about forestry and the environment. . . . An aroused and vocal public has resulted in a re-education of industrial leaders and politicians, the passage of legislation to reduce environmental pollution, degradation, and loss of biological diversity, and in many cases resulted in significant improvements in the quality and sustainability of forest management.

Weed control and vegetation management are no exception to this generalization. In fact, weed control tactics, especially herbicide spraying in forests, represent some of the earliest conflicts over land management policies in the United States and Canada. One such episode in the history of weed science, the public debate about 2,4,5-T use and safety, will be discussed later.

Who Is the "Public" and What Do They Think?

Mater (1977) describes the public as a "large amorphous group, any part of which may suddenly coalesce into action," that is, a powerful lobby. She also suggests that the public is a multiplicity of publics that sometimes blend their efforts and sometimes challenge one another. There is considerable discussion in the literature on this subject of how the majority of people differ in viewpoint from their more vocal counterparts. Some believe the general public is more moderate (perhaps even ambivalent) about most issues than the "extremist" public who vocalize their concerns. Others consider the vocal minority to be representatives of a larger societal issue. Perrin et al. (1993), in a survey about public attitudes toward herbicide use in forests, conclude that,

> It may be wishful thinking on the part of those who believe that the "silent majority" feels differently regarding forest management or herbicides than those who are more outspoken.

They observed that four out of five respondents (81 percent) to their survey (Figure 11.3) said that chemicals used in forest management pose a hazard to human health and the environment, that is, only 19 percent approved of aerial herbicide application. Two-thirds of their respondents also said that the forest industry should be restricted to reduce its environmental impact, in spite of potentially damaging economic consequences. Perrin et al. (1993) concluded that the bulk of the general public, along with "activists," support greater attention to environmental concerns and does not feel the use of chemicals is consistent with that position.

King (1991), in a national survey in the United States, commissioned by the American Farm Bureau, reports similar findings to those of Perrin et al. (1993)

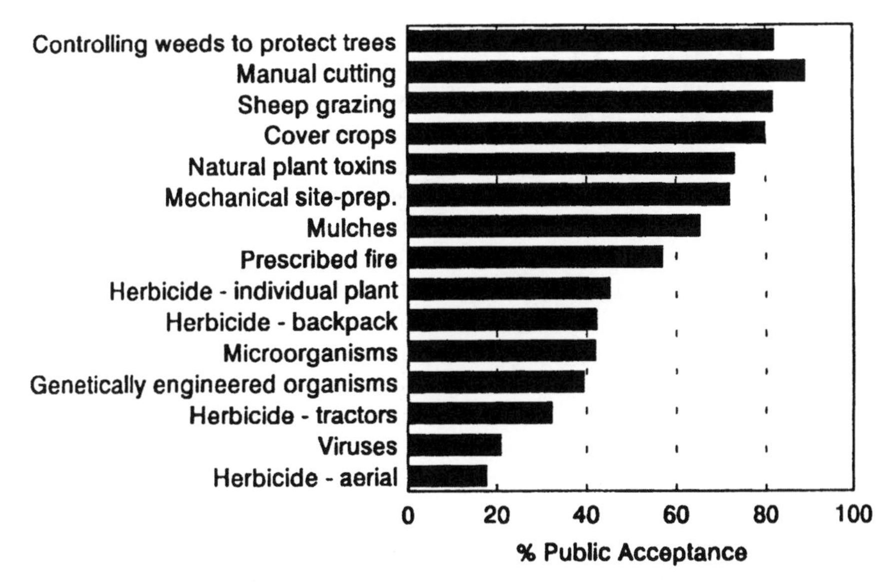

FIGURE 11.3 Acceptability of a number of vegetation management alternatives based on a 1994 survey of 2000 people in Ontario, Canada. (From Wagner and Buse, 1995; Ontario Ministry of Forests.)

about public attitudes to food safety, agricultural chemicals, and on-farm uses of them. A key finding of King's study was that 89 percent of the respondents were more concerned about pesticides than about all other food issues, and 63 percent said that "the dangers to human health of pesticides outweighed their benefits in protecting crops." The existing evidence demonstrates rather convincingly the general public's opposition to herbicide use, but not necessarily to weed control. However, there does not appear to be any in-depth exploration of the reasons for these views, how firmly they are held, or under what circumstances herbicide uses are tolerated (Perrin et al., 1993).

Issue Education versus Resolution

Perrin et al. (1993) indicate that many of their survey informants who work in the agricultural and forest industries point to the limited knowledge of the general public regarding vegetation management and herbicide-related issues. "Inform them and their attitudes can be changed" is the solution. The logic of this argument is hard to refute, except that, based on observation and the literature, one-sided public education programs do not work. Furthermore, Perrin et al. (1993) indicate that "education" is often used as a euphemism for indoctrination as a way to make the public understand, with little or no attempt to also understand public sentiment, knowledge, or values about an issue. For example, Mater (1977) says the statement, "we have to educate the public," when translated says, "we must manipulate the public to understand the situation as we see it." She noted in 1992, while speaking to the forest products industry about itself, that the public does not have to "understand" anything and even if and when it does understand, agreement with a particular point of view does not necessarily follow.

Public education or public relations campaigns to curry public favor or turn public sentiment around from disapproval about any land management practice, including weed control, have been singularly unsuccessful. Again, forestry serves as an example. Mater (1992) observed that, in the United States, the timber industry has made a large investment in public relations and education, with little success in changing the negative public perception about forestry. In Sweden, Breton and Tremblay (1990) point out that the massive amount of money spent by the forestry sector in making and defending the case for herbicides, using a variety of media, had no effect and led to a ban on those products. Similarly, education activities by industry and the U.S. Forest Service in numerous settings have had almost no impact on public opinion about herbicide uses. If anything, Perrin et al. (1993) note, these educational attempts destroyed whatever credibility the forest industry had and firmed up public opposition to current practices. Obviously, public feelings about weeds and their control are related to other important factors, particularly people's underlying attitudes and values.

There is now a general movement in forestry away from the use of education or promotion to solve social conflicts and toward approaches that can be described as conflict resolution. These approaches involve identifying the beliefs, values, interests, concerns, and desired benefits of various segments of the general population. Then an attempt is made to show how what one has to offer can be of benefit to another person, rather than trying to "sell" a belief or preconceived message. There is much written on this subject of natural resource conflict resolution (Wilson and Morren, 1990; Checkland, 1981) and the serious student of this subject should consult this literature. Perrin et al. (1993) summed up the difference in approaches in this way: "In education we say, 'if only they knew more about this,' while in conflict resolution we say, 'if only we knew more about them.' "

ASSESSMENTS OF SOCIAL GAINS OR LOSSES FROM WEED CONTROL

Value systems include the way people think about activities (what we do and how we do it) and technologies (what we do it with). They are actually the underlying ethical framework or foundation through which we reference our own and other people's actions. Although weed control is only a small part of land management decisions, it is important to examine where it is located within this larger value-based context. Castle (1990) believes that there are four fundamental value systems that influence decisions about the management of natural resources, including agriculture:

- *Material well-being.* Any activity or technology should be of benefit or utility to society. Implicit in this belief is that society will be better off with rather than without the new tool, tactic, approach, or activity.
- *Sanctity of nature.* An activity should not proceed if risk or damage to the environment is likely to result. Non-intervention is a key ingredient of this belief system, which is known as environmentalism, although some human involvement is recognized and accepted. While human interventions into nature are accepted as necessary, they should be minimally disruptive so as to avoid adverse consequences.
- *Individual rights.* Individual liberty and property entitlement are the primary concerns of this value system. The marketplace is often posed as the best way for society to accept or reject activities, through the products produced. Government interventions are often seen as a problem, not a solution.
- *Justice as fairness.* All people have equal use of the earth's resources and benefits from them should be distributed equitably. The issue is not bounded by time or geography. Thus, effects of activities and technologies on other people in the world and future generations is of concern.

Each of these value systems raises different questions about our efforts to manage land, raise crops, or manipulate vegetation (Radosevich and Ghersa, 1992). However, the value system most accepted by Western cultures seems to be the first one in the above list, material well-being. Explicit in that value system is the idea that people will be better off with rather than without a particular tool or activity. Nevertheless, people usually do not view such benefits identically. One approach to assess the relative benefits and costs in such situations is *cost-benefit* or, more recently, *risk-benefit* analysis. For example, risk-benefit analysis is used during both herbicide registration and cancellation proceedings, where the potential benefits of a product in terms of weed control are weighed against its potential risks to human health and harm to the environment to determine whether there is a net benefit overall. However, risk-benefit analysis is not the only way or only set of criteria through which to make such judgments.

An Example of Cost-Benefit Analysis

The prevalent way to make assessments of net benefit is to express all values in common economic terms, that is, a cost-benefit ratio. However, even such an analysis is often not straightforward, as the following example, provided by Auld et al. (1987), indicates.

Serrated tussock (*Nassella trichotoma*) is a naturalized weed of Australia that covers over 600,000 hectares of grazing land in the southeastern part of that country. The species is perennial but is propagated mainly by wind-borne panicles of seed. If unchecked it dominates grassland areas, reducing animal carrying capacity from 2 to less than 0.5 sheep per hectare. Long-term control of infestations can be achieved by pasture improvements that prevent reinvasion (Campbell, 1977), although frequent spraying of previously infested areas is common. The various options to control the species and protect tussock-free areas from invasion and reinvasion were examined by simulation and evaluated using a discounted cash flow analysis.

Vere et al. (1980) assumed that the main economic impact from serrated tussock control would come from increased wool production. Given the low elasticity of demand for wool, they found that, at the industry-wide level, the main benefit of serrated tussock control would be to increase consumer's surplus, and that producer's surplus would fall. The overall change in social benefits resulting from serrated tussock control was estimated to be approximately $60 million per year. Using pasture improvement to control serrated tussock would cost $34.2 million when discounted at ten percent over ten years. Vere et al. (1980) predicted a national social benefit/cost ratio of 11 : 1, a very favorable ratio from an economic viewpoint. However, Edwards and Freebairn (1982) pointed out that almost all of Australia's wool is exported. Therefore, it would be foreign consumers who would benefit from tussock control through lower prices, not Australians. Nonetheless, after modifying the Vere et al. estimates to allow for these criticisms, the benefit/cost ratio remained high. Edwards and Freebairn concluded that some producer's surplus would result from improved wool production. This work demonstrates some of the difficulties, mainly associated with economic assumptions, that are involved in applying benefit/cost analysis to the social assessment of weed control (Auld et al., 1987).

Assessing Risk

Many of the values associated with weeds (Chapter 1) and the technology of weed control (Chapters 8, 9, and 10) are a matter of human perception, or are economically intangible. This makes assessment of social benefits from weeds, or the activities to control them, difficult on a purely economic basis. Although no form of weed control is exempt from such debates, herbicides are particularly vulnerable to them because there are more obvious externalities that can result from herbicide use. Impacts to non-target areas and human safety are of particular concern, as noted in Chapter 10 and later portions of this chapter.

Auld et al. (1987) provide some enlightenment about how such externalities may be taken into account and perhaps corrected (Figure 11.4). They suggest that farmers and herbicide producers guided only by their individual profits may use or promote herbicide treatments that either impose a net loss to society or could be modified to the advantage of society. For example, curve OABC in Figure 11.4 represents the profits realized by farmers from increasing the intensity of herbicide use, either by increased frequency of application or by higher rates. They maximize profits when X_2 amount of herbicide is applied. However, suppose that others in society are damaged by such high rates or frequencies of herbicide application (for example, because of diminished water

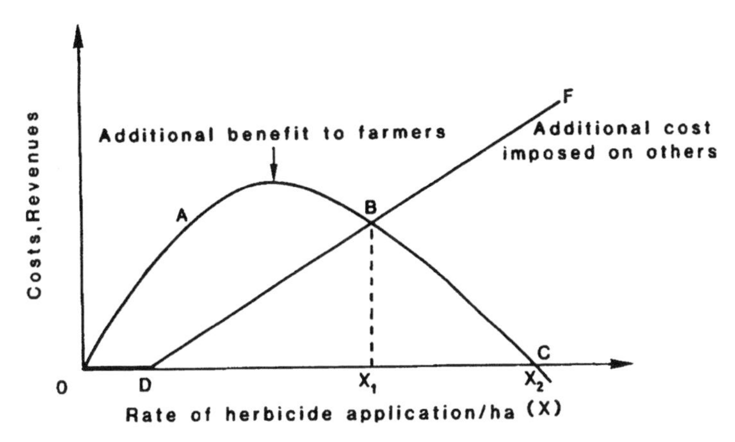

FIGURE 11.4 From a social point of view, account should be taken of losses imposed on others by herbicide use. In the above case, farmers would maximize their benefits by applying X_2 of the herbicide per hectare, whereas losses imposed on others (DBF) would suggest that an application rate of X_1 is socially optimal. (From Auld et al., 1987.)

quality or increased risks to human health, as in the discussion about triazines and 2,45-T later) so that restitution or restoration must be made. Perhaps even morbidity is increased. Should these things occur then the additional externality costs would be represented by curve DBF and the optimal herbicide rate or frequency of application would be represented by X_1. Above a total application rate of X_1 the extra gains to farmers are insufficient to compensate for the extra externality costs or damages imposed on them.

This discussion, so far, ignores the question of how risk and uncertainty from the use of and exposure to herbicides are assessed (see Chapter 9, pp. 398–400 and Chapter 10, pp. 472–494). It also does not address whether the risks are voluntary or involuntary. Auld et al. (1987) suggest that if risks, for example from exposure, are voluntary they can be taken into account without undue difficulty. If, however, individuals are subjected to greater risk than perceived as desirable, or are subjected to such risk against their will or without their knowledge, difficulties of assessment arise because those individuals are made worse off by such risk and are not fully compensated for it.

Information about the overall social and environmental costs of herbicides is extremely limited. One attempt at such an assessment has been conducted by the U.S. Environmental Protection Agency (see discussion of atrazine, this chapter). Pimentel et al. (1978) indicate that risks from all pesticides, in order of economic importance, are

- reduced natural enemies and pesticide resistance
- human pesticide poisonings
- cost of governmental pesticide pollution controls
- honey bee poisoning and reduced pollination

- loss of trees and crops by herbicide drift
- animal pesticide poisonings and contaminated livestock products
- fishery and wildlife losses

It should be pointed out that the externality costs for herbicides could be lower than for insecticides, especially as far as risks from human poisonings are concerned (based solely on LD_{50} studies). However, risks to fish and wildlife may be greater for herbicides because of the higher potential to damage vegetative habitat using these tools.

There are several ways that society, primarily through governmental regulation, can pass on externality costs to herbicide users and producers. These include requiring long periods of testing and establishing minimum acceptable environmental and human health standards, restrictions and cancellations of uses, and stringent labeling requirements. A recent strategy imposed by some states in the United States is a sales tax on all herbicide purchases, with the funds generated being used for research on herbicide alternatives and programs for remediation of adverse environmental impacts.

FARMER INCOME, EMPLOYMENT, AND INCOME DISTRIBUTION FROM WEEDING

According to Altieri and Liebman (1988), weed control is one of the most labor-intensive aspects of tropical agriculture and one of the most chemically intensive aspects of temperate agriculture. This observation is, no doubt, generated from the abundance of labor in many tropical nations and the absence of it for food production in most temperate countries. Akobundu (1987) indicates that at least 50 percent of a farmer's time is taken up in manual weeding during crop production in many Third World nations. He suggests that much of the drudgery associated with subsistence farming in the tropics centers around the peasant farmer and his (her) manual weeding effort, leaving them or their family little time for recreation, creative thinking, or other activities (Akobundu, 1987). Holm (1971), referring to the use of human labor for weeding in tropical agriculture, indicated that more energy is expended on weeding crops than any other human task. In a much quoted speech (Holm, 1971) also commented on the social plight of these people by saying,

> In traveling across the world, one may have the impression that half of the world's men and women are in the fields, stooped, moving slowly and silently, weeding . . . these faceless people without identity, symbolize the great mass of humanity that spends a lifetime in weeding"

This viewpoint by many weed scientists was again substantiated by Moody (1995). Alstrom (1990) agrees with the assessments of Altieri and Liebman (1988) and Akobundu (1987) about the intensive labor requirements for weeding in tropical agriculture, but takes exception to the implications of Holm's comments and indicates that there are two fundamental issues that involve hand weeding, especially in the tropics. The issues identified by Alstrom are:

- low crop production and farmer income as a result of weed infestation
- labor, employment, and income distribution

In further disagreement with Holm, Alstrom (1990) indicates that in many areas of the world the use of human labor in agriculture can be a desirable social objective.

Crop Production and Farmer Income

As we have seen in Chapter 5, the value of timely weed control can greatly increase the output of many cropping systems, whether these systems are in tropical or temperate regions. As such, fully productive systems can be expected to increase overall employment by creating further opportunities in food or product handling and distribution. Alstrom (1990) points out that the contribution of improved weed control to the economy of a country, as a whole, is not only the direct effect of increased net income to the farmer, but also results from the generation of a multiplier effect spreading through all economic activities affected by increased agricultural production. He also suggests that the multiplier effect from increased agricultural production will be greatest when manual labor is used for weed control because, in this case, both a direct and indirect effect on human employment can be achieved. If, however, tools that substitute for human labor are used, then only the indirect effects of added production are realized. This contention (Alstrom, 1990), of course, requires a readily available source of human labor and their willingness to weed. It also requires that farmers and laborers be compensated adequately for their efforts.

Labor, Employment, and Income Distribution

Poverty in a region keeps wages low, which results in the slow introduction of labor-substituting methods, including mechanized and chemical weed control. Alstrom (1990) and Mollison (1988) also point out that it is often economically strong groups inside and outside a region that gain by such technological transfers, because hired laborers are often replaced by high technology inputs even when the additional profits to farmers/landowners are not very significant and the technology change has a net negative effect on the region's social economy (Alstrom 1990). However, generalizations about the effect of herbicide use on the welfare of a country are not always clear because herbicide inputs are usually accompanied by added inputs of other production factors, as well. For example, a much quoted survey of Philippine farmers by deDatta and Barker (1977) indicated that as the number of farmers using herbicides increased from 14 to 61 percent, labor increased from 5 to 15 man-days per hectare and total farm labor increased from 64 to 91 man-days per hectare. This study demonstrates that increased technology to control weeds does not necessarily result in greater human unemployment, owing to the overall effects that improved crop production can have on regional economies.

Nevertheless, Alstrom (1990) indicates that the social consequences of the seemingly specialized and technical activity of chemical weed control can be painful for certain weak and disadvantaged groups of people. For example, Shetty (1980) maintains that for rainfed crops in the Indian semi-arid tropics, herbicide use cannot be advocated because its introduction has no clear cost advantage and could result in decreased income opportunities for one of the most disadvantaged labor groups in India—landless female laborers. Most economists concerned with the problems of developing nations emphasize that massive unemployment among rural populations and concomitant migration to urban areas is one of the greatest problems facing the Third World (Moody, 1988; Alstrom, 1990). Such economic stagnation exacerbates the even greater problem of increasing human population in these countries.

It thus seems questionable to introduce herbicides or other labor-substituting technologies into the agricultural systems of Third World nations. As Alstrom (1990) states:

> In many countries like India there is widespread underemployment resulting in hardships and a weak bargaining position of agricultural laborers. In many cases, they and not the landowners bear the brunt of hard physical labor in weeding operations and they are the ones compelled to migrate to city slums, if and when the use of herbicides reduces employment.

He also points out, however, that it is these same people who will benefit if economic growth is strong enough to create a net increase in the demand for labor, which then could contribute to a more egalitarian distribution of incomes.

WHEN HERBICIDES MIGHT BE USED TO INCREASE INCOME AND SOCIAL WELFARE IN DEVELOPING COUNTRIES

Alstrom (1990) states that, "because of social equity arguments, the use of herbicides in hot climate, peasant agriculture should not be advocated as a general rule, but their use might be agronomically and socially justified under certain conditions." These conditions include:

- When labor is unavailable during critical times of weed-crop competition resulting in low farm income because of yield reduction.
- During periods when weather conditions prevent other cultural practices from occurring in a timely manner, for example, in the sticky vertisol soils of central India during the beginning of monsoon season.
- Under some cropping environments that are unusually difficult to keep free of weeds, for example, dryland rice crops grown in hilly poor-soil regions with medium rainfall.

- In areas with permanent infestations of some perennial weeds such as *Cynodon dactylon, Saccharum spontaneum*, or *Imperata cylindrica*.
- During expansion of areas under production, for example, in lowland humid tropics where both shifting cultivators and semi-subsistence farmers often are forced to restrict cultivated area as a result of intensive weed infestation. Hence, yield and income are small.
- During adoption of new intensified cropping systems. Older, more extensive systems of cropping might have avoided serious weed problems, but if human population growth and the resulting low food per capita availability forces intensification, herbicides might be necessary to cope with new or different weed problems.

Clearly, the need for improved weed control as a way to improve crop production in tropical agriculture is well demonstrated. It is the means and the social outcomes of the use of such tools that are still to be ascertained.

SHIFTING PATTERNS IN AGRICULTURE AND WEED MANAGEMENT

Weed scientists, like many other scientists and land managers in applied natural resource disciplines, are engaged in an ethical debate with other members of society about values and perceptions of food and fiber production. The focus of this debate is often on the tools and tactics (means) used to grow crops, produce wood products, and manage grazing lands. Although the debate is well formed in all areas of natural resource management, it is especially well developed among the agricultural community, perhaps because of agriculture's almost exclusive emphasis, until recently, on marketable yield and production efficiency as societal goals (ends). Periodically, it is important to examine such goals to determine if they have changed and if current technologies continue to meet societal expectations. In the following pages, the role of weed management is examined within the context of an evolving agriculture. It is recognized, however, that other land management disciplines, such as forestry and rangeland management, also evolve and we hope the following discussion will be equally helpful to people in these other areas.

Early Agroecosystems

Agricultural systems began geographically at different times, and have developed at different rates. For example, today agricultural systems in some parts of the world are well developed (e.g., much of North America and western Europe) and backed up with substantial modern technological paraphernalia, while elsewhere (e.g., much of Africa and South America) modern agricultural systems are just beginning. In its early stages, agriculture was, no doubt, tightly coupled with the productivity of ecosystems, as humans competed for food with several other guilds of organisms such as insects, mammals, and mollusks. This competition for subsistence created negative feedbacks (see cybernetics, Chapter 2) that limited human social and agricultural development. The acquisition of food was the primary focus of activity, and the amount of cultivated area was determined by the level of energy that could be transformed into human and animal labor to grow and harvest crops (Merchant, 1980).

Biome level processes (Chapter 2), such as precipitation and radiation patterns, limited crop species that could be established and "weeds" probably grew primarily in open niches in disturbed areas and openings in the natural vegetation. Patchy landscapes emerged with cultivated land, grazed stubble, and fallow and less disturbed ecotones likely coexisting with natural areas (Ghersa et al., 1994a). As successional age increased, the land became more and more infested with weeds, until cultivation finally had to be abandoned and the land used exclusively for grazing or hunting (Naveh and Lieberman, 1990). In a sense, the weed communities behaved as nomads, shifting in concert with different degrees of human disturbance (Holzner and Immonen, 1982). This nomadic character of plants and people in early agriculture helped preserve soil fertility (Brady, 1990) and caused a diverse landscape that probably impeded the development of specialized pests and diseases (Ghersa et al., 1994a). Paradoxically, however, this diversity also fostered unpredictable outbreaks of insect pests and new weeds (Vrijenhoek, 1985), as some colonizers exploited the interfaces between different habitats (Way, 1977). The development of weedy syndromes created the need for better weed management. Improvements probably started with burning and hand weeding and progressed to more sophisticated cultivation implements (Figure 2.2). This coevolution of ecological changes and technological innovations, including new methods of weed management, led to modern agroecosystems (Ghersa et al., 1994a).

Modern Agroecosystems

Significant agroecological modifications, involving horse-plowing, rotations, larger farms and additions of new crops, were introduced during the medieval agricultural revolution in western Europe. Modern agriculture probably began in the late seventeenth century with capitalism (Merchant, 1980), which changed farming from a cyclic process (where input is approximately equal to output) to a more open system in which materials and energy (e.g., fossil fuels) pass from an external supplier to an external buyer, and where money and energy put into the system often exceed money and energy coming out (Lewontin, 1982). In addition to the development of capitalism, the technological and social changes resulting from industrialization substantially altered relationships between humans and other components of the agroecosystem. Feedbacks to the social system (individuals, family, government, and religious, scientific, and educational systems) began to come from political, economic, and academic institutions rather than from daily farming interactions with the environment. In other words, guidance to farmers came from sources external to the locality, and coevolution between the ecosystem and land management became less predominant (Ghersa et al., 1994a).

Agriculture is no longer limited strictly by characteristics of the biome. In fact, the use of information, expressed as power and technology, allows farms to flourish in deserts, former rainforests, drained wetlands, and even on the reclaimed ocean floor. Technology, moreover, has allowed farmers to correct for deficiencies in soil nutrients, disturb enormous areas of land, and modify topography to create a gigantic, nearly homogeneous habitat suitable for a relatively small, but productive, group of crop species (see Chapter 1). This use of technology stabilized agriculture over the short term and increased crop harvests per unit area (Hall et al., 1992; Way, 1977), while also creating a simplified agroecosystem of early seral communities (see Chapter 2). The general properties of this simplified ecosystem can be inferred from net ecosystem production and elemental output rates (Figure 11.5), which are closely correlated. In other words, after disturbance, decomposition of organic matter usually exceeds its production so that net ecosystem production becomes negative (Figure 11.5). Thus, a farming system maintained under continuous disturbance will deteriorate unless it is subsidized with external inputs such as synthetic fertilizers, manure, or irrigation. Deterioration can result from accelerated erosion (Pimentel et al., 1976), pollution by excess fertilization and chemical residues (Tivey, 1990), or reduction in the biotic activity of the soil (Woodmansee, 1984).

AN EXAMPLE OF DETERIORATION

It has been argued that, in intensively farmed areas, an equilibrium has been achieved between nutrient gains and losses (Figure 11.5) and, in some cases, gains may even exceed losses (Tivey, 1990). Because of the intensive soil disturbance and the seasonal interruption of the system, however, these areas probably do not retain all of their natural feedback loops and therefore are more likely to lose nutrients and organic matter than gain them. Many man-

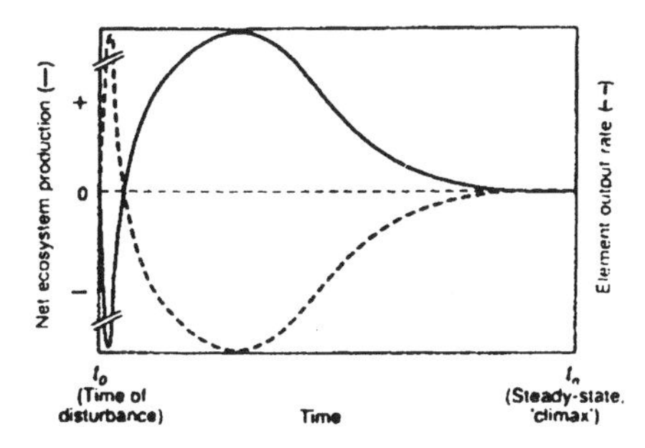

FIGURE 11.5 Patterns of change in net ecosystem production (solid line) and element output rate (broken line), assuming constant input rates, during secondary succession. (From Miles, 1987.)

agement practices have, in fact, coevolved with a deterioration of the resource base. An example of this process occurs in the Rolling Pampa of Argentina.

The Rolling Pampa is a relatively new agricultural area, but one that, nevertheless, has experienced the plow for more than half a century. A typical (traditional) crop sequence is a 3–4 year native perennial grass and forb ley which is used to fatten cattle and restore soil fertility, followed by maize for 10–12 years, after which the area is returned to the ley. With this sequence, soil fertility is maintained at high levels and maize stands range from 70,000 to 120,000 plants per hectare (Ghersa et al., 1985). Both high levels of fertility and high density of crop stands allow good maize yields even with relatively high infestation levels of perennial weeds. In contrast, where rotation with grass-forb leys is not performed, more than 20 years of continuous maize production has caused nutrient depletion, soil erosion, and the resulting maize plant density to be reduced to 40,00 to 50,000 plants per hectare—largely to accommodate poor soil fertility (Andrade et al., 1992). In the deteriorated system, perennial weeds dramatically reduce maize yields (Ghersa et al., 1985) and crop production has shifted to a wheat-soybean rotation (Hall, 1992). This shift in cropping pattern helps improve the nitrogen budget through legume symbiotic nitrogen fixation, but also increases both soil cultivation and herbicide load to control a different spectrum of perennial weed species (Ghersa et al., 1994a).

This example demonstrates that the effects of weed competition on crop yields increase as resource base deterioration forces weed and crop niches to overlap to greater and greater extents (Ghersa and Martinez-Ghersa, 1991a,b). For many agroecosystems, overall availability of resources is either stable (e.g., maize with grass-forb leys) or continually decreases (continuous maize or wheat-soybean rotation) as cropping intensity increases. This deteriorated soil condition often exacerbates the need to control weeds and to adopt new methods or technologies of crop production.

WEED OCCURRENCE AND THE DETERIORATING SOIL BASE

Recent surveys indicate that the use of herbicides increased twice as much as did any other agricultural input between 1950 and 1980 in the United States (National Research Council, 1989). Soil cultivation also increased during that time, so the increased chemical input was apparently not a direct replacement for tillage, although some replacement practices such as no-tillage or minimum tillage systems occurred during that time. Lipsey and Steiner (1984) suggest that increased advertising may account for some of the marked rise in herbicide sales. However, such sustained demand for herbicides also may be due to farmers attempting to compensate for the deterioration of their soil resource. In many cases, crop yields are more sensitive to weed removal than to the addition of fertilizer (Appleby et al., 1976; Hall et al., 1992), especially if the limiting resource cannot be identified readily (Levins, 1986). The elimination of weeds under such circumstances has appeared more reasonable to agronomists and farmers than trying to replace unknown nutrient losses.

Role of Human Institutions in Modern Agroecosystems

Unlike earlier agricultural methods, in which the agroecosystem and its management were tightly coupled, modern weed control techniques are increasingly derived independent of the biological system—that is, generated primarily from feedbacks in human-driven institutions. In early agricultural systems, for example, hand weeding would not take place in areas infested with thorns or woody plants, fire or grazing being the most likely replacement in such situations. Modern management tactics, including weed control, are now just as likely to result from political, scientific, or technological factors (Ruttan, 1982) as biological ones. The problem is that such changes in management come from sources outside the biological hierarchy (Figures 1.14 and 2.1) but they affect all levels of organization within the agroecosystem. The soybean example in Argentina discussed above illustrates this point.

Soybean was introduced into Argentina in 1980 because of political, economic, scientific, and technological forces largely extraneous to the agroecosystem there (Hall et al., 1992). The area devoted to soybean production expanded rapidly to include almost the entire area occupied by summer crops. Thus a mixed system of maize or wheat that was rotated with pastures was replaced with continuous consecutive crops of soybean and wheat. The result is that, in a single decade, losses in soil fertility and concomitant shifts in weed, insect, and disease populations have caused dramatic increases in use of energy, fertilizers, and pesticides to maintain crop production at its former high levels (Ghersa and Martinez-Ghersa, 1991a).

In the United States, social factors and human institutions have had an effect on the agroecosystem similar to that in Argentina. For example, modern agriculture in the United States, encouraged by long-term increases in wages relative to the prices of land and machinery (Ruttan, 1982), has replaced manual labor with machines or chemicals. In other words, technological and social innovations (e.g., commodity subsidies) imposed primarily to solve problems outside the agroecosystem (e.g., a readily available urban workforce) have now contributed to new problems within the agroecosystem (e.g., soil erosion, herbicide resistance, nonpoint pollution, loss of farms), as well as to significant social dilemmas (e.g., unemployment, unequal income distribution and welfare, and crime).

ROLE OF VALUE SYSTEMS IN AGRICULTURE AND NATURAL RESOURCE MANAGEMENT

Value systems play an important role in maintaining existing or developing new ways of thinking in agriculture and most other natural resource areas, because innovations in technology originate with a set of ethical values that tend to justify existing tools or the need for new approaches (Ferre, 1988). Thus, a technology can easily become rooted in the supposition that its underlying values are either universal or take priority over other values in society. A common assumption, for example, related to weed control is that pest (weed) control increases food abundance (material well-being) with negligible harmful affects to either the environment (sanctity of nature) or people (justice as fairness). This assumption places greater value on the amount of food produced than on either the system that produces it or the quality of the product. The assumption also fails to ask who benefits, except in a limited way. Scientific studies have shown, however, that many activities meant to control pests often have caused at least as many problems through harmed wildlife, contaminated soil, watershed erosion, residues in food or water, and pest resistance (Mortimer, 1984; Tivey, 1990) as they have solved by limiting losses due to pests (Pimentel, 1986).

Below are three examples of how social values influence weed management.

The 2,4,5-T Controversy

> 2,4,5-T was the best tool of all; yet 2,4,5-T had a great fall. All of the facts midst all of the din couldn't put "T" back together again (Walstad and Dost, 1986).

In the mid-1960s the first significant social debate over the use and public safety of a tool used exclusively for weed control began. The controversy centered on 2,4,5-T (2,4,5-trichlorophenoxy acetic acid), then the predominant herbicide used for shrub and tree suppression in young forest tree plantations. This debate and subsequent litigations lasted for two decades until early in 1985, when the U.S. Environmental Protection Agency (EPA) canceled the registration and therefore the use of 2,4,5-T in the United States. Thus ended one of the most controversial and turbulent times for forestry in the United States and Canada, until recently when the social debate about clear-cut logging in those countries began to erupt.

The 2,4,5-T dispute also was a milestone in the history of Weed Science because never before had the discipline's assumptions about the societal good and safety of its tools met with such intense social skepticism, activism, litigation, and even violence. It is quite possible that the social awareness about herbicide use generated through the 2,4,5-T controversy still lingers in the disciplines of weed science and forestry. The resulting credibility gaps are now being carried over into many other activities of those disciplines.

A HISTORY OF DISPUTE

In 1964, a study initiated by the National Cancer Institute suggested that there were concerns about the public safety of 2,4,5-T by pointing out that the herbicide was a possible teratogen (a material causing fetal abnormalities). Additional stories and allegations about the safety of 2,4,5-T continued to surface for the next several years, fueled, perhaps, by the fact that 2,4,5-T was 50 percent of the active ingredient in Agent Orange—a defoliant used by the US. Army during the

Vietnam War. By 1970, enough toxicological evidence had accumulated, however, to prompt a halt to military applications of the material and to initiate administrative proceedings, primarily by the EPA, to curtail domestic uses, that is, suspend registration for use. Throughout the 1970s increasing attention was given to the very toxic 2,3,7,8-tetrachlorodibenzo-*p*-dioxin (TCDD), found to contaminate 2,4,5-T during its manufacture. Extensive toxicological studies were performed over these years by private, governmental, and industrial scientists that confirmed the teratogenic effects of the contaminant, TCDD, and, in some cases, the purest form of 2,4,5-T available. These data were supplemented by additional research on the mechanisms of toxicity of the two chemicals and clinical reports of human exposures through industrial accidents, incidences of exposure during forestry uses, and application or ingestion of the herbicide by volunteers. Arguments about the economic necessity for the chemical were also advanced. These reports, while sometimes providing contradictory information, created doubt about the relevancy of laboratory studies in relation to real-world exposure levels and user needs for the herbicide.

In 1979, following a still controversial study on human miscarriages immediately after forestry applications of 2,4,5-T in the Alsea basin of Oregon, the EPA issued an emergency suspension of all uses of 2,4,5-T for forestry, rights-of-way, and pastures. Curiously, however, they did not revoke registrations for uses of the herbicide in rice and rangelands. Public sentiment against the herbicide's use continued through the 1970s and early 1980s (Figure 11.6).

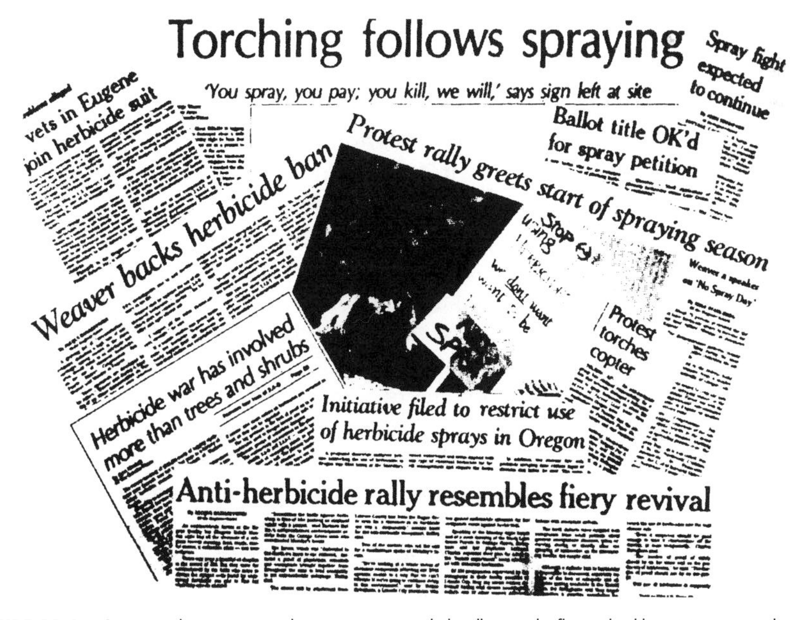

FIGURE 11.6 Campaigns by environmental interest groups made headlines and influenced public opinion against the use of 2,4,5-T. (From Walstad and Dost, 1986.)

Manufacturer and EPA attempts to negotiate a settlement to retain some registrations broke down in 1983, which led to eventual cancellation in 1985. Litigation over pesticide uses in National Forests and other federal lands, originating from the National Policy Act of 1969 (NEPA), also occurred during this same time period. This litigation effectively blocked all herbicide applications on federal lands and resulted in a complete herbicide ban that was maintained until only a few years ago.

LESSONS FROM THE CONTROVERSY

McGee and Levy (1988) point out that public policy decisions that integrally involve scientific issues have become increasingly polarized over the last three decades. They also indicate that herbicide use in forest management embodies most of the important features of such disputes. They especially noted the virulence with which each group contested the other in this particular case study, suggesting the depth to which scientific, political, social, and philosophical issues were intertwined in the dispute.

Walstad and Dost (1986), following the herbicide's cancellation, proposed the following six steps as lessons for responding to public concerns about herbicide applications:

- Establish communication networks and information programs well in advance of operations. Such preparation will provide an atmosphere for interaction within the community and questions and sensitive areas will surface before it is too late to deal with them effectively. Guided tours and door-to-door visits are among the effective ways to communicate with local inhabitants.
- Identify sources of reliable information and expertise. Universities and regulatory agencies tend to be particularly good sources, provided the individuals are well qualified and objective.
- Establish linkages with the media before operations begin. Responsible news reporters can assist land managers in explaining activities to the public. Cooperating with the news media is essential.
- When working with pesticides, encourage prompt, thorough, and impartial investigation of alleged exposure incidents. Such action not only dispels false or exaggerated claims of injury but also ensures that proper attention is given to bona fide incidents of pesticide exposure.
- Ensure consistent and firm enforcement by regulatory officials whenever infractions involving pesticide use occur. Unless enforcement agencies are vigorous and fair, they will not earn public confidence.
- Execute pest control operations with extreme care and proficiency. Detailed planning, thorough training, clear instructions, adequate supervision, and careful implementation are essential. Security and safety precautions protect equipment, materials, personnel, [and the environment].

Most of the steps outlined above are concerned with education about pesticides, establishing and maintaining communication with the public, or assuring conscientious use of the chemicals. But is that enough? McGee and Levy (1988) suggest not. They indicate that increased understanding of the perspectives of both opponents and proponents of herbicide spraying is in order. For example, they suggest that pro-spray representatives, be they chemists, toxicologists, foresters or ecologists, must go beyond assurances of expertise or data to which the public has no access. On the other hand, they also indicate that members of the anti-pesticide movement must learn what is reasonable to expect from any science, especially one as complicated as toxicology, and should stop using scare tactics to drum up support (McGee and Levy, 1988), if reflective deliberation is to occur. At the same time, "pros" must not ridicule those with fears, because "as we understand herbicide toxicology now, neither safety nor risk is absolute and neither total fear nor complete lack of it is rational." Finally, these authors suggest to scientists that some issues, though they involve questions that can be stated in scientific terms, are trans-scientific. In other words, they are simply beyond the ability of science to answer and must be addressed in the policy and political arenas as part of the ever ongoing discussion about what constitutes "the public good."

There is now little doubt that science, technology, and social values played important roles in the activities and decisions surrounding the use and eventual removal of 2,4,5-T. Nor is there much doubt that this episode in the history of Weed Science left its mark on the discipline, especially on those involved directly in the controversy. It was also one of the "battles" that led the way toward much of the public concern, mistrust, and opposition to pesticides. This concern is taken up in the next example.

Atrazine Use and Water Quality

In 1993, the U.S. EPA completed an economic benefits–environmental risks assessment to determine the consequences of a national ban on atrazine and the entire group of triazine herbicides. Atrazine and the other triazine herbicides were chosen for this analysis because they are the most widely used herbicides in the corn-growing region of the United States and coincidentally the most commonly detected group of pesticides in surface and ground water. These herbicides constitute over 12 percent of the total pesticide use in the United States. Thus, a ban on these chemicals could result in a substantial reduction of the overall pesticide load in the agricultural environment. But such a ban could also result in substantial loss in producer income or societal net benefit.

METHODOLOGY

The EPA used an ecological-economic modeling approach, called megamodeling, to accomplish their analysis of the impacts of atrazine and other triazine herbicides on the economic and environmental well-being of the midwestern

United States (see Chapter 10, pp. 488–490). The tool they developed consisted of four interactive components (evaluation criteria, agricultural decision, fate and transport, and policy specification) and eight interfaced models (Table 11.1; see also Figure 10.31), each of which generates statistical response functions. The response functions summarize the relationships between chemical concen-

TABLE 11.1 Models that Comprise CEEPES[a]

Acronym	*Name*	*Citation*	*Comment*
	Economic Decision/Production Models		
RAMS	Resource Adjustment Modeling System	Bouzaher et al. (1992)	Linear programming model that simulates profit-maximizing behavior
WISH	Weather Impact Simulation for Herbicide	Bouzaher et al. (1992)	Systems approach model for weed control strategies
ALMANAC	Agricultural Land Management Alternatives with Numerical Assessment Criteria	Jones and O'Toole (1986)	Process that simulates crop growth and weed competition
AGSIM	Agricultural Sector Integration Model	Taylor (1987)	Econometric model of national & international markets for 10 agricultural sectors
	Environmental Fate & Transport Components		
RUSTIC	Risk of Unsaturated/Saturated Transport and Transformation of Chemical Concentrations	Dean et al. (1989)	Model that partitions mass of pesticides into various media. Groundwater model for pesticides (root zone, vadose zone)
STREAM	Steam Transport and Agricultural Runoff of Pesticides for Exposure: a Methodology	Donigian et al. (1986)	Surface water model for pesticides
BLAYER	Boundary Layer Model	McCorcle (1988)	Atmospheric transport model
PAL	Point, Area, and Line Source	Peterson et al. (1987)	Short-range air transport model

Source: EPA, 1993.

[a]Citations reported in original source; EPA, 1993.

trations within an environment and a set of variables such as weather, soil conditions, tillage, and chemical properties that define the transport and fate of the chemical according to complex process models (EPA, 1993). The final phase of the analysis, aggregation, summarizes economic benefits and environmental risk indicators to allow comparison of policy scenarios (EPA, 1993).

EPA recognized several key problems in constructing the megamodel, called CEEPES (Comprehensive Environmental Economic Policy Evaluation System), so that it would encompass a large region such as the midwestern United States. These include (1) the wide range of variation in temporal and spatial scales of the different models, which required special interfaces; (2) difficulties in aggregating field scale model output to large geographic areas; (3) the lack of adequate calibration and validation data for some models; (4) the lack of detailed data on chemical applications, yields, and producer risks; and (5) the need for diverse soil and weather sources. Thus, CEEPES, according to EPA (1993), is most appropriate for screening level analyses and comparison of policy alternatives.

FINDINGS

The baseline amount of atrazine use estimated by CEEPES was 42 million pounds (active ingredient) for both corn and sorghum, which was less than that estimated by either Gianessi and Puffer (1991) (46 million pounds) or NAPIAP (1992) (54.5 million pounds). In addition, the following observations were made for two possible policy scenarios: (1) a ban on atrazine use only and (2) a ban on all triazine herbicides (also see discussion in Chapter 10, pp. 488–490).

Economic impact. For both an atrazine and a triazine ban, corn acreage and corn yields were predicted to decline, with soybean acreage and soybean yields expected to increase. Corn acreage would be expected to decline by three percent from the baseline of almost 73 million acres with either an atrazine or triazine ban, while soybean acreage would probably increase by about four percent over its 4.4 million acre baseline if either policy scenario occurred. Declines in corn yields by 2.8 and 4.1 percent also would be expected for atrazine and triazine bans, respectively, owing to a shift in weed control practices to other herbicides or increased tillage. The costs of weed control would be expected to increase by $6 to $8 per acre. It also was projected that both long- and short-term national economic welfare (changes in producer income, domestic consumer effect, and government outlay) would decline by $365 million for an atrazine ban and $526 million under a triazine ban (Figure 10.32). Under the atrazine ban crop producers in the midwestern U.S. Corn Belt would bear a large share of this economic burden because producer income in the region would be reduced by $234 million. For the triazine ban, however, some of the losses in the Corn Belt would be offset by higher corn prices and a loss in producer income of $168 million was projected.

Input substitutions. In the baseline estimate of atrazine use, over 65 percent of the corn acreage was found to be treated with a strategy of weed control involving atrazine. Under an atrazine ban over 57 percent of the corn land would still be treated with a strategy of weed control involving a triazine herbicide. With the triazine ban, however, 27 percent of the land would be treated with some form of cultivation, while over 50 percent of the corn acreage would be treated with a non-triazine herbicide. Total herbicide usage under the triazine ban was projected to increase by nearly 50 percent, apparently owing to higher rates and more acres of application of less effective chemicals.

Environmental impacts. Projections of environmental impacts complete the picture of societal impacts of atrazine and triazine bans. Much of this analysis has already been presented in Chapter 10 (see displacement into ground water). EPA used established environmental benchmark concentrations (standards) and projected herbicide concentrations to examine the potential for surface and ground water contamination under the baseline and two policy scenarios (Table 10.5). The EPA concluded from this analysis that a high level of surface loading of soils with other triazine herbicides would increase under an atrazine ban, resulting in chemical concentrations exceeding water quality benchmarks. Generally, this problem was reduced under a total triazine ban. They also noted that, for ground water, all concentrations were projected to be well below the long-term benchmarks under both the atrazine and the triazine ban that were being projected. Exposure values for aquatic vegetation, an indicator of other environmental impacts from corn production under the two scenarios, were always quite high for many herbicides—often twenty-fold greater than EPA standards. Soil erosion was projected to increase under the bans because of the amount of corn production that would shift necessarily from reduced till and no-till systems (see Chapter 8 for more detail) to more conventional tillage operations.

DETERMINATION OF RISK

In deciding whether or not the risks of atrazine use out-weigh its social benefits, EPA must determine not only the magnitude of these risks and benefits, but also whether or not curtailment of its use under different policy options leads to a net environmental gain or loss (EPA, 1993). There appears to be little doubt that substantial benefit arises from the use of atrazine and hardship would arise if it were banned. It also appears that water quality and public safety benchmarks are being exceeded by current weed control practices using the herbicide. For surface water, this problem would apparently be exacerbated by an atrazine ban but alleviated if all triazines were removed from use. Ground water quality would improve under both scenarios.

The question now is, what are the relative weightings that must be placed on the economic, human safety (water quality benchmarks), and environmental considerations regarding atrazine use? This question of relative importance will structure the societal debate now forming over continued atrazine use or curtailment. In 1995, EPA called for a two-year special public review of atrazine and the triazine herbicides. Over this time period, testimonies will, no doubt, be given and further analyses performed. The results of the review could result in little or no action by EPA or cancellation of the products' use registrations. It also could result in prolonged litigation if proceedings are not perceived as lawful and fair. Regardless of the outcome, it is a process that is important—a process that, although cumbersome and troublesome at times, advances the common good.

Herbicide-Resistant Crops

Recent advances in molecular genetics, biochemistry, and cell physiology have opened the possibility of creating herbicide-resistant crops (HRC). These products have been touted as the solution to many environmental problems associated with modern crop production, and have been described as powerful tools to increase options in crop production (Benbrook and Moses, 1986; Goodman 1987; also see Chapter 3). Although the potential benefits of HRCs through improved crop production are readily apparent, there is also concern that their release may occur without much forethought about their impact on agroecosystems, the broader landscape, or rural and urban economies and cultures (Dekker and Comstock, 1992). Some HRCs have already been released (Beversdorf and Hume, 1984) and many others are expected during the current decade. Radosevich et al. (1992) indicate that the concerns or questions about HRCs can be summarized into three categories: technological, biological, and ethical.

TECHNOLOGICAL CONCERNS ABOUT HRCS

HRCs are the latest advance in the current technology to control weeds (see Figure 3.7). They represent a continued refinement of the already existing technology that relies on the development and use of herbicides. Benefits from increased herbicide tolerance in formerly susceptible crop species should result from this technology. For example, higher levels of weed control efficiency are an obvious outcome of HRC use, although improved crop yields and farmer income as a result of the technology may be problematic (see pages 210–215 for more discussion). HRCs also could increase the use of crop rotations by providing a tolerant crop if phytotoxic levels of a herbicide exist in soil from the crop before. Thus, HRCs could provide an option to reduce continuous cropping and prevent many harmful side-effects that accompany overuse of a single chemical product, for example, herbicide resistance, herbicide residues in soil or surface and ground water (see atrazine discussion above). On the other hand, if more chemical is applied as a result of HRCs (for example, to all crops in a rotation), impacts on soil and water quality could be worsened.

The strategies used to market a herbicide and its HRC are primarily technological questions because both the longevity and utility of the combination are influenced negatively by weed resistance to the herbicide. It is uncertain whether HRC use will decrease the incidence of herbicide resistance in weeds or actually enhance it (Maxwell et al., 1990). Whether either happens depends on how HRCs are used in a cropping system. The potential for herbicide resistance could be reduced if HRCs are used to break crop-herbicide cultural monocultures and therefore increase rotations (diversity) among different crops or classes of chemicals. The opposite is true if HRCs are used to fortify market exclusivity of a chemical in crops within a region.

Radosevich et al. (1992) make the point that HRCs are an extension of the existing weed control technology that relies, in some regions, almost exclusively on herbicides. The probable result of such a tactic is similar or increased intensity of herbicide use (for example, from shifts in market share of a herbicide as HRCs become available), not less. Similar or greater patterns of herbicide use will not relieve environmental concerns, unless HRCs provide a marketing advantage to more environmentally benign chemicals.

BIOLOGICAL CONCERNS ABOUT HRCS

Most of the biological concerns about HRCs originate from the paucity of information about the genetics, breeding systems, and population dynamics of most weed species (Chapter 3), and therefore the uncertainty about possible escape of herbicide resistance into susceptible plant populations (Radosevich et al., 1992). Four primary questions exist about the biological risks of HRC release into the environment:

What is the risk of an HRC becoming a super-weed in another crop? This question is relevant because even casual observations of agricultural fields demonstrate that volunteer plants of the preceding crop sometimes can be major weeds in the next crop. Will sensitivity to another herbicide be engineered into HRCs so their control is possible? Who will bear the cost of control if these plants escape?

What is the likelihood of herbicide resistant weed populations evolving when HRCs are used? Weed species are extremely variable genetically, being able to adapt to almost any selective force applied to them (Harper, 1977; Gressel and Segel, 1978; Maxwell et al., 1990). Gressel and Segel (1978) and Maxwell et al. (1990) indicate that selection pressure from persistent or frequently used herbicides is the major factor that causes resistance to evolve in formerly susceptible plant species. As discussed earlier, HRCs could either increase or decrease the selective pressure of a herbicide, depending on the marketing strategies employed. Weed resistance can be prevented and managed (Maxwell et al., 1990), but much better understanding and reliance on biological processes than is currently practiced is necessary to accomplish such a goal.

Will genes inserted into HRCs be further transferred by the plants themselves? If so, what will be the extent and consequence of such genetic dispersion? The biological process of concern here is *introgression*, that is, hybridization among distinct plant populations or species and subsequent backcrossing. Test systems indicate that such genetic exchanges among wild, weed, and crop plants already occur (Doebley, 1990; Ellstrand and Hoffman, 1990; Hoffman, 1990; Wilson, 1990). The incidence of shattercane (*Sorghum bicolor*), a weedy relative of sorghum, demonstrates the potential for this process to have economic as well as biological significance. Because introgression can occur reciprocally between crop plants and their non-crop relatives, could HRCs revert to susceptible forms? Can non-weed relatives become important weeds in HRCs? How would the movement of genes that cause herbicide tolerance influence the diversity of other plant populations? Are traits other than herbicide tolerance associated with HRCs? If so, how are these traits moved and what would be the consequences to other plant populations?

What is the risk of further reducing the complexity of the agroecosystem? Ecosystems include a complex spectrum of organisms that interact in different ways, yet studies to describe and define the nature of these relationships in agriculture are rare. Further impoverishment of biological complexity can compromise the stability of agroecosystems and reduce their resilience to disturbance (an issue considered in more depth later). An impoverished, low-diversity ecosystem also provides optimal conditions for unhampered growth of weeds, insects, and diseases because many ecological niches are not filled by other organisms (Chapter 2). HRCs, through increased herbicide effectiveness, could further reduce plant diversity, allowing almost unlimited growth of any weed species that survives or adapts to those treatments.

The issue in every case is whether these questions should be answered before rather than after the introduction and release of HRCs to the environment.

ETHICAL QUESTIONS ABOUT HRCS

Comstock (1989), in his discussion about genetically engineered herbicide resistance, distinguishes clearly between the roles of science and ethics in such debates. According to Comstock, science is primarily *descriptive* and explanatory, whereas ethics is *prescriptive*. Dekker and Comstock (1992) identify four areas of concern about the moral dimension of HRCs.

What is the medical and environmental safety of the final product? Will herbicide-resistant potato, tomato, or maize be safe for humans, or might toxic residues remain in or on the crops? Will toxic residues accumulate in tissues of fish in streams collecting run-off from HRC fields? Given the magnitude of the ecological questions we now face (problems such as soil erosion, surface and ground water pollution, destruction of habitat) shouldn't less environmentally taxing methods of growing food be sought (Dekker and Comstock, 1992)?

Will HRC technology further concentrate power in the hands of a few companies? Might the introduction of HRCs allow a few chemical companies to further strengthen their hold on an industry that is already controlled by a relatively few corporations, forcing American farmers to pay inflated prices for seed or chemicals? Couch (1990) and Mollison (1988) point out that almost all new technologies serve to strengthen the dominant system. According to Mollison (1988),

> The tragic reality is that very few sustainable systems are designed or applied by those who hold power, for to let people arrange for their own food, energy, and shelter, is to lose economic and political control over them.

The economic power of the chemical industry that markets HRCs deserves more discussion (Dekker and Comstock, 1992).

Will farmers be displaced by HRCs? HRCs might increase the productivity and efficiency of a farmer's time, but what does that mean for farm and rural economies that are already unstable? Dekker and Comstock (1992) point out that time-saving technologies have been substituted for farmers and laborers for over two hundred years in the United States. They ask, "is this a trend we want to continue? Is it socially desirable for rural poverty to exist? Will this technology continue and contribute to the long-term trend of farm foreclosures in America?" On the other hand, these authors also ask if HRCs might not make marginal farmers more productive, helping them to compete better with foreign and corporate competitors, and thus revitalizing rather than destroying the rural economy.

Will HRCs enhance who we are as a people, our cultural identity? According to Dekker and Comstock (1992), HRCs might make American agriculture more dependent on chemical and capital-intensive practices. Is this the direction we as a people want to go? If we follow such a course, do we render our food supply vulnerable to a single virulent organism or resistant weed? Do we want to encourage exploitative attitudes toward nature (Dekker and Comstock, 1992)?

CONCLUSIONS ABOUT HRCS AND OTHER AGRICULTURAL TECHNOLOGIES

Weeds, or specifically decisions about weed control, involve a web of biological, economic, political and social factors. It is, therefore, almost impossible to discuss scientifically the introduction of a new tool for weed control without considering each factor separately and collectively. For this reason, our discussion about HRCs is incomplete, and perhaps it always will be.

In many respects, HRCs are like most other technological innovations of modern agriculture. The National Research Council (1989) points out that the financial viability of many farms and rural communities in the United States has declined over the last several decades as input costs have risen and crop prices and land values have fallen. The consequences of farming on environmental quality and biological integrity also have become increasingly important to the public, policymakers, and farmers, especially over the last decade (National Research Council, 1989). Lewontin (1982) likens the present-day agricultural system to a treadmill in which each new technological innovation requires farmers to use more chemicals, energy, or money to remain productive. Unfortunately, the costs of such inputs in relation to the value of outputs derived from them make it difficult for some farmers to remain in business (Chapter 5, pages 210–215). These technological inputs, such as HRCs, also raise legitimate concerns about the social costs to maintain or restore environmental and biological integrity. The ethical issue raised from such technologies, for example HRCs, is "Who benefits from their use and who is harmed?"

Many farmers have taken steps to reduce the costs and adverse environmental impacts of their operations. Some have improved conventional techniques and practices, while others have adopted alternatives to the costly trend of seemingly ever-increasing inputs. Farmers who have adopted alternatives try to take greater advantage of natural processes and beneficial on-farm biological interactions, reduce off-farm purchases and improve the efficiency of their operations (National Research Council, 1989). Many of these processes and approaches have been discussed throughout this text.

It is hard to give unqualified support to HRCs at this time. Aside from some potential marketing advantages to companies with specific herbicides, both potential benefits and risks from HRCs are uncertain. Until such uncertainties are eased, there will be legitimate questions about herbicide-resistant crops and many other technologies of modern agriculture that must be resolved over the coming decade.

WEED MANAGEMENT IN POST-INDUSTRIAL SOCIETY

Current discussions about change in science, technology, and society itself suggest a new era in which old, albeit successful, concepts based on Cartesian assumptions of reductionism, linearity, and objectivity are yielding to a post-industrial way of thinking that emphasizes holism, circuitry, and connections (Bohm, 1980; Ferre, 1982, 1988; Pauly, 1987; CIESIN, 1992). The seeds of this shift were recognized as early as the late nineteenth century (Steiner, 1983) and were developed in science by the mid-twentieth century (Pauly, 1987). Is it possible to take a different look at weeds and their management from this new perspective?

Weeds pose a dilemma for agriculturists, foresters, and range managers because weeds directly reduce crop yields by causing a decline in crop performance (Chapter 5) or quality, but weed control is often difficult and expensive, and sometimes it creates other undesirable environmental and social side-effects. Designing weed management strategies now requires working through this dilemma from the perspective of how weeds, agroecosystems, and human institutions have evolved, and will continue to evolve together.

As discussed in Chapter 2, the human social and ecological systems that comprise agroecosystems are often represented as nested hierarchies of function and scale (Figure 2.1). The hierarchy in each system can be determined by measuring differences in rates of various processes at each scale or level (Chapter 2). Processes for an ecological system include growth, respiration, and primary productivity, while processes of the social system include adoption of new technology, cultural invasion, and education. The rates of such processes help define the boundaries within each system and are functional rather than arbitrary (O'Neil et al.; 1986). Generally, a hierarchical system is arrayed according to levels of organization, from simple to complex—for example, organism, population, guild, community, and biome in Figure 2.1. For the most part, higher levels are larger, have more components, and their process rates are slower than lower levels. Traits selected at one level of biological organization (scale) can affect evolution occurring at other levels (Stanley, 1989).

Relationship of Agroecosystems to Weed Management

Weed control practices have effects other than to reduce competition and to change a weed flora. For example, some management practices can force radical changes on an entire agroecosystem or discipline (see 2,4,5-T and atrazine discussions above). We know that herbicides and cultivation reduce weed competition, but these tools also cause loss of water quality and soil fertility, in some instances, through leaching or erosion. Weeding also may stimulate a feedback response that increases the probability of invasion by new weeds that are aliens or were derived through new species formation (Rejmanek, 1989). Such a feedback could explain why relative and absolute abundance of the weed flora increased steadily from 1900 to 1980 (Forcella and Harvey, 1983), despite the enormous investments to control weeds over that time period.

The integrated pest management (IPM) movement of the last decade developed an awareness of the positive and negative feedbacks among differing organisms in an agricultural food web (Pimentel, 1986; Levins, 1986; Bird, 1994). However, most integrated pest management studies have, until recently, only focused on a few species at a time (e.g., one crop, one weed, one insect, etc.) and simultaneous evaluation of more than two or three members of a food web has been rare (Duffey and Bloem, 1986; Adams, 1995). Current efforts in IPM could be extended, using systems theory, to examine links among species (May, 1981) and human institutions. Levins (1986) used three general models of pest management, which he called *industrial, IPM*, and *ecological agriculture* (Table 2.2), to illustrate this point. However, only the ecological agriculture model, which contains some elements of the recent sustainable agriculture movement (National Research Council, 1989; Bird, 1994), requires an understanding of fundamental processes in ecology and human institutions (Figure 2.1) to manage pests and weeds.

Perhaps the study of multilevel interactions, as suggested by Levins (1986), will assist efforts to recouple the ecological and social components of agroecosystems (Figure 1.14, Table 2.2). But as far as weeds are concerned, it is not enough simply to use herbicide-resistant crops or to replace herbicides with cultural approaches to control weeds (e.g., crop rotation, weed-suppressive crops, intercrops, or biological controls) to achieve the coevolution of weed management and healthy ecosystems. The recoupling of ecological and social systems in modern agriculture requires the recognition within our social institutions (businesses, governments, scientific and educational organizations) that such coevolution exists, and is desirable. It also requires institutional change to minimize over consumption of resources, optimize labor and energy inputs that are necessary for crop production, and maximize the use of biotic interactions and social values when designing new weed management strategies. This is the seemingly formidable, yet challenging, task facing Weed Scientists.

SUMMARY

Most chapters of this text have been concerned with weeds on individual farms or fields. However, weeds and the methods to control them often extend beyond individual fields and farmers to broader aspects of society. Weeds can sometimes spread over vast areas and affect many people. Weed control also can have widespread benefits and costs to land managers and others in society. Externalities exist when the consequences of weeds or weed control extend beyond the area of infestation or domain of the individual person making a management decision. Since such consequences can be felt or perceived differently, conflicts can arise among people affected differentially by weed control decisions and tactics. Usually, it is best to resolve such conflict, rather than try to teach, educate, or indoctrinate others who have an alternate point of view. The controversies over 2,4,5-T use, influence of atrazine on water quality, and release of herbicide-resistant crops are three examples of how society interacts with and affects weed management, and how biological and social factors are tightly coupled in agriculture.

Value systems are the way people think about activities and technologies. There are four fundamental value systems that affect natural resource management, including agriculture: (1) material well being, (2) sanctity of nature, (3) individual rights, and (4) justice as fairness. Each of these value systems raises different questions about human efforts to raise crops and manage the land, including control of weeds. In Western societies, the most common way to assess differences among value systems is to use a risk-benefit analysis, but this method assumes all things are quantifiable and ignores those aspects of an issue that are not. Important societal issues involving weeds and weed control are farmers' income, employment, and income distribution among laborers. Also of concern is power, control, and the question of who benefits from the various technologies used to control weeds. It is important that all such issues be explored when trying to assess the value of weed control.

It is reasonably certain that agriculture and natural resource management have evolved along with societal needs, wants, and expectations. The three examples discussed in this chapter explore how Weed Science, as a scientific and management-oriented discipline, accommodates such change. It is also possible that some of the modern management practices used to control weeds have coevolved to maintain productivity, but have contributed to a deterioration in the soil resource base at the same time. In the future, Weed Scientists will be called upon to develop new methods of weed management that consider environmental, ethical, and societal concerns, as well as those of agricultural production and efficiency.

Epilogue: What Is a Weed?

Weeds have become such a predominant and pernicious aspect of food production that in many places of the world, especially developing nations, farmers are simply people with hoes. Weeding, for many farmers, is the chore that overshadows other, more satisfying, aspects of farming. It is often the most important task, besides planting, that they do to assure crop growth. The tools used for weed control range from hatchets and machetes to plows, disks, harrows, and chemicals. Even other plants, animals, and microbes are now used to control weeds. No matter what tool is chosen, it is always with the expectation that reducing weeds will increase crop productivity, an assumption proven correct by scores of experiments conducted in almost every cropping system of the world. Even so, the incidence of weeds in the world's cropland has not declined nor has any weed species ever gone extinct in spite of the monumental effort to control weeds each growing season. In fact, the abundance of both new and existing weed species increased during the last half century. Thus, weed control works only from the single, limited perspective of improved crop production. From another perspective, weed control tactics have actually increased the prevalence of weeds, while billions of people still spend their lives hoeing and millions of dollars are spent every year for chemicals to kill unwanted plants.

Our book begins with a quote from Alden Crafts, one of the pioneers of modern chemical weed control. Crafts writes, in his first book (Robbins et al., 1942), "In the beginning there were no weeds," acknowledging that plants became weeds as agriculture evolved and suggesting that many weed species could be the product of the way we grow food now. If so, weeds are a human artifact with their origin identical to that of the hoes, plows, and herbicides now used to kill them. Aldo Leopold, the natural historian and author, carries this idea further in an unfinished essay titled "What is a weed? [1943]." Leopold takes exception to the authors of *The Weeds of Iowa* (Bulletin No. 4, 715p.,1926) who list many of the native plants of Iowa as weeds: black-eyed Susan, partridge pea, flowering spurge, prairie goldenrod, wild rose, blue verbain, chicory, peppermint. Leopold points out that most of the plant species listed in the Bulletin have substantial value to wildlife, soil cover, and soil fertility, or in enhancing the beauty of a place. He grants that some plants "do enormous damage to cropland," but also points out that most weed problems arise from overgrazing, soil exhaustion, or the needless disturbance of more advanced successional stages of vegetation. Leopold also exhorts us to consider that "to live in harmony with [all] plants is, or should be, the ideal of good agriculture."

Perhaps because weeds are a human construct, most descriptions of them reflect some pattern of human behavior. Weeds usually are portrayed negatively as being aggressive, competitive, or invasive—thieves who rob our crops, and therefore us, of needed nutrients. Such descriptions of weeds are necessary if the intent is to kill them, reduce their abundance, or eradicate them entirely. These descriptions also trivialize any positive characteristics these plants might possess. For example, aggression on a football team or competitiveness in a corporation are desirable attributes, and such traits in individuals are often sought by employers. The botanist and natural historian E.J. Salisbury (1942b) points out that even thieves can have admirable qualities when not engaged in nefarious activities. A thief might be a generous benefactor to the poor or a loving husband and father, while a weed in a different environment could be a delightful wildflower or the habitat for other beneficial organisms. Invasiveness, in contrast to competition and aggression, is almost always detrimental. An invading, dispersing species, whether it is a microbe, plant, animal, or human, is viewed as a threat and measures are usually taken to contain it. Antibiotics are used to reduce the effects of invading microbes in our bodies. Fights, feuds, and wars are the outcome of human attempts to protect property from the dispersion of other people. Weed control tactics slow the scatter of weeds into farmers' fields. However, even an invading weed can be helpful, if it slows the erosion of worn out, degraded, or misused land.

Some biological traits of weeds are also similar to those of people. For example, the fecundity of weeds is a serious problem for farmers because weed populations can increase exponentially in a short time, making the land they occupy unproductive and worthless. Similarly, human fecundity and the concomitant overuse of the world's resources are now two of the major problems facing humankind. Weeds, of course, are adaptable, perhaps as adaptable as humans. It is probably this characteristic that is most disconcerting, because these organisms adapt to us. They evolve in response to the very tools, tactics, and practices that we use to kill them. They are vegetation's equivalent to the housefly and wharf rat, products of our own making. Neither wild nor domesticated, none of these organisms can exist now without the presence or activities of people to support them. Perhaps, then, we should learn more about ourselves and our motivations, as well as about the characteristics of weeds, in order to attain the harmony in agriculture to which Leopold refers.

It is an unfortunate fact that almost every innovation in agriculture, including the technology to control weeds, drives rural people from the land. This process, which often substitutes pesticides, fertilizers, energy, and money for human work, can be desirable when it creates labor for urban jobs. However, in many instances, such "efficiency" displaces rural people to urban slums and to even greater poverty. The paradox, as far as weeds are concerned, is that hoeing, while a predominant source of human drudgery, is also a major user of hand labor and employment in developing nations. In developed countries, on the other hand, the continual quest for ever higher yields has become a technological treadmill. Farmer debt, loss of land, and ultimately, displacement to cities result because of the high input costs of chemicals and energy to produce food, coupled with low prices from overproduction. Today less than two percent of the American population are farmers. Serious questions now must be answered about the stability and environmental quality of a system of food production that relies so much on off-farm inputs of chemicals and energy and so little on farmer involvement and an understanding of natural, biological processes.

We chose to examine the ecology of weeds and crops in our book because of our belief that a better, more economical, and more environmentally sound agriculture will result from an understanding of biology and the environment in which plants grow. Nonetheless, farmers will still be faced with weedy fields, the paradox of hoeing will still exist in developing nations, and the dilemma of rural displacement with each new turn of the technological treadmill will continue. It seems to us that the agricultural crisis described in the preceding paragraph, of which weeds are only a symptom, originates from the view of ourselves as managers, manipulators, and dominators of nature. From this perspective, Weed Science is inextricably bound to technology as the only effective way to solve problems and science becomes simply a tool to understand and manipulate nature in different ways, sometimes using new tactics but often only to counteract the problems caused by older technologies. As an example, farmers bothered by declining crop yields look for solutions in weed control. These "solutions" beget more problems, such as soil erosion, chemical contamination of food and water, and the displacement of rural people. Now we look for some better technology, one that incorporates a larger understanding of environment and biology. Unfortunately, we don't yet know what new questions will arise from this technology, but they probably will stem from our very definition of weeds as a problem—a particular discrete problem unconnected to other problems of society.

We would like to imagine a better scenario, a different possibility in which people base their decisions on a deep understanding that they are part of an interacting web of life in which every action causes a suite of reactions. Under this scenario, human beings would need to be more humble, more accountable to nature, adapting to what exists rather than the other way around. With this viewpoint, it is unlikely that weeds as we know them today would even exist, because under such a scenario these plants would simply be incorporated into the "normal" agricultural cycles of production and grazing. However, barring a sudden shift in human consciousness, society seems well entrenched in the current scenario. So, what can be done to resolve the dilemma of weeding and avoid the treadmill?

Do not oppose progress. Think broadly and critically about the consequences of your actions, whether in science or in production. Wonder who will benefit from these actions; think of them as seeds, and imagine where they will disperse, where they will grow. If they are likely to enhance only yourself or a few rich and powerful people, or if they push small farmers off their land, or displace indigenous or other self-sufficient people, or degrade the environment, injure animals, or cause unpredictable economic effects, then consider alternatives that do not. Weigh alternatives carefully. Technology for its own sake is not progress. Remember that weeds are plants whose virtues have not yet been discovered.

References

Abrahamson, W.G. 1980. Demography and vegetative reproduction. Pp. 89–106 in O.T. Solbrig (Ed.), *Demography and Evolution in Plant Populations*. University of California Press, Berkeley.

Adams, C.E. 1995. The role of IPM in a safe, healthy, plentiful food supply. Pp. 25–34 in *Second National Integrated Pest Management Symposium/Workshop Proceedings*. Las Vegas, Nevada. April 19–22, 1994.

Adkins, S. 1995. *Weed Science*. Lecture guide. University of Queensland, Brisbane, Australia.

Akesson, N.B. and S.V. Walton, Jr. 1989. Aircraft. Pp. 155–162 in California Weed Conference. *Principles of Weed Control in California*. 2nd Ed. Thompson Publications, Fresno, CA.

Akey, W.C., T.W. Jurik, and J. Dekker. 1990. Competition for light between velvetleaf (*Abutilon theophrasti*) and soybean (*Glycine max*). *Weed Res.* 30:403–411.

Akobundu, I.O. 1987. *Weed Science in the Tropics: Principles and Practices*. John Wiley & Sons, New York.

Aldrich, R.J. 1984. *Weed-Crop Ecology. Principles in Weed Management*. Breton Publishers, North Scituate, MA.

Alex, J.F. 1967. Competition between *Setaria viridis* (green foxtail) and wheat seeded at two rates. Pp. 286–287 in *Res. Rep., Can. Nat. Weed Comm., West. Sect.*

Allard, R.W. 1965. Genetic systems associated with colonizing ability in predominantly self-pollinated species. Pp. 49–75 in H.G. Baker and G.L. Stebbins (Eds.), *Genetics of Colonizing Species*. Academic Press, New York.

Allen, M.F. 1990. *The Ecology of Mycorrhizae*. Cambridge University Press, New York.

Alm, D.M., E.W. Stoller, and L.M. Wax. 1993. An index model for predicting seed germination and emergence rates. *Weed Technol.* 7:560–569.

Alstrom, S. 1990. *Fundamentals of Weed Management in Hot Climate Peasant Agriculture.* Crop Production Science, Uppsala, Sweden.

Altieri, M.A. and M. Liebman (Eds.). 1988. *Weed Management in Agroecosystems: Ecological Approaches.* CRC Press, Inc., Boca Raton, FL.

Altieri, M.A., A. van Schoonhoven, and J.A. Doll. 1977. The ecological role of weeds in insect pest management systems: A review illustrated by bean (*Phaseolus vulgaris*) cropping systems. *PANS (Pest Artic. News Summ.)* 23:195.

Altom, J.D. and J.F. Stritzke. 1973. Degradation of dicamba, picloram, and four phenoxy herbicides in soils. *Weed Sci.* 21:556–560.

Anderson, R.N. 1968. *Germination and Establishment of Weeds for Experimental Purposes.* Humphrey Press, Geneva, NY.

Anderson, W.P. 1977. *Weeds Science: Principles.* West Publishing Co., St. Paul, MN.

Anderson, W.P. 1996. *Weed Science: Principles and Applications.* 3rd Ed. West Publishing Co., St. Paul, MN.

Andow, D.A. 1983. Effects of agricultural diversity on insect populations. Chapter 5 in W. Lockeretz (Ed.), *Environmentally Sound Agriculture.* Praeger Publishers, New York.

Andow, D.A. 1988. Management of weeds for insect manipulation in agroecosystems. Pp. 265–301 in M.A. Altieri and M. Liebman (Eds.), *Weed Management in Agroecosystems: Ecological Approaches.* CRC Press, Inc., Boca Raton, FL.

Andrade, F.H., F.A. Margiotta, R.M. Martinez, P. Hieland, S. Uhart, A. Grilo, and M. Frugone. 1992. Densidad de platas en maiz. Boletin Tecnico No. 108, Estacion Experimental Balcarce, INTA, Balcarce, Argentina.

Appleby, A.P. 1994. *Weed Control Text Supplement.* Department of Crop and Soil Science, Oregon State University, Corvallis, OR.

Appleby, A.P., P.D. Olson, and D.R. Colbert. 1976. Winter wheat yield reduction from interference by Italian ryegrass. *Agron. J.* 68:463–466.

Ashton, F.M. and A.S. Crafts. 1973. *Mode of Action of Herbicides.* John Wiley & Sons, New York.

Ashton, F.M. and A.S. Crafts. 1981. *Mode of Action of Herbicides.* 2nd Ed. John Wiley & Sons, New York.

Ashton, F.M. and W.A. Harvey. 1971. *Selective Chemical Weed Control.* Univ. of Calif. Circ. 558.

Ashton, F.M. and T.J. Monaco. 1991. *Weed Science: Principles and Practices.* 3rd Ed. John Wiley & Sons, New York.

Audus, L.J. (Ed.). 1964. *The Physiology and Biochemistry of Herbicides.* Academic Press, London.

Audus, L.J. (Ed.). 1976. *Herbicides: Physiology, Biochemistry, Ecology.* 2nd Ed. Vol. 2. Academic Press, London.

Auld, B.A. 1984. Weed distribution. Pp. 173–176 in *Proc. 7th Austral. Weeds Conf.*, Vol. 1.

Auld, B.A. and R.G. Coote. 1990. Invade: Towards the simulation of plant spread. *Agric. Ecosys. Environ.* 30:121–128.

Auld, B.A. and C.A. Tisdell. 1988. Influence of spatial distribution of weeds on crop yield loss. *Plant Prot. Q.* 3:31.

Auld, B.A., K.M. Menz, and C.A. Tisdell. 1987. *Weed Control Economics.* Academic Press, London.

Austin, M.P. 1985. Continuum concept, ordination methods, and niche theory. *Annu. Rev. Ecol. Syst.* 16:39–61.

Avers, C.J. 1986. *Molecular Cell Biology.* Addison-Wesley, Reading, MA.

Bachthaler, G. 1967. Changes in arable weed infestation with modern crop husbandry techniques. *Abstr. 6th Int. Congr. Plant Prot.*, Vienna. Pp. 167–168.

Bailey, G.W. and J.L. White. 1964. Review of adsorption and desorption of organic pesticides by soil colloids, with implications concerning pesticide bioactivity. *J. Agr. Food Chem.* 12:324–332.

Bailey, L.H. and E.Z. Bailey. 1941. *Hortus the Second.* Macmillan, New York.

Baker, D.N. 1965. Effects of certain environmental factors on net assimilation in cotton. *Crop Sci.* 5:53–56.

Baker, D.N. and R.B. Musgrave. 1964. Photosynthesis under field conditions. V. Further plant chamber studies of the effects of light on corn (*Zea mays*). *Crop Sci.* 4:127–131.

Baker, H.G. 1974. The evolution of weeds. *Annu. Rev. Ecol. Syst.* 5:1–24.

Balke, N.E. 1985. Herbicide effects on membrane functions. Pp. 113–139 in S.O. Duke (Ed.), *Weed Physiology, Vol. 2, Herbicide Physiology.* CRC Press, Inc., Boca Raton, FL.

Ballaré, C.L., A.L. Scopel, C.M. Ghersa, and R.A. Sánchez. 1987a. The demography of *Datura ferox* (L.) in soybean crops. *Weed Res.* 27:91–102.

Ballaré, C.L., A.L. Scopel, C.M. Ghersa, and R.A. Sánchez. 1987b. The population ecology of *Datura ferox* in soybean crops: A simulation approach incorporating seed dispersal. *Agric., Ecosyst., Environ.* 19:177–188.

Ballaré, C.L., A.L. Scopel, C.M. Ghersa, and R.A. Sánchez. 1988. The fate of *Datura ferox* seeds in the soil as affected by cultivation, depth of burial and degree of maturity. *Ann. Appl. Biol.* 112:337–345.

Ballaré, C.L., A.L. Scopel, and R.A. Sánchez. 1990. Far-red radiation reflected from adjacent leaves: An early signal of competition in plant canopies. *Science* 247:329–332.

Ballaré, C.L., A.L. Scopel, and R.A. Sánchez. 1991. Photocontrol of stem elongation in plant neighborhoods: Effects of photon fluence rate under natural conditions of radiation. *Plant Cell Environ.* 14:57–65.

Ballaré, C.L., A.L. Scopel, R.A. Sánchez, and S.R. Radosevich. 1992. Photomorphogenic processes in the agricultural environment. *Photochem. Photobiol.* 56:777–778.

Barbour, M.G., J.H. Burk, and W.D. Pitts. 1980. *Terrestrial Plant Ecology.* Benjamin Cummings, Menlo Park, CA.

Barbour, M.G., J.H. Burk, and W.D. Pitts. 1987. *Terrestrial Plant Ecology.* 2nd Ed. Benjamin Cummings, Menlo Park, CA.

Barnes, J.P., A.R. Putnam, and B.A. Burke. 1986. Allelopathic activity of rye (*Secale cereale* L.). Pp. 271–286 in A.R. Putnam and C.S. Tang (Eds.), *The Science of Allelopathy.* John Wiley & Sons, New York.

Barrentine, W.L. 1974. Common cocklebur competition in soybeans. *Weed Sci.* 22:600–603.

Barrett, S.C.H. 1982. Genetic variation in weeds. Pp. 73–98 in R. Charudattan and H.L. Walker (Eds.), *Biological Control of Weeds with Plant Pathogens.* John Wiley & Sons, New York.

Barrett, S.C.H. 1983. Crop mimicry in weeds. *Econ. Bot.* 37:255–282.

Barrett, S.C.H. 1988. Genetics and evolution of agricultural weeds. Pp. 57–75 in M.A. Altieri and M. Liebman (Eds.), *Weed Management in Agroecosystems: Ecological Approaches.* CRC Press, Inc., Boca Raton, FL.

Barrett, S.C.H. and D.E. Seaman. 1980. The weed flora of California rice fields. *Aquat. Bot.* 9:351–376.

Bartels, P.G. 1985. Effects of herbicides on chloroplast and cellular development. Pp. 63–90 in S.O. Duke (Ed.), *Weed Physiology, Vol. 2, Herbicide Physiology.* CRC Press, Inc., Boca Raton, FL.

Baskin, J.M. and C.C. Baskin. 1989. Physiology of dormancy and germination in relation to seed bank ecology. Pp. 53–66 in Leck M.A., V.T. Parker, and R.L. Simpson (Eds.), *Ecology of Soil Seed Banks.* Academic Press, San Diego, CA.

Bazzaz, F.A. and R.W. Carlson. 1982. Photosynthetic acclimation to variability in the light environment of early and late successional plants. *Oecologia* 54:313–316.

Beal, W.J. 1911. The vitality of seeds buried in the soil. *Proc. Sec. Promot. Agric. Sci.* 31:21–23.

Beck, K.G. 1990. Weed management in rangelands and pastures. *Science in Action.* Colorado State University No. 3. 105, Fort Collins, CO.

Beckwith, E.G. 1854. *Report of explorations for a route for the Pacific railroad of the line of the forty-first parallel of north latitude.* In Vol. II, House Rep. Ex. Doc. No. 91, 33rd Congress, 2nd session.

Begon, M. and M. Mortimer. 1986. *Population Ecology. A Unified Study of Animals and Plants.* Blackwell Scientific Publications, London.

Benbrook, C.M. and P.B. Moses. 1986. Engineering crops. P. 57 in C.M. Benbrook and P.B. Moses (Eds.), *Herbicide Resistance: Environmental and Economic Issues.* Staff paper, Board on Agriculture, National Research Council, Washington, D.C.

Benech Arnold R.L., C.M. Ghersa, R.A. Sánchez, and A. Garcia Fernandez. 1988. The role of fluctuating temperatures in germination and establishment of *Sorghum halepense* (L.) Pers. I. Inhibition of germination under leaf canopies. *Func. Ecol.* 2:311–318.

Benech Arnold R.L., C.M. Ghersa, R.A. Sánchez, and P. Insausti. 1990. A mathematical model to predict *Sorghum halepense* (L.) Pers. seedling emergence in relation to soil temperature. *Weed Res.* 30:91–99.

Benech Arnold R.L. and R.A. Sánchez. 1994. Modeling weed seed germination. Pp. 545–566 in J. Kigel and G. Galili (Eds.), *Seed Development and Germination.* Marcel Dekker, Inc., New York.

Berglund, D.R. and J.D. Nalewaja. 1971. Competition between soybeans and wild mustard. *Weed Sci. Soc. Am. Abstr.*, No. 221.

Beversdorf, W.D. and D.J. Hume. 1984. OAC Triton Spring Rapeseed. *Can. J. Plant Sci.* 64:1007–1009.

Bewley, J.D. and M. Black. 1994. *Seeds: Physiology of Development and Germination.* 2nd Ed. Plenum Press, New York.

Beyer, E.M., Jr., M.J. Duffy, J.V. Hay, and D.D. Schlueter. 1988. Sulfonylueas. Pp. 117–189 in P.C. Kearney and D.D. Kaufman (Eds.), *Herbicides—Chemistry, Degradation, and Mode of Action*. Marcel Dekker, New York.

Billings, W.D. 1990. *Bromus tectorum*, a biotic cause of ecosystem impoverishment in the Great Basin. Pp. 301–322 in G.M. Woodwell (Ed.), *The Earth in Transition: Patterns and Processes of Biotic Impoverishment*. Cambridge University Press, New York.

Bird, G.W. 1994. Integrated pest management in a sustainable agriculture. Pp. 35–41 in *Second National Integrated Pest Management Symposium/Workshop Proceedings*. Las Vegas, Nevada. April 19–22, 1994.

Black, C.C., Jr. 1973. Photosynthetic carbon fixation in relation to net CO_2 uptake. *Annu. Rev. Plant Physiol.* 24:258–286.

Black, C.C., Jr. 1985. Effects of herbicides on photosynthesis. Pp. 1–36 in S.O. Duke (Ed.), *Weed Physiology, Vol. 2, Herbicide Physiology*. CRC Press, Inc., Boca Raton, FL.

Black, C.C., Jr., T.M. Chen, and R.H. Brown. 1969. Biochemical basis for plant competition. *Weed Sci.* 17:338–344.

Black, J.N. 1958. Competition between plants of different initial seed sizes in swards of subterranean clover (*Trifolium subterraneum* L.) with particular reference to leaf area and the light microclimate. *Aust. J. Agric. Res.* 9:299–318.

Blackman, V.H. 1919. The compound interest law and plant growth. *Ann. Bot. (London)* 33:353.

Bleasdale, J.K.A. 1960. Studies on plant competition. Pp. 133–142 in J.L. Harper (Ed.), *The Biology of Weeds*. Blackwell Sci. Publ., Oxford, UK.

Bleasdale, J.K.A. 1967. Systematic designs for spacing experiments. *Exp. Agric.* 3:73–85.

Bode, L.E. 1987. Spray application technology. Pp. 85–121 in C.G. McWhorter and M.R. Gebhardt (Eds.), *Methods of Applying Herbicides*. Monograph series of the Weed Science Society of America, Champaign, IL.

Bohm, D. 1980. *Wholeness and the Implicate Order*. Routledge & Kegan Paul, London.

Bonner, J. and A.W. Galston. 1952. *Principles of Plant Physiology*. W.H. Freeman and Co., San Francisco, CA.

Brady, N.C. 1990. *The Nature and Properties of Soils*. 10th Ed. Macmillan, New York.

Brazelton, R.W. and N.B. Akesson. 1989. Safety and regulatory aspects. Pp. 199–213 in California Weed Conference. *Principles of Weed Control in California*. 2nd Ed. Thompson Publications, Fresno, CA.

Brenchley, W.E. 1920. *Weeds of Farm Land*. Longmans, Green, London.

Brenchley, W.E. and K. Warington. 1930. The weed seed population of arable soil. I. Numerical estimation of viable seeds and observations on their natural dormancy. *J. Ecol.* 18:235–272.

Brenchley, W.E. and K. Warington. 1933. The weed seed population of arable soil. II. Influence of crop, soil, and methods of cultivation upon the relative abundance of viable seeds. *J. Ecol.* 21:103–127.

Brenchley, W.E. and K. Warington. 1945. The influence of periodic fallowing on the prevalence of viable weed seeds in arable soil. *Ann. Appl. Biol.* 32:285–296.

Brender, E.V. 1952. From forest to farm to forest again. *Am. For.* 58:24, 25, 40, 41, 43.

Breton, P. and H. Tremblay. 1990. Le secteur forestier Suedois. Rapport de mision d'etude. Forest Canada. Region du Quebec.

Breymeyer, A. and J.M. Meilillo. 1991. The effects of climate change on production and decomposition in coniferous forests and grasslands. *Ecol. Appl.* 1:111.

Bridges, D.C. 1994. Impact of weeds on human endeavors. *Weed Technol.* 8:392–395.

Bridges, D.C. and R.L. Anderson. 1992. Crop losses due to weeds in the United States. Pp. 1–74 in D.C. Bridges (Ed.), *Crop Losses Due to Weeds in the United States—1992*. Weed Science Society of America, Champaign, IL.

Briggs, L.J. and H.L. Shantz. 1913a. The water requirement of plants. I. Investigations in the Great Plains in 1910 and 1911. *Bur. Plant Ind. Bull.* No. 284, U.S. Department of Agriculture, Washington, D.C.

Briggs, L.J. and H.L. Shantz. 1913b. The water requirement of plants. II. A review of the literature. *Bur. Plant Ind. Bull.* No. 285, U.S. Department of Agriculture, Washington, D.C.

Bucha, H.C. and C.W. Todd. 1951. 3(p-chlorophenyl)-1, 1-dimethylurea-A new herbicide. *Science* 114:493–494.

Burdon, J.J. 1987. *Diseases and Plant Population Biology*. Cambridge University Press, Cambridge, UK.

Burkholder, P.R. 1952. Cooperation and conflict among primitive organisms. *Am. Sci.* 40:601–631.

Burnside, O.C. 1992. Rationale for developing herbicide-resistant crops. *Weed Technol.* 6:621-625.

Campbell, M.H. 1977. Assessing the area and distribution of serrated tussock (*Nassella trichotoma*), St. John's wort (*Hypericum perforatum* var. *angustifolium*) and sifton bush (*Cassinia arcuata*) in New South Wales. N.S.W. Department Agric. Tech. Bull. No.18.

Carlson, H.L. and J.E. Hill. 1985. Wild oat (*Avena fatua*) competition with spring wheat: Effects of nitrogen fertilization. *Weed Sci.* 34:29–33.

Casal, J.J. and H. Smith. 1989. The function, action and adaptive significance of phytochrome in light-grown plants. *Plant Cell Environ.* 12:855–862.

Caseley, J.C., G.W. Cussans, and R.K. Atkin (Eds.). 1991. *Herbicide Resistance in Weeds and Crops*. Butterworth Heinemann, Oxford, UK.

Castle, E. 1990. *Toward a Philosophy of Natural Resource Management: A Case for Pluralism.* Unpublished special report. College of Forestry, Oregon State University, Corvallis, OR.

Caswell, H. 1989. *Matrix Population Models. Construction, Analysis and Interpretation*. Sinauer Associates, Inc., Publishers, Sunderland, MA.

Cavers, P.B. and D.L. Benoit. 1989. Seed banks in arable lands. Pp. 309–328 in Leck M.A., V.T. Parker, and R.L. Simpson (Eds.), *Ecology of Soil Seed Banks*. Academic Press, San Diego, CA.

Cavers, P.B., R.H. Groves, and P.E. Kaye. 1995. Seed population dynamics of *Onopordum* over 1 year in southern New South Wales. *J. Appl. Ecol.* 32:425–433.

Chabot, B.F., T.W. Jurik, and J.F. Chabot. 1979. Influence of instantaneous and integrated light-flux density on leaf anatomy and photosynthesis. *Am. J. Bot.* 66:940–945.

Champness, S.S. and K. Morris. 1948. The population of buried viable weed seeds in relation to contrasting pasture and soil types. *J. Ecol.* 36:49–173.
Chancellor, R.J. 1964. Emergence of weed seedlings in the field and the effects of different frequencies of cultivation. Pp. 599–606 in *Proc. 7th Brit. Weed Control Conf.*
Chancellor, R.J. and N.C.B. Peters. 1970. Seed production by *Avena fatua* populations in various crops. Pp. 7–11 in *Proc. 10th Brit. Weed Control Conf.*
Chancellor, R.J. and N.C.B. Peters. 1972. Germination periodicity, plant survival and seed production in populations of *Avena fatua* L. growing in spring barley. Pp. 7–11 in *Proc. 11th Brit. Weed Control Conf.*
Charudattan, R. and C.J. DeLoach, Jr. 1988. Management of pathogens and insects for weed control in agroecosystems. Pp. 245–264 in M.A. Altieri and M. Liebman (Eds.), *Weed Management in Agroecosystems: Ecological Approaches*. CRC Press, Inc., Boca Raton, FL.
Checkland, P. 1981. *Systems Thinking, Systems Practice*. John Wiley & Sons, New York.
Chepil, W.S. 1946. Germination of weed seeds. I. Longevity, periodicity of germination and vitality of seeds in cultivated soil. *Sci. Agric.* 26:307–346.
Chiariello, N.R., H.A. Mooney, and K. Williams. 1991. Growth, carbon allocation and cost of plant tissues. Pp. 327–365 in R.W. Pearcy, J.R. Ehleringer, H.A. Mooney, and P.W. Rundel (Eds.), *Plant Physiological Ecology. Field Methods and Instrumentation*. Chapman and Hall, New York.
Chisaka, H. 1977. Weed damage to crops: Yield loss due to weed competition. Pp. 1–16 in J.D. Fryer and S. Matsunaka (Eds.), *Integrated Control of Weeds*. University of Tokyo Press, Tokyo.
CIESIN. 1992. *Pathways of Understanding: The Interactions of Humanity and Global Environmental Change*. CIESIN, University Center, MI.
Clausen, J., D.D. Keck, and W.M. Hiesey. 1940. Experimental studies on the nature of species. I. Effect of varied environments on Western North American plants. *Carnegie Inst. Wash. Publ.*
Coble, H.D. and D.A. Mortensen. 1991. The threshold concept and its application to weed science. *Weed Technol.* 6:191–195.
Cole, E.C. and M. Newton. 1987. Fifth-year responses of Douglas fir to crowding and nonconiferous competition. *Can. J. For. Res.* 17:181–186.
Comstock, G. 1989. Genetically engineered herbicide resistance, Part 1. *J. Agric. Ethics* 2:263–306.
Conard, S.G. and S.R. Radosevich. 1979. Ecological fitness of *Senecio vulgaris* and *Amaranthus retroflexus* biotypes susceptible or resistant to atrazine. *J. Appl. Ecol.* 16:171–177.
Concannon, J.A. 1987. The effects of density and proportion of spring wheat and *Lolium multiflorum* Lam. M.S. thesis. Oregon State University, Corvallis, OR.
Connell, J.H. and R.O. Slatyer. 1977. Mechanisms of succession in natural communities and their role in community stability and organization. *Am. Nat.* 111:1119–1144.
Cook, R. 1980. The biology of seeds in the soil. Pp 107–129 in O.T. Solbrig (Ed.), *Demography and Evolution in Plant Populations*. University of California Press, Berkeley.

Corbet, P.S. 1981. Non-entomological impediments to the adoption of integrated pest management. *Prot. Ecol.* 3:183–202.

Couch, M.L. 1990. Debating the responsiblilities of plant scientists in the decade of the environment. *Plant Cell* 2:275–277.

Cousens, R. 1985. A simple model relating yield loss to weed density. *Ann. Appl. Biol.* 107:239–252.

Cousens, R. 1991. Aspects of the design and interpretation of competition (interference) experiments. *Weed Technol.* 5:664–673.

Cousens, R. and M. Mortimer. 1995. *Dynamics of Weed Populations*. Cambridge University Press, New York.

Cousens, R., N.C.B. Peters, and C.J. Marshall. 1984. Models of yield loss-weed density relationships. Pp. 367–374 in *7th Int. Symp. Weed Biol. Ecol. Syst.* Paris.

Crafts, A.S. 1975. *Modern Weed Control*. University of California Press, Berkeley.

Crawley, M.J. 1987. What makes a community invasible? Pp. 429–454 in A.J. Gray, M.J. Crawley, and P.J. Edwards (Eds.), *Colonization, Succession and Stability*. Blackwell Scientific Publications, Oxford, UK.

Crop Protection Chemicals Reference. 1986. Chemical and Pharmaceutical Publishing Company, Paris, and John Wiley & Sons, New York.

Cruzz, M., J.L. White, and J.D. Russell. 1968. Montmorillonite-s-triazine interactions. *Israel J. Chem.* 6:315–323.

Cudney, D.W., L.S. Jordan, and A.E. Hall. 1991. Effect of wild oat (*Avena fatua*) infestations on light interception and growth rate of wheat (*Triticum aestivum*). *Weed Sci.* 39:175–179.

Cussans, G.W. 1978. The problem of volunteer crops and some possible means of their control. P. 915 in *Proc. Brit. Crop Prot. Conf.—Weeds*. British Crop Protection Council, Brighton, England.

Cussans, G.W. 1987. Weed management in cropping systems. Pp. 337–347 in *Proc. 8th Aust. Weeds Conf.*

Dale, M.P. and D.R. Causton. 1992. The ecophysiology of *Veronica chamaedrys, V. montana* and *V. officinalis*. I. Light quality and light quantity. *J. Ecol.* 80:483–492.

Daniel, T.W., J.A. Helms, and F.S. Baker. 1979. *Principles of Silviculture*. 2nd Ed. McGraw-Hill Book Co., New York.

Darr, S., V. Souza-Machado, and C.J. Arntzen. 1981. Uniparental inheritance of a chloroplast photosystem II polypeptide controlling herbicide binding. *Biochim. Biophys. Acta* 634:219–228.

Davis, R.G., A.F. Wiese, and J.L. Pafford. 1965. Root moisture extraction profiles of various weeds. *Weeds* 13:98–100.

Dawson, J.H. 1965. Competition between irrigated sugarbeets and annual weeds. *Weeds* 13:245–249.

Dawson, J.H. 1970. Time and duration of weed infestation in relation to weed-crop competition. *South. Weed Sci. Soc.* 23:13–25.

De Datta, S.K. and R. Barker. 1977. Economic evaluation of modern weed control techniques in rice. Pp. 205–228 in J.D. Fryer and S. Matsunaka (Eds.), *Integrated Control of Weeds*. University of Tokyo Press, Tokyo.

Dekker, J. and G. Comstock. 1992. Ethical and environmental considerations in the release of herbicide resistant crops. *Agric. Hum. Values* 9:1–13.

Deregibus, V.A., J.J. Casal, E.J. Jacobo, D. Gibson, M. Kauffman, and A.M. Rodriguez. 1994. Evidence that heavy grazing may promote the germination of *Lolium multiflorum* seeds via phytochrome-mediated perception of high red/far red ratios. *Func. Ecol.* 8:536–542.

Devine, M.D., S.O. Duke, and C. Fedtke. 1993. *Physiology of Herbicide Action.* Prentice-Hall, Inc., Englewood Cliffs, NJ.

De Wet, J.M.J. and J.R. Harlan. 1975. Weeds and domesticates: Evolution in the man-made habitat. *Econ. Bot.* 29:99–107.

di Castri, F. 1989. History of biological invasions with special emphasis on the Old World. P. 26 in A. Drake, H.A. Mooney, F. di Castri, R. Groves, F.J. Kruger, M. Rejmanek, and M. Williamson (Eds.), *Biological Invasions: a Global Perspective.* John Wiley & Sons, New York.

Dilworth, M.J. and A.R. Glenn (Eds.). 1991. *Biology and Biochemistry of Nitrogen Fixation.* Elsevier Sci. Publ. Co., New York.

DiTomaso, J.M. 1995. Approaches for improving crop competitiveness through the manipulation of fertilization strategies. *Weed Sci.* 43:491–497.

Dobzhansky, T. 1970. *Genetics and the Evolutionary Process.* Columbia University Press, New York.

Doebley, J. 1990. Molecular evidence for gene flow among *Zea* species. *BioScience* 40:443–448.

Donald, C.M. 1961. Competition for light in crops and pastures. Pp. 282–313 in *Mechanisms in Biological Competition.* Soc. Exp. Biol. Symposia 15. Academic Press, New York.

Duffey, S.S. and K.A. Bloem. 1986. Plant defense-herbivore-parasite interactions and biological control. Pp. 135–184 in M. Kogan (Ed.), *Ecological Theory and Integrated Pest Management Practice.* John Wiley & Sons, New York.

Duke, S.O. 1985. *Weed Physiology, Vol. 2, Herbicide Physiology.* CRC Press, Inc., Boca Raton, FL.

Duke, S.O. 1986. Microbially produced phytotoxins as herbicides—A perspective. Pp. 287–304 in A.R. Putnam and C.S. Tang (Eds.), *The Science of Allelopathy.* John Wiley & Sons, New York.

Duvel, J.W.T. 1903. Seeds buried in soil. *Science* 17:872–873.

Dyer, W.E., F.D. Hess, J.S. Holt, and S.O. Duke. 1993. Potential benefits and risks of herbicide-resistant crops produced by biotechnology. *Hort. Rev.* 15:365–408.

Eddington, G.E. and W.W. Robbins. 1920. Irrigation water as a factor in the dissemination of weed seeds. *Colo. Exp. Sta. Bull.* 253:25.

Edwards, G.W. and J.W. Freebairn. 1982. The social benefits from an increase in part of an industry. *Rev. Mark. Agric. Econ.* 50:193–210.

Ehrendorfer, F. 1980. Polyploidy and distribution. Pp. 45–60 in W.H. Lewis (Ed.), *Polyploidy: Biological Relevance.* Plenum Press, New York.

Einhellig, F.A. 1986. Mechanisms and modes of action of allelochemicals. Pp. 171–188 in A.R. Putnam and C.S. Tang (Eds.), *The Science of Allelopathy.* John Wiley & Sons, New York.

Ellstrand, N.C. and C.A. Hoffman. 1990. Hybridization as an avenue of escape for engineered genes: Strategies for risk reduction. *BioScience* 40:438–442.

Emerson, R.W. 1878. *Fortune of the Republic.* Houghton and Osgood, Boston.

EPA. 1993. *Agricultural Atrazine Use and Water Quality: A CEEPES Analysis of Policy Options.* U.S. Environmental Protection Agency, Office of Program and Policy Evaluation, Water and Agricultural Policy Division, Agricultural Policy Branch.

Esau, K. 1965. *Plant Anatomy.* 2nd Ed. John Wiley & Sons, New York.

Evans, G.C. 1972. *The Quantitative Analysis of Plant Growth.* University of California Press, Berkeley.

Feltner, K.C. 1967. Light requirements for weed seed germination. P. 64 in *Pro. North Central Weed Control Conf.* USA.

Fenner, M. 1985. *Seed Ecology.* Chapman and Hall, London, UK.

Fenner, M. 1994. Ecology of seed banks. Pp. 507–528 in J. Kigel and G. Galili (Eds.), *Seed Development and Germination.* Marcel Dekker Inc., New York.

Fernández Méndez H., C.M. Ghersa, and E.H. Satorre. 1983. El comportamiento de las semillas de *Sorghum halepense* (L.) Pers. en relacion con la poblacion de rizomas. *Rev. Fac. Agron. UBA* 4:227–231.

Ferre, F. 1982. *The Return to Cosmology: Postmodern Science and the Theology of Nature.* University of California Press, Berkeley.

Ferre, F. 1988. *Philosophy of Technology.* Prentice Hall, Englewood Cliffs, NJ.

Firbank, L.G. and A.R. Watkinson. 1985. On the analysis of competition within two-species mixtures of plants. *Ecology* 22:503–517.

Fischer, B.B., A.H. Lange, J. McCaskill, B. Crampton, and B. Talbraham. 1978. *Growers' Weed Identification Handbook.* No. 4030. Div. Agric. Sci., Univ. Calif., Berkeley, CA.

Fischer, N.H. 1986. The function of mono and sesquiterpenes as plant germination and growth regulators. Pp. 203–218 in A.R. Putnam and C.S. Tang (Eds.), *The Science of Allelopathy.* John Wiley & Sons, New York.

Fischer, R.A. and R.E. Miles. 1973. The role of spatial pattern in the competition between crop plants and weeds. A theoretical analysis. *Math. Biosci.* 18:335–350.

Fisher, H.H. and A.E. Deutch. 1985. *Lever-operated Sprayers: A Practical Scrutiny and Assessment of Features, Components, and Operation—Implications for Purchasers, Users and Manufacturers.* International Plant Protection Center, Oregon State University, Corvallis, OR.

Fisher, R.A. 1920. Some remarks on the methods formulated in a recent article on the quantitative analysis of plant growth. *Ann. Appl. Biol.* 7:376–392.

Fitter, A.H. and R.K.M. Hay. 1987. *Environmental Physiology of Plants.* 2nd Ed. Academic Press, New York.

Fitzherbert, A. 1523. *Boke of Husbandry.* In E.S. Salisbury. 1961. *Weeds and Aliens.* Collins, London.

Forcella, F. and S.J. Harvey. 1983. Relative abundance in an alien weed flora. *Oecologia* 59:292–295.

Francis, C.A., C.A. Flor, and S.R. Temple. 1976. Adapting varieties for intercropping systems in the tropics. Pp. 235–253 in M. Stelly (Ed.), *Multiple Cropping.* American Society of Agronomy Special Publication No. 27.

Froud-Williams, R.J., R.J. Chancellor, and D.S.H. Drennan. 1983. Influence of cultivation regime upon buried weed seeds in arable cropping systems. *J. Appl. Ecol.* 20:199–208.

Fryer, J.D. 1982. Weed control practices and changing weed problems. In J.M. Thresh (Ed.), *Pests, Pathogens and Vegetation*. Pitman, Boston.

Fryer, J.D. and R.J. Chancellor. 1979. Evidence of changing weed populations in arable land. Pp. 958–964 in *Proc. 14th Brit. Weed Control Conf.*

Gajíc, D., S. Malenčić, M. Vrbaški and S. Vrbaški. 1976. Fragm. Herb. Jugoslavica 63:121 in E.L. Rice, 1986. Allelopathic growth stimulation. Pp. 23–42 in A.R. Putnam and C.S. Tang (Eds.), *The Science of Allelopathy*. John Wiley & Sons, New York.

Gashwiler, J.S. 1967. Conifer seed survival in a western Oregon clearcut. *Ecology*. 48:431–433.

Gaudet, C.L. and P.A. Keddy. 1988. A comparative approach to predicting competitive ability from plant traits. *Nature* 334:242–243.

Gause, G.F. 1934. *The Struggle for Existence*. Williams and Wilkins, Baltimore, MD.

Gebhardt, M.R. and C.G. McWhorter. 1987. Introduction to herbicide application technology. Pp.1–8 in C.G. McWhorter and M.R. Gebhardt (Eds.), *Methods of Applying Herbicides*. Monograph series of the Weed Science Society of America. Weed Science Society of America, Champaign, IL.

Geddes, R.D., H.D. Scott, and L.R. Oliver. 1979. Growth and water use by common cocklebur (*Xanthium pensylvanicum*) and soybean (*Glycine max*) under field conditions. *Weed Sci.* 27:206–212.

Gerdemann, J.W. 1968. Vesicular-arbuscular mycorrhizae and plant growth. *Annu. Rev. Phytopathol.* 6:397–418.

Ghersa, C.M. and J.S. Holt. 1995. Using phenology prediction in weed management. *Weed Res.* 35:461–470.

Ghersa, C.M. and M.A. Martínez-Ghersa. 1991a. Cambios ecologicos en los agrosistemas de la pampa ondulada: Efectos de la introduccion de la Soja. *In-vest. Cien.* 2:182–188.

Ghersa, C.M. and M.A. Martínez-Ghersa. 1991b. A field method for predicting yield losses in maize crops caused by *Sorghum halepense*. *Weed Technol.* 5:279–285.

Ghersa, C.M. and M.L. Roush. 1993. Searching for solutions to weed problems. Do we study competition or dispersion? *BioScience* 43:104–109.

Ghersa, C.M., R.J.C. Leon, and Y.A. Soriano. 1985. Efecto del sorgo de Alepo sobre la produccion de soja, de maiz y de las maleas presentes en estos cultivos. *Rev. Fac. Agron. UBA* 6:123–129.

Ghersa, C.M., M.A. Martínez-Ghersa, E.H. Satorre, M.L. Van Esso, and G. Chichotky. 1993. Seed dispersal, distribution and recruitment of seedlings of *Sorghum halepense* (L.) Pers. *Weed Res.* 33:79–88.

Ghersa, C.M., M.L. Roush, S.R. Radosevich, and S.M. Cordray. 1994a. Coevolution of agroecosystems and weed management. *BioScience* 44:85–94.

Ghersa, C.M., M.A. Martínez-Ghersa, J.J. Casal, M. Kaufmann, M.L. Roush, and V.A. Deregibus. 1994b. Effect of light treatments on winter wheat and Italian ryegrass (*Lolium multiflorum*) competition. *Weed Technol.* 8:37–45.

Ghersa, C.M., M.A. Martínez-Ghersa, T.G. Brewer, and M.L. Roush. 1994c. Selection pressures for diclofop-methyl resistance and germination time of Italian ryegrass. *Agron. J.* 86:823–828.

Gianessi, L.P. and C.A. Puffer. 1991. *Herbicide Use in the United States: National Summary Report.* Resources for the Future, Washington, D.C.

Gjerstad, D.H. and B.L. Barber. 1987. Forest vegetation problems in the South. Pp. 55–76 in J.D. Walstad and P.J. Kuch (Eds.), *Forest Vegetation Management for Conifer Production.* John Wiley & Sons, New York.

Glass, E.H. 1975. Integrated pest management: Rationale, potential, needs and improvement. Entomol. Soc. Am., Spec. Publ. Pp. 75–102.

Gliessman, S.R. and M.A. Altieri. 1982. Polyculture cropping has advantages. *Calif. Agric.* 36:14–16.

Goldberg, D.E. 1990. Components of resource competition in plant communities. Pp. 27–49 in J.B. Grace and D. Tilman (Eds.), *Perspectives on Plant Competition.* Academic Press, New York.

Goldberg, D.E. and P.A. Werner. 1983. Equivalence of competitors in plant communities: A null hypothesis and a field experimental approach. *Am. J. Bot.* 70:1098–1104.

Goodman, R.M. 1987. Future potential, problems, and practicalities of herbicide-tolerant crops from genetic engineering. *Weed Sci.* 35:28–31.

Gordon, D.R. and K.J. Rice. 1992. Partitioning of space and water between two California annual grassland species. *Am. J. Bot.* 79:967–976.

Gowgani, G., F.A. Holmes and F.O. Colbert. 1989. Herbicides. Pp. 173–197 in California Weed Conference. *Principles of Weed Control in California.* 2nd Ed. Thompson Publications, Fresno, CA.

Grace, J.B. 1990. On the relationship between plant traits and competitive ability. Pp. 51–65 in J.B. Grace and D. Tilman (Eds.), *Perspectives on Plant Competition.* Academic Press, San Diego, CA.

Grace, J.B. 1991. A clarification of the debate between Grime and Tilman. *Func. Ecol.* 5:583–587.

Grace, J.B. and D. Tilman. 1990. *Perspectives on Plant Competition.* Academic Press, San Diego, CA.

Graham, P.L., J.L. Steiner, and A.F. Weise. 1988. Light absorption and competition in mixed sorghum-pigweed communities. *Agron. J.* 80:415–418.

Gray, A. 1879. The predominance and pertinacity of weeds. *Am. J. Sci.* 118:161–167.

Gray, B., M. Drake, and W.G. Colby. 1953. Potassium competition in grass-legume associations as a function of root cation exchange capacity. *Soil Sci. Soc. Am. Proc.* 17:235–239.

Gressel, J. 1991. Why get resistance? It can be prevented or delayed. Pp. 1–26 in J.C. Caseley, G.W. Cussans, and R.K. Atkin (Eds.), *Herbicide Resistance in Weeds and Crops.* Butterworth-Heinemann, Oxford, UK.

Gressel, J. and L.A. Segel. 1978. The paucity of plants evolving genetic resistance to herbicides: Possible reasons and implications. *J. Theor. Biol.* 75:349–372.

Gressel, J. and L.A. Segel. 1982. Interrelating factors controlling the rate of appearance of resistance: The outlook for the future. Pp. 325–347 in H.M. LeBaron and J. Gressel (Eds.), *Herbicide Resistance in Plants.* John Wiley & Sons, New York.

Gressel, J. and L.A. Segel. 1990. Modeling the effectiveness of herbicide rotations and mixtures as strategies to delay or preclude resistance. *Weed Technol.* 4:186–198.

Griffiths, A.J.F., J.H. Miller, D.T. Suzuki, R.C. Lewontin, and W.M. Gelbart. 1993. *An Introduction to Genetic Analysis*. W.H. Freeman and Company, New York.

Grime, J.P. 1974. Vegetation classification by reference to strategies. *Nature* 250:26–31.

Grime, J.P. 1977. Evidence for the existence of three primary strategies in plants and its relevance to ecological and evolutionary theory. *Am. Nat.* 111:1169–1194.

Grime, J.P. 1979. *Plant Strategies and Vegetation Processes*. John Wiley & Sons, New York.

Grime, J.P. 1989. Seed banks in ecological perspective. In M.A. Leck, V.T. Parker, and R.L. Simpson (Eds.), *Ecology of Soil Seed Banks*. Academic Press, San Diego, CA.

Grime, J.P. and S.H. Hillier. 1981. Predictions based upon the laboratory characteristics of seeds. P. 6 in *Annual Report 1981*. Unit of Comparative Plant Ecology (NERC). University of Sheffield, Sheffield, England.

Grime, J.P. and R. Hunt. 1975. Relative growth rate: Its range and adaptive significance in a local flora. *J. Ecol.* 63:393–422.

Gronwald, J.W., C.V. Eberlein, K.J. Betts, K.M. Rosow, and N.J. Ehlke. 1989. Diclofop resistance in a biotype of Italian ryegrass. P. 115 in *Proc. Annu. Meet. Am. Soc. Plant Physiol.* Toronto, Canada.

Guenzi, W.D. and T. McCalla. 1962. Inhibitions of germination and seedling development by crop residues. *Proc. Soil Sci. Soc. Am.* 26:456–458.

Gutterman, Y. 1992. Maternal effects on seeds during development. Pp. 27–59 in M. Fenner (Ed.), *Seeds: The Ecology of Regeneration in Plant Communities*. C.A.B. International, UK.

Haas, H. and J.C. Streibig. 1982. Changing patterns of weed distribution as a result of herbicide use and other agronomic factors. Pp. 57–80 in H.M. LeBaron and J. Gressel (Eds.), *Herbicide Resistance in Plants*. John Wiley & Sons, New York.

Hakansson, S. 1988. Behovet av ogr sbek mpning—bed mningsgrunder och prognosm jligheter. Pp. 21–31 in *Ogr s- och v xtskyddskonfer-enserna. Gemensam och v xtskyddsdel*. Uppsala, Sweden.

Hall, A.J., C.M. Rebella, C.M. Ghersa, and P.M. Culot. 1992. Field crop systems of the Pampas. Pp. 413–450 in C.J. Pearson (Ed.), *Ecosystems of the World. Field Crop Ecosystems*. Elsevier, New York.

Halliwell, B. 1984. *Chloroplast Metabolism. The Structure and Function of Chloroplasts in Green Leaf Cells*. Rev. Ed. Clarendon Press, Oxford, UK.

Hamaker, J.W. 1976. Mathematical prediction of cumulative levels of pesticides in soil. Pp. 122–131 in *Organic Pesticides in the Environment*. Adv. in Chem. Series 60. Amer. Chem. Soc., Washington, DC.

Hanf, M. 1944. Der Einfluss des Bodens auf Keimen und Auflaufen. *Beih. 2. Bot. Zentralb.* 57: Abst. A, 405–425.

Hanson, J.B. and F.W. Slife. 1969. Role of RNA metabolism in the action of auxin-herbicides. *Resid Rev.* 25:59–67.

Harper, J.L. 1956. The evolution of weeds in relation to the resistance to herbicides. *Proc. 3rd Brit. Weed Control Conf.* 1:179–188.

Harper, J.L. 1957. Ecological aspects of weed control. *Outlook Agric.* 1:197–205.

Harper, J.L. 1964a. The nature and consequence of interference amongst plants. Pp. 465–482 in Genetics Today. *Proc. XI Internat. Cong. Genet.* 2:465–482.

Harper, J.L. 1964b. The individual in the population. *J. Ecol.* 52: (Suppl.) 149–158.

Harper, J.L. 1977. *The Population Biology of Plants*. Academic Press, London, UK.

Harper, J.L. 1981. The concept of population in modular organisms. Pp. 53–77 in *Theoretical Ecology: Principles and Applications*. 2nd Ed. Blackwell Scientific Publications, Oxford, UK.

Harper, J.L. 1984. *Modules, Branches and the Capture of Resources*. Yale University Press, New Haven, CT.

Harper, J.L. and A.D. Bell. 1979. The population dynamics of growth forms in organisms with modular construction. Pp. 29–52 in R.M. Anderson, B.D. Turner, and R.L. Taylor (Eds.), *Population Dynamics*. Blackwell Scientific Publications, Oxford, UK.

Harper, J.L. and J. White. 1971. The dynamics of plant populations. Pp. 41–63 in *Proc. Adv. Study Inst. Dynamics Numbers Popul. (Oosterbeek 1970)*.

Harper, J.L., J.T. Williams, and G.R. Sagar. 1965. The behaviour of seeds in soil. Part 1. The heterogeneity of soil surfaces and its role in determining the establishment of plants from seed. *J. Ecol.* 53:273–286.

Harper, R.M. 1944. Preliminary report on the weeds of Alabama. *Bull. Geol. Surv. Ala.* 53:275.

Harrington, J.F. 1972. Seed storage and longevity. Pp. 145–245 in T.T. Kozlowski (Ed.), *Seed Biology*. Academic Press, New York.

Harrison, H.F., Jr. 1992. Developing herbicide-tolerant crop cultivars: Introduction. *Weed Technol.* 6:613–614.

Hartl, D.L. and A.G. Clark. 1989. *Principles of Population Genetics*. 2nd Ed. Sinauer Associates, Inc., Sunderlund, MA.

Hashem, A. 1991. Effect of density, proportion, and spatial arrangement on the competition of winter wheat and Italian ryegrass (*Lolium multiflorum* Lam.). Ph.D. thesis, Oregon State University, Corvallis, OR.

Hashem, A., S.R. Radosevich, and M.L. Roush. 1996. Effect of proximity factors on competition between winter wheat and Italian ryegrass (*Lolium multiflorum* Lam.). *Weed Sci.* In Press.

Hegazy, A.K. 1994. Trade-off between sexual and vegetative reproduction of the weedy *Heliotropium curassavicum*. *J. Arid Environ.* 27:209–220.

Heiser, C.B. and T.W. Whitaker. 1948. Chromosome number, polyploidy, and growth habit in California weeds. *Am. J. Bot.* 35:179–186.

Hess, F.D. 1985. Herbicide absorption and translocation and their relationship to plant tolerances and susceptibility. Pp. 191–121 in S.O. Duke (Ed.), *Weed Physiology, Vol. 2, Herbicide Physiology*. CRC Press, Inc., Boca Raton, FL.

Hess, F.D. 1987. Herbicide effects on the cell cycle of meristematic plant cells. *Rev. Weed Sci.* 3:183–203.

Hess, F.D. 1994. Research needs in weed science. *Weed Technol.* 8:408–409.

Hess, F.D. and D.E. Bayer. 1974. The effect of trifluralin on the ultrastructure of dividing cells of the root meristem of cotton (*Gossypium hirsutum* L. 'Acala 4–42'). *J. Cell. Sci.* 15:429–441.

Hewson, R.T. and H.A. Roberts. 1973a. Effects of weed competition for different periods on the growth and yield of red beet. *J. Hort. Sci.* 48:281–292.

Hewson, R.T. and H.A. Roberts. 1973b. Some effects of weed competition on the growth of onions. *J. Hort. Sci.* 48:51–57.
Hewson, R.T., H.A. Roberts, and W. Bond. 1973. Weed competition in spring sown broad beans. *Hort. Res.* 13:25–32.
Hilu, K.W. 1993. Polyploidy and the evolution of domesticated plants. *Am. J. Bot.* 80:1494–1499.
Hipkins, M.F. 1984. Photosynthesis. Pp. 219–248 in M.B. Wilkins (Ed.), *Advanced Plant Physiology*. Pitman, London, UK.
Hitchcock, A.S. and G.L. Clothier. 1898. *Kans. Exp. Sta. Bot. Bull.* 80:113–169.
Hoffman, C.A. 1990. Ecological risks of genetic engineering of crop plants. *BioScience* 40:434–437.
Holling, C.S. (Ed.). 1978. *Adaptive Environmental Assessment and Management*. John Wiley & Sons, New York.
Holm, L.G. 1971. The role of weeds in human affairs. *Weed Sci.* 19:485–490.
Holm, L.G. 1978. Some characteristics of weed problems in two worlds. Pp. 3–12 in *Proc. West. Soc. Weed Sci.*.
Holm, L.G., D.L. Plucknett, J.V. Pancho, and J.P. Herberger. 1977. *The World's Worst Weeds: Distribution and Biology*. University Press of Hawaii, Honolulu.
Holmes, M.G. and H. Smith. 1975. The function of phytochrome in plants growing in the natural environment. *Nature* 254:512–514.
Holt, J.S. 1988. Reduced growth, competitiveness, and photosynthetic efficiency of triazine-resistant *Senecio vulgaris* from California. *J. Appl. Ecol.* 25:307–318.
Holt, J.S. 1991. Applications of physiological ecology to weed science. *Weed Sci.* 39:521–528.
Holt, J.S. 1992. History of identification of herbicide-resistant weeds. *Weed Technol.* 6:615–620.
Holt, J.S. 1995. Plant responses to light: A potential tool for weed management. *Weed Sci.* 43:474–482.
Holt, J.S. and H.M. LeBaron. 1990. Significance and distribution of herbicide resistance. *Weed Technol.* 4:141–149.
Holt, J.S. and D.R. Orcutt. 1991. Functional relationships of growth and competitiveness in perennial weeds and cotton (*Gossypium hirsutum*). *Weed Sci.* 39:575–584.
Holt, J.S. and S.R. Radosevich. 1983. Differential growth of two common groundsel (*Senecio vulgaris*) biotypes. *Weed Sci.* 31:112–120.
Holt, J.S., A.J. Stemler, and S.R. Radosevich. 1981. Differential light responses of photosynthesis by triazine-resistant and triazine-susceptible *Senecio vulgaris* biotypes. *Plant Physiol.* 67:744–748.
Holt, J.S., S.R. Radosevich, and A.J. Stemler. 1983. Differential efficiency of photosynthetic oxygen evolution in flashing light in triazine-resistant and triazine-susceptible biotypes of *Senecio vulgaris* L. *Biochim. et Biophys. Acta* 722:245–255.
Holt, J.S., S.B. Powles and J.A.M. Holtum. 1993. Mechanisms and agronomic aspects of herbicide resistance. *Annu. Rev. Plant Physiol. Plant Mol. Biol.* 44:203–229.
Holzner, W. and R. Immonen. 1982. Europe: An overview. Pp. 203–226 in W. Holzner and M. Numata (Eds.), *Biology and Ecology of Weeds*. Dr W. Junk Publishers, The Hague.

Holzner, W. and M. Numata. 1982. *Biology and Ecology of Weeds*. Dr W. Junk Publishers, The Hague.
Holzner, W., I. Hayashi, and J. Glauninger. 1982. Reproductive strategy of annual agrestals. Pp. 111–121 in W. Holzner and M. Numata (Eds.), *Biology and Ecology of Weeds*. Dr W. Junk Publishers, The Hague.
Horn, H.S. 1971. *The Adaptive Geometry of Trees*. Princeton University Press, Princeton, NJ.
Horwith, B. 1985. A role for intercropping in modern agriculture. *BioScience* 35:286.
Hoyer, G.E. and D. Belz. 1984. Stump sprouting related to time of cutting red alder. DNR Rep. 45. State of Washington Department of Natural Resources, Olympia, WA.
Hsiao, T.C. 1973. Plant responses to water stress. *Annu. Rev. Plant Physiol.* 24:519–570.
Hull, H.M., D.G. Davis, and G.E. Stoltenburg. 1982. Action of adjuvants on plant surfaces. Pp. 26–27 in *Adjuvants for Herbicides*, Weed Science Society of America, Champaign, IL.
Hull, R.J. and O.A. Leonard. 1964. Physiological aspects of parasitism in mistletoes (*Arceuthobium* and *Phoradendron*). I. The carbohydrate nutrition of mistletoe. *Plant Physiol.* 39:996–1007.
Hunt, R. 1978. *Plant Growth Analysis*. Edward Arnold, London, UK.
Hunt, R. 1982. *Plant Growth Curves. The Functional Approach to Plant Growth Analysis*. University Park Press, Baltimore, MD.
Hunt, R. and I.T. Parsons. 1974. A computer program for deriving growth functions in plant growth analysis. *J. Appl. Ecol.* 11:297–307.
Hyde, E.O.C. 1954. The function of the hilum in some Papilionaceae in relation to the ripening of the seed and the permeability of the testa. *Ann. Bot. N.S.* 18:241–256.
Ince, J.W. 1915. Fertility and weeds. *N.D. Bull.* 112:233–247.
Ivy, H.W. and R.S. Baker. 1972. Prickly sida control and competition in cotton. *Weed Sci.* 20:137–139.
Jasieniuk, M., A.L. Brule-Babel, and I.N. Morrison. 1996. The evolution and genetics of herbicide resistance in weeds. *Weed Sci.* 44:176–193.
Jensen, P.K. 1991. Udnyttelse af ukrudtsfro's behov for lysinduktion. Pp. 215–230 in Dansk plantevaernskonference, Ukrudt. Tidsskrift for Planteavls Specialserie, Beretning nr. S2110, Statens Planteavlsforsog. Lyngby, Denmark.
Jolliffe, P.A., A.N. Nimjas, and V.C. Runeckles. 1984. A reinterpretation of yield relationships in replacement series experiments. *J. Appl. Ecol.* 21:227–243.
Karssen, C.M. 1970a. The light promoted germination of *Chenopodium album* L. III. The effect of the photoperiod during growth and development of the plants on the dormancy of the produced seeds. *Acta Bot. Neerl.* 19:81–94.
Karssen, C.M. 1970b. The light promoted germination of *Chenopodium album* L. VI. P_{fr} requirements during different stages of the germination process. *Acta Bot. Neerl.* 19:297–312.
Kays, S. and J.L. Harper. 1974. The regulation of plant and tiller density in a grass sward. *J. Ecol.* 62:97–105.

Kearney, P.C. and Kaufman, D.D. (Eds.). 1988. *Herbicides: Chemistry, Degradation and Mode of Action*. 2nd Ed. Vol. 1 and 2. Marcel Dekker, New York.

Kellman, M.C. 1974a. Preliminary seed budgets for two plant communities in coastal British Columbia. *J. Biogeog.* 1:123–133.

Kellman, M.C. 1974b. The viable weed seed content of some tropical agricultural soils. *J. Appl. Ecol.* 2:669–677.

Kennedy, P.B. 1903. Summer ranges of eastern Nevada sheep. *Nevada Agric. Exp. Sta. Bull.* 55.

Kidd, F. and C. West. 1919. Physiological pre-determination: The influence of the physiological condition of the seed upon the subsequent growth and upon the yield. IV. Review of the literature. Chapter III. *Ann. Appl. Biol.* 5:220–251.

Kimmins, J.P. 1991. The future of the forested landscapes of Canada. *For. Chron.* 67:14–18.

King, G.A. and R.P. Neilson. 1992. The transient response of vegetation to climate change: A potential source of CO_2 to the atmosphere. *Water, Air Soil Poll.* 64:365–383.

King, J. 1991. A matter of public confidence—Consumers' concerns about pesticide residues unjustified. *Agric. Eng.* 72(4):16–18.

King, J.J. 1966. *Weeds of the World: Biology and Control*. Interscience, New York. Pp. 1–48.

Kivilaan, A. and R.S. Bandurski. 1973. The ninety-year period for Dr. W.J. Beal's seed viability experiment. *Am. J. Bot.* 60:140–145.

Kivilaan, A. and R.S. Bandurski. 1981. The one-hundred year period for Dr. W.J. Beal's seed viability experiment. *Am. J. Bot.* 68:1290–1292.

Klemmedson, J.O. and J.G. Smith. 1964. Cheatgrass (*Bromus tectorum* L.). *Bot. Rev.* 30:226–291.

Knake, E.I. and F.W. Slife. 1962. Competition of *Setaria faberi* with corn and soybeans. *Weeds* 10:26–29.

Knowe, S.A., R.G. Shula, and S.R. Radosevich. 1995. Regional vegetation management model for Douglas-fir and associated vegetation. Pp. 131–133 in *Popular Summaries from Second International Conference on Forest Vegetation Management*. Forest Research Institute, Rotorua, New Zealand, March 20–24, 1995.

Koide, R.T., R.H. Robichaux, S.R. Morse, and C.M. Smith. 1991. Plant water status, hydraulic resistance and capacitance. Pp. 161–183 in R.W. Pearcy, J.R. Ehleringer, H.A. Mooney, and P.W. Rundel (Eds.), *Plant Physiological Ecology. Field Methods and Instrumentation*. Chapman and Hall, New York.

Kropff, M.J. 1988. Modeling the effects of weeds on crop production. *Weed Res.* 28:465–471.

Kropff, M.J. and L.A.P. Lotz. 1992. Optimization of weed management systems: The role of ecological models of interplant competition. *Weed Technol.* 6:462–470.

Kropff, M.J. and C.J.T. Spitters. 1991. A simple model of crop loss by weed competition from early observations on relative leaf area of the weeds. *Weed Res.* 31:97–105.

Kropff, M.J. and C.J.T. Spitters. 1992. An eco-physiological model for interspecific competition, applied to the influence of *Chenopodium album* L. on sugar beet. I. Model description and parameterization. *Weed Res.* 32:437–450.

Kropff, M.J. and H.H. van Laar (Eds.). 1993. *Modelling Crop-Weed Interactions.* CAB International, Wallingford, UK.

Kropff, M.J., F.J.H. Vossen, C.J.T. Spitters, and W. de Groot. 1984. Competition between a maize crop and a natural population of *Echinochloa crus-galli* (L.). *Neth. J. Agric. Sci.* 32:324–327.

Kropff, M.J., C.J.T. Spitters, B.J. Schnieders, W. Joenje, and W. de Groots. 1992. An eco-physiological model for interspecific competition, applied to the influence of *Chenopodium album* L. on sugar beet. II. Model evaluation. *Weed Res.* 32:451–463.

Kropff, M.J., L.A.P. Lotz, S.E. Weaver, H.J. Bos, J. Wallinga, and T. Migo. 1995. A two parameter model for prediction of crop loss by weed competition from early observations of relative leaf area of the weeds. *Ann. Appl. Biol.* 126:329–346.

Kuijt, J. 1969. *The Biology of Parasitic Flowering Plants.* University of California Press, Berkeley, CA.

Lamberts, M.L. 1980. Intercropping with potatoes. M.S. thesis. Cornell University, Ithaca, NY.

Lang, A.L., J.W. Pendleton, and G.H. Dungan. 1956. Influence of population and nitrogen level on yield and protein contents of nine corn hybrids. *Agron. J.* 48:284–289.

Lawton, J.H. 1994. What do species do in ecosystems? *Oikos* 71:367–374.

Leather, G.R. and F.A. Einhellig. 1986. Bioassays in the Study of Allelopathy. Pp. 133–146 in A.R. Putnam and C.S. Tang (Eds.), *The Science of Allelopathy,* John Wiley & Sons, New York.

LeBaron, H.M. 1991. Distribution and seriousness of herbicide-resistant weed infestations worldwide. Pp. 27–43 in J.C. Caseley, G.W. Cussans, and R.K. Atkin (Eds.), *Herbicide Resistance in Weeds and Crops.* Butterworth-Heinemann, Oxford, UK.

LeBaron, H.M. and J. Gressel (Eds.). 1982. *Herbicide Resistance in Plants.* John Wiley & Sons, New York.

Leck, M.A., V.T. Parker, and R.L. Simpson (Eds.). 1989. *Ecology of Soil Seed Banks.* Academic Press, San Diego, CA.

Lehninger, A.L., D.L. Nelson, and M.M. Cox. 1993. *Principles of Biochemistry.* 2nd Ed. Worth Publishers, New York.

Leon, R.J.C. and A. Suero. 1962. Las comunidades de malezas en los maizales y su valor indicador. *Rev. Arg. Agron.* 29:23–28.

Leopold, A. 1943. What is a weed? *The Weed Flora of Iowa*, Bulletin no. 4. Pp. 306–309 in S.L. Hader and J.B. Callicott (Eds.). 1991. *The River of the Mother of God and Other Essays by Aldo Leopold.* University of Wisconsin Press, Madison, WI.

Leslie, P.H. 1943. On the use of matrices in certain population mathematics. *Biometrika* 33:183–212.

LeStrange, M. and J. Hill. 1983. Crop management practices affect barnyardgrass competition in rice. P. 115 in Proc. 35th Annual *California Weed Conf.*, San Jose, CA.

Levin, D.A. and H.W. Kerster. 1974. Gene flow in seed plants. *Evol. Biol.* 7:139–220.

Levins, R. 1986. Perspectives in integrated pest management. Pp. 1–18 in M. Kogan (Ed.), *Ecological Theory and Integrated Pest Management Practice*. John Wiley & Sons, New York.

Levins, R. and J.H. Vandermeer. 1994. The agroecosystem embedded in a complex ecological community. Pp. 341–362 in C.R. Carroll, J.H. Vandermeer, and P.M. Rosset (Eds.), *Agroecology*. McGraw-Hill Publishing Company, New York.

Lewis, D.H. 1974. Micro-organisms and plants: The evolution of parasitism and mutualism. Pp. 367–392 in M.J. Carlile and J.J. Skehel (Eds.), *Evolution in the Microbial World*. Cambridge University Press, New York.

Lewis, J. 1973. Longevity of crop and weed seed: Survival after 20 years in the soil. *Weed Res.* 13:179–191.

Lewontin, R. 1982. Agricultural research and the penetration of capital. *Sci. People* 1:12–17.

Lewontin, R.C. 1974. *The Genetic Basis of Evolutionary Change*. Columbia University Press, New York.

Liebl, R. and A.D. Worsham. 1987. Interference of Italian ryegrass (*Lolium multiflorum*) in wheat (*Triticum aestivum*). *Weed Sci.* 35:819–823.

Liebman, M. 1986. Ecological suppression of weeds in intercropping systems: Experiments with barley, pea, and mustard. Ph.D. dissertation. Department of Botany, University of California, Berkeley.

Liebman, M. 1988. Ecological suppression of weeds in intercropping systems: A review. Pp. 197–212 in M.A. Altieri and M. Liebman (Eds.), *Weed Management in Agroecosystems: Ecological Approaches*. CRC Press, Inc., Boca Raton, FL.

Lippert, R.D. and H.H. Hopkins. 1950. Study of viable seeds in various habitats in mixed prairies. *Trans. Kans. Acad. Sci.* 53:355–364.

Lippold, K.W. 1995. Protecting our green earth: How to manage global warming through environmentally sound farming and preservation of the world's forests. Enquete Commission "*Protecting the Earth's Atmosphere*" of the German Bundestag (Ed.), Economica Verlag. 683 pp.

Lipsey, R.G. and P.O. Steiner. 1984. *Economics*. 7th Ed. Harper and Row, New York.

Lotka, A.J. 1925. *Elements of Physical Biology*. Williams and Wilkins, Baltimore, MD.

Louda, S.M. 1989. Predation in the dynamics of seed regeneration. Pp. 25–51 in M.A. Leck, V.T. Parker, and R.L. Simpson (Eds.), *Ecology of Soil Seed Banks*. Academic Press, San Diego, CA.

Lueschen, W.E. and R.N. Anderson. 1980. Longevity of velvetleaf seeds in soil under agricultural practices. *Weed Sci.* 28:341–346.

Lybecker, D.W., E.E. Schweizer, and R.P. King. 1991. Weed management decisions in corn based on bioeconomic modeling. *Weed Sci.* 39:124–129.

Lynn, D.G. 1985. The involvement of allelochemicals in the host selection of parasitic angiosperm. Pp. 55–81 in A.C. Thompson (Ed.), *The Chemistry of Allelopathy: Biochemical Interactions Among Plants*. American Chemical Society, Washington, D.C.

MacArthur, R.H. 1962. Generalized theorems of natural selection. *Proc. Natl. Acad. Sci.* 48:1893–1897.

Mack, R.N. 1981. Invasion of *Bromus tectorum* L. into western North America: An ecological chronicle. *Agro-Ecosystems* 7:145–165.

Mack, R.N. 1984. Invaders at home on the range. *Nat. Hist.* February 1984:40–47.

Mack, R.N. 1986. Alien plant invasion into the intermountain West: A case history. Pp. 191–213 in H.A. Mooney and J.A. Drake (Eds.), *Ecology of Biological Invasions of North America and Hawaii*, Ecological Studies 58, Springer-Verlag, New York.

MacLeod, D.G. 1962. Some anatomical and physiological observations on two species of *Cuscuta. Trans. Bot. Soc. Edin.* 39:302–315.

Maddy, K.T., F. Schneider, and S. Edmiston. 1989. Regulation and registration. Pp. 214–225 in California Weed Conference, *Principles of Weed Control in California.* 2nd Ed. Thompson Publications, Fresno, CA.

Major, J. and W.T. Pyott. 1966. Buried viable seed in California bunchgrass sites and their bearing on the definition of a flora. *Veg. Acta Geobot.* 13:253–282.

Marshall, D.R. and S.K. Jain. 1967. Cohabitation and relative abundance of two species of wild oats. *Ecology* 48:656–659.

Mason, J.M. 1932. Weed Survey of the Prairie Provinces. *Dom. Can. Natl. Res. Council Rep.* 26. p. 34.

Mater, J. 1977. *Citizens Involved: Handle with Care! A Forest Industry Guide to Working with the Public.* Timber Press, Forest Grove, OR.

Mater, J. 1992. A paradigm shift for marketing the forest industry—From public relations to research. 46th Annual meeting, June, 1992, Forest Products Research Society, Charleston, South Carolina.

Maxwell, B.D. and C.M. Ghersa. 1992. The influence of weed seed dispersion versus the effect of competition on crop yield. *Weed Technol.* 6:196–204.

Maxwell, B.D., M.V. Wilson, and S.R. Radosevich. 1988. Population modeling approach for evaluating leafy spurge (*Euphorbia esula*) development and control. *Weed Technol.* 2:132–138.

Maxwell, B.D., M.L. Roush, and S.R. Radosevich. 1990. Predicting the evolution and dynamics of herbicide resistance in weed populations. *Weed Technol.* 4:2–13.

May, R.M. 1981. Models for two interacting populations. Pp. 78–104 in R. May (Ed.) *Theoretical Ecology: Principles and Applications.* Blackwell Scientific, Oxford, UK.

McCloskey, W.B. and J.S. Holt. 1990. Triazine resistance in *Senecio vulgaris* parental and nearly isonuclear backcrossed biotypes is correlated with reduced productivity. *Plant Physiol.* 92:954–962.

McCloskey, W.B. and J.S. Holt. 1991. Effect of growth temperature on biomass production of nearly isonuclear triazine-resistant and -susceptible common groundsel (*Senecio vulgaris* L.). *Plant Cell Environ.* 14:699–705.

McGee, A.B. and E. Levy. 1988. Herbicide use in forestry: Communication and information gaps. *J. Environ. Manag.* 26:111–126.

McGiffen, M.E., Jr., J.B. Masiunas, and J.D. Hesketh. 1992. Competition for light between tomatoes and nightshades (*Solanum nigrum* or *S. ptycanthum*). *Weed Sci.* 40:220–226.

McHenry, W.B. and R.F. Norris. 1972. *Study Guide for Agricultural Pest Control Advisers on Weed Control.* University of California Publication 4050, Berkeley, CA.

McWhorter, C.G. and M.R. Gebhardt. (Eds.) 1987. *Methods of Applying Herbicides*. Monograph series of the Weed Science Society of America. Weed Science Society of America, Champaign, IL.

Merchant, C. 1980. *The Death of Nature*. Harper and Row, New York.

Methy, M., P. Alpert, and J. Roy. 1990. Effects of light quality and quantity on growth of the clonal plant *Eichhornia crassipes*. *Oecologia* 84:265–271.

Miles, J. 1987. Vegetation succession: Past and present perceptions. Pp. 1–30 in A.J. Gray, M.J. Crawley, and P.J. Edwards (Eds.), *Colonization, Succession, and Stability*. Blackwell Scientific, Oxford, UK.

Millar, C.E., L.M. Turk, and H.D. Foth. 1965. *Fundamentals of Soil Science*. 4th Ed. John Wiley & Sons, New York.

Miller, T.E. and P.A. Werner. 1987. Competitive effects and responses in plants. *Ecology* 68:1201–1210.

Molisch, H. 1937. *Der Einfluss einer Pflanze auf die andere-Allelopathie*. Fischer, Jena.

Mollison, B. 1988. *Permaculture: A Designer's Manual*. Tagan, Tyalgum, Australia.

Moody, K. 1988. Developing appropriate weed management strategies for small-scale farmers. Pp. 319–330 in M.A. Altieri and M. Liebman (Eds.), *Weed Management in Agroecosystems: Ecological Approaches*. CRC Press, Inc., Boca Raton, FL.

Moody, K. 1995. Weed control—a different perspective. Weed Science Society of America 35th Annual Meeting, Jan 30–Feb 2, 1995, Seattle, WA.

Moody, K. and S.V.R. Shetty. 1981. Weed management in intercropping systems. Pp. 229–237 in *Proc. Int. Workshop Intercroppings*, ICRISAT, 10–13 Jan. 1979. Patancheru, A.P., India.

Moreland, D.E. 1985. Effects of herbicides on respiration. Pp. 37–61 in S.O. Duke (Ed.), *Weed Physiology, Vol. 2, Herbicide Physiology*. CRC Press, Inc., Boca Raton, FL.

Moreno, R.A. and R.D. Hart. 1979. Intercropping with cassava in Central America. In E. Weber, B. Nestel, and M. Campbell (Eds.), *Intercropping with Cassava*. IDRC, Ottawa, Canada.

Mortensen, D.A., G.A. Johnson, and L.J. Young. 1993. Weed distribution in agricultural fields. Pp. 113–124 in P. Robert and R.H. Rust (Eds.), *Site Specific Crop Management*. ASA-CSSA-SSSA, 677 South Sedge Rd., Madison, WI.

Mortimer, A.M. 1983. On weed demography. Pp. 3–41 in W.W. Fletcher (Ed.), *Recent Advances in Weed Research*. Commonwealth Agricultural Bureau, Farnham Royal, UK.

Mortimer, A.M. 1984. Population ecology and weed science. Pp. 363–388 in R. Dirzo and J. Sarukhan (Eds.), *Perspectives on Plant Population Ecology*. Sinauer, Sunderland, MA.

Musselman, L.J. 1982. Parasitic weeds in arable land. Pp. 175–186 in W. Holzner and M. Numata (Eds.), *Biology and Ecology of Weeds*. Dr W. Junk Publishers, The Hague.

Musselman, L.J. (Ed.). 1987. *Parasitic Weeds in Agriculture*. CRC Press, Inc., Boca Raton, FL.

NAPIAP. 1992. *The Effects of Restricting or Banning Atrazine Use to Reduce Surface Water Contamination in the Upper Mississippi River Basin*. National Agricultural Pesticide Impact Assessment Program, Washington, D.C.

National Research Council. 1989. *Alternative Agriculture*. National Academy Press. Washington, D.C.

Naveh, Z. 1980. Landscape ecology as a scientific and educational tool for teaching the total human ecosystem. Pp. 149–163 in T.S. Bakshi and Z. Naveh (Eds.), *Environmental Education: Principles, Methods and Applications*. Plenum Press, New York.

Naveh, Z. and A.S. Lieberman. 1990. *Landscape Ecology: Theory and Applications*. Springer-Verlag, New York.

Naylor, R.E.L. 1970a. The prediction of blackgrass infestations. *Weed Res.* 10:296–299.

Naylor, R.E.L. 1970b. Weed predictive indices. Pp. 26–29 in *Proc. 10th Brit. Weed Control Conf.*

Neill, R.L. and E.L. Rice. 1971. Possible role of *Ambrosia psilostachya* on pattern and succession in old-fields. *Am. Midland Nat.* 86:344.

Nelder, J.A. 1962. New kinds of systematic designs for spacing studies. *Biometrics* 18:283–307.

Nicollier, G.F., D.F. Pope, and A.C. Thompson. 1985. Pp. 207–218 in A.C. Thompson (Ed.), *The Chemistry of Allelopathy*. American Chemical Society, Washington, D.C.

Nobel, P.S., I.N. Forseth, and S.P. Long. 1993. Canopy structure and light interception. Pp. 79–90 in D.O. Hall, J.M.O. Scurlock, H.R. Bolhàr-Nordenkampf, R.C. Leegood, and S.P. Long (Eds.), *Photosynthesis and Production in a Changing Environment: A Field and Laboratory Manual*. Chapman and Hall, New York.

Norris, R.F. 1982. Interactions between weeds and other pests in the agroecosystem. Pp. 343–406 in J.L. Hatfield and I.J. Thomason (Eds.), *Biometeorology in Integrated Pest Management*. Academic Press, New York.

Norris, R.F. 1992. Case history for weed competition/population ecology: barnyardgrass (*Echinochloa crus-galli*) in sugarbeets (*Beta vulgaris*). *Weed Technol.* 6:220–222.

Norris, R.F. and G.W. Fick. 1981. Consortium for Integrated Pest Management. Unpublished mimeographed document.

Novoplansky, A., D. Cohen, and T. Sachs. 1990. How portulaca seedlings avoid their neighbors. *Oecologia* 82:490–493.

O'Donovan, J.T., R.E. Blackshaw, K.N. Harker, D. Derksen, and A.G. Thomas. 1995. Relative seed germination and growth of triallate/difenzoquat susceptible and resistant wild oat (*Avena fatua*) populations. P. 136 in *Proc. Internat. Symp. Weed Crop Resist. Herb.*, Cordoba, Spain, April 1995.

Odum, E.P. 1971. *Fundamentals of Ecology*, 3rd Ed. Saunders, Philadelphia.

Odum, S. 1965. Germination of ancient seeds: Floristical observations and experiments with archaeologically dated soil samples. *Dan. Bot. Ark.* 24:2.

Odum, W. 1974. Seeds in ruderal soils, their longevity and contribution to the flora of disturbed ground in Denmark. Pp. 1131–1141 in *Proc. 12th Brit. Weed Control Conf.*

Ogden, J. 1970. Plant population structure and productivity. *Proc. N.Z. Ecol. Soc.* 17:1–9.

Oliver, L.R. and M.M. Schreiber. 1974. Competition for CO_2 in heteroculture. *Weed Sci.* 22:125–131.

Oliver, W.W. and R.F. Powers. 1978. Growth models for ponderosa pine. I. Yield of unthinned plantations in northern California. USDA, *For. Ser. Research Paper* PSW-133.

O'Neill, R.V., D.L. de Angelis, J.B. Waide, and T.F. Allen. 1986. *A Hierarchical Concept of Ecosystems.* Princeton University Press, Princeton, NJ.

Oosting, H.J. and M.E. Humphreys. 1940. Buried viable seeds in a successional series of old field and forest soils. *Bull. Torrey Bot. Club* 57:253–273.

Pacala, S.W. and J.A. Silander, Jr. 1990. Field test of neighborhood population dynamic models of two annual weed species. *Ecol. Monogr.* 60:113–134.

Palmblad, I.G. 1968. Competition in experimental populations of weeds with emphasis on the regulation of population size. *Ecology* 49:26–34.

Patrick, Z.A. 1971. Phytotoxic substances associated with the decomposition in soil of plant residues. *Phytopathology* 53:152–161.

Patterson, D.T. 1982. Effect of light and temperature on weed/crop growth and competition. Pp. 407–421 in J.L. Hatfield and I.J. Thomason (Eds.), *Biometeorology in Integrated Pest Management.* Academic Press, New York.

Patterson, D.T. 1985. Comparative ecophysiology of weeds and crops. Pp. 101–129 in S.O. Duke (Ed.), *Weed Physiology, Vol. 1, Reproduction and Ecophysiology.* CRC Press, Inc., Boca Raton, FL.

Patterson, D.T. 1986. Responses of soybean (*Glycine max*) and three C_4 grass weeds to CO_2 enrichment during drought. *Weed Sci.* 34:203–210.

Patterson, D.T. 1995. Effects of environmental stress on weed/crop interactions. *Weed Sci.* 43:483–490.

Patterson, D.T. and E.P. Flint. 1980. Potential effects of global atmospheric CO_2 enrichment on the growth and competitiveness of C_3 and C_4 weed and crop plants. *Weed Sci.* 28:71–75.

Patterson, D.T. and E.P. Flint. 1983. Comparative water relations, photosynthesis, and growth of soybean (*Glycine max*) and seven associated weeds. *Weed Sci.* 31:318–323.

Patterson, D.T., C.R. Meyers, E.P. Flint, and P.C. Quimby, Jr. 1979. Temperature responses and potential distribution of itchgrass (*Rottboellia exaltata*) in the United States. *Weed Sci.* 27:77–82.

Pauly, P.J. 1987. *Controlling Life: Jacque Loeb and the Engineering Ideal in Biology.* University of California Press, Berkeley.

Pavlychenko, T.K. 1937a. Quantitative study of the entire root system of weed and crop plants under field conditions. *Ecology* 18:62–79.

Pavlychenko, T.K. 1937b. The soil washing method in quantitative root study. *Can. J. Res.* 15:33–57.

Pavlychenko, T.K. 1940. Investigations relating to weed control in Western Canada. Imp. Bur. Pastures and Forage Crops, *Herbage Publ. Serv. Bull.* 27 Aberystwyth, Wales.

Pearcy, R.W. 1990. Sunflecks and photosynthesis in plant canopies. *Ann. Rev. of Plant Physiol. Plant Molec. Biol.* 41:421–453.

Pearcy, R.W., N. Tumosa, and K. Williams. 1981. Relationship between growth, photosynthesis, and competitive interactions for a C_3 and a C_4 plant. *Oecologia* 48:371–376.

Pearcy, R.W., J.S. Roden, and J.A. Gamon. 1990. Sunfleck dynamics in relation to canopy structure in a soybean (*Glycine max* (L.) Merr.) canopy. *Agric. For. Meteorol.* 52:359–372.

Perlin, J. 1989. *A Forest Journey: The Role of Wood in the Development of Civilization.* Harvard University Press, Cambridge, MA.

Perrin, B., K. Wiele, and L. Iles. 1993. *Public attitudes towards herbicides and their implications for the public involvement strategy for the vegetation management alternatives program.* VMAP Technical Report 93–04. Queen's Printer, Toronto, Canada.

Pfister, K., S.R. Radosevich, and C.J. Arntzen. 1979. Modification of herbicide binding to photosystem II in two biotypes of *Senecio vulgaris* L. *Plant Physiol.* 64:995–999.

Phillips, D.A. 1980. Efficiency of symbiotic nitrogen fixation in legumes. *Annu. Rev. Plant Physiol.* 31:29–49.

Pianka, E.R. 1970. On r- and K-selection. *Am. Nat.* 104:592–597.

Pianka, E.R. 1994. *Evolutionary Ecology.* 5th Ed. Harper Collins, New York.

Pickett, S.T.A. and F.A. Bazzaz. 1978. Organization of an assemblage of early successional species on a soil moisture gradient. *Ecology* 59:1248–1255.

Pimentel, D. 1986. Acroecology and economics. Pp. 299–319 in M. Kogan (Ed.), *Ecological Theory and Integrated Pest Management Practice.* John Wiley & Sons, New York.

Pimentel, D.E., C. Terhune, R. Dyson-Hudson, S. Rochereau, R. Samis, E. Smith, D. Denman, D. Reifschnieder, and M. Shepard. 1976. Land degradation: Effects on food and energy resources. *Science* 194:149–155.

Pimentel, D.E., J. Krummel, D. Gallhan, J. Hough, A. Merrill, I. Schreiner, P. Vittum, F. Koziol, E. Back, D. Yen, and S. Fiance. 1978. Benefits and costs of pesticide use in U.S. food production. *BioScience* 28:772, 778–784.

Pokorny, R. 1941. Some chlorophenoxyacetic acids. *J. Am. Chem. Soc.* 63:1768.

Poorter, H. 1990. Interspecific variation in relative growth rate: On ecological causes and physiological consequences. Pp. 45–68 in H. Lambers, M.L. Cambridge, H. Konings, and T.L. Pons (Eds.), *Causes and Consequences of Variation in Growth Rate and Productivity of Higher Plants.* SPB Academic Publishing, The Hague.

Poorter, H. and C. Remkes. 1990. Leaf area ratio and net assimilation rate of 24 wild species differing in relative growth rate. *Oecologia* 83:553–559.

Popay, A.I. and E.H. Roberts. 1970a. Ecology of *Capsella bursa-pastoris* and *Senecio vulgaris* in relation to germination behavior. *J. Ecol.* 58:123–139.

Popay, A.I. and E.H. Roberts. 1970b. Factors involved in the dormancy and germination of *Capsella bursa-pastoris* and *Senecio vulgaris. J. Ecol.* 58:103–122.

Powles, S.B. and J.A.M. Holtum (Eds.). 1994. *Herbicide Resistance in Plants: Biology and Biochemistry.* Lewis Publishers, Boca Raton, FL.

Price, S.C., K.M. Shumaker, A.L. Kahler, R.W. Allard, and J.E. Hill. 1984 Estimates of population differentiation obtained from enzyme polymorphisms and quantitative characters. *J. Hered.* 75:141–142.

Probert, R.J. 1992. The role of temperature in germination ecophysiology. Pp. 285–325 in M. Fenner (Ed.). *Seeds: The Ecology of Regeneration in Plant Communities.* C.A.B. International, Wallingford, UK.

Puckridge, D.W. 1968. Photosynthesis of wheat under field conditions. II. Effect of defoliation on the carbon dioxide uptake of the community. *Aust. J. Agric. Res.* 20:623–634.

Putnam, A.R. and W.B. Duke. 1978. Allelopathy in agroecosystems. *Annu. Rev. Phytopathol.* 16:431–451.

Putnam, A.R. and C.S. Tang. 1986. *The Science of Allelopathy.* John Wiley & Sons, New York.

Putnam, A.R. and L.A. Weston. 1986. Adverse impacts of allelopathy in agricultural systems. Pp. 43–56 in A.R. Putnam and C.S. Tang (Eds.), *The Science of Allelopathy.* John Wiley & Sons, New York.

Putwain, P.D. and A.M. Mortimer. 1989. The resistance of weeds to herbicides: Rational approaches for containment of a growing problem. Pp. 285–294 in *Proceedings of the 1989 Brighton Crop Protection Conference—Weeds.* BCPC Publications, Surrey, UK.

Putwain, P.D. and A.M. Mortimer. 1995. Evolution and spread of herbicide resistant weeds. The role of selection pressure. P. 115 in *Proc. Internat. Symp. Weed Crop Resist. Herb.* Cordoba, Spain.

Qasem, J.R. 1992. Nutrient accumulation by weeds and their associated vegetable crops. *J. Hort. Sci.* 67:189–195.

Radosevich, S.R. 1987. Methods to study interactions among crops and weeds. *Weed Technol.* 1:190–198.

Radosevich, S.R. and S.G. Conard. 1981. Interactions among weeds, other pests, and conifers in forest regeneration. Pp. 453–476 in J.L. Hatfield and I.L. Thomason (Eds.), *Biometeorology in Integrated Pest Management.* Academic Press, New York.

Radosevich, S.R. and C.M. Ghersa. 1992. Weeds, crops, and herbicides: A modern-day "neckriddle." *Weed Technol.* 6:788–795.

Radosevich, S.R. and J.S. Holt. 1982. Physiological responses and fitness of susceptible and resistant weed biotypes to triazine herbicides. Pp. 163–183 in H.M. LeBaron and J. Gressel (Eds.), *Herbicide Resistance in Plants.* John Wiley & Sons, New York.

Radosevich, S.R. and M.L. Roush. 1990. The role of competition in agriculture. Pp. 341–363 in J.B. Grace and D. Tilman (Eds.), *Perspectives on Plant Competition.* Academic Press, San Diego, CA.

Radosevich, S.R. and R. Shula. 1994. Implementation of weed control in IPM. Pp. 58–70 in *Pesticide Risk Reduction and Strategic Planning Forum.* Pest Management Alternatives Office, Val-Morin, Quebec, Canada.

Radosevich, S.R., N.L. Smith, and F. Kegel. 1975. Johnsongrass control in field corn. *West. Soc. Weed Sci. Prog. Rep.* Pp. 62–64.

Radosevich, S.R., C.M. Ghersa, and G. Comstock. 1992. Concerns a weed scientist might have about herbicide-tolerant crops. *Weed Technol.* 6:635–639.

Raven, P.H., R.F. Evert, and S.E. Eichhorn. 1992. *Biology of Plants*. 5th Ed. Worth Publishers, Inc., New York.

Regnier, E.E., M.E. Salvucci, and E.W. Stoller. 1988. Photosynthesis and growth responses to irradiance in soybean (*Glycine max*) and three broadleaf weeds. *Weed Sci.* 36:487–496.

Reineke, L.H. 1933. Perfecting a stand-density index for even-aged forests. *J. Agric. Res.* 46:627–638.

Rejmanek, M. 1989. Invasibility of plant communities. Pp. 369–388 in J.A. Drake, H.A. Mooney, F. di Castri, R. Groves, F.J. Kruger, M. Rejmanek, and M. Williamson (Eds.), *Biological Invasions: A Global Perspective*. John Wiley & Sons, New York.

Rejmanek, M., G.R. Robinson, and E. Rejmankova. 1989. Weed-crop competition: Experimental designs and models for data analysis. *Weed Sci.* 37:276–284.

Rice, E.L. 1984. *Allelopathy*. 2nd Ed. Academic Press, Orlando, FL.

Rizvi, S.J.H. and V. Rizvi. 1992. *Allelopathy: Basic and Applied Aspects*. Chapman & Hall, London.

Robbins, W.W., M.K. Bellue, and W.S. Ball. 1941. *Weeds of California*. California State Dept. of Agriculture.

Robbins, W.W., A.S. Crafts, and R.N. Raynor. 1942. *Weed Control*. McGraw-Hill, New York.

Robbins, W.W., M.K. Bellue, and W.S. Ball. 1951. *Weeds of California*. Univ. of California, Davis, CA.

Roberts, E.H. 1972a. Cytological, genetical and metabolic changes associated with loss of viability. Pp. 253–306 in E.H. Roberts (Ed.), *Viability of Seeds*. Syracuse University Press, Syracuse, NY.

Roberts, E.H. 1972b. Storage environment and the control of viability. Pp. 14–58 in E.H. Roberts (Ed.), *Viability of Seeds*. Syracuse University Press, Syracuse, NY.

Roberts, H.A. and P.M. Feast. 1972. Emergence and longevity of seeds of annual weeds in cultivated and undisturbed soil. *J. Appl. Ecol.* 10:133–143.

Roberts, H.A. and P.M. Feast. 1973. Changes in the number of viable weed seeds in soil under different regimes. *Weed Res.* 13:298–303.

Roberts, H.A. and M.E. Potter. 1980. Emergence patterns of weed seedlings in relation to cultivation and rainfall. *Weed Res.* 20:377–386.

Roberts, H.A. and M.E. Ricketts. 1979. Quantitative relationship between the weed flora after cultivation and the seed population in the soil. *Weed Res.* 19:269–275.

Roberts, H.A., W. Bond, and R.T. Hewson. 1976. Weed competition in drilled summer cabbage. *Ann. Appl. Biol.* 84:91–95.

Roberts, H.A., R.T. Hewson, and M.E. Ricketts. 1977. Weed competition in drilled summer lettuce. *Hort. Res.* 17:39–45.

Robertson, J.H. 1954. Half-century changes of northern Nevada ranges. *J. Range Manag.* 7:117–121.

Rodgers, N.K., G.A. Buchanan, and W.C. Johnson. 1976. Influence of row spacing on weed competition with cotton. *Weed Sci.* 24:410–413.

Rosenthal, S.S., D.M. Maddox, and K. Brunetti. 1989. Biological control methods. Pp. 77–99 in California Weed Conference, *Principles of Weed Control in California*. 2nd Ed. Thomson Publications, Fresno, CA.

Ross, M.A. 1968. The establishment of seedlings and the development of patterns in grassland. Ph.D. thesis, University of Wales, Bangor, UK.

Ross, M.A. and C.A. Lembi. 1985. *Applied Weed Science*. Burgess Publishing Company, Minneapolis, MN.

Roush, M.L. 1988. Models of a four-species annual weed community: Growth, competition, and community dynamics. Ph.D. thesis, Oregon State University, Corvallis, OR.

Roush, M.L. and S.R. Radosevich. 1985. Relationships between growth and competitiveness of four annual weeds. *J. Appl. Ecol.* 22:895–905.

Roush, M.L. and S.R. Radosevich. 1987. A weed community model of germination, growth, and competition of annual weed species. *Abstr. Weed Sci. Soc. Am.* 27:147.

Roush, M.L., S.R. Radosevich, R.G. Wagner, B.O. Maxwell, and T.D. Petersen. 1989a. A comparison of methods for measuring effects of density and proportion in plant competion experiments. *Weed Sci.* 37:268–275.

Roush, M.L., N. Jordon, and J.S. Holt. 1989b. Ecological basis for weed management in IPM. Pp. 137–157 in *Proc. National Integrated Pest Management Symposium/Workshop*, Las Vegas, NV. Communications Services, New York State Agricultural Experiment Station, Cornell Univ., Geneva, NY.

Roush, M.L., S.R. Radosevich, and B.D. Maxwell. 1990. Future outlook for herbicide resistance research. *Weed Technol.* 4:208–214.

Ruttan, V.W. 1982. *Agricultural Research Policy*. University of Minnesota Press, Minneapolis, MN.

Sagar, G.R. 1970. Factors controlling the size of plant populations. Pp. 965–979 in *Proc. 10th Brit. Weed Control Conf.*

Sagar, G.R. 1974. On the ecology of weed control. Pp. 42–56 in D. Price-Jones and M.E. Solomon (Eds.), *Biology in Pest and Disease Control*. Blackwell Scientific Publications, Oxford, UK.

Sagar, G.R. and A.M. Mortimer. 1976. An approach to the study of the population dynamics of plants with special reference to weeds. *Ann. Appl. Biol.* 1:1–47.

Salisbury, E.J. 1942a. *The Reproductive Capacity of Plants*. G. Bell, London.

Salisbury, E.J. 1942b. The weed problem. *Proc. R. Inst. Gr. Brit.* 31:1–15.

Salisbury, E.J. 1961. *Weeds and Aliens*. Collins, London.

Salisbury, F.B. and C.W. Ross. 1978. *Plant Physiology*. 2nd Ed. Wadsworth Publishing Co., Belmont, CA.

Salisbury, F.B. and C.W. Ross. 1992. *Plant Physiology*. 4th Ed. Wadsworth Publishing Co., Belmont, CA.

Sarukhan, J. and M. Gadgil. 1974. Studies on plant demography: *Ranunculus repens* L., *R. bulbosus* L. and *R. acris* L. III. A mathematical model incorporating multiple modes of reproduction. *J. Ecol.* 62:921–936.

Sauer, J. and G. Struik. 1964. A possible ecological relation between soil disturbance, light flash, and seed germination. *Ecology* 45:884–886.

Schmitt, J. and R.D. Wulff. 1993. Light spectral quality, phytochrome and plant competition. *Trends Ecol. Evol.* 8:47–51.

Scholander, P.F., H.T. Hammel, E.A. Hemmingsen, and E.D. Bradstreet. 1964. Hydrostatic pressure and osmotic potential in leaves of mangroves and some other plants. *Proc. Natl. Acad. Sci.* 52:119–125.

Schoner, C.A., R.F. Norris, and W. Chilcote. 1978. Yellow foxtail (*Setaria lutescens*) biotype studies: Growth and morphological characteristics. *Weed Sci.* 26:632–636.

Schultz, E.D. and H.A. Mooney. 1993. *Biodiversity and Ecosystem Function.* Springer-Verlag, Berlin.

Schulz, A., F. Wengenmayer, and H.M. Goodman. 1990. Genetic engineering of herbicide resistance in higher plants. *CRC Crit. Rev. Plant Sci.* 9:1–15.

Scopel, A.L., C.L. Ballaré, and C.M. Ghersa. 1988. Role of seed reproduction in the population ecology of *Sorghum halepense* in maize crops. *J. Appl. Ecol.* 25:951–962.

Scopel, A.L., C.L. Ballaré, and R.A. Sánchez. 1991. Induction of extreme light sensitivity in buried weed seeds and its role in the perception of soil cultivations. *Plant Cell Environ.* 14:501–508.

Scopel, A.L., C.L. Ballaré, and S.R. Radosevich. 1994. Photostimulation of seed germination during soil tillage. *New Phytol.* 126:145–152.

Scott, H.D. and R.D. Geddes. 1979. Plant water stress of soybean (*Glycine max*) and common cocklebur (*Xanthium pensylvanicum*): A comparison under field conditions. *Weed Sci.* 27:285–289.

Searle, S.R. 1966. *Matrix Algebra for the Biological Sciences.* John Wiley & Sons, New York.

Selman, M. 1970. The population dynamics of *Avena fatua* (wild oats) in continuous spring barley—desirable frequency of spraying with triallate. Pp. 1176–1188 in *Proc. 10th Brit. Weed Control Conf.*

Shainsky, L.J. and S.R. Radosevich. 1991. Analysis of yield-density relationships in experimental stands of Douglas-fir and red alder seedlings. *For. Sci.* 37:574–592.

Shainsky, L.J. and S.R. Radosevich. 1992. Mechanisms of competition between Douglas-fir and red alder seedlings. *Ecology* 73:30–45.

Shantz, H.L. and L.N. Piemeisel. 1927. The water requirements of plants at Akron, Colorado. *J. Agric. Res.* 34:1093–1189.

Sheldon, J.C. 1974. The behavior of seed in soil. III. The influence of seed morphology and the behavior of seedlings on the establishment of plants from surface lying seeds. *J. Ecol.* 62:47–66.

Sheldon, J.C. and F.M. Burrows. 1973. The dispersal effectiveness of the achene pappus units of selected Compositae in steady winds with convection. *New Phytol.* 72:665–675.

Shetty, S.V.R. 1980. Some agro-economical aspects of improved weed management systems in Indian Semi-Arid Tropics. Pp. 899–910 in *Proc. Brit. Crop Prot. Conf.—Weeds.*

Shimabukuro, R.H. 1985. Detoxication of herbicides. Pp. 215–240 in S.O. Duke (Ed.), *Weed Physiology, Vol. 2, Herbicide Physiology.* CRC Press, Inc., Boca Raton, FL.

Shimabukuro, R.H., G.L. Lamoureux, and D.S. Frear. 1982. Pesticide metabolism in plants: Reactions and mechanisms. Pp. 21–66 in F. Matsumura and C.R. Krishna Murti (Eds.), *Biological Degradation of Pesticides.* Plenum, New York.

Shribbs, J.M., D.W. Lybecker, and E.E. Schweizer. 1990a. Bioeconomic weed management models for sugarbeet (*Beta vulgaris*) production. *Weed Sci.* 38:436–444.

Shribbs, J.M., E.E. Schweizer, L. Hergert, and D.W. Lybecker. 1990b. Validation of four bioeconomic weed management models for sugarbeet (*Beta vulgaris*) production. *Weed Sci.* 38:445–451.

Shuttleworth, C.L. 1973. The case for reducing wild oats in commercial grain. In *Let's Clean Up on Wild Oats: Proceedings, Action Proposals and Programs.* Agriculture Canada and United Grain Growers Limited Special Seminar.

Silverton, J.W. 1987. *Introduction to Plant Population Ecology.* 2nd Ed. John Wiley & Sons, New York.

Silvertown, J.W. and J. Lovett Doust. 1993. *Introduction to Plant Population Biology.* Blackwell Scientific Publications, London.

Simpson, G.M. 1990. Timing in dormancy. Pp. 195–231 in *Seed Dormancy in Grasses.* Cambridge University Press, Cambridge, UK.

Singh, K. 1968. Thermoresponse of *Portulaca oleracea* seeds. *Curr. Sci.* 37:506–507.

Slabaugh, W.H. and T.D. Parsons. 1966. Chaps. 10 and 27 in *General Chemistry.* John Wiley & Sons, New York.

Slatyer, R.O. 1967. *Plant-Water Relationships.* Academic Press, New York.

Smith, D.M. 1986. *The Practice of Silviculture.* John Wiley & Sons. New York.

Smith, H. 1982. Light quality, photoperception, and plant strategy. *Annu. Rev. Plant Physiol.* 33:481–518.

Smith, H., J.J. Casal, and G.M. Jackson. 1990. Reflection signals and the perception by phytochrome of the proximity of neighboring vegetation. *Plant Cell Environ.* 13:73–78.

Smith, R.J., Jr. 1968. Weed competition in rice. *Weed Sci.* 16:252–254.

Solangaarachchi, S.M. and J.L. Harper. 1987. The effect of canopy filtered light on the growth of white clover *Trifolium repens. Oecologia* 72:372–376.

Solbrig, O.T. (Ed.). 1980. *Demography and Evolution in Plant Populations.* Botanical Monographs 15. University of California Press, Berkeley, CA.

Soriano, A. 1971. Aspectos ritmicos o ciclicos del dinamismo de la comunidad vegetal. Pp. 441–445 in R.H. Mejia y J.A. Moguilevsky (Eds.), *Recientes Adelantos en Biologia.*

Souza Machado, V., J.D. Bandeen, G.R. Stephenson, and P. Lavigne. 1978. Uniparental inheritance of chloroplast atrazine tolerance in *Brassica campestris. Can. J. Plant Sci.* 58:977.

Spitters, C.J.T. 1983a. An alternative approach to the analysis of mixed cropping experiments. 1. Estimation of competition effects. *Neth. J. Agric. Sci.* 31:1–11.

Spitters, C.J.T. 1983b. An alternative approach to the analysis of mixed cropping experiments. 2. Marketable yield. *Neth. J. Agric. Sci.* 31:143–155.

Spitters, C.J.T. 1989. Weeds: Population dynamics, germination and competition. Pp. 182–216 in R. Rabingge, S.A. Ward, and H.H. van Laar (Eds.), *Simulation and Systems Management in Crop Protection.* Vol. 32. Simulation Monographs, Pudoc, Wageningen, The Netherlands.

Spitters, C.J.T. and R. Aerts. 1983. Simulation of competition for light and water in crop-weed associations. *Asp. Appl. Biol.* 4:467–483.

Spitters, C.J.T. and J.P. Van den Bergh. 1982. Competion between crops and weeds: A system approach. Pp. 137–146 in W. Holzner and M. Numata (Eds.), *Biology and Ecology of Weeds.* Dr W. Junk Publishers, The Hague.

Stacey, G., R.H. Burris, and H.J. Evans (Eds.). 1992. *Biological Nitrogen Fixation*. Chapman and Hall, New York.

Stallings, G.P., D.C. Thill, C.A. Mallory-Smith, and L. Lass. 1995. Plant movement and seed dispersal of Russian thistle (*Salsola iberica*). *Weed Sci.* 43:63–69.

Stanley, S.M. 1989. Fossils, macroevolution, and theoretical ecology. Pp. 125–134 in J. Roughgarden, R.M. May, and S.A. Levin (Eds.), *Perspectives in Ecological Theory*. Princeton University Press, Princeton, NJ.

Steiner, R. 1983. *The Boundaries of Natural Science*. Translated from the German text (1969). Anthroposophic Press, Spring Valley, NY.

Stevens, O.A. 1954. Weed Seed Facts. *N.D. Agric. Coll. Ext. Cir.* A-128, 4 pp.

Stevens, O.A. 1957. Weights of seeds and numbers per plant. *Weeds* 5:46–55.

Stewart, R.E., L.L. Gross, and B.H. Honkala. 1984. Effects of competing vegetation on forest trees: A bibliography with abstracts. USDA Forest Service, *Gen. Tech. Rep.* WO-43. Washington, D.C. 260 pp.

Stigliani, L. and C. Resina. 1993. SELOMA: Expert system for weed management in herbicide-intensive crops. *Weed Technol.* 7:550–559.

Stoller, E.W. and R.A. Myers. 1989. Response of soybeans (*Glycine max*) and four broadleaf weeds to reduced irradiance. *Weed Sci.* 37:570–574.

Stowe, A.E. and J.S. Holt. 1988. Comparison of triazine-resistant and -susceptible biotypes of *Senecio vulgaris* and their F_1 hybrids. *Plant Physiol.* 87:183–189.

Strauss, S.H., W.H. Rottmann, A.M. Brunner, and L.A. Sheppard. 1995. Genetic engineering of reproductive sterility in forest trees. *Mol. Breed.* 1:5–26.

Streibig, J.C. 1988. Herbicide bioassay. *Weed Res.* 28:479.

Streibig, J.C., M. Rudemo, and J.E. Jensen. 1993. Dose-response curves and statistical models. P. 29 in J.C. Streibig and P. Kudsk (Eds.), *Herbicide Bioassays*. CRC Press, Boca Raton, FL.

Stroud, D. and H.M. Kempen. 1989. Wick/Wiper. Pp. 148–150 in *California Weed Conference. Principles of Weed Control in California*. 2nd Ed. Thompson Publishing, Fresno, CA.

Suarez, S., R.J.C. Leon, C.M. Ghersa, and S. Burkhart. 1995. Cambios floristicos en las comunidades de malezas de maiz relacionados con el deterioro del ambiente. *Pro. Primeras Jornadas Cientificas del Medio Amionto*. Montevideo, Uraguay. 8–10 November. II, p. 48.

Sutton, R.F. 1985. Vegetation management in Canadian forestry. Great Lakes Forest Research Centre, Canadian Forestry Service, Information Rep. 0-X-369, Sault Ste. Marie, Canada.

Swift, M.J. and J.M. Anderson. 1993. Biodiversity and ecosystem function in agricultural systems. Pp. 15–38 in E.D. Schulze and H.A. Mooney (Eds.), *Biodiversity and Ecosystem Function*. Springer-Verlag, Berlin.

Taiz, L. and E. Zeiger. 1991. *Plant Physiology*. The Benjamin/Cummings Publishing Company, Redwood City, CA.

Tang, C. and C. Young. 1982. Collection and identification of allelopathic compounds from the undisturbed root systems of bigalta limpograss (*Hemarthria altissima*). *Plant Physiol.* 69:155–160.

Taylorson, R.B. and S.B. Hendricks. 1977. Dormancy in seeds. *Annu. Rev. Plant Physiol.* 28:331–354.

Terminology Committee of the Weed Science Society of America. 1956. Terminology Committee Report—WSSA. *Weeds* 4:278–287.

Teyker, R.H., H.D. Hoelzer, and R.A. Liebl. 1991. Maize and pigweed response to nitrogen supply and form. *Plant Soil* 135:287–292.

Thomas, W.L., Jr. (Ed.). 1956. *Man's Role in Changing the Face of the Earth*. An international symposium under the co-chairmanship of C. Sauer, M. Bates, and L. Mumford. Sponsored by the Wenner-Gren Foundation for Anthropological Research. University of Chicago Press, Chicago, IL.

Thompson, K. and J.P. Grime. 1979. Seasonal variation in the seed banks of herbaceous species in ten contrasting habitats. *J. Ecol.* 67:893–921.

Thurston, J.M. 1961. The effect of depth of burying and frequency of cultivation on survival and germination of seeds of wild oats (*Avena fatua* L. and *Avena ludoviciana*). *Weed Res.* 1:19–31.

Thurston, J.M. 1964. Weed studies in winter wheat. Pp. 592–598 in *Proc. 7th Brit. Weed Control Conf.*

Tilman, G.D. 1982. *Resource Competition and Community Structure*. Princeton Univ. Press, Princeton, NJ.

Tilman, G.D. 1985. The resource-ratio hypothesis of plant succession. *Am. Nat.* 125:827–852.

Tilman, G.D. 1988. *Plant Strategies and the Dynamics and Structure of Plant Communities*. Princeton Monographs, Princeton, NJ.

Tilman, G.D. 1990. Mechanisms of plant competition for nutrients: The elements of a predictive theory of competition. Pp. 117–141 in J.B. Grace and D. Tilman (Eds.), *Perspectives in Plant Competition*. Academic Press, New York.

Tivey, J. 1990. *Agricultural Ecology*. Longman, New York.

Toole, E.H. and E. Brown. 1946. Final results of the Duvel buried seed experiment. *J. Agric. Res.* 72:201–210.

Tremmel, D.C. and F.A. Bazzaz. 1993. How neighbor canopy architecture affects target plant performance. *Ecology* 74:2114–2124.

Trenbath, B.R. 1976. Plant interactions in mixed crop communities. Pp. 129–170 in R.I. Papendick, P.A. Sanchez, and G.B. Triplett (Eds.), *Multiple Cropping*. ASA Special Publication No. 27. Am. Soc. Agron., Madison, WI.

Tuesca, D., E.C. Puricelli, and J.C. Papa. 1995. Changes in the weed community in contrasting tillage systems. *Actas del XII Congreso Latinoamericano de Malezas, Montevideo (Uruguay)*.

Tukey, H.B., Jr. 1966. Leaching of metabolites from above-ground plant parts and its implications. *Bull. Torrey Bot. Club* 93:385–401.

Tull, J. 1733. The horse hoeing husbandry. In E.J. Salisbury, 1961. *Weeds and Aliens*. Collins, London.

Vandermeer, J. 1989. *Ecology of Intercropping*. Cambridge University Press, Cambridge, UK.

Van Esso, M.L. and C.M. Ghersa. 1989. Dynamics of *Sorghum halepense* (L.) Pers. seeds in the soil of an uncultivated field. *Can. J. Bot.* 67:940–944.

Van Esso, M.L., C.M. Ghersa, and A. Soriano. 1986. Cultivation effects on the dynamics of a Johnsongrass seed population in the soil. *Soil Tillage Res.* 6:325–335.

Vengris, J. 1956. Weeds—Robbers of our farms. *Better Crops* 40:9–12, 46–47.

Vengris, J., W.G. Colby, and M. Drake. 1955. Plant nutrient competition between weeds and corn. *Agron. J.* 47:213–216.

Vere, D.T., J.A. Sinden, and M.H. Campbell. 1980. Social benefits of serrated tussock control in New South Wales. *Rev. Mark. Agric. Econ.* 48:123–138.

Villiers, T.A. 1972. Cytological studies in dormancy. II. Pathological aging changes during prolonged dormancy and recovery upon dormancy release. *New Phytol.* 71:145–152.

Villiers, T.A. 1974. Seed aging: chromosome stability and extended viability of seeds stored fully imbibed. *Plant Physiol.* 53:875–878.

Vitousek, P.M. and L.R. Walker. 1987. Colonization, succession and resource availability: Ecosystem-level interactions. Pp. 207–224 in A.J. Gray, M.J. Crawley, and P.J. Edwards (Eds.), *Colonization, Succession and Stability*. Blackwell Scientific Publications, London, UK.

Volterra, V. 1926. Fluctuations in the abundance of a species considered mathematically. *Nature* 118:558–560.

Vrijenhoek, R.C. 1985. Animal population genetics and disturbance: The effects of local extinctions and recolonizations on heterozygosity and fitness. Pp. 266–286 in S.T.A. Pickett and P.S. White (Eds.), *Ecology of Natural Disturbance and Patch Dynamics*. Academic Press, San Diego, CA.

Wagner, R.G. and L. Buse. 1995. Public and professional perspectives about forest vegetation management and risk. *Veg. Manag. Alt. Prog.* 4:1–12.

Wagner, R.G., T.D. Petersen, D.W. Ross, and S.R. Radosevich. 1989. Competition thresholds for the survival and growth of ponderosa pine seedlings associated with woody and herbaceous vegetation. *New For.* 3:151–170.

Waller, G.R., D. Kumari, J. Friedman, N. Friedman, and C. Chou. 1986. Caffeine autotoxicity in *Coffea arabica* L. Pp. 243–269 in A.R. Putnam and C.S. Tang (Eds.), *The Science of Allelopathy*. John Wiley & Sons, New York.

Walstad, J.D. and F.N. Dost. 1986. All the king's horses and all the king's men: The lessons of 2,4,5-T. *J. For.* 84:28–33.

Walstad J.D. and P.J. Kuch. 1987. *Forest Vegetation Management for Conifer Production*. John Wiley & Sons, New York.

Walstad, J.D., M. Newton, and R.J. Boyd. 1987. Forest vegetation problems in the Northwest. Pp. 157–202 in J.D. Walstad and P.J. Kuch (Eds.), *Forest Vegetation Management for Conifer Production*. John Wiley & Sons, New York.

Warwick, S.I. 1990. Genetic variation in weeds—with particular reference to Canadian agricultural weeds. Pp. 3–18 in S. Kawano (Ed.), *Biological Approaches and Evolutionary Trends in Plants*. Academic Press, San Diego, CA.

Warwick, S.I. and L. Black. 1980. Uniparental inheritance of atrazine resistance in *Chenopodium album* L. *Can. J. Plant Sci.* 60:751.

Warwick, S.I. and L.D. Black. 1994. Relative fitness of herbicide-resistant and susceptible biotypes of weeds. *Phytoprotection* 75:37–49.

Watson, A.K. 1985. Integrated management of leafy spurge. Pp. 93–104 in A.K. Watson (Ed.), *Leafy Spurge*. Weed Sci. Soc. Am., Champaign, IL.

Watson, S. 1871. *Botany*. U.S. Geological Exploration of the Fortieth Parallel. Vol. 5, Washington, D.C.

Way, M.F. 1977. Pest and disease status in mixed stands vs. monocultures: The relevance of ecosystem stability. Pp. 127–138 in J.M. Cherrett and G.R. Sagar (Eds.),

Origins of Pest, Parasite, Disease and Weed Problems. Blackwell Scientific, Oxford, UK.

Weatherspoon, D.M. and E.E. Schweizer. 1971. Competition between sugarbeets and five densities of kochia. *Weed Sci.* 19:125–128.

Weed Science Society of America. 1994. *Herbicide Handbook.* 7th Ed. Champaign, IL.

Weier, T.E., C.R. Stocking, M.G. Barbour, and T.L. Rost. 1982. *Botany. An Introduction to Plant Biology.* John Wiley & Sons, New York.

Welbank, P.J. 1961. A study of nitrogen and other factors in competition with *Agropyron repens* (L.) Beauv. *Ann. Bot. (N.S.)* 25:116–137.

Weldon, C.W. and W.L. Slausen. 1986. The intensity of competition versus its importance: An overlooked distinction and some implications. *Q. Rev. Biol.* 61:23–44.

Wesson, G. and P.F. Wareing. 1969a. The induction of light sensitivity in weed seeds by burial. *J. Exp. Bot.* 20:413–425.

Wesson, G. and P.F. Wareing. 1969b. The role of light in the germination of naturally occurring populations of buried weed seeds. *J. Exp. Bot.* 20:402–413.

Westra, P.H. and D.L. Wyse. 1980. Growth and development of quackgrass (*Agropyron repens*) biotypes. *Weed Sci.* 29:44–53.

White, D.E. 1979. Physiological adaptations in two ecotypes of Canada thistle (*Cirsium arvense* (L) Scop.). Master's thesis, University of California, Davis, CA.

Whitson, T.D. Ed. 1991. *Weeds of the West.* Western Society of Weed Science, in cooperation with the Western United States Land Grant Universities Cooperative Extension Services.

Whittaker, R.H. 1975. *Communities and Ecosystems.* 2nd Ed. Macmillan, New York.

Wiens, D. 1978. Mimicry in plants. *Evol. Biol.* 11:365–403.

Wilen, C.A., J.S. Holt, and W.B. McCloskey. 1996. Development of a predictive degree-day model for yellow nutsedge (*Cyperus esculentus*). *Weed Science.* In press.

Wilkerson, G.G., S.A. Modena, and H.D. Coble. 1991. HERB: Decision model for postemergence weed control in soybean. *Agron. J.* 83:413–417.

William, R.D. 1981. Complementary interactions between weeds, weed control practices, and pests in horticultural cropping systems. *Hort. Sci.* 16:508–513.

Williams, J. 1963. Biological flora of the British Isles: *Chenopodium album* L. *J. Ecol.* 51:711–725.

Williams, J.T. and J.L. Harper. 1965. Seed polymorphism and germination. I. The influence of nitrates and low temperatures on the germination of *Chenopodium album. Weed Res.* 5:141–150.

Williams, W.A. 1963. Competition for light between annual species of *Trifolium* during the vegetative phase. *Ecology* 44:475–485.

Williamson, M.H. 1991. Are communities ever stable? Pp. 353–372 in A.J. Gray, M.J. Crawley, and P.J. Edwards (Eds.), *Colonization, Succession and Stability.* Blackwell Scientific Publications, London, UK.

Wilson, B.J. 1972. Studies on the fate of *Avena fatua* seeds on cereal stubble as influenced by Autumn treatment. Pp. 242–247 in *Proc. 11th Brit. Weed Control Conf.*

Wilson, H.D. 1990. Gene flow in squash species. *BioScience* 40:449–455.

Wilson, K. and G.E.B. Morren, Jr. 1990. *Systems Approaches for Improvement in Agriculture and Resource Management.* Macmillan Publishing Company, New York.

Wilson, P.W. 1940. *The Biochemistry of Symbiotic Nitrogen Fixation.* University of Wisconsin Press, Madison, WI.

Wilson, R.G. 1988. Biology of weed seeds in the soil. Pp. 25–39 in M.A. Altieri and M. Liebman (Eds.), *Weed Management in Agroecosystems: Ecological Approaches.* CRC Press, Inc., Boca Raton, FL.

Wilson, R.G. and P. Westra. 1991. Wild proso millet (*Panicum miliaceum*) interference in corn (*Zea mays*). *Weed Sci.* 39:217–220.

Wit, C.T. de. 1960. On competition. *Versl. landbouwkd. onderz.* No. 66.8.

Woodmansee, R.G. 1984. Comparative nutrient cycles of natural and agricultural ecosystems: A step toward principles. Pp. 145–156 in R. Lowrance, B.R. Stinner, and G.J. House (Eds.), *Agricultural Ecosystems: Unifying Concepts.* John Wiley & Sons, New York.

Woodwell, G.M. (Ed.). 1986. *The Earth in Transition: Patterns and Processes of Biotic Impoverishment.* Cambridge University Press, New York.

Woolley J.T. and E.W. Stoller. 1978. Light penetration and light-induced seed germination in soil. *Plant Physiol.* 61:597–600.

Yadav, R.L. 1982. Minimizing nitrate-nitrogen leaching by parallel multiple cropping in long-duration row crops. *Exp. Agric.* 18:37–42.

Yoda, K., T. Kira, H. Ogawa, and K. Hozumi. 1963. Self-thinning in overcrowded pure stands under cultivated and natural conditions. *J. Biol. Osaka City Univ.* 14:107–129.

Young, J.A. and R.A. Evans. 1976. Response of weed populations to human manipulations of the natural environment. *Weed Sci.* 24:186–190.

Zimdahl, R.L. 1980. *Weed Crop Competition: A Review.* International Plant Protection Center, Corvallis, OR.

Zimdahl, R.L. 1988. The concept and application of the critical weed-free period. Pp. 145–155 in M.A. Altieri and M. Liebman (Eds.), *Weed Management in Agroecosystems: Ecological Approaches.* CRC Press, Inc., Boca Raton, FL.

Zimdahl, R.L. 1991. *Weed Science: A Plea for Thought.* Colorado State University, Fort Collins, CO.

Zimdahl, R.L. 1993. *Fundamentals of Weed Science.* Academic Press, San Diego, CA.

Zimmerman, C.A. 1976. Growth characteristics of weediness in *Portulaca oleracea* L. *Ecology* 57:964–974.

Appendix

1

An Example of a Demographic Model Using Difference Equations*

Example. Suppose that the hypothetical plant population described on pages 111–112 consists of 1000 individuals at t_1 and these individuals are distributed among the age groups as follows:

$$\begin{aligned} t_1a_0 &= 750 \\ t_1a_1 &= 100 \\ t_1a_2 &= 100 \\ t_1a_3 &= 50 \end{aligned}$$

Suppose, further, that the age-specific birth and survival-rates are:

Age	B	p
0	0	0.1
1	5	0.6
2	15	0.3
3	10	0.0

Then, for instance, no offspring are born to adults aged a_0, while adults in age group a_2 are the most fecund. Also, only 10 percent of the individuals aged a_0 survive to the next time-step and become incorporated into a_1, whereas none

* See pp. 108–112.

of the individuals in a_3 survive ($p = 0$). The number of individuals in each age class at time t_2 will become:

$$
\begin{aligned}
t_2a_0 &= (750 \times 0) + (100 \times 5) + (100 \times 15) + (50 \times 10) &= 2500 \\
t_2a_1 &= 750 \times 0.1 &= 75 \\
t_2a_2 &= 100 \times 0.6 &= 60 \\
t_2a_3 &= 100 \times 0.3 &= \underline{30} \\
& & 2665
\end{aligned}
$$

The population increased by 1665 individuals during this timestep and the age distribution changed from:

$$
\begin{aligned}
a_0 &= 750 \text{ to } 2500 \\
a_1 &= 100 \text{ to } 75 \\
a_2 &= 100 \text{ to } 60 \\
a_3 &= 50 \text{ to } 30
\end{aligned}
$$

It is important to realize that these equations assume that individuals reproduce before they die. Taking age group a_2, for instance, the value 100 was used in equation 4.2 to compute the number of births ($100 \times 15 = 1500$), but then used in equation 4.5 to compute the number of a_2 survivors into age group a_3 ($100 \times 0.3 = 30$). Such an assumption may not always be true in nature.

Appendix

2

An Example Using a Matrix Model*

The calculations in Appendix 1 can be expressed much more compactly using matrix algebra. A matrix is simply a group, table or array of numbers. For instance, the numbers for B and p in the example in Appendix 1 could be described as the following matrix:

$$\begin{bmatrix} 0 & 0.1 \\ 5 & 0.6 \\ 15 & 0.3 \\ 10 & 0.0 \end{bmatrix}$$

The fact that this is a matrix is signified by surrounding the numbers with square brackets. Conventionally, matrices are symbolized by a letter in bold face, say **X**. The numbers of rows and columns that comprise a matrix may vary, and matrices may consist of only a single column. The initial age structure of the previous example population may be written in this way:

$$\begin{bmatrix} 750 \\ 100 \\ 100 \\ 50 \end{bmatrix}$$

* See pg. 112.

and symbolized as t_1 **A**. The age distribution at t_2, t_2 **A**, can then be written as:

$$\begin{bmatrix} 2500 \\ 75 \\ 60 \\ 30 \end{bmatrix}$$

The two matrices t_1 **A** and t_2 **A** are technically called column vectors, indicating that the matrices are, in fact, just one column of figures.

In Appendix 1, the hypothetical population changed from t_1 **A** to t_2 **A** by using recurrence equations. To complete the matrix model, we have to construct a suitable matrix to enable t_1 **A** to become t_2 **A** by multiplication. The construction of this matrix is determined by the rules of matrix multiplication, and would appear as:

$$\begin{bmatrix} 0 & 5 & 15 & 10 \\ 0.1 & 0 & 0 & 0 \\ 0 & 0.6 & 0 & 0 \\ 0 & 0 & 0.3 & 0 \end{bmatrix} = \mathbf{T}$$

Note that matrix **T** is square, and that all the age-specific survival and birth data have been entered in particular positions, writing all the other numbers (matrix elements) as zeroes. The multiplication of the initial age-distributed population occurs according to the rules of matrix algebra. Each individual element in each row of **T** is multiplied with the corresponding element in t_1 **A**; the first with the first, the second with the second, and so on. These pairwise multiplications are then summed, and the sum is entered as the appropriate element in a new column vector. Thus, the first row of the square matrix leads to the first element of the new vector, the second leads to the second, and so on. This is illustrated as:

$$\begin{bmatrix} 0 & 5 & 15 & 10 \\ 0.1 & 0 & 0 & 0 \\ 0 & 0.6 & 0 & 0 \\ 0 & 0 & 0.3 & 0 \end{bmatrix} \times \begin{bmatrix} 750 \\ 100 \\ 100 \\ 50 \end{bmatrix} = \begin{bmatrix} (0)(750) + (5)(100) + (15)(100) + (10)(50) \\ (0.1)(750) + (0)(100) + (0)(100) + (0)(50) \\ (0)(750) + (0.6)(100) + (0)(100) + (0)(50) \\ (0)(750) + (0)(100) + (0.3)(100) + (0)(50) \end{bmatrix} = \begin{bmatrix} 2500 \\ 75 \\ 60 \\ 30 \end{bmatrix}$$

The positions of the zeroes in the matrix **T** are critical because they reduce terms in each row of the multiplication to zero, giving rise in t_2 **A** to the age-structure that was shown previously in Appendix 1.

Index